Foundations in Signal Processing, Communications and Networking

Volume 8

Series Editors

Wolfgang Utschick
Holger Boche
Rudolf Mathar

For further volumes:
http://www.springer.com/series/7603

Raphael Hunger

Analysis and Transceiver Design for the MIMO Broadcast Channel

Springer

Raphael Hunger
Technische Universität München
Fachgebiet Methoden der Signalverarbeitung
Munich
Germany

ISSN 1863-8538 ISSN 1863-8546 (electronic)
ISBN 978-3-642-31691-3 ISBN 978-3-642-31692-0 (eBook)
DOI 10.1007/978-3-642-31692-0
Springer Heidelberg New York Dordrecht London

Library of Congress Control Number: 2012942717

Printed on acid-free paper

Springer is part of Springer Science+Business Media (www.springer.com)

Contents

Abstract

Due to the asymmetry between the amount of data traffic in the downlink and uplink direction of nowadays and future wireless networks, a proper design of the transceivers in the broadcast channel is inevitable in order to satisfy the users' demands on data rate and transmission quality. This book deals with the optimization-based joint design of the transmit and receive filters in a MIMO broadcast channel in which the user terminals may be equipped with several antenna elements. Furthermore, the maximum performance of the system in the high power regime as well as the set of all feasible quality-of-service requirements is analyzed. First, a fundamental duality is derived that holds between the MIMO broadcast channel and a virtual MIMO multiple access channel. This duality construct allows for the efficient solution of problems originally posed in the broadcast channel in the dual domain where a possibly hidden convexity can often be revealed. Afterwards, the optimum solution is converted back to the original domain. Different kinds of dualities are derived like the multi-level mean square error duality and the Shannon rate duality, both for the case of purely linear transceivers and for the case of more complex transceivers utilizing dirty paper coding. On the basis of the established duality result, the gradient-projection algorithm is introduced as a tool to solve constrained optimization problems to global optimality under certain conditions. We present a complete description of the algorithm for precoder-based and covariance-based optimizations and derive a sophisticated step-size adaptation. The gradient-projection tool is then applied to solving the weighted sum rate maximization problem which is a central optimization that arises in any network utility maximization as well as in quality-of-service based optimizations and the design of schedulers. Thanks to the duality result, this problem can be optimally solved when dirty paper coding is used in the broadcast channel. For linear transceivers, we derive a combinatorial algorithm that achieves the global optimum in most of the cases. The iterative nature of the described algorithms makes a simple characterization of the obtained performance very difficult. In the high power regime, however, this changes due to the fact that the weighted sum rate utility converges to an affine asymptote in the logarithmic power domain irrespective of whether dirty paper coding is applied or not. We find closed-form expressions for these asymptotes given a fixed channel realization and extend the results also to the case of semicorrelated Rayleigh fading. The so-obtained results allow for a quantification of the asymptotic rate loss that linear transceivers have to face with respect to dirty paper coding. Depending on the antenna configuration, the high power analysis remains accurate at moderately large transmit powers which renders a quick estimate of the achievable performance possible. In the last part, we answer the fundamental question of feasibility in quality-of-service-based optimizations with inelastic traffic that features strict delay constraints. Under the assumption of linear transceivers, not every set of quality-of-service requirements might be feasible making the power minimization problem with given lower bound constraints on the rate for example infeasible in these cases. Before running any power minimization algorithm, it is therefore of prime importance to verify that the constraint set is not empty. We derive a complete description of the quality-of-service feasibility region for arbitrary channel matrices and come to the surprising result that the feasibility problem is independent of the channel matrix as long as some mild rank conditions are satisfied.

Chapter 1
Introduction

The broadcast channel is an integral entity of many wireless network topologies and denotes the downlink transmission from a central base station to several decentralized users. During the formation phase of the first wireless networks, only few subscribers were joining the network and voice telephony was the dominant service offered. Due to the consequently small rate requirements, the network load was expected to be small such that it was possible to employ orthogonal multiple access techniques to serve the different users in the broadcast channel. However, the increasing number of subscribers and the arising amount of multimedia data led to the insight that orthogonal access schemes do not fully exploit the available degrees of freedom that are offered by the scarce resources power and frequency bandwidth. One step to cope with the higher network load is to allow for some interference in a controlled way instead of completely suppressing it, a method which enhances the efficiency of the system. Obviously, higher data rates can also be offered by using a broader frequency band for the communication but due to regulations or the high expenses, this is not always possible. A second step to meet the increased demand for high data rate communication of a growing number of subscribers is the application of multiple antennas which enhance the spectral efficiency without having to spend more power or bandwidth. Due to the asymmetry of the data traffic in the uplink and the downlink, multi-antenna element base stations play a central role in the broadcast channel as the amount of downlink traffic exceeds the uplink traffic by far.

The huge demand for high data rate communication in the downlink translates itself to the need for an efficient transceiver design in the broadcast channel. As the maximum system performance can only be attained if the receivers and the transmitter are optimized *jointly*, some kind of channel state information must be available at the transmitter. In case of a multi-antenna base station, this information is not only required for the power control, but also for the design of the precoding matrix which has crucial impact on the achievable performance. While accurate channel state information can be obtained relatively easy at the receivers, this is quite challenging at the transmitter side. However, upcoming (and partly also hitherto existing) mobile communication standards include some feedback of the channel quality and/or (quantized) channel direction information from the users to the base station which then chooses its transmission strategy from a finite set of operation

R. Hunger, *Analysis and Transceiver Design for the MIMO Broadcast Channel*, Foundations in Signal Processing, Communications and Networking 8,
DOI: 10.1007/978-3-642-31692-0_1, © Springer-Verlag Berlin Heidelberg 2013

modes such that it fits best to the current interference situation. Although perfect channel state information will never be available at the transmitter, the tendency is to have more accurate information. It is to be expected that future wireless networks will turn their attention on the availability of very detailed channel state information also at the transmitter by applying more and more sophisticated feedback algorithms. Hence, it is important to know the maximum performance of a wireless broadcast channel under the assumption of perfect channel state information. In addition, knowing the optimum performance also allows for a reasonable assessment of the performance reduction that one has to face given imperfect channel state information.

Throughout this book, the limiting case of perfect channel state information is considered and the entire focus is laid on the broadcast channel as most of the data traffic is transmitted in the downlink direction. Therefore, a proper transceiver design is of central importance for present and future wireless networks in particular when multi-antenna terminals are involved. Under the assumption of perfect channel state information, this book deals with the analysis and the transceiver design in a MIMO broadcast channel with multi-antenna terminals. The elementary problem of maximizing a weighted sum of user rates is covered in detail since it represents the core of any network utility optimization. More precisely, we present efficient algorithms that solve the weighted sum rate maximization problem both for the case when the optimal (but computationally complex) transmission scheme termed dirty paper coding is applied and for the simpler case of linear transceivers. The transceiver design according to these algorithms is based on a fundamental duality result which relates the MIMO broadcast channel to a virtual MIMO multiple access channel. Instead of handling the optimizations directly in the broadcast channel, they are solved in the dual domain where they turn out to have a more favorable structure such that a hidden convexity of the original problem can often be revealed.

Besides the transceiver design, we analyze the maximum system performance in the high power regime which often allows for simple closed-form expressions of the maximum achievable rates that are accurate even at moderately large transmit powers. These results are derived both for a fixed channel realization as well as for semi-correlated Rayleigh fading. Finally, quality-of-service based optimizations are considered since they represent the second class of important optimizations in the broadcast channel. A frequently arising problem setting is the minimization of the required transmit power given some requirements for the users' transmission rate or quality. We derive the complete set of requirements that are feasible when linear transceivers are applied.

1.1 Overview

In addition to the introduction, this book is divided into eight chapters. The following overview briefly subsumes their content and reflects the main results derived therein.

- *Chapter 2: System Models*
  First, the discrete-time baseband model of the multi-user MIMO broadcast channel

is introduced since it serves as the mathematical description of the downlink in a cellular system for example. Different data are transmitted by the base station to the individual users which may be equipped with several antenna elements each, and a frequency-flat radio channel is assumed. Evidently, the performance of the system crucially depends on the choice of the precoding matrices and the receive filters. By means of these variables, the given setup can be optimized according to some desired requirements. Apart from Chaps. 8 and 9, an average sum power constraint is imposed on the transmitted symbol vector in order to model the limited amount of the resource power. Obviously, this constraint limits the set of admissible precoding matrices and therefore has to be considered during the optimization of the system.
While many problem settings share the unpleasant property of being hard to handle directly in the broadcast channel, they surprisingly often turn out to have a much simpler mathematical structure, e.g., convexity, in the dual multiple access channel. This multiple access channel can approximately be regarded as the same system setup with reversed signal flow, i.e., the users send their data to the base station (uplink). Due to the same *sum* power constraint, this virtual construct differs from a conventional multiple access channel with individual power constraints per user and it will only be used as a vehicle to solve problems in the broadcast channel. The duality relation between the MIMO broadcast channel and this dual MIMO multiple access channel will be derived in Chaps. 3 and 4.

- *Chapter 3: Dualities for the MIMO BC and the MIMO MAC with Linear Transceivers*
For the case of linear transceivers, the stream-based mean square error (MSE) duality between the broadcast channel and the dual multiple access channel is a well-investigated result in the literature. In this chapter, we generalize the stream-based variant to the multi-level MSE duality which features three different levels of resolution and can thus be tailored to the problem setting at hand. While the hitherto existing stream-based duality preserves the MSE of every data stream, we introduce two additional levels in which either the total sum MSE over all users is preserved during the duality conversion, or the sum MSE per user. Due to this application-specific property, the level with the smallest complexity can be applied for the problem setting at hand. For example, if only the total sum MSE shall be minimized in the broadcast channel, then the low-complexity level-one duality which only preserves this total sum MSE should be chosen since the resulting conversion of the filter matrices from the MAC to the BC is extremely simple. In general, duality between the two domains means that the same MSE tuples can be achieved both in the BC and the dual MAC. Instead of solving an MSE-based optimization directly in the BC where it is formulated, it can equivalently be solved in the dual domain, where it often features a simpler mathematical structure. The sum MSE minimization, for example, is a convex problem in the dual MAC but not (directly) in the BC. Optimality in one domain translates itself to optimality in the other domain. Besides the equivalence of the two domains, the duality also delivers conversion rules from one domain to the other, i.e., the optimal solution in the MAC can conveniently be transformed back to the BC.

Besides the extended MSE duality, a novel rate duality for linear transceivers is presented which fully exploits the gains that multi-antenna terminals offer. So far, rate dualities were treating the individual streams of a user separately and thus failed to achieve the maximum rate that joint decoding of these streams would offer. For *separate* stream decoding, the data stream of a user treats all other streams that belong to him as self-interference which reduces the attainable rates. The key idea to overcome this self-interference is to apply unitary decorrelation filters as the first stage of the precoding filters. Since the rate expressions under joint decoding only depend on the transmit covariance matrices and not on the precoding filters themselves, these unitary decorrelation matrices do not have any impact on the achievable rate of any user when joint decoding is applied per user. However, they can be chosen such that the intra-user self-interference is completely removed, thus, making the data rates of joint decoding achievable with *separate* decoding. In combination with the unitary decorrelations, any stream-based signal-to-interference-and-noise ratio (SINR) duality can be used to end up with a rate duality which preserves the gain of joint decoding. In particular, our level-three MSE duality also keeps the SINRs constant during the conversion such that the combination of these two parts forms the novel rate duality. The main advantage offered by the new duality is that rate-based optimization problems in the broadcast channel with linear transceivers can now equivalently be solved in the dual MAC where the arising rate expressions are simpler. Moreover, there is no need to split the optimization variables into unit-norm precoders, receive filters, and power allocation since the Shannon rate expressions only depend on the transmit covariance matrices of the users, over which one can conveniently optimize. Besides the rate gain, the computational complexity can thus be reduced, see the linear transceiver design in Chap. 7.

- *Chapter 4: Rate Duality with Nonlinear Interference Cancelation*

  The maximum potential of the broadcast channel usually cannot be obtained by means of linear transceivers. Instead, a rate pair on the boundary of the capacity region in general requires the application of dirty paper coding, a nonlinear technique in which the interference that is caused by the data of one user to some other users is partly treated as known side information which can be canceled by means of coding. Unfortunately, the rate expressions of the users in this kind of system are not concave in the transmit covariance matrices making rate-based optimizations with dirty paper coding hard to tackle directly in the broadcast channel. Nonetheless, there is again a dual multiple access channel that features the same capacity region under a sum power constraint. In this dual MAC, successive decoding is applied in which the already decoded codewords are subtracted from the received signal making the decoding of the remaining users' codewords easier. Additionally, the rate expressions are concave in the MAC transmit covariance matrices making rate-based optimizations solvable in this dual domain.

  Before the already existing duality for systems with dirty paper coding is revisited, Chap. 4 first recapitulates the capacity region of the broadcast channel with dirty paper coding and the multiple access channel with successive decoding. Then, the novel rate-duality for linear transceivers is adopted to the case of interference cancelation. In turn, the unitary decorrelation filters play the central role of the duality. In contrast to the existing rate dualities for dirty paper coding, the presented one has a very low computational complexity and features a reduced encoding delay. This results from the fact that due to the removal of the total intra-user interference, the individual data streams of a user can be encoded in parallel and not serially. Furthermore, the presented duality gives hints how to implement the cancelation of a *vector*-valued side information by relating it to the scalar case thanks to the decorrelation filters.

- *Chapter 5: Matrix-Based Gradient-Projection Algorithm*
  Due to the limited resource power, the optimization of the transmit and receive filters usually has to cope with the upper bound on the radiated power. When optimizing with respect to the transmit covariance matrices instead of the precoding filters, their positive definiteness also has to be ensured. As a consequence, we have to deal with (inequality-)*constrained* optimizations making standard approaches like Cauchy's steepest ascent algorithm not applicable. In Chap. 5, the gradient-projection algorithm is presented as a tool to find a stationary point of a utility maximization problem with sufficiently 'simple' constraints. Since moving towards the direction of the gradient usually leaves the constraint set, the so-obtained variables need to be projected back to meet all constraints. An integral part of the gradient-projection algorithm is thus the orthogonal projection onto the constraint set. We derive this projection for the two relevant types of constraints that arise during broadcast channel optimizations. In the first case, the precoding matrices have an upper bounded Frobenius norm due to the limited resource power but no other structural limitation. In the second case, when optimizing with respect to the transmit covariance matrices, not only the sum of their traces is limited but also their positive definiteness structure has to be ensured. For both cases, we come up with closed form expressions for the projection. Besides the projection onto the constraint set, the orthogonal projection onto the tangent cone plays a major role especially during the adaptation of the step-size. We use Armijo's rule for the choice of the step-size and show that convergence to a stationary point of the optimization can be guaranteed. Such a detailed analysis including the tangent cone projection is presented for the first time in the literature. A specialty which is worthwhile to mention is that we derive the *matrix*-valued variant of the gradient projection algorithm with complex-valued arguments. As a consequence, the Wirtinger calculus must be applied as the utilities are not holomorphic functions. A brief introduction to the Wirtinger calculus is therefore part of the appendix.
  Using the gradient-projection algorithm, the constrained maximization problem with an arbitrary concave utility can consequently be solved to global optimality. In Chap. 6, this algorithm is applied to the weighted sum rate maximization in a MIMO broadcast channel with dirty paper coding whereas Chap. 7 deals with the optimization of linear transceivers.
- *Chapter 6: MIMO BC Transceiver Design with Interference Cancelation*
  An optimization problem of central importance in the broadcast channel is the weighted sum rate maximization. This is due to the fact that many quality-of-service based optimizations and basically any network utility maximization

problem require the solution of a weighted sum rate maximization as a subproblem. The weights are usually determined in a higher layer by a master program. Besides, the weighted sum rate maximization arises in the context of scheduling and is for example part of the proportional fair scheduler. By varying the weight vector, any rate tuple on the boundary of the broadcast channel capacity region can be synthesized using the gradient-projection algorithm.
While the weighted sum rate utility is nonconcave in the broadcast channel (with dirty paper coding) which practically prevents from finding the globally optimum solution directly in the BC, this utility becomes concave when switching to the dual MIMO multiple access channel. Thanks to the duality between these two domains shown in Chap. 4, the weighted sum rate maximization problem can alternatively be solved in the dual domain to global optimality. Choosing the weighted sum rate as the utility, the gradient-projection algorithm from Chap. 5 is applied. More precisely, both the precoder-based and the covariance-based approach are simulated and evaluated in terms of their speed of convergence. In addition, we relate the presented algorithm to existing ones in the literature. It turns out that the covariance-based approach features the highest speed of convergence in terms of the number of iterations. However, the orthogonal projection of the covariances onto the constraint set is computationally more expensive than the projection of the precoders. The latter approach, however, requires a larger number of iterations in order to achieve the same accuracy. When the search direction is chosen as the projection of the Wirtinger derivative onto the tangent cone, this corresponds to the direction of the steepest ascent with the orthogonal projection taken into consideration. When the projection of the Wirtinger derivative onto the constraint set is chosen as search direction instead of the derivative itself, a higher speed of convergence can be observed for the precoder-based gradient-projection algorithm. For the covariance-based approach, the two search directions lead to the same result in the majority of cases.

- *Chapter 7: Linear Transceiver Design for the MIMO BC*
Despite being the optimum transmission scheme, dirty paper coding is very difficult to implement in practice. In Chap. 7, two possible implementations (typical sequences approach and modulo-lattice additive noise channel approach) of dirty paper coding are presented from which one can conclude that the computational complexity is extremely high especially for the receivers. To overcome this high burden, linear transceivers are an attractive alternative with drastically smaller complexity but obviously reduced performance. Chapter 7 therefore deals with the weighted sum rate maximization in the broadcast channel without dirty paper coding and purely linear transceivers instead. As for dirty paper coding, the weighted sum rate utility is again nonconcave in the broadcast channel. For linear transceivers, however, it remains nonconcave even in the dual multiple access channel. Nonetheless, we tackle the problem in this dual MAC due to the better mathematical structure (like common inverses in the rate expressions) in order to reduce the computational complexity. The nonconcavity of the utility manifests itself also in the existence of several stationary points with *different* utilities such that the gradient-projection algorithm from Chap. 5 will probably fail to achieve

the global maximum. As a remedy, we propose a combinatorial stream allocation algorithm which allocates a certain number of data streams to every user. Then, we exploit the fact that the precoder-based variant of the gradient-projection algorithm cannot increase the number of active streams per user during the iterations—it may only reduce it. This way, we manage to cover many different local optima and the global optimum is likely to be amongst them such that despite the nonconcavity of the problem, the global optimum is achieved with high probability. Besides the combinatorial approach which basically probes all reasonable stream initializations for the users, a greedy variant is introduced as well. There, the number of active streams of the users at the initialization is never reduced but only increased. As a consequence, drastically less stream configurations have to be tested without a noticeable performance reduction. Again, we investigate the speed of convergence for a fixed stream initialization and compare the obtained weighted sum rate utilities to those achieved by available algorithms in the literature. The superior performance of both presented algorithms is confirmed to hold for any value of the maximum sum power.

- *Chapter 8: Asymptotic High Power Analysis of the MIMO BC*
  The weighted sum rate analysis in Chaps. 6 and 7 revealed that the gradient-projection algorithm from Chap. 5 always achieves the global optimum in case of dirty paper coding and most of the times in case of linear transceivers without non-linear interference cancelation. However, since no closed-form solutions are available, the optimum covariance matrices and the maximum utility are the outcome of an iterative algorithm. This makes the prospective performance hard to quantify in advance without running the algorithm. In the high power regime when the maximum available transmit power becomes large, the maximum weighted sum rate utility can be shown to converge to an affine asymptote in the logarithmic power domain irrespective of whether dirty paper coding is applied or not. Chapter 8 covers the high power analysis of the maximum weighted sum rate utility for an instantaneous channel realization and the ergodic analysis for the case of semi-correlated Rayleigh fading.
  The high power asymptote can be characterized by its slope (which corresponds to the multiplexing gain in case of equal weights) and the rate shift. While the slope is usually only a function of the available degrees of freedom, the rate shift also depends on the channel realization or the fading model. We start with the asymptotic analysis for a fixed channel realization. If the base station is equipped with at least as many transmit antennas as the user terminals have in sum, the high power analysis becomes tractable. Again, all investigations are carried out in the dual MIMO MAC. A first, rather surprising observation is that cooperation of the decentralized receivers in the broadcast channel does not increase the achievable sum rate if dirty paper coding is applied. This observation results from the fact that the base station has enough degrees of freedom and that the worst case noise distribution is indeed uncorrelated amongst the users in this case. A second result is the exact quantification of the loss of the weighted sum rate utility that linear transceivers have to face compared to dirty paper coding. For a base station with

less antennas than the user terminals have in sum, the asymptotic analysis of the sum capacity remains difficult. However, we are able to show that in case of linear transceivers, the number of active data streams must match the rank of the channel matrix. This means that at most as many streams may be transmitted as the base station has antenna elements. A rigorous proof for this statement is derived for the first time although many publications made use of this assumption without proof. When dirty paper coding is employed, we can show that it may be necessary to have more active streams than the base station has antenna elements in order to achieve the sum capacity. In particular, for a two-antenna base station, we substantiate our conjecture that up to four (but not more) streams may be necessary in order to achieve the sum capacity.

Thanks to the closed-form expressions for the maximum asymptotic weighted sum rate given a fixed channel realization (for linear transceivers, these results were so far not available), channel-averaged results can quickly be obtained for any fading model by means of simulations. For the case of semi-correlated Rayleigh fading with correlations at the terminals, even closed-form expressions are possible and derived in the second part of Chap. 8, again both for the case of linear transceivers and dirty paper coding. For their derivation, the Wishart and inverse Wishart distributions play a major role.

- *Chapter 9: Description of the Quality of Service Feasibility Region*

The optimizations discussed in the previous chapters primarily focused on the weighted sum rate as the utility to maximize. Given some limitation on the maximum sum power, the gradient-projection algorithm determines the optimum covariance matrices and the user subset that must be served in order to maximize the utility. Depending on the choice of the weights, the available transmit power, and the transmission strategy, individual users might not receive any data at the given time instant. Some scheduler would later increase the priority of this user in order to satisfy his needs in terms of data rate. While this procedure is appropriate for elastic traffic, it is not applicable for inelastic traffic with time-critical delay requirements. There, certain quality-of-service (QoS) requirements must be guaranteed instantaneously and not on a larger time-scale. Chapter 9 answers the fundamental question, which QoS targets are feasible when *linear* transceivers are applied without time sharing. For a system with dirty paper coding, any requirements are feasible, but as mentioned before, the implementation of dirty paper coding features tremendous complexity especially when the known side-information has a large variance. For linear transceivers, not every QoS requirement can be ensured. Hence, before running an iterative algorithm which minimizes the required transmit power subject to given lower bounds on the quality-of-service of all users, it must be ensured that the constraint set is nonempty, i.e., that the required QoS targets are feasible.

First, feasibility is investigated for the case of single-antenna terminals. By means of the MSE duality from Chap. 3, the feasibility problem is in turn tackled in the dual vector MAC. While most of the contributions in this field use the SINR as the QoS metric, we show that a very simple description of the feasibility region is

possible in terms of the minimum mean square error. More precisely, the closure of the feasible MMSE region is shown to be a polytope. In case of the regular channels scenario, which requires the channel matrix to satisfy some very mild rank conditions, this polytope is given by the individual box constraints which limit the MMSE of every user to the closed interval between zero and one *and* by a lower bound on the sum MMSE of all users. This lower bound is given by the difference of the number of users and the number of antennas at the base station. Any MMSE tuple in the interior of this polytope can be achieved with finite sum power and the feasibility region is even independent of the channel matrix as long as it fulfills the very mild rank conditions. Interestingly, the inability of the users in the broadcast channel to cooperate does not reduce the set of feasible MMSE requirements compared to the case when the users were allowed to cooperate and form a virtual point-to-point MIMO system. However, cooperation clearly reduces the required transmit power to achieve the targets, but it does not render targets feasible that are infeasible without cooperation. When the channel matrix does not satisfy the regular channels condition, additional half-space constraint further crop the polytope. As a byproduct, we solved a long-standing problem in the signal-to-interference (SIR) balancing where the noise component is not considered. The maximum common scaling factor which scales the relative SIR requirements now directly follows from the polytope structure of the MMSE feasibility region.
In the second part of Chap. 9, the feasibility problem is extended to the MIMO broadcast channel with multi-antenna terminals. For the QoS metric, the Shannon data rate is now chosen. Interestingly, single-stream transmission is sufficient to clarify the question whether given rate requirements are feasible or not. Due to this circumstance, the feasibility problem in the MIMO BC boils down to a feasibility test in the vector BC with single-antenna terminals, for which the feasibility region was characterized in the first part. As a consequence, the feasibility test of given rate requirements is again very simple: the corresponding MMSEs only have to be tested to exceed the lower bound in sum.

Finally, some proofs that would have led to an impaired reading fluency have been moved to the appendix, where one can also find a short introduction to the Wirtinger calculus.

1.2 Used Symbols and Operators

The following list describes the meaning of the symbols and operators that will be used throughout this book:

$\lvert\cdot\rvert$	Absolute Value
$\mathbf{\lvert\cdot\rvert}$	Determinant
$\Vert\cdot\Vert_1$	Manhattan Norm (L^1 norm)
$\Vert\cdot\Vert_2$	Euclidean Norm (L^2 norm)
$\Vert\cdot\Vert_{\mathrm{F}}$	Frobenius Norm
$\lceil\cdot\rceil$	Ceiling Function
$\odot$	Element-Wise Schur Product
$(\cdot)^*$	Complex Conjugate
$(\cdot)^{\mathrm{T}}$	Transposition
$(\cdot)^{\mathrm{H}}$	Hermitian Transposition
$\frac{\partial_{\mathrm{w}}}{\partial_{\mathrm{w}}(\cdot)}$	Wirtinger Derivative
$[\cdot]_{k,j}$	Scalar Element in the k-th Row and j-th Column
$\boldsymbol{A} \succ \boldsymbol{B}$	$\boldsymbol{A}-\boldsymbol{B}$ is Positive Definite
$\boldsymbol{A} \succeq \boldsymbol{B}$	$\boldsymbol{A}-\boldsymbol{B}$ is Positive Semi-Definite
$a \sim b$	a Has the Same Distribution as b
$k!$	k Factorial, i.e., $k! = k\cdot(k-1)\cdot\ldots\cdot 2\cdot 1$
$\mathrm{blockdiag}\{\cdot\}$	Generates a Block-Diagonal Matrix with the Arguments as Diagonal Blocks
$\mathrm{cl}(\cdot)$	Closure of a Set
$\mathrm{co}\{\cdot\}$	Convex Hull Operator
$\mathrm{diag}\{\cdot\}$	Generates a Diagonal Matrix with the Arguments as the Diagonal Entries
e	Euler's constant, i.e., $\mathrm{e}\approx 2.71$
$\mathrm{E}[\cdot]$	Expectation
$\mathrm{Im}(\cdot)$	Imaginary Part Operator
$\mathrm{int}(\cdot)$	Interior of a Set
mod	Modulo
$\mathrm{rank}(\cdot)$	Rank of a Matrix
$\mathrm{Re}(\cdot)$	Real Part Operator
$\mathrm{tr}(\cdot)$	Trace
$\mathrm{vec}(\cdot)$	Operator that Stacks the Columns of a Matrix
$\emptyset$	Empty Set
$\mathbb{C}$	Set of Complex Numbers
$\mathcal{CN}(\boldsymbol{a},\boldsymbol{B})$	Circularly Symmetric Complex Gaussian Distribution with Mean $\mathbf{a}$ and Covariance Matrix $\boldsymbol{B}$
$\mathcal{N}(\boldsymbol{a},\boldsymbol{B})$	Real Gaussian Distribution with Mean $\boldsymbol{a}$ and Covariance Matrix $\boldsymbol{B}$
$\mathbb{N}_0$	Set of Natural Numbers Including the Zero
$\mathbb{R}$	Set of Real Numbers
$\mathbb{R}_+$	Set of Positive Real Numbers
$\mathbb{R}_{+,0}$	Set of Nonnegative Real Numbers
$\mathbb{S}^N$	Set of $N\times N$ Positive Semi-Definite Matrices
$\mathbb{S}_+^N$	Set of $N\times N$ Positive Definite Matrices
$\mathbf{0}$	All-Zeros Matrix
$\mathbf{I}_M$	$M\times M$ Identity Matrix
1_M	All-Ones Vector of Dimension M
$\mathcal{H}(\cdot)$	Differential Entropy
$I(a;b)$	Mutual Information Between a and b

1.3 Abbreviations

For a compact notation, below phrases will frequently be replaced by their corresponding abbreviations:

AWGN	Additive White Gaussian Noise
BC	Broadcast Channel
BpCU	Bits per Channel Use
CDMA	Code Division Multiple Access
CSI	Channel State Information
KKT	Karush-Kuhn-Tucker
MAC	Multiple Access Channel
MF	Matched Filter
MIMO	Multiple-Input Multiple-Output
MLAN	Modulo-Lattice Additive Noise
MMSE	Minimum Mean Square Error
MSE	Mean Square Error
OSI	Open Systems Interconnections
QoS	Quality of Service
SINR	Signal-to-Interference-and-Noise Ratio
SIR	Signal-to-Interference Ratio
SISO	Single-Input Single-Output
TDD	Time-Division Duplex
TDMA	Time-Division Multiple Access
WSR	Weighted Sum Rate

Chapter 2
System Models

2.1 MIMO Broadcast Channel System Model

In the MIMO broadcast channel, a centralized transmitter serves K decentralized users who cannot cooperate and therefore have to process and decode their received signals on their own. Each of the K users asks for data that is specific to him thus constituting a type of information theoretic broadcast channel [1]. This is in contrast to the conventional radio, TV, or satellite broadcast where common data is sent to all receivers. The considered scenario with different data for the users as depicted in Fig. 2.1 can be used to model the downlink of a cellular network in a single radio cell but also for wired communication like the digital subscriber line (e.g., [2, 3]). The number of antennas deployed at the base station shall be denoted by N and the kth terminal is equipped with M_k antennas where $k \in \{1, \ldots, K\}$. The information-bearing symbol vector $\boldsymbol{s}_k \in \mathbb{C}^{L_k}$ that is dedicated to user k is modeled as an L_k-dimensional complex-valued zero-mean random variable with identity covariance matrix $\mathrm{E}[\boldsymbol{s}_k \boldsymbol{s}_k^H] = \mathbf{I}_{L_k} \forall k \in \{1, \ldots, K\}$. Here, L_k is the number of streams assigned to user k. When needed, the distribution of the symbol vectors will be specified further in the respective sections and chapters. In addition, the symbol vectors of all K users are pairwise independent. The modulated symbol vector $\boldsymbol{s}_k$ is then filtered by the precoding matrix $\boldsymbol{P}_k \in \mathbb{C}^{N \times L_k}$ which maps the L_k data streams to the transmit signal of the N transmit antennas in the discrete-time system model. This way, the overall transmit signal

$$\boldsymbol{x} = \sum_{k=1}^{K} \boldsymbol{P}_k \boldsymbol{s}_k \in \mathbb{C}^N \tag{2.1}$$

is generated by summing up all the precoded symbol vectors and then radiated from the N antennas leading to a symbol-averaged dissipated transmit power

$$P_{\text{diss}}^{\text{BC}} = \mathrm{E}\big[\|\boldsymbol{x}\|_2^2\big] = \sum_{k=1}^{K} \|\boldsymbol{P}_k\|_{\text{F}}^2. \tag{2.2}$$

R. Hunger, *Analysis and Transceiver Design for the MIMO Broadcast Channel*,
Foundations in Signal Processing, Communications and Networking 8,
DOI: 10.1007/978-3-642-31692-0_2,

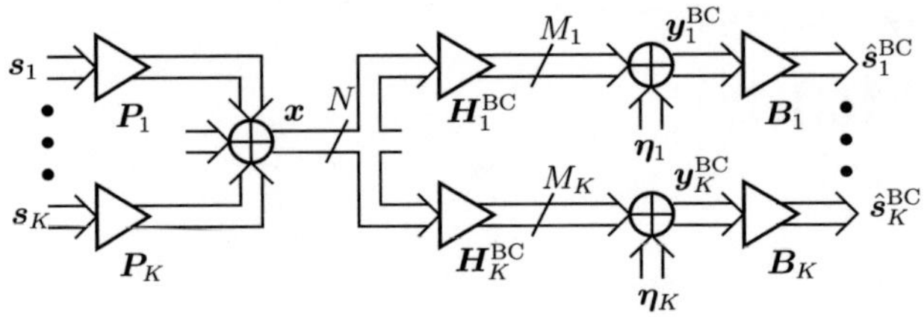

Fig. 2.1 System model for the K user MIMO broadcast channel

Apart from the feasibility analysis with given quality of service requirements in Chap. 9, we impose a sum power constraint on all antennas as upper bound on the totally radiated power:

$$P_{\text{diss}}^{\text{BC}} \leq P_{\text{max}}. \tag{2.3}$$

We assume a frequency-flat physical channel between the transmitter and the K receivers. In the discrete-time system model, the radio propagation from the base station to the kth receiver is described by the channel matrix $\boldsymbol{H}_k^{BC} \in \mathbb{C}^{M_k \times N}$ whose entry in the ith row and the jth column defines the channel coefficient describing the propagation from the jth antenna at the transmitter to the ith antenna at receiver k. Every channel coefficient subsumes the effects of the pulse shaping filter, the physical channel involving the path loss and fading effects, the analog receiver stage, the up- and downconversion, and the sampling with analog-to-digital conversion. At the kth receiver, circularly symmetric complex Gaussian noise $\boldsymbol{\eta}_k$ is added, which we assume to be zero-mean with covariance matrix $\boldsymbol{C}_{\boldsymbol{\eta}_k} := \text{E}[\boldsymbol{\eta}_k \boldsymbol{\eta}_k^{\text{H}}]$. This perturbation contains both the thermal noise and inter-cell interference from surrounding cells and it is assumed to be independent of all symbol vectors $\boldsymbol{s}_1, \ldots, \boldsymbol{s}_K$. Hence, the received signal at the kth user reads as

$$\boldsymbol{y}_k^{\text{BC}} = \boldsymbol{H}_k^{\text{BC}} \boldsymbol{x} + \boldsymbol{\eta}_k = \boldsymbol{H}_k^{\text{BC}} \sum_{i=1}^{K} \boldsymbol{P}_i \boldsymbol{s}_i + \boldsymbol{\eta}_k \tag{2.4}$$

and this signal is filtered by the $L_k \times M_k$ receive filter $\boldsymbol{B}_k$ to generate the symbol estimate

$$\hat{\boldsymbol{s}}_k^{\text{BC}} = \boldsymbol{B}_k \boldsymbol{y}_k^{\text{BC}} = \boldsymbol{B}_k \boldsymbol{H}_k^{\text{BC}} \boldsymbol{x} + \boldsymbol{B}_k \boldsymbol{\eta}_k \tag{2.5}$$

which is the input of the decoder. As long as the receive filters $\boldsymbol{B}_1, \ldots, \boldsymbol{B}_K$ generate a sufficient statistic and do not destroy any information, they do not have an impact on the mutual information. Note that we implicitly assume perfect coding, i.e., the symbol vectors $\boldsymbol{s}_1, \ldots, \boldsymbol{s}_K$ refer to the *encoded* symbols.

Throughout this book, perfect channel state information (CSI) is assumed to be available at both the transmitter and all receivers. While receiver-side CSI consisting of the *precoded* channel matrices can be obtained with reasonable accuracy by transmitting dedicated and common pilot symbols to the receivers either in a time-multiplexed or superimposed pilot scheme (see for example [4, 5] for the single-user scenario and [6] for the multi-user scenario), the situation is more complicated at the transmitter side. Depending on the coherence time of the channel, accurate transmitter-side CSI may be hard to obtain. In time-division duplex (TDD) systems, the reciprocity of the channel can be exploited to use the uplink channel estimates for the transmitter-side channel state given a proper calibration in signal space, i.e., a proper modeling and estimation of the radio frequency circuitry impulse responses [7]. Outdating due to a time-variant scenario can be overcome by means of a robust precoder design including channel prediction, see for example [8, 9, 10, 11]. Another possibility to obtain CSI at the transmitter which is also applicable in non-TDD systems is feedback from the receivers, e.g. [12, 13]. There, analog or quantized receiver-side CSI is fed back to the transmitter. Although perfect transmitter and receiver-side CSI will never be available, its assumption makes sense for the following two reasons. First, it is necessary to know how the idealized system without any kind of imperfections performs in order to have an upper bound on the realizable performance and in order to be able to evaluate the efficiency of robust algorithms taking into account imperfect CSI. Second, such robust transceiver designs are often based on their idealized counterparts which assume perfect CSI. A degradation of the CSI may for example become manifest in regularized inverses in robust transceivers, see [9, 10] amongst others.

2.2 Dual MIMO Multiple Access Channel System Model

By switching the roles of transmitter and receivers, the dual multiple access channel arises with reversed signal flow. The K decentralized users send their data to the centralized receiver and the same symbol vectors $\boldsymbol{s}_1, \ldots, \boldsymbol{s}_K$ shall be transmitted as in the original BC. But now, the precoder $\boldsymbol{T}_k \in \mathbb{C}^{M_k \times L_k}$ of user k maps the L_k-dimensional symbol vector $\boldsymbol{s}_k$ to his M_k antennas, where the precoded signal $\boldsymbol{T}_k \boldsymbol{s}_k$ propagates over the channel described by the matrix $\boldsymbol{H}_k \in \mathbb{C}^{N \times M_k}$, see Fig. 2.2. In principle, the MAC channel matrices $\boldsymbol{H}_1, \ldots, \boldsymbol{H}_K$ correspond to the Hermitian channel matrices $\boldsymbol{H}_1^{\mathrm{BC,H}}, \ldots, \boldsymbol{H}_K^{\mathrm{BC,H}}$ in the BC, but they are also influenced by the noise covariance matrices $\boldsymbol{C}_{\eta_1}, \ldots, \boldsymbol{C}_{\eta_K}$ as will be seen in Chaps. 3 and 4. For a shorter notation, we omit the superscript $(\cdot)^{\mathrm{MAC}}$ for the MAC channel matrices and the MAC symbol estimates. If all K noise covariance matrices in the BC are scaled identities, i.e., $\boldsymbol{C}_{\eta_k} = \sigma_{\eta_k}^2 \boldsymbol{I}_{M_k}\ \forall k$, the dual MIMO MAC can be seen as a system with reversed signal flow having the same channel transfer function as the original broadcast channel. However, imposing the same upper bound $P_{\mathrm{diss}}^{\mathrm{MAC}} \leq P_{\max}$ as in the BC on the *sum* power

$$P_{\mathrm{diss}}^{\mathrm{MAC}} = \sum_{k=1}^{K} \mathrm{E}\big[\|\boldsymbol{T}_k \boldsymbol{s}_k\|_2^2\big] = \sum_{k=1}^{K} \|\boldsymbol{T}_k\|_{\mathrm{F}}^2 \tag{2.6}$$

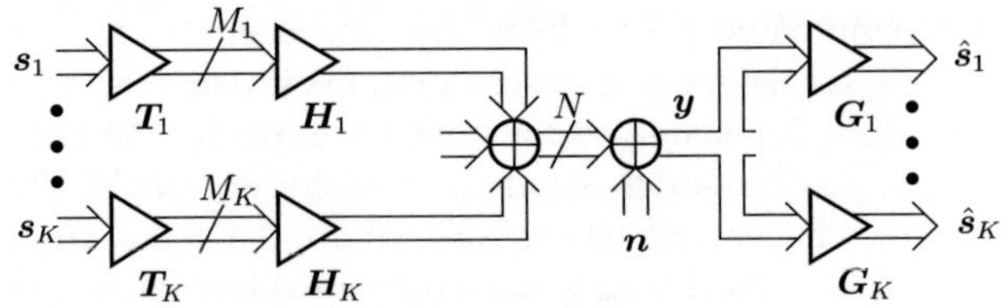

Fig. 2.2 System model for the K user MIMO multiple access channel

dissipated in the MAC distinguishes this dual MIMO MAC from a conventional MIMO MAC where individual power constraints for every user are imposed since cooperation and power sharing between the users is not possible. All precoded symbol vectors are superimposed by the wireless channel and constitute the received signal

$$y = \sum_{k=1}^{K} \boldsymbol{H}_k \boldsymbol{T}_k \boldsymbol{s}_k + \boldsymbol{n} \in \mathbb{C}^N \tag{2.7}$$

where the additive white Gaussian noise $\boldsymbol{n} \sim \mathcal{CN}(\mathbf{0}, \mathbf{I}_N)$ can be chosen to be zero-mean with identity covariance matrix. By means of the receive filter $\boldsymbol{G}_k \in \mathbb{C}^{L_k \times N}$, the symbol estimate for user k reads as

$$\hat{\boldsymbol{s}}_k = \boldsymbol{G}_k \boldsymbol{y} = \boldsymbol{G}_k \sum_{i=1}^{K} \boldsymbol{H}_i \boldsymbol{T}_i \boldsymbol{s}_i + \boldsymbol{G}_k \boldsymbol{n} \in \mathbb{C}^{L_k} \tag{2.8}$$

which can then be passed to the detector.

Most parts of the analysis and the transceiver designs presented in this book are done in the dual MIMO MAC instead of the original MIMO BC. The equivalence between these two domains is shown in Chaps. 3 and 4. Thus, problem settings in the BC can equivalently be solved in the dual MAC.

Chapter 3
Dualities for the MIMO BC and the MIMO MAC with Linear Transceivers

3.1 Motivation

In the past few years, dualities were successfully employed as the linking element between the MAC and the BC and thus, they have gained in importance in the signal processing and information theory communities. If a domain (e.g., rate, mean square error, or signal-to-interference-and-noise ratio domain) in the BC is dual to the corresponding one in the MAC, then both regions coincide and every problem setting and analysis can be investigated and performed either in the BC or in the dual MAC. Consequently, any conclusion drawn in the MAC remains valid in the BC and vice versa. Optimality in one system translates itself into optimality in the other one as well. Besides the capability of proving the congruency of two regions, dualities also deliver explicit conversion formulas how to switch from one domain to the other. While the dual system has some fundamental properties in common with the original system such that the two underlying regions coincide, it additionally features some attributes that differ and hence can be exploited.

Among those attributes may for example be a better mathematical structure. Using the mean square error (MMSE) duality, the hidden convexity [14] of the sum MSE minimization problem in the MIMO BC with linear transceivers under a sum power constraint could be revealed, see [15, 16, 17, 18]. Similarly, the weighted sum rate maximization problem in the sum power constrained MIMO BC with dirty paper coding turns out to be convex in the dual MAC [19, 20], see Chap. 6. Finally, the rate duality between the MIMO MAC and the MIMO BC with joint stream decoding and linear transceivers [21] renders possible the analysis of the asymptotic MIMO BC in the high power regime, cf. [22, 23, 24] and Chap. 8. These benefits have their origin in the property that the channel and precoders indices in the MAC are always aligned, i.e., the precoders $\boldsymbol{T}_k$ of user k is always linked to the channel matrix $\boldsymbol{H}_k$ of the *same* user k, see (2.7). This leads to the fact that a common inverse is involved in the MSE or rate expressions of different users, cf. Sect. 7.4 for example. In the BC, however, this indices alignment does not hold, see (2.4).

R. Hunger, *Analysis and Transceiver Design for the MIMO Broadcast Channel*, Foundations in Signal Processing, Communications and Networking 8,
DOI: 10.1007/978-3-642-31692-0_3,

Another advantage of tackling a BC problem in its dual MAC is the reduced complexity. In general, the K equalizers in the BC and the single equalizer in its dual MAC do not have to be computed explicitly since they usually turn out to be the minimum mean square error (MMSE) equalizers or do not come into play at all as long as the equalizer generates a sufficient statistic and does not destroy any information. Therefore, it usually suffices to determine the optimum precoders(s) which is mostly done in an iterative fashion. In the original BC, the precoders $\boldsymbol{P}_1, \ldots, \boldsymbol{P}_K$ consist of up to $N \sum_{k=1}^{K} M_k$ complex-valued coefficients whereas the K precoders in the dual MAC have only up to $\sum_{k=1}^{K} M_k^2$ coefficients in sum, which for most practical scenarios is smaller than $N \sum_{k=1}^{K} M_k$. Therefore, the optimization in the dual MAC is with respect to a smaller number of optimization variables than in the BC yielding a reduced computational complexity.

3.2 Mean Square Error Duality for Linear Transceivers

3.2.1 Literature Overview

Due to the fact that the third kind of the MSE duality derived in the following subsections not only preserves the stream-wise MSEs but also the signal-to-interference-and-noise ratios (SINRs), we include the evolution of the SINR dualities into this literature overview as well. Furthermore, the first MSE duality was deduced from the SINR duality making the origins of the SINR duality worthwhile to summarize.

Early work in this field focused on the signal-to-interference ratio (SIR) neglecting the noise portion, see for example [25, 26, 27]. The first SINR duality concept taking the noise portion explicitly into account was introduced 1998 by Rashid-Farrokhi et al. [28] in the context of power minimization under minimum SINR requirements. In [29] Visotsky et al. constructed a virtual uplink channel by normalizing the users' channels. However, an explicit duality framework had not been presented yet, since duality aspects were proven only for simultaneously optimum beamformers and power allocation. Boche et al. proved the uplink-downlink SINR duality for general unit-norm beamforming vectors [30, 31], whose power allocations have to fulfill a balancing of certain SINR ratios, i.e., the SINR targets have to be set to the currently achieved SINRs. At the same time, Viswanath et al. derived the SINR duality between the MAC and the BC from the duality of the point-to-point MIMO system [32]. Up to that time, all dualities were applicable only for single antenna receivers. This restriction was eliminated by Tse et al. in [33], who extended the SINR duality to single-stream transmission for multi-antenna receivers/transmitters. An interesting property that was observed already in the first contributions on SINR duality is, that when applying a set of unit-norm beamforming vectors in the downlink with an arbitrary power allocation, the achieved SINR tuple in the downlink can also be achieved in the dual uplink by means of *the same set* of performers as receive filters, but with a *different* power allocation at the single antenna transmitters in the dual uplink. This property was exploited by the authors in [34] to find the SINR duality with the lowest complexity currently available. This SINR duality in [34] turns out to be a byproduct of the third kind of the MSE duality presented in Sects. 3.2.3 and 3.2.4. With the SINR duality established, the sum power minimization problem subject to minimum SINR requirements in the BC as well as the balancing problem with given relative SINRs could be solved, see for example [30, 35, 36, 37, 38].

Many of the existing MSE dualities for linear transceivers were deduced on the basis of the SINR duality, e.g., [39, 16, 37]. As a consequence, such MSE dualities fail for setups with data streams that are switched off, i.e., the respective SINRs are zero. While this configuration does not arise in SINR-based optimizations, it can be observed that individual data streams or even users are switched off in MSE-based optimizations when the noise variance at the receiver is too large or when the channel gains are too small, see for example [40, 41]. A second property that is inherited from the SINR duality is the maintenance of every *stream's* MSE which depending on the problem that shall be solved by applying the MSE duality may be far too complex and hence a waste of computational resources, see Sect. 3.2.6. For many optimization problems, preserving either the sum MSE per user or even the sum MSE of all users suffices due to a user-wise formulation of QoS constraints for example. Hence, preserving the MSEs per data stream is not necessary. The first user-wise MSE duality was proposed in [38] without SINR detouring which was later extended to become applicable to frequency selective scenarios for example by treating precoders, receivers, and channels as bounded operators in a Hilbert space [42]. In a similar way, Shi et al. directly derived an MSE-based duality in [17]. Finally, a duality where single-sided deactivated data streams are allowed was derived in [18]. There, switched-off data streams at the transmitter where the respective receive filter is active are allowed as well as active data streams with switched off receive filters. Such a setup might arise in an intermediate step of an iterative algorithm, where the duality conversion is one part of an iteration.

3.2.2 Multi-Level Structure and MSE Expressions

As discussed in Sect. 3.2.1, the state-of-the-art MSE duality preserves the MSEs per data stream when switching from the MIMO BC to the MIMO MAC and vice versa. The MSE duality presented in this book consists of three different kinds corresponding to three different resolutions of the MSE conservation. Thus, it features an application specific property and depending on the problem setting under investigation, one out of the three resolutions can be chosen to simplify or solve the problem with the minimum possible computational complexity.

In the first and likewise simplest level, the overall sum MSE of all $L := \sum_{k=1}^{K} L_k$ data streams of all K users in the system does not change during the duality transformation. This level of the duality is the optimum choice when minimizing the overall sum MSE in the MIMO BC, see Sect. 7.4. The two other levels do not need to be applied when the overall sum MSE shall be minimized although they basically could.

Only a single scalar (φ for the MAC to BC conversion and ψ for the BC to MAC conversion) has to be computed and applied for the duality conversion of the first level. Note that this first level of the duality cannot be deduced from the SINR duality since any SINR duality naturally exhibits a stream-wise formulation, whereas only the overall sum MSE is of interest here. Also note that the other two levels of duality can be applied to the minimization of the overall sum MSE without any change of the optimality in the overall sum MSE metric. However, the other two levels of MSE duality have a much larger computational complexity.

The second level of the duality does not only preserve the overall sum MSE during the duality transformations, but also the K individual sum MSEs of the data streams belonging to the K users. Therefore, it features a finer resolution than the level one duality which however implies that instead of only a single scalar, K scalars ($\varphi_1, \ldots, \varphi_K$ for the MAC to BC conversion and $\psi_1, \ldots, \psi_K$ for the BC to MAC conversion) have to be determined from a linear system of equations in general. These K scalars then link the receive filters in the BC to the transmit filters in the MAC and vice versa. The application specific nature of the level two duality can be observed in the weighted sum MSE minimization in the MIMO BC [18], where K weights are combined with the K sum MSEs of the K users and the weighted sum is then minimized. From this user-wise formulation, it is obvious that the duality of level two has to be applied, since the level one duality does not offer the required user-wise resolution whereas the level three duality entails a too fine resolution with unnecessary computational complexity. At a Karush-Kuhn-Tucker (KKT) point of the weighted sum MSE minimization in the dual MAC, the K scalars that establish the duality transformation from the dual MAC to the BC turn out to be a byproduct of the optimization. Thus, they need not be computed from a linear system of equations as they would have to be when transforming arbitrary MSE tuples from the MAC to the BC. The ratios of these K squared scalars are completely determined by the user weights at a KKT point such that only a single scalar has to be computed in the end which sets the absolute value of the scalars $\varphi_1, \ldots, \varphi_K$ and $\psi_1, \ldots, \psi_K$, respectively. Thus, even the level two duality conversion can be achieved with very small complexity [18].

The highest resolution is obtained by the level three duality where in addition to the K sum MSEs of the K users, the individual MSEs of the L transmitted data streams are conserved during the duality transform. In principle, this level three duality is similar to the existing layer-wise MSE dualities in [16, 15] and [17]. This level of duality becomes attractive if one focuses on rate optimization problems where instead of a joint stream decoding at the users in the BC, separate stream decoding is applied (under the assumption of separate coding). This follows from the fact that the level three duality not only leaves the MSEs of the L data streams unchanged during the conversion, but also keeps the SINRs of the L streams constant such that the data rates under separate stream decoding with Gaussian alphabets do not change. However, the obtained data rates per user can be increased if the rate duality from Sect. 3.3 is applied for the conversion from the dual MIMO MAC to the MIMO BC to avoid the self-interference of all streams dedicated to one particular user.

3.2.2.1 MSE Expressions in the MIMO BC

Starting with the level one MSE duality, we first compute the overall sum MSE of all L data streams in the MIMO BC introduced in Sect. 2.1. To this end, we define respectively the stacked symbol vector $\boldsymbol{s}$, its estimate $\hat{\boldsymbol{s}}^{\mathrm{BC}}$, the BC noise contributions $\boldsymbol{\eta}$, the composite precoders matrix $\boldsymbol{P}$, the composite BC channel matrix $\boldsymbol{H}^{\mathrm{BC}}$, and the composite equalizer filter $\boldsymbol{B}$ via

$$\begin{aligned}
\boldsymbol{s} &:= [\boldsymbol{s}_1^{\mathrm{T}}, \ldots, \boldsymbol{s}_K^{\mathrm{T}}]^{\mathrm{T}} \in \mathbb{C}^L \\
\hat{\boldsymbol{s}}^{\mathrm{BC}} &:= [\hat{\boldsymbol{s}}_1^{\mathrm{BC,T}}, \ldots, \hat{\boldsymbol{s}}_K^{\mathrm{BC,T}}]^{\mathrm{T}} \in \mathbb{C}^L \\
\boldsymbol{\eta} &:= [\boldsymbol{\eta}_1^{\mathrm{T}}, \ldots, \boldsymbol{\eta}_K^{\mathrm{T}}]^{\mathrm{T}} \in \mathbb{C}^M \\
\boldsymbol{P} &:= [\boldsymbol{P}_1, \ldots, \boldsymbol{P}_K] \in \mathbb{C}^{N \times L} \\
\boldsymbol{H}^{\mathrm{BC}} &:= [\boldsymbol{H}_1^{\mathrm{BC,T}}, \ldots, \boldsymbol{H}_K^{\mathrm{BC,T}}]^{\mathrm{T}} \in \mathbb{C}^{M \times N} \\
\boldsymbol{B} &:= \mathrm{blockdiag}\{\boldsymbol{B}_k\}_{k=1}^K \in \mathbb{C}^{L \times M}
\end{aligned} \tag{3.1}$$

where $M := \sum_{k=1}^K M_k$ denotes the total number of antennas at the K users. Using (3.1), we can rewrite (2.5) for all K users simultaneously yielding

$$\hat{\boldsymbol{s}}^{\mathrm{BC}} = \boldsymbol{B}\boldsymbol{H}^{\mathrm{BC}}\boldsymbol{P}\boldsymbol{s} + \boldsymbol{B}\boldsymbol{\eta}.$$

Thus, the overall sum MSE $\varepsilon^{\mathrm{BC}} := \mathrm{E}[\|\boldsymbol{s} - \hat{\boldsymbol{s}}^{\mathrm{BC}}\|_2^2]$ reads as

$$\varepsilon^{\mathrm{BC}} = L + \mathrm{tr}\left(\boldsymbol{B}\boldsymbol{H}^{\mathrm{BC}}\boldsymbol{P}\boldsymbol{P}^{\mathrm{H}}\boldsymbol{H}^{\mathrm{BC,H}}\boldsymbol{B}^{\mathrm{H}} + \boldsymbol{B}\boldsymbol{C}_{\boldsymbol{\eta}}\boldsymbol{B}^{\mathrm{H}} - \boldsymbol{B}\boldsymbol{H}^{\mathrm{BC}}\boldsymbol{P} - \boldsymbol{P}^{\mathrm{H}}\boldsymbol{H}^{\mathrm{BC,H}}\boldsymbol{B}^{\mathrm{H}}\right) \tag{3.2}$$

where the covariance matrix $\boldsymbol{C}_{\boldsymbol{\eta}}$ of $\boldsymbol{\eta}$ can be chosen to be block-diagonal irrespective of whether $\boldsymbol{\eta}_1, \ldots, \boldsymbol{\eta}_K$ are correlated or not since $\boldsymbol{B}$ is block-diagonal. Without loss of generality, we may set $\boldsymbol{C}_{\boldsymbol{\eta}} := \mathrm{blockdiag}\{\boldsymbol{C}_{\boldsymbol{\eta}_k}\}_{k=1}^K$.

For the next higher resolution of the level two duality, the individual user MSEs need to be resolved. Using the symbol estimate definition for $\hat{\boldsymbol{s}}_k^{\mathrm{BC}}$ in (2.5), the MSE $\varepsilon_k^{\mathrm{BC}} := \mathrm{E}[\|\boldsymbol{s}_k - \hat{\boldsymbol{s}}_k^{\mathrm{BC}}\|_2^2]$ of user k in the BC can be expressed as

$$\begin{aligned}
\varepsilon_k^{\mathrm{BC}} = L_k + \mathrm{tr}\Bigg(&\boldsymbol{B}_k\boldsymbol{H}_k^{\mathrm{BC}}\sum_{i=1}^K \boldsymbol{P}_i\boldsymbol{P}_i^{\mathrm{H}}\boldsymbol{H}_k^{\mathrm{BC,H}}\boldsymbol{B}_k^{\mathrm{H}} + \boldsymbol{B}_k\boldsymbol{C}_{\boldsymbol{\eta}_k}\boldsymbol{B}_k^{\mathrm{H}} \\
&- \boldsymbol{B}_k\boldsymbol{H}_k^{\mathrm{BC}}\boldsymbol{P}_k - \boldsymbol{P}_k^{\mathrm{H}}\boldsymbol{H}_k^{\mathrm{BC,H}}\boldsymbol{B}_k^{\mathrm{H}}\Bigg).
\end{aligned} \tag{3.3}$$

Obviously, the K individual sum MSEs of the users sum up to the overall sum MSE, i.e., we have

$$\varepsilon^{\mathrm{BC}} = \sum_{k=1}^K \varepsilon_k^{\mathrm{BC}}.$$

When the MSEs are resolved per data stream, the highest resolution is achieved in the level three duality. With $\boldsymbol{b}_{k,j}^{\mathrm{T}}$ denoting the jth row of the equalizer filter $\boldsymbol{B}_k$ corresponding to the equalizer filter for the jth data stream of user k, and $\boldsymbol{p}_{k,j}$ denoting the jth column of the precoders $\boldsymbol{P}_k$, the MSE $\varepsilon_{k,j}^{\mathrm{BC}}$ of the jth stream of user k reads as

$$\begin{aligned}\varepsilon_{k,j}^{\mathrm{BC}} = 1 + \boldsymbol{b}_{k,j}^{\mathrm{T}} \boldsymbol{H}_k^{\mathrm{BC}} \sum_{i=1}^{K} \sum_{\ell=1}^{L_i} \boldsymbol{p}_{i,\ell} \boldsymbol{p}_{i,\ell}^{\mathrm{H}} \boldsymbol{H}_k^{\mathrm{BC,H}} \boldsymbol{b}_{k,j}^{*} + \boldsymbol{b}_{k,j}^{\mathrm{T}} \boldsymbol{C}_{\eta_k} \boldsymbol{b}_{k,j}^{*} \\ - \boldsymbol{b}_{k,j}^{\mathrm{T}} \boldsymbol{H}_k^{\mathrm{BC}} \boldsymbol{p}_{k,j} - \boldsymbol{p}_{k,j}^{\mathrm{H}} \boldsymbol{H}_k^{\mathrm{BC,H}} \boldsymbol{b}_{k,j}^{*}.\end{aligned} \tag{3.4}$$

The MSEs of all L_k streams of user k sum up to his sum MSE, i.e.,

$$\varepsilon_k^{\mathrm{BC}} = \sum_{j=1}^{L_k} \varepsilon_{k,j}^{\mathrm{BC}}.$$

3.2.2.2 MSE Expressions in the Dual MIMO MAC

Equivalently, we now define the overall sum MSE, the K user-wise sum MSEs, and the L stream-wise MSEs in the dual MIMO MAC described in Sect. 2.2. For this purpose, the stacked symbol estimate $\hat{\boldsymbol{s}}$, the composite precoders matrix $\boldsymbol{T}$, the composite channel matrix $\boldsymbol{H}$, and the composite equalizer filter $\boldsymbol{G}$ in this dual MIMO MAC are defined according to

$$\begin{aligned}\hat{\boldsymbol{s}} &:= [\hat{\boldsymbol{s}}_1^{\mathrm{T}}, \ldots, \hat{\boldsymbol{s}}_K^{\mathrm{T}}]^{\mathrm{T}} \in \mathbb{C}^{L} \\ \boldsymbol{T} &:= \mathrm{blockdiag}\{\boldsymbol{T}_k\}_{k=1}^{K} \in \mathbb{C}^{M \times L} \\ \boldsymbol{H} &:= [\boldsymbol{H}_1, \ldots, \boldsymbol{H}_K] \in \mathbb{C}^{N \times M} \\ \boldsymbol{G} &:= [\boldsymbol{G}_1^{\mathrm{T}}, \ldots, \boldsymbol{G}_K^{\mathrm{T}}]^{\mathrm{T}} \in \mathbb{C}^{L \times N}\end{aligned} \tag{3.5}$$

respectively. With these definitions, the stacked symbol estimate vector can be expressed similar to (2.8) as

$$\hat{\boldsymbol{s}} = \boldsymbol{G}\boldsymbol{H}\boldsymbol{T}\boldsymbol{s} + \boldsymbol{G}\boldsymbol{n}$$

inducing the overall sum MSE $\varepsilon^{\mathrm{MAC}} := \mathrm{E}[\|\boldsymbol{s} - \hat{\boldsymbol{s}}\|_2^2]$, which reads as

$$\varepsilon^{\mathrm{MAC}} = L + \mathrm{tr}\left(\boldsymbol{G}\boldsymbol{H}\boldsymbol{T}\boldsymbol{T}^{\mathrm{H}}\boldsymbol{H}^{\mathrm{H}}\boldsymbol{G}^{\mathrm{H}} + \boldsymbol{G}\boldsymbol{G}^{\mathrm{H}} - \boldsymbol{G}\boldsymbol{H}\boldsymbol{T} - \boldsymbol{T}^{\mathrm{H}}\boldsymbol{H}^{\mathrm{H}}\boldsymbol{G}^{\mathrm{H}}\right) \tag{3.6}$$

where we used $\mathrm{E}[\boldsymbol{n}\boldsymbol{n}^{\mathrm{H}}] = \mathbf{I}_N$.

The level two duality resolves the user-wise MSEs and leaves them unchanged during the duality transformation. With the definition of $\hat{\boldsymbol{s}}_k$ in (2.8), the MSE $\varepsilon_k^{\mathrm{MAC}} := \mathrm{E}[\|\boldsymbol{s}_k - \hat{\boldsymbol{s}}_k\|_2^2]$ of user k in the MAC reads as

$$\varepsilon_k^{\mathrm{MAC}} = L_k + \mathrm{tr}\Big(\boldsymbol{G}_k \sum_{i=1}^{K} \boldsymbol{H}_i \boldsymbol{T}_i \boldsymbol{T}_i^{\mathrm{H}} \boldsymbol{H}_i^{\mathrm{H}} \boldsymbol{G}_k^{\mathrm{H}} + \boldsymbol{G}_k \boldsymbol{G}_k^{\mathrm{H}} - \boldsymbol{G}_k \boldsymbol{H}_k \boldsymbol{T}_k - \boldsymbol{T}_k^{\mathrm{H}} \boldsymbol{H}_k^{\mathrm{H}} \boldsymbol{G}_k^{\mathrm{H}}\Big). \tag{3.7}$$

For the highest resolution, the user-wise MSEs are decomposed into the contributions of the respective streams. Let $\boldsymbol{g}_{k,j}^{\mathrm{T}}$ denote the jth row of the equalizer filter $\boldsymbol{G}_k$ and let $\boldsymbol{t}_{k,j}$ denote the jth column of the precoders $\boldsymbol{T}_k$. Then, the MSE $\varepsilon_{k,j}^{\mathrm{MAC}}$ of the jth data stream of user k can be written as

$$\varepsilon_{k,j}^{\mathrm{MAC}} = 1 + \boldsymbol{g}_{k,j}^{\mathrm{T}} \sum_{i=1}^{K} \boldsymbol{H}_i \sum_{\ell=1}^{L_i} \boldsymbol{t}_{i,\ell} \boldsymbol{t}_{i,\ell}^{\mathrm{H}} \boldsymbol{H}_i^{\mathrm{H}} \boldsymbol{g}_{k,j}^{*} + \|\boldsymbol{g}_{k,j}\|_2^2 - \boldsymbol{g}_{k,j}^{\mathrm{T}} \boldsymbol{H}_k \boldsymbol{t}_{k,j} - \boldsymbol{t}_{k,j}^{\mathrm{H}} \boldsymbol{H}_k^{\mathrm{H}} \boldsymbol{g}_{k,j}^{*}. \tag{3.8}$$

In turn, $\varepsilon_{k,1}^{\mathrm{MAC}}, \ldots, \varepsilon_{k,L_k}^{\mathrm{MAC}}$ sum up to $\varepsilon_k^{\mathrm{MAC}}$, and $\varepsilon_1^{\mathrm{MAC}}, \ldots, \varepsilon_K^{\mathrm{MAC}}$ sum up to $\varepsilon^{\mathrm{MAC}}$.

3.2.3 BC to MAC Conversion

In this subsection, the conversion from the MIMO BC to the MIMO MAC is described which, according to the chosen level of the duality, does not change the overall sum MSE, the sum MSEs of the K users, or the MSEs of the L data streams in the system when switching from the BC to the dual MAC. Therefore, one direction of the duality statement is proven here, and this direction allows us to design precoders and receive filters in the dual MIMO MAC featuring the same MSE region as the original MIMO BC has. This first direction of the MSE duality is stated in the following lemma:

Lemma 3.1 *There exist precoding and receive filters in the dual MIMO MAC such that any overall sum MSE, any user-wise MSE tuple, and any stream-wise MSE tuple that is feasible in the MIMO BC described in Sect. 2.1 under linear filtering is also achievable in the dual MIMO MAC from Sect. 2.2 with the same sum power. This constitutes one direction of the MSE duality for linear transceivers which holds for all three levels of duality.*

The proof of Lemma 1 is shown for the three levels of duality separately.

3.2.3.1 Level One BC to MAC Conversion

In the level one duality conversion, the overall sum MSE is preserved, which means that $\varepsilon^{\mathrm{MAC}}$ from (3.6) must be equal to $\varepsilon^{\mathrm{BC}}$ from (3.2). In addition, the same amount of transmit power has to be dissipated in the BC and in the MAC. To achieve both goals, only a single scalar $\psi \in \mathbb{R}_+$ that depends on the noise covariance matrix $\boldsymbol{C}_\eta$ and the composite equalizer filter $\boldsymbol{B}$ has to be determined when we set

$$
\begin{aligned}
\boldsymbol{T} &= \psi \boldsymbol{A}^{\mathrm{H}} \boldsymbol{B}^{\mathrm{H}} \\
\boldsymbol{G} &= \psi^{-1} \boldsymbol{P}^{\mathrm{H}}
\end{aligned} \tag{3.9}
$$

i.e., the equalizer filter $\boldsymbol{G}$ in the dual MAC is a scaled version of the Hermitian precoders $\boldsymbol{P}^{\mathrm{H}}$ in the BC, and the precoders $\boldsymbol{T}$ in the MAC is a transformed version of the Hermitian equalizer filter $\boldsymbol{B}^{\mathrm{H}}$ in the BC. The matrix $\boldsymbol{A} = \mathrm{blockdiag}\{\boldsymbol{A}_k\}_{k=1}^{K} \in \mathbb{C}^{M \times M}$ is block-diagonal and invertible. The two degrees of freedom ψ and $\boldsymbol{A}$ can be used to equate $\varepsilon^{\mathrm{BC}}$ and $\varepsilon^{\mathrm{MAC}}$. Inserting $\boldsymbol{T}$ and $\boldsymbol{G}$ from (3.9) into the overall MAC sum MSE expression in (3.6), we obtain

$$
\varepsilon^{\mathrm{MAC}} = L + \mathrm{tr}\big(\boldsymbol{P}^{\mathrm{H}} \boldsymbol{H} \boldsymbol{A}^{\mathrm{H}} \boldsymbol{B}^{\mathrm{H}} \boldsymbol{B} \boldsymbol{A} \boldsymbol{H}^{\mathrm{H}} \boldsymbol{P} + \psi^{-2} \boldsymbol{P}^{\mathrm{H}} \boldsymbol{P} - \boldsymbol{P}^{\mathrm{H}} \boldsymbol{H} \boldsymbol{A}^{\mathrm{H}} \boldsymbol{B}^{\mathrm{H}} - \boldsymbol{B} \boldsymbol{A} \boldsymbol{H}^{\mathrm{H}} \boldsymbol{P}\big). \tag{3.10}
$$

The two subtrahends in (3.10) match those of $\varepsilon^{\mathrm{BC}}$ in (3.2), if we choose the composite channel matrix $\boldsymbol{H}$ in the dual MIMO MAC such that

$$
\boldsymbol{H} = \boldsymbol{H}^{\mathrm{BC,H}} \boldsymbol{A}^{-\mathrm{H}} \Leftrightarrow \boldsymbol{H}_k = \boldsymbol{H}_k^{\mathrm{BC,H}} \boldsymbol{A}_k^{-\mathrm{H}} \quad \forall k \in \{1, \ldots, K\}. \tag{3.11}
$$

In other words, the channel matrices in the dual MIMO MAC are the Hermitian channel matrices of the MIMO BC transformed by the matrices $\boldsymbol{A}_1, \ldots, \boldsymbol{A}_K$. Since this dual MAC has to match the BC for variable precoding and equalizer filters $\boldsymbol{P}$ and $\boldsymbol{B}$, the matrix $\boldsymbol{A}$ may depend only on the noise covariance matrix $\boldsymbol{C}_\eta$ and the channel $\boldsymbol{H}^{\mathrm{BC}}$. Equating $\varepsilon^{\mathrm{BC}}$ in (3.2) and $\varepsilon^{\mathrm{MAC}}$ in (3.10) for the choice of the composite MAC channel matrix $\boldsymbol{H}$ in (3.11), we get

$$
\mathrm{tr}(\boldsymbol{B} \boldsymbol{C}_\eta \boldsymbol{B}^{\mathrm{H}}) = \psi^{-2} \|\boldsymbol{P}\|_{\mathrm{F}}^2
$$

which in conjunction with (2.2) shows that the positive solution for the duality conversion scalar ψ is given by

$$
\psi = \sqrt{\frac{P_{\mathrm{diss}}^{\mathrm{BC}}}{\mathrm{tr}(\boldsymbol{B} \boldsymbol{C}_\eta \boldsymbol{B}^{\mathrm{H}})}} = \sqrt{\frac{P_{\mathrm{diss}}^{\mathrm{BC}}}{\sum_{k=1}^{K} \mathrm{tr}(\boldsymbol{B}_k \boldsymbol{C}_{\eta_k} \boldsymbol{B}_k^{\mathrm{H}})}} \tag{3.12}
$$

i.e., ψ^2 is the ratio of the totally dissipated power in the BC divided by the overall filtered noise power. While the equality $\varepsilon^{\mathrm{BC}} = \varepsilon^{\mathrm{MAC}}$ can be assured by the choice of ψ, the matrix $\boldsymbol{A}$ can be chosen to ensure that the same overall sum MSE as in the BC can be achieved in the dual MAC using the same amount of transmit power. Using (2.6), (3.9), and (3.12), the dissipated power in the MAC can be expressed as

$$
P_{\mathrm{diss}}^{\mathrm{MAC}} = \|\boldsymbol{T}\|_{\mathrm{F}}^2 = \psi^2 \|\boldsymbol{A}^{\mathrm{H}} \boldsymbol{B}^{\mathrm{H}}\|_{\mathrm{F}}^2 = P_{\mathrm{diss}}^{\mathrm{BC}} \cdot \frac{\mathrm{tr}(\boldsymbol{B} \boldsymbol{A} \boldsymbol{A}^{\mathrm{H}} \boldsymbol{B}^{\mathrm{H}})}{\mathrm{tr}(\boldsymbol{B} \boldsymbol{C}_\eta \boldsymbol{B}^{\mathrm{H}})}.
$$

Equality between $P_{\mathrm{diss}}^{\mathrm{BC}}$ and $P_{\mathrm{diss}}^{\mathrm{MAC}}$ holds, if $\boldsymbol{A}$ is chosen to satisfy

$$\boldsymbol{A}\boldsymbol{A}^{\mathrm{H}} = \boldsymbol{C}_{\eta} \Leftrightarrow \boldsymbol{A}_k\boldsymbol{A}_k^{\mathrm{H}} = \boldsymbol{C}_{\eta_k} \quad \forall k \in \{1, \ldots, K\}$$

which is achieved for any (not necessarily Hermitian for a faster computation) square root matrix

$$\boldsymbol{A}_k = \boldsymbol{C}_{\eta_k}^{\frac{1}{2}} \tag{3.13}$$

that satisfies $\boldsymbol{C}_{\eta_k}^{\frac{1}{2}}\boldsymbol{C}_{\eta_k}^{\frac{1}{2},\mathrm{H}} = \boldsymbol{C}_{\eta_k}$. Together with (3.11), the channel matrices of the dual MIMO MAC have to be chosen such that

$$\boldsymbol{H}_k = \boldsymbol{H}_k^{\mathrm{BC,H}}\boldsymbol{C}_{\eta_k}^{-\frac{1}{2},\mathrm{H}}. \tag{3.14}$$

For uncorrelated noise terms $\boldsymbol{\eta}_1, \ldots, \boldsymbol{\eta}_K$ in the BC with identity covariance matrix, the MAC channels are simply the Hermitian BC channels. With (3.12) and (3.13), the optimum composite precoders $\boldsymbol{T}$ and receive filter $\boldsymbol{G}$ in the MAC can be computed from (3.9).

Synopsis: For any MIMO BC with channel matrices $\boldsymbol{H}_1^{\mathrm{BC}}, \ldots, \boldsymbol{H}_K^{\mathrm{BC}}$ and arbitrary linear transmit and receive filters there exists a dual MIMO MAC with channel matrices

$$\boldsymbol{H}_k = \boldsymbol{H}_k^{\mathrm{BC,H}}\boldsymbol{C}_{\eta_k}^{-\frac{1}{2},\mathrm{H}} \forall k$$

such that if an overall sum MSE in the BC is feasible, it is also achievable in the dual MAC with the same amount of total transmit power. The conversion rules from the BC to the dual MAC for the composite precoding filter $\boldsymbol{T}$ and the composite equalizer filter $\boldsymbol{G}$ in the MAC read as

$$\boldsymbol{T} = \psi\boldsymbol{C}_{\eta}^{\frac{1}{2},\mathrm{H}}\boldsymbol{B}^{\mathrm{H}} \quad \text{and} \quad \boldsymbol{G} = \psi^{-1}\boldsymbol{P}^{\mathrm{H}} \quad \text{with} \quad \psi = \sqrt{\frac{P_{\mathrm{diss}}^{\mathrm{BC}}}{\mathrm{tr}(\boldsymbol{B}\boldsymbol{C}_{\eta}\boldsymbol{B}^{\mathrm{H}})}}.$$

3.2.3.2 Level Two BC to MAC Conversion

The second level of the duality keeps the sum MSEs of the K users constant during the BC to MAC conversion. Equating $\varepsilon_k^{\mathrm{BC}}$ and $\varepsilon_k^{\mathrm{MAC}}$ for all $k \in \{1, \ldots, K\}$ requires K scaling factors $\psi_1, \ldots, \psi_K$ (that can be chosen to be positive) instead of a single one as in the level one duality conversion. Similar to (3.9), the dual MIMO MAC filter parametrization is set to

$$\begin{aligned} \boldsymbol{T}_k &= \psi_k\boldsymbol{A}_k^{\mathrm{H}}\boldsymbol{B}_k^{\mathrm{H}} \\ \boldsymbol{G}_k &= \psi_k^{-1}\boldsymbol{P}_k^{\mathrm{H}} \quad \forall k \in \{1, \ldots, K\}. \end{aligned} \tag{3.15}$$

Substituting these precoding and equalizer filters into the MSE expression of user k in (3.7), we obtain

$$\begin{aligned}\varepsilon_k^{\mathrm{MAC}} &= L_k + \mathrm{tr}\Big(\sum_{i=1}^{K} \frac{\psi_i^2}{\psi_k^2} \boldsymbol{P}_k^{\mathrm{H}} \boldsymbol{H}_i \boldsymbol{A}_i^{\mathrm{H}} \boldsymbol{B}_i^{\mathrm{H}} \boldsymbol{B}_i \boldsymbol{A}_i \boldsymbol{H}_i^{\mathrm{H}} \boldsymbol{P}_k + \psi_k^{-2} \boldsymbol{P}_k^{\mathrm{H}} \boldsymbol{P}_k\Big) \\ &\quad - \mathrm{tr}\big(\boldsymbol{P}_k^{\mathrm{H}} \boldsymbol{H}_k \boldsymbol{A}_k^{\mathrm{H}} \boldsymbol{B}_k^{\mathrm{H}}\big) - \mathrm{tr}\big(\boldsymbol{B}_k \boldsymbol{A}_k \boldsymbol{H}_k^{\mathrm{H}} \boldsymbol{P}_k\big). \end{aligned} \tag{3.16}$$

The two subtrahends in (3.16) and (3.3) are identical, if [cf. (3.11)]

$$\boldsymbol{H}_k = \boldsymbol{H}_k^{\mathrm{BC,H}} \boldsymbol{A}_k^{-\mathrm{H}} \quad \forall k \in \{1, \ldots, K\}. \tag{3.17}$$

For the conservation of the K MSEs, we equate $\varepsilon_k^{\mathrm{BC}}$ in (3.3) with $\varepsilon_k^{\mathrm{MAC}}$ in (3.16) for all k using the dual MAC channels in (3.17) and the identity $\mathrm{tr}(\boldsymbol{X}^{\mathrm{H}}\boldsymbol{X}) = \mathrm{tr}(\boldsymbol{X}\boldsymbol{X}^{\mathrm{H}}) = \|\boldsymbol{X}\|_{\mathrm{F}}^2$:

$$\begin{aligned}\varepsilon_k^{\mathrm{BC}} = \varepsilon_k^{\mathrm{MAC}} \Leftrightarrow \psi_k^2\Big[\sum_{i\neq k} \|\boldsymbol{B}_k \boldsymbol{H}_k^{\mathrm{BC}} \boldsymbol{P}_i\|_{\mathrm{F}}^2 + \mathrm{tr}(\boldsymbol{B}_k \boldsymbol{C}_{\eta_k} \boldsymbol{B}_k^{\mathrm{H}})\Big] \\ - \sum_{i\neq k} \psi_i^2 \|\boldsymbol{B}_i \boldsymbol{H}_i^{\mathrm{BC}} \boldsymbol{P}_k\|_{\mathrm{F}}^2 = \|\boldsymbol{P}_k\|_{\mathrm{F}}^2\end{aligned}$$

Above equation must hold for all users $k \in \{1, \ldots, K\}$, which can conveniently be expressed in matrix vector notation:

$$\boldsymbol{Y}' \cdot \big[\psi_1^2, \ldots, \psi_K^2\big]^{\mathrm{T}} = \big[\|\boldsymbol{P}_1\|_{\mathrm{F}}^2, \ldots, \|\boldsymbol{P}_K\|_{\mathrm{F}}^2\big]^{\mathrm{T}} \tag{3.18}$$

The entries of the $K \times K$ matrix $\boldsymbol{Y}'$ read as

$$[\boldsymbol{Y}']_{k,j} := \begin{cases} \sum_{i\neq k} \|\boldsymbol{B}_k \boldsymbol{H}_k^{\mathrm{BC}} \boldsymbol{P}_i\|_{\mathrm{F}}^2 + \mathrm{tr}(\boldsymbol{B}_k \boldsymbol{C}_{\eta_k} \boldsymbol{B}_k^{\mathrm{H}}) & \text{for } \ j = k \\ \quad - \|\boldsymbol{B}_j \boldsymbol{H}_j^{\mathrm{BC}} \boldsymbol{P}_k\|_{\mathrm{F}}^2 & \text{for } \ j \neq k. \end{cases} \tag{3.19}$$

It is easy to see that $\boldsymbol{Y}'$ is column diagonally dominant [43, Definition 4.5]. If, in addition, none of the equalizer filters $\boldsymbol{B}_1, \ldots, \boldsymbol{B}_K$ is the zero matrix, $\boldsymbol{Y}'$ is even strictly column diagonally dominant, which we assume to be the case (the noise covariance matrices have full rank). Passive users, whose equalizers *and* precoders are zero simultaneously can simply be dropped from (3.19) since they are passive in the dual MAC as well and can be dropped from the system model. For semi-active users to which data is actively transmitted but their equalizer filters are the zero matrix, or who do not get any data at all but have nonzero equalizer filters, see [18]. From the strict column diagonal dominance, it follows that $\boldsymbol{Y}'$ is invertible [43, Corollary 4.8]. Moreover, $\boldsymbol{Y}'$ is real-valued and has only nonpositive off-diagonal entries, so $\boldsymbol{Y}'$ is also a *Z-matrix* [44, p. 132]. Together with the strict column diagonal dominance, being a *Z-matrix* also means being a nonsingular *M-matrix* [44, p. 133], which only has nonnegative entries in its inverse [44, p. 139] with at least one positive element per row (otherwise, the inverse would be rank deficient), see Appendix A1 for a rigorous proof. As the right-hand side of (3.18) is strictly positive when only active users are assumed, the *squared* scaling factors $\psi_1^2, \ldots, \psi_K^2$ in (3.18) that may not be negative have the well-defined, *positive* solution

$$\left[\psi_1^2, \ldots, \psi_K^2\right]^{\mathrm{T}} = \boldsymbol{Y}'^{-1} \cdot \left[\|\boldsymbol{P}_1\|_{\mathrm{F}}^2, \ldots, \|\boldsymbol{P}_K\|_{\mathrm{F}}^2\right]^{\mathrm{T}} \tag{3.20}$$

which means that the conversion from the BC to the MAC can always be achieved without altering the sum MSEs of the K users. Instead of explicitly computing $\boldsymbol{Y}'^{-1}$, an *LU-factorization* [45, Sect. 3.2.5] can also be used to solve (3.18) which is computationally more efficient.

So far, we ensured that the K sum MSEs do not change when switching from the BC to the MAC. It remains to show that the dual MAC achieves these user-wise MSEs using the same amount of transmit power. Summing up the K rows of (3.18), we obtain

$$\begin{aligned}
\mathbf{1}_K^{\mathrm{T}} \boldsymbol{Y}' \cdot \left[\psi_1^2, \ldots, \psi_K^2\right]^{\mathrm{T}} &= \sum_{k=1}^{K} \psi_k^2 \operatorname{tr}(\boldsymbol{B}_k \boldsymbol{C}_{\eta_k} \boldsymbol{B}_k^{\mathrm{H}}) \\
&= \mathbf{1}_K^{\mathrm{T}} \cdot \left[\|\boldsymbol{P}_1\|_{\mathrm{F}}^2, \ldots, \|\boldsymbol{P}_K\|_{\mathrm{F}}^2\right]^{\mathrm{T}} = P_{\mathrm{diss}}^{\mathrm{BC}}
\end{aligned}$$

where we used (3.19) and the definition of $P_{\mathrm{diss}}^{\mathrm{BC}}$ in (2.2). In the dual MAC, the dissipated power $P_{\mathrm{diss}}^{\mathrm{MAC}}$ in (2.6) reads by means of (3.15) as

$$P_{\mathrm{diss}}^{\mathrm{MAC}} = \sum_{k=1}^{K} \operatorname{tr}(\boldsymbol{T}_k^{\mathrm{H}} \boldsymbol{T}_k) = \sum_{k=1}^{K} \psi_k^2 \operatorname{tr}(\boldsymbol{B}_k \boldsymbol{A}_k \boldsymbol{A}_k^{\mathrm{H}} \boldsymbol{B}_k^{\mathrm{H}}).$$

Equality between $P_{\mathrm{diss}}^{\mathrm{MAC}}$ and $P_{\mathrm{diss}}^{\mathrm{BC}}$ holds, if

$$\boldsymbol{A}_k \boldsymbol{A}_k^{\mathrm{H}} = \boldsymbol{C}_{\eta_k} \Rightarrow \boldsymbol{A}_k = \boldsymbol{C}_{\eta_k}^{\frac{1}{2}} \quad \forall k. \tag{3.21}$$

Hence, the channel matrices of the dual MAC are always constructed in the same way irrespective of the chosen level of duality, which allows us to state the following corollary:

Corollary 3.1 *If the noise covariance matrix in the dual MIMO MAC is chosen as the identity matrix, the MAC channel matrices $\boldsymbol{H}_1, \ldots, \boldsymbol{H}_K$ have to be constructed from the MIMO BC channel matrices $\boldsymbol{H}_1^{BC}, \ldots, \boldsymbol{H}_K^{BC}$ and the MIMO BC noise covariance matrices $\boldsymbol{C}_{\eta_1}, \ldots, \boldsymbol{C}_{\eta_K}$ such that*

$$\boldsymbol{H}_k = \boldsymbol{H}_k^{BC,H} \boldsymbol{C}_{\eta_k}^{-\frac{1}{2},H} \quad \forall k.$$

This result also holds for the level three duality.

Synopsis: Any user-wise MSE tuple that is feasible in a MIMO BC can also be achieved in the dual MIMO MAC with the same sum power. If the BC channel matrices are denoted by $\boldsymbol{H}_1^{\mathrm{BC}}, \ldots, \boldsymbol{H}_K^{\mathrm{BC}}$, the MAC channels have to be set to

$$\boldsymbol{H}_k = \boldsymbol{H}_k^{\mathrm{BC,H}} \boldsymbol{C}_{\eta_k}^{-\frac{1}{2},\mathrm{H}} \quad \forall k$$

if the noise covariance matrix in the MAC is an identity matrix. Furthermore, the MAC precoding and equalizer filters read as

$$\boldsymbol{T}_k = \psi_k \boldsymbol{C}_{\eta_k}^{\frac{1}{2},\mathrm{H}} \boldsymbol{B}_k^{\mathrm{H}} \quad \text{and} \quad \boldsymbol{G}_k = \psi_k^{-1} \boldsymbol{P}_k^{\mathrm{H}} \quad \forall k$$

where $\psi_1, \ldots, \psi_K$ are the positive roots of (3.20). The noise covariance matrices have to be included into the MAC channel and precoders definition to ensure the power conservation.

3.2.3.3 Level Three BC to MAC Conversion

The third level of the MSE duality does not only preserve the user-wise MSEs, but also the MSE of every single stream in the system making it thus attractive for optimizations with data stream based QoS requirements, see for example [36, 37, 46, 47]. Let the precoding filters of user k in the dual MIMO MAC and the MIMO BC be composed by the columns

$$\boldsymbol{T}_k = \left[\boldsymbol{t}_{k,1}, \ldots, \boldsymbol{t}_{k,L_k}\right] \quad \text{and} \quad \boldsymbol{P}_k = \left[\boldsymbol{p}_{k,1}, \ldots, \boldsymbol{p}_{k,L_k}\right] \tag{3.22}$$

and let the respective equalizer filters be composed by the rows

$$\boldsymbol{G}_k = \begin{bmatrix} \boldsymbol{g}_{k,1}^{\mathrm{T}} \\ \vdots \\ \boldsymbol{g}_{k,L_k}^{\mathrm{T}} \end{bmatrix} \quad \text{and} \quad \boldsymbol{B}_k = \begin{bmatrix} \boldsymbol{b}_{k,1}^{\mathrm{T}} \\ \vdots \\ \boldsymbol{b}_{k,L_k}^{\mathrm{T}} \end{bmatrix}. \tag{3.23}$$

For the dual MIMO MAC, we choose the parametrization

$$\begin{aligned} \boldsymbol{t}_{k,j} &= \psi_{k,j} \boldsymbol{A}_k^{\mathrm{H}} \boldsymbol{b}_{k,j}^* \\ \boldsymbol{g}_{k,j}^{\mathrm{T}} &= \psi_{k,j}^{-1} \boldsymbol{p}_{k,j}^{\mathrm{H}} \quad \forall k \, \forall j \end{aligned} \tag{3.24}$$

i.e., we associate L_k real-valued scalars $\psi_{k,1}, \ldots, \psi_{k,L_k}$ to user k, one for each data stream. Given the parametrization in (3.24), the MSE of the jth stream of user k from (3.8) in the MAC can be expressed as

$$\varepsilon_{k,j}^{\mathrm{MAC}} = 1 + \sum_{i=1}^{K}\sum_{\ell=1}^{L_i} \frac{\psi_{i,\ell}^2}{\psi_{k,j}^2} |\boldsymbol{b}_{i,\ell}^{\mathrm{T}} \boldsymbol{A}_i \boldsymbol{H}_i^{\mathrm{H}} \boldsymbol{p}_{k,j}|^2 + \psi_{k,j}^{-2} \|\boldsymbol{p}_{k,j}\|_2^2$$
$$- \boldsymbol{p}_{k,j}^{\mathrm{H}} \boldsymbol{H}_k \boldsymbol{A}_k^{\mathrm{H}} \boldsymbol{b}_{k,j}^{*} - \boldsymbol{b}_{k,j}^{\mathrm{T}} \boldsymbol{A}_k \boldsymbol{H}_k^{\mathrm{H}} \boldsymbol{p}_{k,j}. \tag{3.25}$$

Equating $\varepsilon_{k,j}^{\mathrm{MAC}}$ from (3.25) with $\varepsilon_{k,j}^{\mathrm{BC}}$ in (3.4) and using the MAC channels from (3.17) yields

$$\psi_{k,j}^2 \Big[\sum_{\substack{i=1 \\ (i,\ell)\neq(k,j)}}^{K} \sum_{\ell=1}^{L_i} |\boldsymbol{b}_{k,j}^{\mathrm{T}} \boldsymbol{H}_k^{\mathrm{BC}} \boldsymbol{p}_{i,\ell}|^2 + \boldsymbol{b}_{k,j}^{\mathrm{T}} \boldsymbol{C}_{\eta_k} \boldsymbol{b}_{k,j}^{*} \Big] - \sum_{\substack{i=1 \\ (i,\ell)\neq(k,j)}}^{K} \sum_{\ell=1}^{L_i} \psi_{i,\ell}^2 |\boldsymbol{b}_{i,\ell}^{\mathrm{T}} \boldsymbol{H}_i^{\mathrm{BC}} \boldsymbol{p}_{k,j}|^2$$
$$= \|\boldsymbol{p}_{k,j}\|_2^2. \tag{3.26}$$

When all L MSEs of the K users are equated, this can conveniently be expressed in matrix vector notation in analogy to (3.18):

$$\boldsymbol{Y} \cdot \left[\psi_{1,1}^2, \ldots, \psi_{K,L_K}^2\right]^{\mathrm{T}} = \left[\|\boldsymbol{p}_{1,1}\|_2^2, \ldots, \|\boldsymbol{p}_{K,L_K}\|_2^2\right]^{\mathrm{T}} \tag{3.27}$$

The $L \times L$ matrix $\boldsymbol{Y}$ reads as

$$\boldsymbol{Y} := (\boldsymbol{B}\boldsymbol{C}_{\eta}\boldsymbol{B}^{\mathrm{H}}) \odot \mathbf{I}_L + \mathrm{diag}\{\mathbf{1}_L^{\mathrm{T}} \boldsymbol{Y}_0\} - \boldsymbol{Y}_0$$

with '$\odot$' denoting the element-wise Schur product and $\boldsymbol{Y}_0$ is defined as

$$\boldsymbol{Y}_0 := \left[(\boldsymbol{B}\boldsymbol{H}^{\mathrm{BC}}\boldsymbol{P}) \odot (\boldsymbol{B}\boldsymbol{H}^{\mathrm{BC}}\boldsymbol{P})^{*} \right]^{\mathrm{T}}. \tag{3.28}$$

For the same reasoning as for $\boldsymbol{Y}'$ in the level two BC to MAC conversion, $\boldsymbol{Y}$ is an *M-matrix*, if $\boldsymbol{b}_{k,j} \neq \mathbf{0}$ for all users k and all streams j. Again, passive streams can simply be dropped from (3.27) and semi-active streams are handled in [18]. The *M-matrix* property of $\boldsymbol{Y}$ ensures that there is a well-defined, positive solution to the L squared scaling factors $\psi_{1,1}^2, \ldots, \psi_{K,L_K}^2$:

$$\left[\psi_{1,1}^2, \ldots, \psi_{K,L_K}^2\right]^{\mathrm{T}} = \boldsymbol{Y}^{-1} \cdot \left[\|\boldsymbol{p}_{1,1}\|_2^2, \ldots, \|\boldsymbol{p}_{K,L_K}\|_2^2\right]^{\mathrm{T}}. \tag{3.29}$$

To see that the dissipated sum power is conserved when we choose $\boldsymbol{A}_1, \ldots, \boldsymbol{A}_K$ and thus implicitly the channel matrices in the MAC according to Corollary 1, we sum up all rows of (3.27):

$$\mathbf{1}_L^{\mathrm{T}} \boldsymbol{Y} \cdot \left[\psi_{1,1}^2, \ldots, \psi_{K,L_K}^2\right]^{\mathrm{T}} = \mathbf{1}_L^{\mathrm{T}} \cdot \left[\|\boldsymbol{p}_{1,1}\|_2^2, \ldots, \|\boldsymbol{p}_{K,L_K}\|_2^2\right]^{\mathrm{T}}$$
$$\sum_{k=1}^{K}\sum_{j=1}^{L_k} \psi_{k,j}^2 \boldsymbol{b}_{k,j}^{\mathrm{T}} \boldsymbol{C}_{\eta_k} \boldsymbol{b}_{k,j}^{*} = P_{\mathrm{diss}}^{\mathrm{BC}}.$$

The dissipated power in the MAC reads as [cf. (3.24)]

$$P_{\text{diss}}^{\text{MAC}} = \sum_{k=1}^{K}\sum_{j=1}^{L_k} \|\boldsymbol{t}_{k,j}\|_2^2 = \sum_{k=1}^{K}\sum_{j=1}^{L_k} \psi_{k,j}^2 \boldsymbol{b}_{k,j}^{\mathrm{T}} \boldsymbol{A}_k \boldsymbol{A}_k^{\mathrm{H}} \boldsymbol{b}_{k,j}^*$$

which matches $P_{\text{diss}}^{\text{BC}}$, if $\boldsymbol{A}_1, \ldots, \boldsymbol{A}_K$ are chosen as in (3.21). Summing up, the conversion rules from the MIMO BC to the MIMO MAC read as

$$\begin{aligned} \boldsymbol{t}_{k,j} &= \psi_{k,j} \boldsymbol{C}_{\eta_k}^{\frac{1}{2},\mathrm{H}} \boldsymbol{b}_{k,j}^* \\ \boldsymbol{g}_{k,j}^{\mathrm{T}} &= \psi_{k,j}^{-1} \boldsymbol{p}_{k,j}^{\mathrm{H}} \quad \forall k \ \forall j \end{aligned} \tag{3.30}$$

where $\psi_{1,1}, \ldots, \psi_{K,L_K}$ are the positive roots of (3.29).

Besides the individual MSEs of all L data streams in the system, the presented level three duality conversion from the BC to the MAC also preserves the SINRs of these streams as we will show in the following. For that purpose, we define the SINRs of the jth data stream belonging to user k in the BC and MAC as the ratio of the expected desired signal power and the sum of the expected noise and interference power, where the expectation is taken with respect to the noise and the symbols:

$$\begin{aligned} \text{SINR}_{k,j}^{\text{BC}}(\boldsymbol{P}, \boldsymbol{b}_{k,j}) &:= \frac{|\boldsymbol{b}_{k,j}^{\mathrm{T}} \boldsymbol{H}_k^{\text{BC}} \boldsymbol{p}_{k,j}|^2}{\sum_{i=1}^{K} \sum_{\ell=1}^{L_i}{}_{(i,\ell)\neq(k,j)} |\boldsymbol{b}_{k,j}^{\mathrm{T}} \boldsymbol{H}_k^{\text{BC}} \boldsymbol{p}_{i,\ell}|^2 + \boldsymbol{b}_{k,j}^{\mathrm{T}} \boldsymbol{C}_{\eta_k} \boldsymbol{b}_{k,j}^*} \\ \text{SINR}_{k,j}^{\text{MAC}}(\boldsymbol{T}, \boldsymbol{g}_{k,j}) &:= \frac{|\boldsymbol{g}_{k,j}^{\mathrm{T}} \boldsymbol{H}_k \boldsymbol{t}_{k,j}|^2}{\sum_{i=1}^{K} \sum_{\ell=1}^{L_i}{}_{(i,\ell)\neq(k,j)} |\boldsymbol{g}_{k,j}^{\mathrm{T}} \boldsymbol{H}_i \boldsymbol{t}_{i,\ell}|^2 + \|\boldsymbol{g}_{k,j}\|_2^2}. \end{aligned} \tag{3.31}$$

Inserting $\boldsymbol{t}_{k,j}$ and $\boldsymbol{g}_{k,j}^{\mathrm{T}}$ from (3.30) and the MAC-BC channel relationship from Corollary 1 into (3.31), we see that the two numerators are equivalent. Equating the two denominators in addition leads to (3.26), so $\text{SINR}_{k,j}^{\text{BC}}$ and $\text{SINR}_{k,j}^{\text{MAC}}$ are the same if the two MSEs $\varepsilon_{k,j}^{\text{BC}}$ and $\varepsilon_{k,j}^{\text{MAC}}$ are the same. Hence, our MSE-based level-three duality can also be used to solve SINR-based problems in the BC.

Synopsis: Any stream-wise MSE tuple and any SINR tuple that is feasible in the MIMO BC with linear filtering can also be achieved in the dual MIMO MAC with linear transceivers and the same transmit power. If the noise covariance matrix in this dual MAC is the identity matrix, the MAC channels $\boldsymbol{H}_1, \ldots, \boldsymbol{H}_K$ follow from the BC channels via

$$\boldsymbol{H}_k = \boldsymbol{H}_k^{\text{BC},\mathrm{H}} \boldsymbol{C}_{\eta_k}^{-\frac{1}{2},\mathrm{H}} \quad \forall k.$$

The precoding and equalizer filter for the jth data stream of user k in the dual MAC read as

$$t_{k,j} = \psi_{k,j} C_{\eta_k}^{\frac{1}{2},\mathrm{H}} b_{k,j}^*$$
$$g_{k,j}^{\mathrm{T}} = \psi_{k,j}^{-1} p_{k,j}^{\mathrm{H}}$$

where the scaling factors $\psi_{1,1}, \ldots, \psi_{K,L_K}$ are the positive roots of (3.29) and $p_{k,j}$ and $b_{k,j}^{\mathrm{T}}$, respectively, are the precoding and equalizer filter in the BC.

3.2.4 MAC to BC Conversion

The MSE-based BC↔MAC duality is always applied as a tool to solve a problem stated in the BC. In Sect. 3.2.3, it was shown that a dual MIMO MAC can be constructed that features at least the same overall, user-wise, and stream-wise MSE region. Apart from representing one direction of the duality proof, the conversion rules from the BC to the MAC are usually not applied in practice since the problem setting can completely be solved in the dual MAC and then be converted back to the original BC given the conversion rules deduced now. Nonetheless, some iterative algorithms exist that repeatedly switch between the BC and the MAC such that the BC to MAC conversion really has to be implemented, see for example [47, 48].

In this subsection, we derive the converse, i.e., the MAC to BC direction of the duality proof for all three levels, which is stated in the following lemma:

Lemma 3.2 *For any set of linear transceivers in the dual MIMO MAC, there exist linear precoding and equalizer filters in the MIMO BC such that the same overall sum MSE, the same user-wise MSEs, and the same stream-wise MSEs can be achieved with the same amount of transmit power. This constitutes the other direction of the MSE duality for all three levels.*

Throughout this subsection, we assume that the relationship between the MIMO MAC and the MIMO BC channels is given by Corollary 1, i.e.,

$$H_k = H_k^{\mathrm{BC,H}} C_{\eta_k}^{-\frac{1}{2},\mathrm{H}} \quad \forall k \in \{1, \ldots, K\}.$$

3.2.4.1 Level One MAC to BC Conversion

For the first level of the MAC to BC conversion, a precoder P and an equalizer G from the dual MAC are converted to the BC without a change of the overall sum MSE using the BC parametrization

$$\begin{aligned} P &= \varphi G^{\mathrm{H}} \\ B &= \varphi^{-1} T^{\mathrm{H}} A^{-1}. \end{aligned} \tag{3.32}$$

It will become clear below why the matrix $\boldsymbol{A}$ arises at the equalizer filter $\boldsymbol{B}$ and not at the precoding filter as in the parametrization for the BC to MAC conversion in (3.9). Using the representation in (3.32), the overall sum MSE $\varepsilon^{\mathrm{BC}}$ from (3.2) reads as

$$\varepsilon^{\mathrm{BC}} = L + \mathrm{tr}\big(\boldsymbol{T}^{\mathrm{H}}\boldsymbol{A}^{-1}\boldsymbol{H}^{\mathrm{BC}}\boldsymbol{G}^{\mathrm{H}}\boldsymbol{G}\boldsymbol{H}^{\mathrm{BC,H}}\boldsymbol{A}^{-\mathrm{H}}\boldsymbol{T} + \varphi^{-2}\boldsymbol{T}^{\mathrm{H}}\boldsymbol{A}^{-1}\boldsymbol{C}_{\eta}\boldsymbol{A}^{-\mathrm{H}}\boldsymbol{T}\big)$$
$$-\mathrm{tr}\big(\boldsymbol{T}^{\mathrm{H}}\boldsymbol{A}^{-1}\boldsymbol{H}^{\mathrm{BC}}\boldsymbol{G}^{\mathrm{H}}\big) - \mathrm{tr}\big(\boldsymbol{G}\boldsymbol{H}^{\mathrm{BC,H}}\boldsymbol{A}^{-\mathrm{H}}\boldsymbol{T}\big). \tag{3.33}$$

The two subtrahends of $\varepsilon^{\mathrm{BC}}$ in (3.33) automatically match those of $\varepsilon^{\mathrm{MAC}}$ in (3.6) given the MAC-BC channel relationship in Corollary 1 if $\boldsymbol{A}$ is chosen as supplied before in (3.11), i.e., $\boldsymbol{A}\boldsymbol{A}^{\mathrm{H}} = \boldsymbol{C}_{\eta}$. This match requires the matrix $\boldsymbol{A}$ to be linked to the receive filter $\boldsymbol{B}$ in the original BC and not to the precoders $\boldsymbol{P}$. In addition, equating $\varepsilon^{\mathrm{BC}}$ and $\varepsilon^{\mathrm{MAC}}$ yields

$$\varepsilon^{\mathrm{BC}} = \varepsilon^{\mathrm{MAC}} \Leftrightarrow \varphi^{-2}\|\boldsymbol{T}\|_{\mathrm{F}}^2 = \|\boldsymbol{G}\|_{\mathrm{F}}^2.$$

Using the dissipated power definition $P_{\mathrm{diss}}^{\mathrm{MAC}} = \|\boldsymbol{T}\|_{\mathrm{F}}^2$ in the MAC, the positive solution of the scaling factor φ is

$$\varphi = \frac{\sqrt{P_{\mathrm{diss}}^{\mathrm{MAC}}}}{\|\boldsymbol{G}\|_{\mathrm{F}}}.$$

In combination with the parametrization for $\boldsymbol{P}$ in (3.32), the dissipated power in the BC is exactly the same as in the dual MAC:

$$P_{\mathrm{diss}}^{\mathrm{BC}} = \|\boldsymbol{P}\|_{\mathrm{F}}^2 = \varphi^2\|\boldsymbol{G}\|_{\mathrm{F}}^2 = P_{\mathrm{diss}}^{\mathrm{MAC}}.$$

Synopsis: Let $\boldsymbol{T}$ and $\boldsymbol{G}$ denote the composite precoding and equalizer filter in the dual MAC. Then, the composite precoding filter $\boldsymbol{P}$ and the composite equalizer filter $\boldsymbol{B}$ in the BC achieve the same overall sum MSE using the same amount of power $P_{\mathrm{diss}}^{\mathrm{MAC}}$, if

$$\boldsymbol{P} = \sqrt{P_{\mathrm{diss}}^{\mathrm{MAC}}}\frac{\boldsymbol{G}^{\mathrm{H}}}{\|\boldsymbol{G}\|_{\mathrm{F}}} \quad \text{and} \quad \boldsymbol{B} = \frac{\|\boldsymbol{G}\|_{\mathrm{F}}}{\sqrt{P_{\mathrm{diss}}^{\mathrm{MAC}}}}\boldsymbol{T}^{\mathrm{H}}\boldsymbol{C}_{\eta}^{-\frac{1}{2}}.$$

3.2.4.2 Level Two MAC to BC Conversion

To preserve the user MSEs of the K users, K scaling factors $\varphi_1, \ldots, \varphi_K$ (that may be chosen to be from $\mathbb{R}_+$ each), one for each transmit-receive filter pair, have to be determined. This second level of the MSE duality becomes interesting for problem settings which are formulated user-wise, e.g., the weighted sum MSE minimization where different weights can be chosen for each user in the cost function. Similar to the BC to MAC conversion, we choose the filter parametrization

$$\begin{aligned} \boldsymbol{P}_k &= \varphi_k \boldsymbol{G}_k^{\mathrm{H}} \\ \boldsymbol{B}_k &= \varphi_k^{-1} \boldsymbol{T}_k^{\mathrm{H}} \boldsymbol{A}_k^{-1} \quad \forall k \in \{1, \ldots, K\}. \end{aligned} \tag{3.34}$$

Plugging these representations into the BC MSE expression of user k in (3.3), we obtain

$$\begin{aligned} \varepsilon_k^{\mathrm{BC}} = L_k &+ \operatorname{tr}\Big(\sum_{i=1}^{K} \frac{\varphi_i^2}{\varphi_k^2} \boldsymbol{T}_k^{\mathrm{H}} \boldsymbol{A}_k^{-1} \boldsymbol{H}_k^{\mathrm{BC}} \boldsymbol{G}_i^{\mathrm{H}} \boldsymbol{G}_i \boldsymbol{H}_k^{\mathrm{BC,H}} \boldsymbol{A}_k^{-\mathrm{H}} \boldsymbol{T}_k \\ &+ \varphi_k^{-2} \boldsymbol{T}_k^{\mathrm{H}} \boldsymbol{A}_k^{-1} \boldsymbol{C}_{\eta_k} \boldsymbol{A}_k^{-\mathrm{H}} \boldsymbol{T}_k \Big) - \operatorname{tr}\big(\boldsymbol{T}_k^{\mathrm{H}} \boldsymbol{A}_k^{-1} \boldsymbol{H}_k^{\mathrm{BC}} \boldsymbol{G}_k^{\mathrm{H}}\big) \\ &- \operatorname{tr}\big(\boldsymbol{G}_k \boldsymbol{H}_k^{\mathrm{BC,H}} \boldsymbol{A}_k^{-\mathrm{H}} \boldsymbol{T}_k\big) \end{aligned}$$

where the two subtrahends again match those of $\varepsilon_k^{\mathrm{MAC}}$ in (3.7) due to the MAC-BC channel relationship in Corollary 1 and the choice of $\boldsymbol{A}_k$ from (3.21). Equating $\varepsilon_k^{\mathrm{BC}}$ with $\varepsilon_k^{\mathrm{MAC}}$ from (3.7) yields

$$\varepsilon_k^{\mathrm{BC}} = \varepsilon_k^{\mathrm{MAC}} \Leftrightarrow \varphi_k^2 \Big[\sum_{i \neq k} \|\boldsymbol{G}_k \boldsymbol{H}_i \boldsymbol{T}_i\|_{\mathrm{F}}^2 + \|\boldsymbol{G}_k\|_{\mathrm{F}}^2 \Big] - \sum_{i \neq k} \varphi_i^2 \|\boldsymbol{G}_i \boldsymbol{H}_k \boldsymbol{T}_k\|_{\mathrm{F}}^2 = \|\boldsymbol{T}_k\|_{\mathrm{F}}^2.$$

Since this equality has to hold for all users $k \in \{1, \ldots, K\}$, we end up with the following linear system of equations for the squared scaling factors $\varphi_1^2, \ldots, \varphi_K^2$:

$$\boldsymbol{Z}' \cdot \big[\varphi_1^2, \ldots, \varphi_K^2\big]^{\mathrm{T}} = \big[\|\boldsymbol{T}_1\|_{\mathrm{F}}^2, \ldots, \|\boldsymbol{T}_K\|_{\mathrm{F}}^2\big]^{\mathrm{T}}. \tag{3.35}$$

The entries of the $K \times K$ matrix $\boldsymbol{Z}'$ read as

$$[\boldsymbol{Z}']_{k,j} := \begin{cases} \sum_{i \neq k} \|\boldsymbol{G}_k \boldsymbol{H}_i \boldsymbol{T}_i\|_{\mathrm{F}}^2 + \|\boldsymbol{G}_k\|_{\mathrm{F}}^2 & \text{for } j = k \\ -\|\boldsymbol{G}_j \boldsymbol{H}_k \boldsymbol{T}_k\|_{\mathrm{F}}^2 & \text{for } j \neq k. \end{cases} \tag{3.36}$$

In analogy to the BC to MAC conversion with user-wise MSE conservation, the matrix $\boldsymbol{Z}'$ is column diagonally dominant [43, Definition 4.5]. In addition, it is strictly column diagonally dominant if none the K receive filters $\boldsymbol{G}_1, \ldots, \boldsymbol{G}_K$ is the zero matrix, which we assume to be true here. A completely passive user whose transmit-receive filter pair vanishes in the MAC can be dropped from the duality conversion since he is completely passive in the BC as well. Here, we assume that all users have nonvanishing transmit and receive filters. For semi-active users, see [18]. Due to its nonpositive off-diagonal elements, the matrix $\boldsymbol{Z}'$ is a *Z-matrix* [44, p. 132], which in combination with the strict column diagonal dominance implies that $\boldsymbol{Z}'$ is also a nonsingular *M-matrix* [44, p. 133], whose inverse has only nonnegative entries [44, p. 132] with at least one positive entry per row. Given the positive right-hand side of (3.35) from the K active users, a well-defined, positive solution to the *squared* real-valued scalars $\varphi_1^2, \ldots, \varphi_K^2$ exists:

$$\left[\varphi_1^2, \ldots, \varphi_K^2\right]^{\mathrm{T}} = \boldsymbol{Z}'^{-1} \cdot \left[\|\boldsymbol{T}_1\|_{\mathrm{F}}^2, \ldots, \|\boldsymbol{T}_K\|_{\mathrm{F}}^2\right]^{\mathrm{T}}. \tag{3.37}$$

The scalars $\varphi_1, \ldots, \varphi_K$ can then be chosen as the positive roots of above equation and the same user-wise MSEs are achieved in the BC as in the dual MAC. Summing up all rows of (3.35) yields

$$\mathbf{1}_K^{\mathrm{T}} \boldsymbol{Z}' \cdot \left[\varphi_1^2, \ldots, \varphi_K^2\right]^{\mathrm{T}} = \sum_{k=1}^{K} \varphi_k^2 \|\boldsymbol{G}_k\|_{\mathrm{F}}^2 = \mathbf{1}_k^{\mathrm{T}} \cdot \left[\|\boldsymbol{T}_1\|_{\mathrm{F}}^2, \ldots, \|\boldsymbol{T}_K\|_{\mathrm{F}}^2\right]^{\mathrm{T}} = P_{\mathrm{diss}}^{\mathrm{MAC}}$$

where we used (2.6) for the definition of $P_{\mathrm{diss}}^{\mathrm{MAC}}$. Noticing that the BC transmit filter $\boldsymbol{P}_k$ is linked to the MAC receive filter $\boldsymbol{G}_k$ via $\boldsymbol{P}_k = \varphi_k \boldsymbol{G}_k^{\mathrm{H}}$ from (3.34), above equation can be written as [cf. 2.2)]

$$P_{\mathrm{diss}}^{\mathrm{MAC}} = \sum_{k=1}^{K} \varphi_k^2 \|\boldsymbol{G}_k\|_{\mathrm{F}}^2 = \sum_{k=1}^{K} \|\boldsymbol{P}_k\|_{\mathrm{F}}^2 = P_{\mathrm{diss}}^{\mathrm{BC}}.$$

So the same amount of power is dissipated in the original BC as in the dual MAC.

Synopsis: Any user-wise MSE tuple that is feasible in the dual MIMO MAC can also be achieved in the MIMO BC with the same sum power. The BC precoding and receive filters read as

$$\boldsymbol{P}_k = \varphi_k \boldsymbol{G}_k^{\mathrm{H}} \quad \text{and} \quad \boldsymbol{B}_k = \varphi_k^{-1} \boldsymbol{T}_k^{\mathrm{H}} \boldsymbol{C}_{\eta_k}^{-\frac{1}{2}} \quad \forall k$$

where the scaling factors $\varphi_1, \ldots, \varphi_K$ are the positive roots of (3.37).

3.2.4.3 Level Three MAC to BC Conversion

Given the transmit and receive filter partitioning in (3.22) and (3.23), the MIMO BC filter parametrization that preserves the stream-wise MSEs of all L data streams can be written as

$$\begin{aligned} \boldsymbol{p}_{k,j} &= \varphi_{k,j}\boldsymbol{g}_{k,j}^*, \\ \boldsymbol{b}_{k,j}^{\mathrm{T}} &= \varphi_{k,j}^{-1}\boldsymbol{t}_{k,j}^{\mathrm{H}}\boldsymbol{A}_k^{-1} \quad \forall k\,\forall j \end{aligned} \tag{3.38}$$

where the scaling factors $\varphi_{1,1},\ldots,\varphi_{K,L_K}$ are again positive. When these BC filters are inserted into the MSE expression for the jth stream of user k in (3.4), we get

$$\begin{aligned} \varepsilon_{k,j}^{\mathrm{BC}} = 1 + \sum_{i=1}^{K}\sum_{\ell=1}^{L_i} \frac{\varphi_{i,\ell}^2}{\varphi_{k,j}^2} |\boldsymbol{g}_{i,\ell}^{\mathrm{T}}\boldsymbol{H}_k^{\mathrm{BC,H}}\boldsymbol{A}_k^{-\mathrm{H}}\boldsymbol{t}_{k,j}|^2 + \varphi_{k,j}^{-2}\boldsymbol{t}_{k,j}^{\mathrm{H}}\boldsymbol{A}_k^{-1}\boldsymbol{C}_{\eta_k}\boldsymbol{A}_k^{-\mathrm{H}}\boldsymbol{t}_{k,j} \\ - \boldsymbol{t}_{k,j}^{\mathrm{H}}\boldsymbol{A}_k^{-1}\boldsymbol{H}_k^{\mathrm{BC}}\boldsymbol{g}_{k,j}^* - \boldsymbol{g}_{k,j}^{\mathrm{T}}\boldsymbol{H}_k^{\mathrm{BC,H}}\boldsymbol{A}_k^{-\mathrm{H}}\boldsymbol{t}_{k,j} \end{aligned}$$

and the two subtrahends match those of $\varepsilon_{k,j}^{\mathrm{MAC}}$ in (3.8) for $\boldsymbol{A}_k = \boldsymbol{C}_{\eta_k}^{\frac{1}{2}}$. Setting $\varepsilon_{k,j}^{\mathrm{BC}} = \varepsilon_{k,j}^{\mathrm{MAC}}$ delivers

$$\varphi_{k,j}^2\Big[\sum_{\substack{i=1\\(i,\ell)\neq(k,j)}}^{K}\sum_{\ell=1}^{L_i} |\boldsymbol{g}_{k,j}^{\mathrm{T}}\boldsymbol{H}_i\boldsymbol{t}_{i,\ell}|^2 + \|\boldsymbol{g}_{k,j}\|_2^2\Big] - \sum_{\substack{i=1\\(i,\ell)\neq(k,j)}}^{K}\sum_{\ell=1}^{L_i} \varphi_{i,\ell}^2|\boldsymbol{g}_{i,\ell}^{\mathrm{T}}\boldsymbol{H}_k\boldsymbol{t}_{k,j}|^2 = \|\boldsymbol{t}_{k,j}\|_2^2 \tag{3.39}$$

which can be put into a linear system of equations forcing $\varepsilon_{k,j}^{\mathrm{BC}} = \varepsilon_{k,j}^{\mathrm{MAC}}$ for all k and all j:

$$\boldsymbol{Z}\cdot\big[\varphi_{1,1}^2,\ldots,\varphi_{K,L_K}^2\big]^{\mathrm{T}} = \big[\|\boldsymbol{t}_{1,1}\|_2^2,\ldots,\|\boldsymbol{t}_{K,L_K}\|_2^2\big]^{\mathrm{T}} \tag{3.40}$$

The $L \times L$ matrix $\boldsymbol{Z}$ reads as

$$\boldsymbol{Z} := (\boldsymbol{G}\boldsymbol{G}^{\mathrm{H}}) \odot \mathbf{I}_L + \operatorname{diag}\{\mathbf{1}_L^{\mathrm{T}}\boldsymbol{Z}_0\} - \boldsymbol{Z}_0$$

where $\boldsymbol{Z}_0$ is defined as

$$\boldsymbol{Z}_0 := \Big[(\boldsymbol{G}\boldsymbol{H}\boldsymbol{T}) \odot (\boldsymbol{G}\boldsymbol{H}\boldsymbol{T})^*\Big]^{\mathrm{T}}. \tag{3.41}$$

Due to the *M-matrix* property of $\boldsymbol{Z}$, a well-defined solution for $\varphi_{1,1}^2,\ldots,\varphi_{K,L_K}^2$ exists:

$$\big[\varphi_{1,1}^2,\ldots,\varphi_{K,L_K}^2\big]^{\mathrm{T}} = \boldsymbol{Z}^{-1}\cdot\big[\|\boldsymbol{t}_{1,1}\|_2^2,\ldots,\|\boldsymbol{t}_{K,L_K}\|_2^2\big]^{\mathrm{T}}. \tag{3.42}$$

Summing up all rows of (3.40) and using (3.38), we see that the same sum power $P_{\mathrm{diss}}^{\mathrm{BC}}$ is dissipated in the BC as in the MAC.

Besides the conservation of the stream-wise MSEs, the level three conversion from the MAC to the BC also keeps the SINRs defined in (3.31) constant. Equality of the numerators of $\mathrm{SINR}_{k,j}^{\mathrm{MAC}}$ and $\mathrm{SINR}_{k,j}^{\mathrm{BC}}$ in (3.31) follows from (3.38) and equality of the denominators implies (3.39).

Synopsis: Any stream-wise MSE and SINR tuple that is feasible in the dual MIMO MAC with linear filtering can also be achieved in the MIMO BC with linear transceivers using the same sum power, if the transmit and receive filter for stream j of user k are chosen as

$$\begin{aligned} \boldsymbol{p}_{k,j} &= \varphi_{k,j}\boldsymbol{g}_{k,j}^{*} \\ \boldsymbol{b}_{k,j}^{\mathrm{T}} &= \varphi_{k,j}^{-1}\boldsymbol{t}_{k,j}^{\mathrm{H}}\boldsymbol{C}_{\eta_k}^{-\frac{1}{2}} \end{aligned}$$

for all users and all streams with $\varphi_{1,1},\ldots,\varphi_{K,L_K}$ following from (3.42).

3.2.5 Duality of BC and MAC

Combining Lemma 1 from Sect. 3.2.3 and Lemma 2 from Sect. 3.2.4, the complete MSE duality is established:

Theorem 3.1 *The MIMO BC and the dual MIMO MAC share the same overall, user-wise, and stream-wise MSE region under the same sum power constraint. If the noise covariance matrix in the dual MAC is chosen to be the identity matrix, the channel matrices in the dual MAC result from the channel matrices and the noise covariance matrices in the BC as shown in Corollary 1.*

As a consequence of Theorem 1, it is possible to tackle an MSE-based problem in the dual MIMO MAC that is originally stated in the MIMO BC where it usually has a complicated structure due to the inexistent indices alignment of the precoders and the channel matrices (see the discussion in Sect. 3.1). Afterwards, the solution in the dual MIMO MAC can be converted back to the BC by means of the conversion rules derived in Sect. 3.2.4.

3.2.6 Advantages of the Presented Duality Compared to Hitherto Existing Ones

Due to its multi-level structure, the presented duality is tailored to match the properties of the underlying problem setting and is thus application-specific. Hitherto existing dualities were all based on the stream-wise MSE conservation irrespective of which MSE formulation the problem at hand had. This is accompanied by a waste of computational complexity. For the overall sum MSE minimization problem tackled in [15, 16, 17] for example, it would suffice to apply the very simple level one duality where instead of solving an $L \times L$ system of equations, only a single scalar has to be determined, cf. (3.32). In almost the same manner, the level two duality suffices to handle optimizations that are formulated on a user-wise MSE basis, e.g., [49].

Another advantage of the presented MSE duality is its direct nature without detouring. Given transmit and receive filters in the BC (dual MAC) which achieve an arbitrary MSE tuple, our duality directly generates transmit and receive filters in the dual MAC (BC) which achieve the same MSE tuple without actually having to know which MSE tuple is achieved. This is in contrast to other dualities, where first the currently achieved MSE tuple has to be computed and afterwards, the power allocation is computed that achieves this particular MSE tuple in the dual system.

3.3 Rate Duality for Linear Transceivers Under Gaussian Signaling

Another, probably even more important metric is the Shannon capacity that is defined as the maximum rate of transmission over a system with arbitrarily small error probability [50, 51]. Given a proper encoding and decoding scheme, the frequency of error can be made as small as desired by increasing the length of the encoding sequences. For a particular distribution of the channel input $P_1 s_1$ in a single-input single-output (SISO) system

$$y_1 = H_1^{\mathrm{BC}} P_1 s_1 + \eta_1 \tag{3.43}$$

which is a special case of the MIMO BC in Sect. 2.1, Shannon showed that the rate of transmission R_{SISO} corresponds to the difference of the input entropy and the equivocation:

$$R_{\mathrm{SISO}} = \mathcal{H}(y_1) - \mathcal{H}(y_1|s_1)$$

Here, $\mathcal{H}(\cdot)$ denotes the differential entropy function. Since we assumed the noise η_1 to be independent of the (precoded) data symbol, the rate of error-free transmission simplifies to

$$R_{\mathrm{SISO}} = \mathcal{H}(y_1) - \mathcal{H}(\eta_1) \tag{3.44}$$

i.e., the difference between the output entropy and the noise entropy. The Shannon capacity C_{SISO} is now defined as the maximum transmission rate R_{SISO} when optimizing with respect to the input distribution of $P_1 s_1$ under some power constraint. For Gaussian noise η_1 (which is the worst distribution given the noise power c_{η_1}) and an average power constraint as in (2.2)–(2.3), the optimum input distribution must also be Gaussian and achieves the capacity

$$C_{\mathrm{SISO}} := \log_2\left(1 + \frac{|H_1^{\mathrm{BC}}|^2 \cdot P_{\max}}{c_{\eta_1}}\right) \tag{3.45}$$

in bits per second and per unit bandwidth. Throughout this section, we therefore assume the symbol vectors $\boldsymbol{s}_1, \ldots, \boldsymbol{s}_K$ to be Gaussian with identity covariance matrix.

This capacity concept of the SISO system was extended to a point-to-point MIMO link fifty years later by Telatar in [52]. Again, the point-to-point link can be seen as a single-user broadcast channel, where the received signal reads as

$$\boldsymbol{y}_1^{\mathrm{BC}} = \boldsymbol{H}_1^{\mathrm{BC}} \boldsymbol{P}_1 \boldsymbol{s}_1 + \boldsymbol{\eta}_1$$

cf. Sect. 2.1. For such a link, the mutual information reads in analogy to (3.44) as [52]

$$I_{\mathrm{MIMO}}(\boldsymbol{P}_1) := \mathcal{H}(\boldsymbol{y}_1^{\mathrm{BC}}) - \mathcal{H}(\boldsymbol{\eta}_1) = \log_2 \left| \mathbf{I}_N + \boldsymbol{H}_1^{\mathrm{BC,H}} \boldsymbol{C}_{\eta_1}^{-1} \boldsymbol{H}_1^{\mathrm{BC}} \boldsymbol{P}_1 \boldsymbol{P}_1^{\mathrm{H}} \right| \quad (3.46)$$

and the capacity C_{MIMO} is the maximum mutual information $I_{\mathrm{MIMO}}(\boldsymbol{P}_1)$ when optimizing the input distribution $\boldsymbol{P}_1 \boldsymbol{s}_1$ with respect to $\boldsymbol{P}_1$ under some power constraint. In turn, by increasing the length of the encoding sequences, the error probability can be made arbitrarily small while transmitting at a rate lower than C_{MIMO}.

For a single data stream, the transmission rate R_{SISO} depends on the SINR via the bijective mapping [cf. (3.45)]

$$R_{\mathrm{SISO}} = \log_2(1 + \mathrm{SINR}). \quad (3.47)$$

Given the SINR definition of the jth data stream of user k in the MIMO BC from (3.31), the transmission rate of this particular stream reads as

$$R_{k,j}^{\mathrm{BC}}(\boldsymbol{P}, \boldsymbol{b}_{k,j}) := \log_2 \left(1 + \mathrm{SINR}_{k,j}^{\mathrm{BC}}(\boldsymbol{P}, \boldsymbol{b}_{k,j})\right). \quad (3.48)$$

Due to this one-to-one mapping, one might think that a rate duality for linear transceivers directly follows from any SINR duality that is applicable to multi-user MIMO systems. In fact, this approach can be found in [53], where the Lagrangian duality [54] is used to establish a rate duality between the MIMO BC and the MIMO MAC. Due to the imposed per-antenna power constraints, the dual MIMO MAC has an uncertain noise covariance matrix making the dual system dependent on the applied filter in the BC which spoils the duality relationship between the two systems. But the authors manage to solve the power minimization problem given minimum SINR requirements per data stream. One could think of any other SINR-based duality that induces such a rate duality for linear transceivers, see for example [55], where the dual network is constructed using the duality framework for linear programming given a weighted sum power constraint. Other examples are [56], where the sum power constrained MIMO interference channel is rather a MIMO BC, and also [16].

However, the stream-wise SINR-based rate dualities do in general not offer the full potential of linear transceivers. The reason for this circumstance is the fact that the SINR-based dualities treat the individual streams of a particular user as virtual users which thus interfere with each other, i.e., the jth stream of user k has to face not only the inter-user interference from all other users, but also intra-user interference from the streams $\{1, \ldots, L_k\} \backslash \{j\}$. This self-interference was inherent to all rate dualities for linear transceivers until the first rate duality preserving the total mutual information between every link in the system [21]. To see the performance loss of the stream-based SINR dualities, we first compute the data rates of the L_k streams of user k when optimized receive filters are applied. Afterwards, we compare the result to the mutual information between the transmitter and receiver k when the interference of all other users is treated as noise.

In case of stream-wise, separate decoding, the data rate of user k can be expressed as [cf. (3.48)]

$$R_k^{\mathrm{BC}}(\boldsymbol{P}, \boldsymbol{B}_k) = \sum_{j=1}^{L_k} R_{k,j}^{\mathrm{BC}}(\boldsymbol{P}, \boldsymbol{b}_{k,j}) = \sum_{j=1}^{L_k} \log_2 \left(1 + \mathrm{SINR}_{k,j}^{\mathrm{BC}}(\boldsymbol{P}, \boldsymbol{b}_{k,j})\right) \tag{3.49}$$

which holds irrespective of the chosen transmit and receive filters as long as the filtered noise, all interfering data streams, and the desired data stream have a Gaussian distribution. Clearly, the SINR of the jth stream belonging to user k only depends on the receive filter $\boldsymbol{b}_{k,j}$ and on no other receive filters, so the receivers can be optimized separately. In Appendix A3, it is shown that the MMSE receive filter

$$\boldsymbol{B}_k^{\mathrm{MMSE}} = \boldsymbol{P}_k^{\mathrm{H}} \boldsymbol{H}_k^{\mathrm{BC,H}} \left(\boldsymbol{H}_k^{\mathrm{BC}} \sum_{i=1}^{K} \boldsymbol{P}_i \boldsymbol{P}_i^{\mathrm{H}} \boldsymbol{H}_k^{\mathrm{BC,H}} + \boldsymbol{C}_{\eta_k}\right)^{-1} \tag{3.50}$$

not only minimizes the MSEs of the L_k data streams, but also maximizes their SINRs and thus also the sum rate $R_k^{\mathrm{BC}}(\boldsymbol{P}, \boldsymbol{B}_k)$. Inserting the jth row $\boldsymbol{b}_{k,j}^{\mathrm{MMSE,T}}$ of $\boldsymbol{B}_k^{\mathrm{MMSE}}$ into the SINR expression (3.31), we find

$$\mathrm{SINR}_{k,j}^{\mathrm{BC}}(\boldsymbol{P}, \boldsymbol{b}_{k,j}^{\mathrm{MMSE}}) = \frac{1}{\left[\mathbf{I} - \boldsymbol{B}_k^{\mathrm{MMSE}} \boldsymbol{H}_k^{\mathrm{BC}} \boldsymbol{P}_k\right]_{j,j}} - 1.$$

For fixed $\boldsymbol{P}$, the SINR-maximizing equalizers lead to the maximum sum rate

$$\begin{aligned} R_k^{\mathrm{BC}}(\boldsymbol{P}) &:= \max_{\boldsymbol{B}_k} R_k^{\mathrm{BC}}(\boldsymbol{P}, \boldsymbol{B}_k) = R_k^{\mathrm{BC}}(\boldsymbol{P}, \boldsymbol{B}_k^{\mathrm{MMSE}}) \\ &= -\log_2 \prod_{j=1}^{L_k} \left[\mathbf{I}_{L_k} - \boldsymbol{B}_k^{\mathrm{MMSE}} \boldsymbol{H}_k^{\mathrm{BC}} \boldsymbol{P}_k\right]_{j,j}. \end{aligned} \tag{3.51}$$

In contrast, the mutual information between the transmitter and the kth receiver in the BC reads in analogy to (3.46) as [52]

$$I_k^{\mathrm{BC}}(\boldsymbol{P}) = \log_2 \left| \mathbf{I}_{M_k} + \left(\boldsymbol{H}_k^{\mathrm{BC}} \sum_{i \neq k} \boldsymbol{P}_i \boldsymbol{P}_i^{\mathrm{H}} \boldsymbol{H}_k^{\mathrm{BC,H}} + \boldsymbol{C}_{\eta_k}\right)^{-1} \boldsymbol{H}_k^{\mathrm{BC}} \boldsymbol{P}_k \boldsymbol{P}_k^{\mathrm{H}} \boldsymbol{H}_k^{\mathrm{BC,H}} \right| \tag{3.52}$$

and represents the maximum transmission rate that can be achieved with arbitrarily small errors given the precoders in $\boldsymbol{P}$ and a joint decoding of all L_k streams. Although the matched filter yields a much more convenient notation for the mutual information in (3.52) and will therefore be used for the actual duality conversion, we reparametrize (3.52) in terms of the MMSE receive filter for a nicer illustration that separate stream decoding is in general inferior to the joint stream decoding. Using the matrix inversion lemma from Appendix A2 and the identity $|\mathbf{I} + \mathbf{C}_1\mathbf{C}_2| = |\mathbf{I} + \mathbf{C}_2\mathbf{C}_1|$ for rectangular $\boldsymbol{C}_1$ and $\boldsymbol{C}_2$ with matching dimensions, we obtain from (3.52)

$$\begin{aligned} I_k^{\mathrm{BC}}(\boldsymbol{P}) &= -\log_2 \left| \mathbf{I}_{L_k} - \boldsymbol{P}_k^{\mathrm{H}} \boldsymbol{H}_k^{\mathrm{BC,H}} \left(\boldsymbol{H}_k^{\mathrm{BC}} \sum_{i=1}^{K} \boldsymbol{P}_i \boldsymbol{P}_i^{\mathrm{H}} \boldsymbol{H}_k^{\mathrm{BC,H}} + \boldsymbol{C}_{\eta_k} \right)^{-1} \boldsymbol{H}_k^{\mathrm{BC}} \boldsymbol{P}_k \right| \\ &= -\log_2 \left| \mathbf{I}_{L_k} - \boldsymbol{B}_k^{\mathrm{MMSE}} \boldsymbol{H}_k^{\mathrm{BC}} \boldsymbol{P}_k \right|. \end{aligned} \tag{3.53}$$

The second line follows from the definition of the MSE receive filter $\boldsymbol{B}_k^{\mathrm{MMSE}}$ in (3.50). From (3.53) we see, that the rate $I_k^{\mathrm{BC}}(\boldsymbol{P})$ of user k with joint decoding is the negative logarithm of a matrix determinant. In case of separate stream decoding, the rate $R_k^{\mathrm{BC}}(\boldsymbol{P})$ in (3.51) is the negative logarithm of the main diagonal entries product of the *same* matrix. According to Hadamard's inequality which is shown in Appendix A4, the product of the main diagonal elements of a positive definite matrix is always larger than or equal to the determinant of this matrix. Because of the *negative* logarithm in (3.51) and (3.53), the possible rate of a user obtained with separate stream decoding is smaller than or equal to the mutual information:

$$R_k^{\mathrm{BC}}(\boldsymbol{P}) \leq I_k^{\mathrm{BC}}(\boldsymbol{P}) \quad \forall \boldsymbol{P} \ \forall k. \tag{3.54}$$

Therefore, stream-wise, SINR-based rate dualities do not offer the full potential of linear transceivers in general.
Taking a closer look at the rate expression for joint stream-decoding in (3.52), we make the following observations:

(1) The mutual information $I_k^{\mathrm{BC}}(\boldsymbol{P})$ of user k only depends on the Gram product $\boldsymbol{P}_k\boldsymbol{P}_k^{\mathrm{H}}$ of his precoders and not on the precoders itself. Therefore, it is invariant to an $L_k \times L_k$ unitary matrix $\boldsymbol{W}_k$ multiplied from the right-hand side to $\boldsymbol{P}_k$ yielding $\bar{\boldsymbol{P}}_k := \boldsymbol{P}_k\boldsymbol{W}_k$.
(2) The mutual information $I_k^{\mathrm{BC}}(\boldsymbol{P})$ of user k only depends on the outer products $\boldsymbol{P}_i\boldsymbol{P}_i^{\mathrm{H}}$ $\forall i \neq k$ of the interfering users' precoders, and not on the precoders themselves. Hence, changing $\boldsymbol{P}_k$ to $\bar{\boldsymbol{P}}_k$ leads to the same interference covariance matrix for other users $i \neq k$.
(3) The transmit power $P_{\mathrm{diss}}^{\mathrm{BC}}$ in the BC again only depends on the outer product of the precoders, see (2.2). The unitarily transformed precoders $\bar{\boldsymbol{P}}_k$ lead to the same dissipated power due to $\|\boldsymbol{P}_k\|_{\mathrm{F}} = \|\bar{\boldsymbol{P}}_k\|_{\mathrm{F}}$.

The key observation for the rate duality of linear transceiver that achieves the BC mutual information expressions in the dual MAC and vice versa is that we have the unitary rotations $\boldsymbol{W}_1, \ldots, \boldsymbol{W}_K$ as additional degrees of freedom. Although $\boldsymbol{W}_k$ does not influence the interference covariance matrix experienced by any other user $i \neq k$, it can be used as a spatial decorrelation filter for the point-to-point link from the transmitter to user k. Updating the precoders to $\bar{\boldsymbol{P}}_k = \boldsymbol{P}_k \boldsymbol{W}_k \ \forall k$ leads to the modified receive filters $\bar{\boldsymbol{B}}_k^{\mathrm{MMSE}} := \boldsymbol{W}_k^{\mathrm{H}} \boldsymbol{B}_k^{\mathrm{MMSE}} \ \forall k$, see (3.50). When $\boldsymbol{W}_k$ is used to rotate the precoders $\boldsymbol{P}_k$ and thus also the receiver $\boldsymbol{B}_k^{\mathrm{MMSE}}$, the mutual information expression (3.53) can be rewritten to

$$I_k^{\mathrm{BC}}(\boldsymbol{P}) = I_k^{\mathrm{BC}}(\bar{\boldsymbol{P}}) = -\log_2 \left| \mathbf{I}_{L_k} - \boldsymbol{W}_k^{\mathrm{H}} \boldsymbol{B}_k^{\mathrm{MMSE}} \boldsymbol{H}_k^{\mathrm{BC}} \boldsymbol{P}_k \boldsymbol{W}_k \right|.$$

Choosing $\boldsymbol{W}_k$ as the unitary eigenbasis of $\boldsymbol{B}_k^{\mathrm{MMSE}} \boldsymbol{H}_k^{\mathrm{BC}} \boldsymbol{P}_k$ for all $k \in \{1, \ldots, K\}$ where $\boldsymbol{B}_k^{\mathrm{MMSE}}$ is taken from (3.50), we can decorrelate and diagonalize every point-to-point link. Doing so, Hadamard's inequality becomes an equality, i.e., (3.54) holds with equality:

$$R_k^{\mathrm{BC}}(\bar{\boldsymbol{P}}) = I_k^{\mathrm{BC}}(\bar{\boldsymbol{P}}) = I_k^{\mathrm{BC}}(\boldsymbol{P}). \tag{3.55}$$

Thus, we can achieve the mutual information that usually requires joint decoding of the L_k streams associated to user k with separate stream decoding since no intra-user interference remains!

Theorem 3.2 *In the K user MIMO BC with linear transceivers, the K mutual information expressions between the transmitter and each receiver can be achieved by separate stream decoding using MSE receivers followed by appropriate decorrelation matrices applied at all transmit and receive filters. This also holds for the MIMO MAC.*

The proof of Theorem 2 for the MIMO MAC follows from the BC to MAC conversion in Sect. 3.3.1 and the MAC to BC conversion in Sect. 3.3.2.

Synopsis The rate duality between the MIMO MAC and the MIMO BC with linear transceivers was so far deduced from the stream-wise SINR duality in the literature. Different streams of a user were treated as virtual users generating self-interference. In general, such a conversion of the precoders to the dual domain does not achieve the mutual information of the original domain in the dual domain and hence fails to offer the full potential of linear transceivers. Introducing unitary rotations applied to all users' data symbols prior to the precoding process allows to achieve the mutual information by separate stream decoding. This way, a novel rate duality for linear transceivers can be established making rate-based optimizations manageable in a convenient way.

3.3.1 BC to MAC Conversion

Starting in the MIMO BC from Sect. 2.1 with given precoders $\boldsymbol{P}_1, \ldots, \boldsymbol{P}_K$ obtaining mutual information expressions as in (3.52), we construct a dual MIMO MAC with precoders $\boldsymbol{T}_1, \ldots, \boldsymbol{T}_K$ achieving *at least* the same mutual information using the same amount of transmit power. In addition, we derive receive filters $\boldsymbol{G}_1, \ldots, \boldsymbol{G}_K$ in this dual MAC for which *exactly* the user-wise BC mutual information is obtained in the dual MAC by means of separate decoding.

This way, one direction of the duality statement is proven, and it will be possible to tackle a rate-based optimization problem stated in the MIMO BC in its dual MIMO MAC which offers better mathematical properties, see Sect. 3.1. Having obtained an expression for the precoders (and the equalizers) in the dual MIMO MAC, this solution can be converted back to the original MIMO BC using the MAC to BC conversion derived later. The BC to MAC conversion of the rate duality is stated in the following lemma:

Lemma 3.3 *For every set of precoders in the MIMO BC leading to certain values for the mutual information of the K users, there exist precoders in a dual MIMO MAC which achieve at least the same mutual information using the same amount of transmit power. In addition, there exist receive filters such that the BC mutual information expressions can exactly be achieved with separate decoding in the dual MAC.*

For the proof, we first assume matched filter (MF) receivers $\boldsymbol{B}_1^{\mathrm{MF}}, \ldots, \boldsymbol{B}_K^{\mathrm{MF}}$, where

$$\boldsymbol{B}_k^{\mathrm{MF}} = \boldsymbol{P}_k^{\mathrm{H}} \boldsymbol{H}_k^{\mathrm{BC,H}} \Big(\boldsymbol{H}_k^{\mathrm{BC}} \sum_{i \neq k} \boldsymbol{P}_i \boldsymbol{P}_i^{\mathrm{H}} \boldsymbol{H}_k^{\mathrm{BC,H}} + \boldsymbol{C}_{\eta_k} \Big)^{-1} \quad \forall k \in \{1, \ldots, K\}. \quad (3.56)$$

Since the rate expressions are easily linked to the SINR (not only to the maximum SINR), the MF receivers allow for a more convenient derivation than the MMSE receivers. In principle, the derivation could also be based on the MSE receivers. However, only the *minimum* MSEs are directly related to the data rate. As the minimum MSEs in the MIMO BC are in general no longer minimum MSEs in the dual MIMO MAC after the duality conversion, the data rates will be expressed in terms of SINRs. Additionally, the matched filter allows for very simple SINR expressions of the uncorrelated data streams.

Making use of the matched filter expression in (3.56), the data rate of user k under joint or successive decoding reads as [cf. (3.52)]

$$I_k^{\mathrm{BC}}(\boldsymbol{P}) = \log_2 \left| \mathbf{I}_{L_k} + \boldsymbol{B}_k^{\mathrm{MF}} \boldsymbol{H}_k^{\mathrm{BC}} \boldsymbol{P}_k \right| \quad \forall k \in \{1, \ldots, K\}. \quad (3.57)$$

Preprocessing the data symbol vectors $\boldsymbol{s}_1, \ldots, \boldsymbol{s}_K$ via right-hand side multiplications of all BC precoders by the unitary decorrelation matrices $\boldsymbol{W}_1, \ldots, \boldsymbol{W}_K$ also leads to modified receive filters. This operation does not change the mutual information in

(3.57) as mentioned before, but allows us to remove the intra-user self-interference and to achieve the mutual information with separate stream decoding:

$$
\begin{aligned}
I_k^{\mathrm{BC}}(\boldsymbol{P}) &= \log_2 \left| \mathbf{I}_{L_k} + \boldsymbol{W}_k^{\mathrm{H}} \boldsymbol{B}_k^{\mathrm{MF}} \boldsymbol{H}_k^{\mathrm{BC}} \boldsymbol{P}_k \boldsymbol{W}_k \right| \\
&= \log_2 \left| \mathbf{I}_{L_k} + \bar{\boldsymbol{B}}_k^{\mathrm{MF}} \boldsymbol{H}_k^{\mathrm{BC}} \bar{\boldsymbol{P}}_k \right| \\
&= \log_2 \prod_{j=1}^{L_k} \left[\mathbf{I}_{L_k} + \bar{\boldsymbol{B}}_k^{\mathrm{MF}} \boldsymbol{H}_k^{\mathrm{BC}} \bar{\boldsymbol{P}}_k \right]_{j,j} \\
&= \sum_{j=1}^{L_k} \log_2 \left(1 + \left[\bar{\boldsymbol{B}}_k^{\mathrm{MF}} \boldsymbol{H}_k^{\mathrm{BC}} \bar{\boldsymbol{P}}_k \right]_{j,j} \right).
\end{aligned} \tag{3.58}
$$

For the second equality, we introduced the rotated precoders

$$
\bar{\boldsymbol{P}}_k := \boldsymbol{P}_k \boldsymbol{W}_k \quad \forall k \in \{1, \ldots, K\} \tag{3.59}
$$

and the matched filter for the rotated precoders

$$
\bar{\boldsymbol{B}}_k^{\mathrm{MF}} := \boldsymbol{W}_k^{\mathrm{H}} \boldsymbol{B}_k^{\mathrm{MF}} \quad \forall k \in \{1, \ldots, K\}. \tag{3.60}
$$

In order to let the third equality in (3.58) hold, the isometry $\boldsymbol{W}_k$ has to be chosen such that

$$
\boldsymbol{W}_k \boldsymbol{\Lambda}_k \boldsymbol{W}_k^{\mathrm{H}} = \boldsymbol{B}_k^{\mathrm{MF}} \boldsymbol{H}_k^{\mathrm{BC}} \boldsymbol{P}_k \quad \forall k \tag{3.61}
$$

holds with unitary $\boldsymbol{W}_k$ and diagonal $\boldsymbol{\Lambda}_k$. Note that $\boldsymbol{W}_k$ also dragonflies $\boldsymbol{B}_k^{\mathrm{MMSE}} \boldsymbol{H}_k^{\mathrm{BC}} \boldsymbol{P}_k$ which can be shown by means of the matrix inversion lemma. For the choice of the decorrelation matrix $\boldsymbol{W}_k$ in (3.61), the intra-user interference is completely removed. As a consequence, the rotated matched filter $\bar{\boldsymbol{B}}_k^{\mathrm{MF}}$ maximizes the SINRs of every of the L_k data streams belonging to user k. According to (3.31), the SINR of the jth data stream without self-interference reads as

$$
\begin{aligned}
\mathrm{SINR}_{k,j}^{\mathrm{BC}}(\bar{\boldsymbol{P}}, \bar{\boldsymbol{b}}_{k,j}^{\mathrm{MF}}) &= \frac{|\bar{\boldsymbol{b}}_{k,j}^{\mathrm{MF,T}} \boldsymbol{H}_k^{\mathrm{BC}} \bar{\boldsymbol{p}}_{k,j}|^2}{\bar{\boldsymbol{b}}_{k,j}^{\mathrm{MF,T}} \boldsymbol{C}_{\eta_k} \bar{\boldsymbol{b}}_{k,j}^{\mathrm{MF}*} + \sum_{i \neq k} \sum_{\ell=1}^{L_i} |\bar{\boldsymbol{b}}_{k,j}^{\mathrm{MF,T}} \boldsymbol{H}_k^{\mathrm{BC}} \bar{\boldsymbol{p}}_{i,\ell}|^2} \\
&= \frac{|\bar{\boldsymbol{b}}_{k,j}^{\mathrm{MF,T}} \boldsymbol{H}_k^{\mathrm{BC}} \bar{\boldsymbol{p}}_{k,j}|^2}{\bar{\boldsymbol{b}}_{k,j}^{\mathrm{MF,T}} \left(\boldsymbol{C}_{\eta_k} + \boldsymbol{H}_k^{\mathrm{BC}} \sum_{i \neq k} \bar{\boldsymbol{P}}_i \bar{\boldsymbol{P}}_i^{\mathrm{H}} \boldsymbol{H}_k^{\mathrm{BC,H}} \right) \bar{\boldsymbol{b}}_{k,j}^{\mathrm{MF}*}} \\
&= \left[\bar{\boldsymbol{B}}_k^{\mathrm{MF}} \boldsymbol{H}_k^{\mathrm{BC}} \bar{\boldsymbol{P}}_k \right]_{j,j}.
\end{aligned} \tag{3.62}
$$

Hence, the mutual information $I_k^{\mathrm{BC}}(\boldsymbol{P})$ in (3.58) can also be expressed in terms of the SINRs with rotated filters and can thus also be achieved with separate stream decoding:

$$I_k^{\mathrm{BC}}(\boldsymbol{P}) = \sum_{j=1}^{L_k} \log_2 \left(1 + \mathrm{SINR}_{k,j}^{\mathrm{BC}}(\bar{\boldsymbol{P}}, \bar{\boldsymbol{b}}_{k,j}^{\mathrm{MF}})\right). \tag{3.63}$$

Having removed the intra-user interference from every user, we can create a dual MIMO MAC that offers *at least* the same data rates per user by simply applying the level three duality conversion on the *rotated* transmit and receive filters $\bar{\boldsymbol{P}}_1, \ldots, \bar{\boldsymbol{P}}_K$ and $\bar{\boldsymbol{B}}_1, \ldots, \bar{\boldsymbol{B}}_K$, respectively, see Sect. 3.2.3.3. In particular, we make use of the SINR conservation in order to be able to express the rates in the dual domain in terms of the SINRs. Indeed, the MSEs are also preserved but the MSEs in the dual MAC cannot easily be mapped to data rate since they are not the *minimum* MSEs in general.

For the dual MAC precoders and receivers, the same parametrization as in (3.24) is chosen:

$$\begin{aligned} \boldsymbol{t}_{k,j} &= \boldsymbol{T}_k \boldsymbol{e}_j = \psi_{k,j} \boldsymbol{C}_{\eta_k}^{\frac{1}{2},\mathrm{H}} \bar{\boldsymbol{b}}_{k,j}^{\mathrm{MF}*} = \psi_{k,j} \boldsymbol{C}_{\eta_k}^{\frac{1}{2},\mathrm{H}} \boldsymbol{B}_k^{\mathrm{MF,H}} \boldsymbol{W}_k \boldsymbol{e}_j \\ \boldsymbol{g}_{k,j}^{\mathrm{T}} &= \boldsymbol{e}_j^{\mathrm{T}} \boldsymbol{G}_k = \psi_{k,j}^{-1} \bar{\boldsymbol{p}}_{k,j}^{\mathrm{H}} = \psi_{k,j}^{-1} \boldsymbol{e}_j^{\mathrm{T}} \boldsymbol{W}_k^{\mathrm{H}} \boldsymbol{P}_k^{\mathrm{H}} \quad \forall k \forall j \end{aligned} \tag{3.64}$$

where $\boldsymbol{e}_j$ denotes the jth canonical unit vector of appropriate dimension. Due to the lack of the intra-user interference, the linear system of equations from which we derive the L squared scaling factors $\psi_{1,1}^2, \ldots, \psi_{K,L_K}^2$ looks slightly different from the one in Sect. 3.2.3.3 making it worthwhile to briefly revisit the duality conversion. When this MAC filter parametrization is inserted into the SINR expression from (3.31), we get

$$\mathrm{SINR}_{k,j}^{\mathrm{MAC}}(\boldsymbol{T}, \boldsymbol{g}_{k,j}) = \frac{|\bar{\boldsymbol{p}}_{k,j}^{\mathrm{H}} \boldsymbol{H}_k^{\mathrm{BC,H}} \bar{\boldsymbol{b}}_{k,j}^{\mathrm{MF}*}|^2}{\sum_{i \neq k} \sum_{\ell=1}^{L_i} \frac{\psi_{i,\ell}^2}{\psi_{k,j}^2} |\bar{\boldsymbol{p}}_{k,j}^{\mathrm{H}} \boldsymbol{H}_i^{\mathrm{BC,H}} \bar{\boldsymbol{b}}_{i,\ell}^{\mathrm{MF}*}|^2 + \psi_{k,j}^{-2} \|\bar{\boldsymbol{p}}_{k,j}\|_2^2} \tag{3.65}$$

where we used $\boldsymbol{H}_k \boldsymbol{C}_{\eta_k}^{\frac{1}{2},\mathrm{H}} = \boldsymbol{H}_k^{\mathrm{BC,H}}$ from Corollary 1 and the diagonal structure of $\boldsymbol{G}_k \boldsymbol{H}_k \boldsymbol{T}_k$ from (3.61) and (3.64) to see that there is no intra-user interference also in the dual MIMO MAC. The two numerators in (3.65) and (3.62) are the same due to the choice of the filters in the dual MAC such that for the equality of the SINRs, we only have to equate the denominators:

$$\begin{aligned} &\psi_{k,j}^2 \cdot \left[\sum_{i \neq k} \sum_{\ell=1}^{L_i} |[\bar{\boldsymbol{B}}_k^{\mathrm{MF}} \boldsymbol{H}_k^{\mathrm{BC}} \bar{\boldsymbol{P}}_i]_{j,\ell}|^2 + [\bar{\boldsymbol{B}}_k^{\mathrm{MF}} \boldsymbol{C}_{\eta_k} \bar{\boldsymbol{B}}_k^{\mathrm{MF,H}}]_{j,j} \right] \\ &\quad - \sum_{i \neq k} \sum_{\ell=1}^{L_i} \psi_{i,\ell}^2 \cdot |[\bar{\boldsymbol{B}}_i^{\mathrm{MF}} \boldsymbol{H}_i^{\mathrm{BC}} \bar{\boldsymbol{P}}_k]_{\ell,j}|^2 = \|\bar{\boldsymbol{p}}_{k,j}\|_2^2. \end{aligned}$$

Since above equation has to hold for all L streams of the K users, we end up with the linear system of equations [cf. (3.27)]

$$\boldsymbol{Y} \cdot \left[\psi_{1,1}^2, \ldots, \psi_{K,L_K}^2\right]^{\mathrm{T}} = \left[\|\bar{\boldsymbol{p}}_{1,1}\|_2^2, \ldots, \|\bar{\boldsymbol{p}}_{K,L_K}\|_2^2\right]^{\mathrm{T}} \tag{3.66}$$

where the matrix $\boldsymbol{Y}$ is defined as

$$\boldsymbol{Y} = \begin{bmatrix} \boldsymbol{Y}_{1,1} & \ldots & \boldsymbol{Y}_{1,K} \\ \vdots & \ddots & \vdots \\ \boldsymbol{Y}_{K,1} & \ldots & \boldsymbol{Y}_{K,K} \end{bmatrix} \in \mathbb{R}^{L\times L}. \tag{3.67}$$

The off-diagonal blocks $\boldsymbol{Y}_{k,i} \in \mathbb{R}^{L_k\times L_i}$ with $k \neq i$ are defined similar to (3.28) via

$$\boldsymbol{Y}_{k,i} = -\left[\left(\bar{\boldsymbol{B}}_i^{\mathrm{MF}} \boldsymbol{H}_i^{\mathrm{BC}} \bar{\boldsymbol{P}}_k\right) \odot \left(\bar{\boldsymbol{B}}_i^{\mathrm{MF}} \boldsymbol{H}_i^{\mathrm{BC}} \bar{\boldsymbol{P}}_k\right)^*\right]^{\mathrm{T}} \quad \forall k \neq i \tag{3.68}$$

whereas the main diagonal blocks are now diagonal matrices defined as

$$\boldsymbol{Y}_{k,k} = \left(\bar{\boldsymbol{B}}_k^{\mathrm{MF}} \boldsymbol{C}_{\eta_k} \bar{\boldsymbol{B}}_k^{\mathrm{MF,H}}\right) \odot \mathbf{I}_{L_k} - \sum_{i\neq k} \mathrm{diag}\{\mathbf{1}_{L_i}^{\mathrm{T}} \boldsymbol{Y}_{i,k}\} \quad \forall k. \tag{3.69}$$

The squared scaling factors result from (3.66) and read as

$$\left[\psi_{1,1}^2, \ldots, \psi_{K,L_K}^2\right]^{\mathrm{T}} = \boldsymbol{Y}^{-1} \cdot \left[\|\bar{\boldsymbol{p}}_{1,1}\|_2^2, \ldots, \|\bar{\boldsymbol{p}}_{K,L_K}\|_2^2\right]^{\mathrm{T}}. \tag{3.70}$$

Again, $\boldsymbol{Y}$ is a nonsingular *M-matrix* as long as $\bar{\boldsymbol{B}}_k^{\mathrm{MF}}$ has no row being the zero-vector for all $k \in \{1, \ldots, K\}$. A zero column in $\bar{\boldsymbol{B}}_k^{\mathrm{MF}}$ arises if $\mathrm{rank}(\boldsymbol{H}_k^{\mathrm{BC}} \boldsymbol{P}_k) < \min\{M_k, L_k\}$, cf. (3.50). This configuration arises for example when a single user transmits more data streams than he has degrees of freedom, i.e., when $L_k > M_k$. But due to the fact that MF receivers are assumed in the BC for the duality transformation, a passive transmit filter in the *rotated* system also leads to an inactive receive filter. Thus, no semi-active streams are present and passive streams can simply be dropped from $\boldsymbol{Y}$ in (3.67), i.e., the zero rows and columns in $\boldsymbol{Y}$ have to be removed. They remain passive in the dual domain as well, i.e., there is a zero column in a MAC precoder and a zero row in a MAC receiver. Accordingly, the respective entries $\psi_{\cdot,\cdot}^2$ and the zeros in the right-hand side of (3.66) have to be removed.

Having removed these entries, the matrix $\boldsymbol{Y}$ becomes a nonsingular *M-matrix* whose inverse has only nonnegative entries with at least one positive entry per row, see Appendix A1. In addition, the shortened right-hand side of (3.66) has only positive entries such that we end up with a positive solution for the shortened left-hand side of (3.70).

To show the power conservation, we may assume without loss of generality that there are no passive streams in the system. This is due to the fact that the rows (and columns) to be removed from $\boldsymbol{Y}$ per passive data stream are zero vectors. Moreover, only zeros would be removed from the right-hand side of (3.66), and every passive precoders in the BC is passive in the dual MAC as well, such that the following considerations are also valid when passive streams are present. Summing up all rows of (3.66) yields

$$\sum_{k=1}^{K}\sum_{j=1}^{L_k}\psi_{k,j}^2\big[\bar{\boldsymbol{B}}_k^{\mathrm{MF}}\boldsymbol{C}_{\eta_k}\bar{\boldsymbol{B}}_k^{\mathrm{MF,H}}\big]_{j,j}=\sum_{k=1}^{K}\|\bar{\boldsymbol{P}}_k\|_{\mathrm{F}}^2=P_{\mathrm{diss}}^{\mathrm{BC}}$$

where we used (2.2) for the last equality. The dissipated power in the dual MAC reads by means of (2.6) and (3.64) as

$$P_{\mathrm{diss}}^{\mathrm{MAC}}=\sum_{k=1}^{K}\|\boldsymbol{T}_k\|_{\mathrm{F}}^2=\sum_{k=1}^{K}\sum_{j=1}^{L_k}\psi_{k,j}^2\big[\bar{\boldsymbol{B}}_k^{\mathrm{MF}}\boldsymbol{C}_{\eta_k}\bar{\boldsymbol{B}}_k^{\mathrm{MF,H}}\big]_{j,j}=P_{\mathrm{diss}}^{\mathrm{BC}}$$

so the same amount of transmit power is dissipated in the dual MAC.

During the conversion from the MIMO BC to the dual MIMO MAC, the SINRs have been preserved:

$$\mathrm{SINR}_{k,j}^{\mathrm{MAC}}(\boldsymbol{T},\boldsymbol{g}_{k,j})=\mathrm{SINR}_{k,j}^{\mathrm{BC}}(\bar{\boldsymbol{P}},\bar{\boldsymbol{b}}_{k,j}^{\mathrm{MF}})\quad\forall k\ \forall j. \tag{3.71}$$

These can be related to the rate of user k in the dual MAC that is achieved by separate stream decoding [cf. (3.49)]:

$$R_k^{\mathrm{MAC}}(\boldsymbol{T},\boldsymbol{G}_k):=\sum_{j=1}^{L_k}R_{k,j}^{\mathrm{MAC}}(\boldsymbol{T},\boldsymbol{g}_{k,j}):=\sum_{j=1}^{L_k}\log_2(1+\mathrm{SINR}_{k,j}^{\mathrm{MAC}}(\boldsymbol{T},\boldsymbol{g}_{k,j})). \tag{3.72}$$

Combining (3.63), (3.71), and (3.72), we obtain

$$R_k^{\mathrm{MAC}}(\boldsymbol{T},\boldsymbol{G}_k)=I_k^{\mathrm{BC}}(\boldsymbol{P})\quad\forall k\in\{1,\ldots,K\}$$

so the dual MAC *exactly* achieves the mutual information in the BC with separate stream decoding when the receive filters $\boldsymbol{G}_1,\ldots,\boldsymbol{G}_K$ are chosen. As mentioned before, the resulting receivers do not maximize the individual SINRs in general. Hence, the data rate in the dual domain may be increased by applying SINR maximizing receivers and afterwards removing the arising intra-user interference via unitary decorrelation filters again. This corresponds to applying the matched filter receivers

$$\boldsymbol{G}_k^{\mathrm{MF}}:=\boldsymbol{T}_k^{\mathrm{H}}\boldsymbol{H}_k^{\mathrm{H}}\Big(\mathbf{I}_N+\sum_{i\neq k}\boldsymbol{H}_i\boldsymbol{T}_i\boldsymbol{T}_i^{\mathrm{H}}\boldsymbol{H}_i^{\mathrm{H}}\Big)^{-1} \tag{3.73}$$

and removing the self-interference again via unitary decorrelation filters $\boldsymbol{V}_1,\ldots,\boldsymbol{V}_K$ for every point-to-point link. Doing so, every user can achieve the mutual information

$$I_k^{\mathrm{MAC}}(\boldsymbol{T}) := \log_2 \left| \mathbf{I}_{L_k} + \boldsymbol{T}_k^{\mathrm{H}} \boldsymbol{H}_k^{\mathrm{H}} \left(\mathbf{I}_N + \sum_{i \neq k} \boldsymbol{H}_i \boldsymbol{T}_i \boldsymbol{T}_i^{\mathrm{H}} \boldsymbol{H}_i^{\mathrm{H}} \right)^{-1} \boldsymbol{H}_k \boldsymbol{T}_k \right| \tag{3.74}$$

in the dual MAC via separate stream decoding. Clearly,

$$R_k^{\mathrm{MAC}}(\boldsymbol{T}, \boldsymbol{G}_k) \leq I_k^{\mathrm{MAC}}(\boldsymbol{T})$$

holds, which implies the inequality

$$I_k^{\mathrm{MAC}}(\boldsymbol{T}) \geq I_k^{\mathrm{BC}}(\boldsymbol{P}) \quad \forall k \in \{1, \ldots, K\}. \tag{3.75}$$

Summing up, the BC to MAC rate duality conversion achieves data rates exactly matching the BC mutual information when the receive filters $\boldsymbol{G}_1, \ldots, \boldsymbol{G}_K$ are applied in the dual MAC and separate stream decoding is used. In addition, the values for the mutual information in the dual MAC are larger than or equal to the ones in the original BC.

Synopsis: The BC to MAC conversion constitutes one direction of the rate duality for linear transceivers. Although it is seldomly applied in practice, it represents one part of the rate duality which enables us to tackle a rate-based optimization problem posed in the MIMO BC in its dual MIMO MAC. Moreover, it shows how the channel matrices and the noise covariance matrix in this dual MIMO MAC have to be chosen. In general, a problem is solved in the dual system and then converted back to the original BC. Nonetheless, the conversion from the MIMO BC to the dual MIMO MAC is obtained by the following steps:

1) Given MIMO BC precoders $\boldsymbol{P}_1, \ldots, \boldsymbol{P}_K$ achieving the mutual information values $I_1^{\mathrm{BC}}(\boldsymbol{P}), \ldots, I_K^{\mathrm{BC}}(\boldsymbol{P})$, one first has to compute the corresponding MF receive filters $\boldsymbol{B}_1^{\mathrm{MF}}, \ldots, \boldsymbol{B}_k^{\mathrm{MF}}$ via (3.56).
2) Afterwards, unitary decorrelation filters $\boldsymbol{W}_1, \ldots, \boldsymbol{W}_K$ have to be determined such that every matrix product $\boldsymbol{B}_k^{\mathrm{MF}} \boldsymbol{H}_k^{\mathrm{BC}} \boldsymbol{P}_k$ is diagonalized:
$$\boldsymbol{W}_k \boldsymbol{\Lambda}_k \boldsymbol{W}_k^{\mathrm{H}} = \boldsymbol{B}_k^{\mathrm{MF}} \boldsymbol{H}_k^{\mathrm{BC}} \boldsymbol{P}_k \quad \forall k.$$
3) The rotated precoders $\bar{\boldsymbol{P}}_k := \boldsymbol{P}_k \boldsymbol{W}_k \forall k$ and receivers $\bar{\boldsymbol{B}}_k^{\mathrm{MF}} := \boldsymbol{W}_k^{\mathrm{H}} \boldsymbol{B}_k^{\mathrm{MF}} \forall k$ are passed to the level three duality from Sect. 3.2.3.3 to generate MAC precoders $\boldsymbol{T}_1, \ldots, \boldsymbol{T}_K$ and MAC receivers $\boldsymbol{G}_1, \ldots, \boldsymbol{G}_K$ which offer the same SINRs per data stream. The structure of these filters follows from (3.64), and the scaling factors $\psi_{1,1}, \ldots, \psi_{K,L_K}$ result from (3.70).
4) Using the receive filter $\boldsymbol{G}_k$, the rate of user k with separate stream decoding matches *exactly* the mutual information expression of the BC:

$$R_k^{\mathrm{MAC}}(\boldsymbol{T}, \boldsymbol{G}_k) = I_k^{\mathrm{BC}}(\boldsymbol{P}) \quad \forall k.$$

However, the dual MAC features *at least* the BC mutual information:

$$I_k^{\mathrm{MAC}}(\boldsymbol{T}) \geq I_k^{\mathrm{BC}}(\boldsymbol{P}) \quad \forall k.$$

This mutual information can also be achieved by separate stream decoding when MF receivers are used and additional unitary decorrelation filters are applied at each transmit and receive filter to decorrelate every point-to-point link, see Theorem 2.

3.3.2 MAC to BC Conversion

This direction of the rate duality for linear transceivers is the one that is actually used to convert filters derived in the dual MIMO MAC back to the original MIMO BC. The BC to MAC conversion is primarily just a vehicle to prove that both rate regions under linear filtering coincide. In this subsection, the following lemma is shown:

Lemma 3.4 *For every set of precoders in the MIMO MAC leading to certain values for the mutual information of the K users, there exist precoders in the MIMO BC which achieve at least the same mutual information using the same amount of transmit power. Moreover, there exist receive filters such that the MAC mutual information expressions can exactly be achieved with separate stream decoding in the BC.*

With the BC to MAC conversion from Sect. 3.3.1 and the MAC to BC conversion in this subsection, the two directions of the duality are established. Hence, by combining Lemma 3 and Lemma 4, we conclude:

Theorem 3.3 *The mutual information regions obtained by linear transceivers in the MIMO BC and the MIMO MAC subject to a sum power constraint coincide.*

The conversion from the dual MIMO MAC to the MIMO BC is very similar to the reverse case discussed in Sect. 3.3.1 but features some slight modifications. Hence, a shortened derivation for the conversion rules is presented.

By means of the MAC MF receiver in (3.73), the mutual information $I_k^{\mathrm{MAC}}(\boldsymbol{T})$ of user k in the MAC can be written as [cf. (3.74)]

$$I_k^{\mathrm{MAC}}(\boldsymbol{T}) = \log_2 \left| \mathbf{I}_{L_k} + \boldsymbol{G}_k^{\mathrm{MF}} \boldsymbol{H}_k \boldsymbol{T}_k \right|$$

where the MAC channels are defined in (3.14). Introducing unitary decorrelation filters $\boldsymbol{V}_1, \ldots, \boldsymbol{V}_K$, the rotated precoders and receivers read as

$$\bar{\boldsymbol{T}}_k := \boldsymbol{T}_k \boldsymbol{V}_k$$
$$\bar{\boldsymbol{G}}_k^{\mathrm{MF}} := \boldsymbol{V}_k^{\mathrm{H}} \boldsymbol{G}_k^{\mathrm{MF}} \quad \forall k$$

where $\boldsymbol{V}_k$ must be chosen such that

$$\boldsymbol{V}_k \boldsymbol{\Xi}_k \boldsymbol{V}_k^{\mathrm{H}} = \boldsymbol{G}_k^{\mathrm{MF}} \boldsymbol{H}_k \boldsymbol{T}_k$$

holds with unitary $\boldsymbol{V}_k$ and diagonal $\boldsymbol{\Xi}_k$. For these rotated filters, the mutual information is achieved via separate stream decoding [cf. (3.72)]:

$$I_k^{\mathrm{MAC}}(\boldsymbol{T}) = R_k^{\mathrm{MAC}}(\bar{\boldsymbol{T}}, \bar{\boldsymbol{G}}_k^{\mathrm{MF}}) \quad \forall k. \tag{3.76}$$

The SINRs induced by the rotated filters are converted to the BC using the level three duality from Sect. 3.2.4.3 and the parametrization [cf. (3.38)]

$$\begin{aligned} \boldsymbol{p}_{k,j} &= \boldsymbol{P}_k \boldsymbol{e}_j = \varphi_{k,j} \boldsymbol{G}_k^{\mathrm{MF,H}} \boldsymbol{V}_k \boldsymbol{e}_j \\ \boldsymbol{b}_{k,j}^{\mathrm{T}} &= \boldsymbol{e}_j^{\mathrm{T}} \boldsymbol{B}_k = \varphi_{k,j}^{-1} \boldsymbol{e}_j^{\mathrm{T}} \boldsymbol{V}_k^{\mathrm{H}} \boldsymbol{T}_k^{\mathrm{H}} \boldsymbol{C}_{\eta_k}^{-\frac{1}{2}}. \end{aligned} \tag{3.77}$$

The squared scaling factors follow from the linear system of equations that arises when the SINRs of all L streams in the BC are equated with those in the dual MAC:

$$\left[\varphi_{1,1}^2, \ldots, \varphi_{K,L_K}^2\right]^{\mathrm{T}} = \boldsymbol{Z}^{-1} \cdot \left[\|\bar{\boldsymbol{t}}_{1,1}\|_2^2, \ldots, \|\bar{\boldsymbol{t}}_{K,L_K}\|_2^2\right]^{\mathrm{T}} \tag{3.78}$$

The $L \times L$ matrix $\boldsymbol{Z}$ reads as

$$\boldsymbol{Z} := \begin{bmatrix} \boldsymbol{Z}_{1,1} & \cdots & \boldsymbol{Z}_{1,K} \\ \vdots & \ddots & \vdots \\ \boldsymbol{Z}_{K,1} & \cdots & \boldsymbol{Z}_{K,K} \end{bmatrix} \tag{3.79}$$

with off-diagonal blocks

$$\boldsymbol{Z}_{k,i} := -\left[\left(\bar{\boldsymbol{G}}_i^{\mathrm{MF}} \boldsymbol{H}_k \bar{\boldsymbol{T}}_k\right) \odot \left(\bar{\boldsymbol{G}}_i^{\mathrm{MF}} \boldsymbol{H}_k \bar{\boldsymbol{T}}_k\right)^*\right]^{\mathrm{T}} \in \mathbb{R}^{L_k \times L_i} \quad \text{for} \quad k \neq i.$$

All diagonal blocks are diagonal matrices, and the kth block reads as

$$\boldsymbol{Z}_{k,k} := \left(\bar{\boldsymbol{G}}_k^{\mathrm{MF}} \bar{\boldsymbol{G}}_k^{\mathrm{MF,H}}\right) \odot \mathbf{I}_{L_k} - \sum_{i \neq k} \operatorname{diag}\{\mathbf{1}_{L_i}^{\mathrm{T}} \boldsymbol{Z}_{i,k}\} \quad \forall k. \tag{3.80}$$

Again, zero rows and columns representing passive streams have to be removed and after this removal, $\boldsymbol{Z}$ is a nonsingular *M-Matrix* (see Appendix A1) such that there is a positive solution for the scaling factors of all active streams. Summing up all rows of (3.78) multiplied by $\boldsymbol{Z}$ from the left-hand side reveals that the resulting precoders in the BC dissipate the same power as their dual MAC counterparts.

Due to the conservation of the SINRs, the rate obtained by user k in the BC using the receiver $\boldsymbol{B}_k$ and separate stream decoding can be expressed as [cf. (3.49)]

$$R_k^{\mathrm{BC}}(\boldsymbol{P}, \boldsymbol{B}_k) = R_k^{\mathrm{MAC}}(\bar{\boldsymbol{T}}, \bar{\boldsymbol{G}}_k^{\mathrm{MF}}) = I_k^{\mathrm{MAC}}(\boldsymbol{T}) \quad \forall k$$

where the second equality corresponds to (3.76). In general, the receivers $\boldsymbol{B}_1, \ldots, \boldsymbol{B}_K$ do not maximize the receive SINRs and diagonalize the matrix product $\boldsymbol{B}_k \boldsymbol{H}_k^{\mathrm{BC}} \boldsymbol{P}_k$ simultaneously. Thus, the precoders and receivers resulting from the duality conversion usually do not achieve the mutual information given the precoders in $\boldsymbol{P}$ from (3.52):

$$I_k^{\mathrm{BC}}(\boldsymbol{P}) \geq R_k^{\mathrm{BC}}(\boldsymbol{P}, \boldsymbol{B}_k) = I_k^{\mathrm{MAC}}(\boldsymbol{T}) \quad \forall k.$$

According to Theorem 2, unitary decorrelation matrices can be found to achieve the mutual information in the BC. If, by chance, the dual MAC precoders are chosen such that the obtained mutual information tuple lies on the boundary of the mutual information region for a given power constraint, then the duality conversion will deliver filter pairs that diagonalize the error covariance matrix and achieve the BC mutual information.

Synopsis: Given precoders $\boldsymbol{T}_1, \ldots, \boldsymbol{T}_K$ in the dual MIMO MAC achieving mutual information expressions $I_1^{\mathrm{MAC}}(\boldsymbol{T}), \ldots, I_K^{\mathrm{MAC}}(\boldsymbol{T})$, it is possible to derive transmit and receive filters $\boldsymbol{P}$ and $\boldsymbol{B}$, respectively, such that the same or larger data rates can be obtained in the MIMO BC using the same amount of transmit power. The duality conversion from the dual MIMO MAC to the MIMO BC works as follows:

1) The MF receive filters $\boldsymbol{G}_1^{\mathrm{MF}}, \ldots, \boldsymbol{G}_K^{\mathrm{MF}}$ are computed via (3.73).
2) By means of unitary decorrelation filters $\boldsymbol{V}_1, \ldots, \boldsymbol{V}_K$, all matrix products $\boldsymbol{G}_k^{\mathrm{MF}} \boldsymbol{H}_k \boldsymbol{T}_k$ are diagonalized:

$$\boldsymbol{V}_k \boldsymbol{\Xi}_k \boldsymbol{V}_k^{\mathrm{H}} = \boldsymbol{G}_k^{\mathrm{MF}} \boldsymbol{H}_k \boldsymbol{T}_k \quad \forall k.$$

3) The rotated precoders $\bar{\boldsymbol{T}}_k := \boldsymbol{T}_k \boldsymbol{V}_k \forall k$ and the rotated MF receive filters $\bar{\boldsymbol{G}}_k^{\mathrm{MF}} := \boldsymbol{V}_k^{\mathrm{H}} \boldsymbol{G}_k^{\mathrm{MF}} \forall k$ achieve the mutual information via separate stream decoding. They are passed to the level three duality to generate BC precoders $\boldsymbol{P}_1, \ldots, \boldsymbol{P}_K$ and MAC receivers $\boldsymbol{B}_1, \ldots, \boldsymbol{B}_K$ which offer the same SINRs per data stream. Their parametrization is given in (3.77) and the scaling factors $\varphi_{1,1}, \ldots, \varphi_{K,L_K}$ follow from (3.78).

4) Using the receive filter $\boldsymbol{B}_k$, the rate of user k with separate decoding matches *exactly* the mutual information expression of the MAC:

$$R_k^{\mathrm{BC}}(\boldsymbol{P}, \boldsymbol{B}_k) = I_k^{\mathrm{MAC}}(\boldsymbol{T}) \quad \forall k.$$

However, the original BC offers *at least* the dual MAC mutual information:

$$I_k^{\mathrm{BC}}(\boldsymbol{P}) \geq I_k^{\mathrm{MAC}}(\boldsymbol{T}).$$

In combination with the result that the mutual information values in the dual MAC after the BC to MAC duality transformation is larger than or equal to those of the original BC, we conclude that the two mutual information regions are identical.

3.3.3 Advantages of the Novel Rate Duality for Linear Transceivers

The presented rate duality for linear transceivers serves as the key element to handle rate-based optimizations in the MIMO BC with multi-antenna terminals under linear filtering without having to sacrifice the full mutual information gain of joint decoding. Basically all state-of-the-art weighted sum rate maximization algorithms for example treat the individual streams of a particular user as virtual users thus generating interference to other streams of the same user. In [48], the authors present an algorithm that frequently switches between the dual MAC and the BC using a stream-wise SINR duality. They exploit the fact that the receive filters that maximize the SINR can be computed independently and serve as the transmit filters after normalization thus making dualities attractive to tackle such a problem. Using an alternating optimization approach they alternate between the receive filter optimization in the dual MAC, the receive filter optimization in the BC, and the power allocation. The latter one follows from a local approximation of the signomial optimization problem by a monomial function and a geometric programming toolbox is applied. The decoupling into different sets of variables to optimize over during the alternating optimization and the geometric programming part lead to a huge complexity. Almost the same alternating optimization approach with frequent MAC to BC changes and vice versa can be found in [57, 58]. But in contrast to [48], the authors do not have to approximate the weighted sum rate utility locally during the power allocation step. They convert the objective of the power optimization directly into a posynomial for integer weighting factors such that the geometric programming approach directly becomes applicable. Again, the stream-wise formulation deters one from using the full potential of the mutual information expressions.

An algorithm targeted at maximizing the weighted sum rate directly in the MIMO BC is given in [59], where again the stream-wise formulation of the sum rate of a user is chosen. Since no duality is applied, the transmit beamformers have to be computed using a second-order-cone-program implying a high computational burden. In addition, the signomial programming part during the power allocation step has a very high complexity. One of the first contributions on rate-based optimizations with multi-antenna receivers in the BC where the mutual information expression $I_k^{\mathrm{BC}}(\boldsymbol{P})$ from (3.52) is chosen instead of the sum rate of the individual streams of a user can be found in [60]. Therein, the authors noticed that $I_k^{\mathrm{BC}}(\boldsymbol{P})$ can be written in terms of the determinant of the MSE error covariance matrix as in (3.53). Given their parametrization in terms of unit norm beamformers and powers per data stream they mention that the resulting weighted sum rate maximization could in principle be approached using sequential quadratic programming. However, due to the involved complexity, they do not suggest to do so and hence, switch to a stream-wise operation (without decorrelation) and apply the stream-wise MSE duality. In the dual domain, a sequential quadratic program is then used for the power allocation of the streams. Another (suboptimum) weighted sum rate maximization algorithm is presented in [61]. Again, the mutual information expressions in the BC are not chosen as rate terms. Instead, the authors reformulate the problem as a weighted sum MSE minimization problem and apply an alternating optimization algorithm.

In the beginning of 2008, Böhnke et al. came up with a very interesting gradient-based weighted sum rate maximization approach using the mutual information expressions from (3.74) in the dual MIMO MAC [62]. The presented algorithm is very similar to the one that we will describe in Sect. 7.3. For the duality conversion, they referred to an early version of our user-wise MSE duality in [38] which however does *not* conserve the mutual information expressions. At that time, their results in the dual MIMO MAC could not be converted back to the BC where the problem had been posed. Their contribution for linear transceivers became meaningful at the end of 2008 when we published the rate duality from Sect. 3.3 in [21].

These examples show that practically all rate-based optimizations under linear filtering where multi-antenna terminals are involved directly start with stream-wise rate formulations instead of the mutual information expressions. In addition, they have a very high computational complexity since optimization packages are required to solve the power allocation step. Some greedy stream-allocation schemes are an exception, see [63, 64]. However they are also based on stream-wise rate expressions.

Besides the advantage that our rate duality allows to use the mutual information expressions instead of the stream-wise sum rates for a user, there is no need for a partitioning of the optimization variables into unit-norm transmit beamformers, unit-norm receive beamformers, and powers of the individual streams. This grouping becomes superfluous and we can either optimize with respect to the total transmit filters $\boldsymbol{T}_1, \ldots, \boldsymbol{T}_K$ of the K users in the dual MAC or with respect to their outer products, i.e., the transmit covariance matrices. The optimum receivers are implicitly taken into account in the mutual information expression. Hence, we are able to use standard gradient-based algorithms and are not reliant on optimization packages. Due to the indices alignment in the dual MIMO MAC, iterative algorithms can be implemented very efficiently.

Using the rate-based duality, we were able to derive a close-to-optimum (weighted) sum rate maximization algorithm [65] that will be derived in Sect. 7.3. Moreover, we were able to further explore the properties of linear multi-antenna transceivers. In [22, 23], we found an asymptotic expression for the maximum possible weighted sum rate in the high power regime for an instantaneous channel realization (see Chap. 8), and the geometry of the asymptotic rate region was analyzed in [24].

Chapter 4
Rate Duality with Nonlinear Interference Cancelation

In this chapter the rate duality for linear transceivers from Sect. 3.3 is extended to systems utilizing nonlinear interference cancelation. To this end, we make use of the same stream decorrelation approach to remove the intra-user interference of every user. Before revisiting the state-of-the-art duality in Sect. 4.3 and deriving the novel rate duality for systems with interference cancelation in Sect. 4.4, we first focus on the capacity regions of the MIMO MAC and the MIMO BC in Sects. 4.1 and 4.2, respectively.

4.1 The Capacity Region of the MIMO MAC

4.1.1 The Two-User MIMO MAC

Let us first consider the $K = 2$ user MIMO MAC with independent Gaussian symbol vectors according to the system model from Sect. 2.2. In addition, let the precoded symbol vector of user k be defined via

$$\boldsymbol{x}_k := \boldsymbol{T}_k \boldsymbol{s}_k \in \mathbb{C}^{M_k} \quad \forall k \in \{1, \ldots, K\}.$$

Under the assumption that the precoded symbol vector $\boldsymbol{x}_2$ of user 2 (and of course his channel matrix $\boldsymbol{H}_2$) is known at the receiving base station such that its influence $\boldsymbol{H}_2 \boldsymbol{x}_2$ can be subtracted from the received signal $\boldsymbol{y}$ defined in (2.7), the maximum achievable rate R_1^{MAC} of user 1 is clearly upper bounded by the mutual information between $\boldsymbol{x}_1$ and $\boldsymbol{y}$ conditioned on $\boldsymbol{x}_2$ (see for example [66, p. 532] for the two-user SISO case):

$$R_1^{\mathrm{MAC}} \leq I(\boldsymbol{x}_1; \boldsymbol{y} | \boldsymbol{x}_2). \tag{4.1}$$

By interchanging user indices and thus conditioning on $\boldsymbol{x}_1$, we arrive at the equivalent upper bound on the rate of user 2:

R. Hunger, *Analysis and Transceiver Design for the MIMO Broadcast Channel*, Foundations in Signal Processing, Communications and Networking 8,
DOI: 10.1007/978-3-642-31692-0_4, © Springer-Verlag Berlin Heidelberg 2013

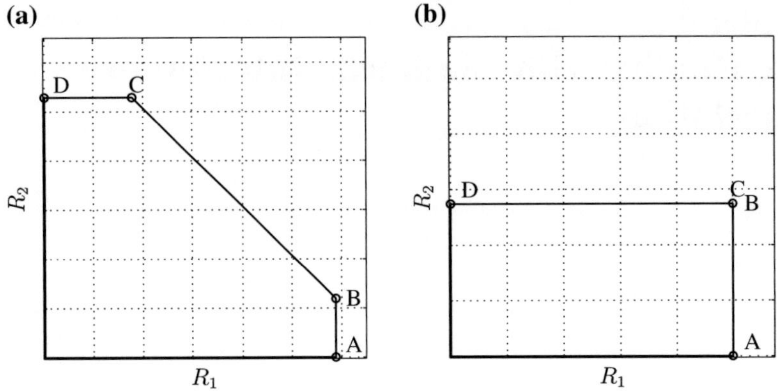

Fig. 4.1 MIMO MAC rate region for fixed transmit covariance matrices $\boldsymbol{Q}_1$ and $\boldsymbol{Q}_2$: **a** standard pentagon, **b** degenerated pentagon

$$R_2^{\mathrm{MAC}} \leq I(\boldsymbol{x}_2; \boldsymbol{y}|\boldsymbol{x}_1). \tag{4.2}$$

From the total mutual information between both senders and the receiver, i.e., when using joint decoding, we find the third upper bound inequality, which limits the sum of the two data rates:

$$R_1^{\mathrm{MAC}} + R_2^{\mathrm{MAC}} \leq I(\boldsymbol{x}_1, \boldsymbol{x}_2; \boldsymbol{y}) \tag{4.3}$$

For given precoders $\boldsymbol{T}_1$ and $\boldsymbol{T}_2$ inducing *fixed* covariance matrices $\boldsymbol{Q}_i := \boldsymbol{T}_i\boldsymbol{T}_i^{\mathrm{H}} \succcurlyeq \boldsymbol{0}$, $i \in \{1, 2\}$, the intersection of rate pairs defined by (4.1)–(4.3) and the nonnegative quadrant $\mathbb{R}^2_{+,0}$ defines a (possibly degenerated) pentagon:

$$\mathcal{R}(\boldsymbol{Q}_1, \boldsymbol{Q}_2) := \left\{ \begin{bmatrix} R_1 \\ R_2 \end{bmatrix} \in \mathbb{R}^2_{+,0} \middle| \begin{array}{r} R_1 \leq I(\boldsymbol{x}_1; \boldsymbol{y}|\boldsymbol{x}_2) = \log_2 \left|\mathbf{I} + \boldsymbol{H}_1\boldsymbol{Q}_1\boldsymbol{H}_1^{\mathrm{H}}\right| \\ R_2 \leq I(\boldsymbol{x}_2; \boldsymbol{y}|\boldsymbol{x}_1) = \log_2 \left|\mathbf{I} + \boldsymbol{H}_2\boldsymbol{Q}_2\boldsymbol{H}_2^{\mathrm{H}}\right| \\ R_1 + R_2 \leq I(\boldsymbol{x}_1, \boldsymbol{x}_2; \boldsymbol{y}) = \log_2 \left|\mathbf{I} + \boldsymbol{H}_1\boldsymbol{Q}_1\boldsymbol{H}_1^{\mathrm{H}} + \boldsymbol{H}_2\boldsymbol{Q}_2\boldsymbol{H}_2^{\mathrm{H}}\right| \end{array} \right\}. \tag{4.4}$$

Note that the mutual information is maximized with Gaussian inputs $\boldsymbol{x}_1$ and $\boldsymbol{x}_2$. Since the symbol vectors and the noise are pairwise independent, the mutual information expressions follow from

$$\begin{aligned} I(\boldsymbol{x}_1; \boldsymbol{y}|\boldsymbol{x}_2) &= \mathcal{H}(\boldsymbol{H}_1\boldsymbol{T}_1\boldsymbol{s}_1 + \boldsymbol{n}) - \mathcal{H}(\boldsymbol{n}) \\ I(\boldsymbol{x}_2; \boldsymbol{y}|\boldsymbol{x}_1) &= \mathcal{H}(\boldsymbol{H}_2\boldsymbol{T}_2\boldsymbol{s}_2 + \boldsymbol{n}) - \mathcal{H}(\boldsymbol{n}) \\ I(\boldsymbol{x}_1, \boldsymbol{x}_2; \boldsymbol{y}) &= \mathcal{H}(\boldsymbol{y}) - \mathcal{H}(\boldsymbol{n}) \end{aligned}$$

in analogy to (3.46). Note that the polytope $\mathcal{R}(\boldsymbol{Q}_1, \boldsymbol{Q}_2)$ is convex by definition for any pair of covariance matrices $\boldsymbol{Q}_i \succcurlyeq \boldsymbol{0}$, $i \in \{1, 2\}$. An example for a typically pentagon-shaped region $\mathcal{R}(\boldsymbol{Q}_1, \boldsymbol{Q}_2)$ is shown in Fig. 4.1a.

An important observation is that the two corner points B and C can be achieved by *successive decoding*, see for example [67, 68]. At point B, user two is decoded first and hence has to face interference from user one. For fixed $\boldsymbol{Q}_1$ and $\boldsymbol{Q}_2$, the maximum possible rate for user two with arbitrarily small error probability thus reads as

$$R_2^{(\mathrm{B})} = \log_2 \left| \mathbf{I} + \left(\mathbf{I} + \boldsymbol{H}_1\boldsymbol{Q}_1\boldsymbol{H}_1^{\mathrm{H}}\right)^{-1}\boldsymbol{H}_2\boldsymbol{Q}_2\boldsymbol{H}_2^{\mathrm{H}} \right|. \tag{4.5}$$

With infinitely long codes, error-free decoding of $\boldsymbol{s}_2$ becomes possible as long as $R_2 \leq R_2^{(\mathrm{B})}$, i.e., for all points on the line segment between A and B, such that the portion $\boldsymbol{H}_2\boldsymbol{T}_2\boldsymbol{s}_2$ in the received signal $\boldsymbol{y}$ can be subtracted. As a consequence, user one can be decoded without perturbation by interference from user two such that the maximum data rate for user one given $\boldsymbol{Q}_1$ reads as

$$R_1^{(\mathrm{B})} = \log_2 \left| \mathbf{I} + \boldsymbol{H}_1\boldsymbol{Q}_1\boldsymbol{H}_1^{\mathrm{H}} \right|. \tag{4.6}$$

For any point on the line segment between A and B that is different from B, user two simply uses a smaller code rate than the one in (4.5). Additionally, $\boldsymbol{Q}_2$ could be scaled to consume less power.

In order to achieve the corner point C, the decoding order is reversed, i.e., user one is decoded first, its signal portion $\boldsymbol{H}_1\boldsymbol{T}_1\boldsymbol{s}_1$ is subtracted from the received signal $\boldsymbol{y}$, and afterwards, user two is decoded. At point C, the rate of user one seeing interference from user two is given by

$$R_1^{(\mathrm{C})} = \log_2 \left| \mathbf{I} + \left(\mathbf{I} + \boldsymbol{H}_2\boldsymbol{Q}_2\boldsymbol{H}_2^{\mathrm{H}}\right)^{-1}\boldsymbol{H}_1\boldsymbol{Q}_1\boldsymbol{H}_1^{\mathrm{H}} \right|. \tag{4.7}$$

The error-free decoded codeword is subtracted from the received signal such that the rate of user two is given by

$$R_2^{(\mathrm{C})} = \log_2 \left| \mathbf{I} + \boldsymbol{H}_2\boldsymbol{Q}_2\boldsymbol{H}_2^{\mathrm{H}} \right|. \tag{4.8}$$

In turn, any point on the line segment connecting the points C and D can be reached by allocating a rate to user one that is smaller than or equal to the one in (4.7). To achieve a point on the line segment connecting B and C, one can either apply time sharing between the two decoding orders or jointly decode the codewords of both users [68]. The third, most elegant, way to achieve a point on this line segment is to apply a rate-splitting approach [69]. There, one of the two users is split into two virtual users to create a three-dimensional polymatroid.[1] A point on the line segment BC then corresponds to a vertex of the three-dimensional polymatroid which can be obtained via successive decoding. In contrast to the time-sharing approach between the vertices B and C, no synchronization in time is required and this approach is drastically less complex than the joint decoding approach.

---

[1] A polymatroid is the extension of a two-dimensional pentagon to three dimensions, see Sect. 4.1.2 for the definition.

Note that the line segment BC has a $-45°$ slope as $R_1^{(B)} + R_2^{(B)} = R_1^{(C)} + R_2^{(C)}$, cf. (4.5)–(4.8). In other words, changing the decoding order of the two users does not change the sum rate. A generalization of this statement to more than two users can be found in Appendix A.7.

The pentagon shape depicted in Fig. 4.1a degenerates to a rectangle[2] if the two points B and C merge, i.e., if changing the decoding order does not change the achieved rate pair. $R_1^{(B)} = R_1^{(C)}$ and $R_2^{(B)} = R_2^{(C)}$ holds if $\boldsymbol{H}_1\boldsymbol{Q}_1\boldsymbol{H}_1^{\mathrm{H}}\boldsymbol{H}_2\boldsymbol{Q}_2\boldsymbol{H}_2^{\mathrm{H}} = \mathbf{0}$. Such a configuration is shown in Fig. 4.1b where the two channel matrices fulfill $\boldsymbol{H}_1^{\mathrm{H}}\boldsymbol{H}_2 = \mathbf{0}$.

4.1.1.1 Individual Power Constraints

In the real MIMO MAC, the users have individual power constraints. Hence, the capacity region is a union of pentagons $\mathcal{R}(\boldsymbol{Q}_1, \boldsymbol{Q}_2)$ (see for example [70]), where the union is taken over all positive semi-definite transmit covariance matrices $\boldsymbol{Q}_1$ and $\boldsymbol{Q}_2$ that fulfill the individual power constraints $\mathrm{tr}(\boldsymbol{Q}_i) \leq P_{\mathrm{max},i}\ \forall i$:

$$\mathcal{C}^{\mathrm{MAC}}(P_{\mathrm{max},1}, P_{\mathrm{max},2}) := \bigcup_{\substack{\boldsymbol{Q}_i \succeq \mathbf{0}\ \forall i\in\{1,2\} \\ \mathrm{tr}(\boldsymbol{Q}_i)\leq P_{\mathrm{max},i}\ \forall i\in\{1,2\}}} \mathcal{R}(\boldsymbol{Q}_1, \boldsymbol{Q}_2). \tag{4.9}$$

Note that the union of pentagons boils down to a single pentagon in the two-user vector MAC with single-antenna users and individual power constraints:

Theorem 4.1 *The capacity region of the two-user vector MAC with individual upper bounds $P_{\mathrm{max},1}$ and $P_{\mathrm{max},2}$ on the dissipated powers is a (possibly degenerated) pentagon, i.e.,*

$$\mathcal{C}^{\mathrm{MAC}}(P_{\mathrm{max},1}, P_{\mathrm{max},2}) = \mathcal{R}(P_{\mathrm{max},1}, P_{\mathrm{max},2}).$$

The union over all pentagons $\mathcal{R}(Q_1, Q_2)$ satisfying $0 \leq Q_i \leq P_{\mathrm{max},i},\ \forall i \in \{1, 2\}$ corresponds to a single pentagon where both transmitters make use of their whole power budget.

For the proof, we make use of the fact that the weighted sum rate is a nondecreasing function in all transmit covariance matrices, see Appendix A.8. In other words, $Q_1 = P_{\mathrm{max},1}$ clearly maximizes $R_1^{(B)}$ from (4.6) and $Q_2 = P_{\mathrm{max},2}$ maximizes $R_2^{(C)}$ from (4.8). In addition, the weighted sum rate is maximized for $Q_1 = P_{\mathrm{max},1}$ and $Q_2 = P_{\mathrm{max},2}$ because the sum rate is a nondecreasing function in Q_1 and Q_2.

In the MIMO MAC with multi-antenna transmitters, however, the union in (4.9) is mandatory because the transmit covariance matrix $\boldsymbol{Q}_i$ that maximizes the rate of user i usually fails to achieve the maximum sum rate. The capacity region of a two user MIMO MAC with individual power constraints is shown in Fig. 4.2. At

[2] Further degenerations to a line segment or a point are not considered here, i.e., $\boldsymbol{H}_i\boldsymbol{Q}_i\boldsymbol{H}_i^{\mathrm{H}} \neq \mathbf{0}\ \forall i \in \{1, 2\}$.

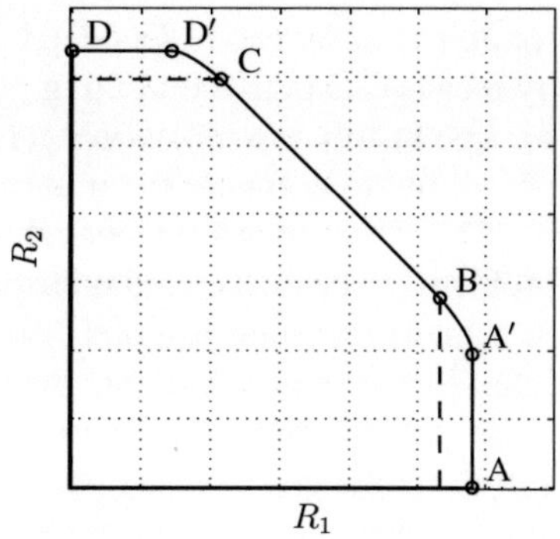

Fig. 4.2 Two user MIMO MAC capacity region with individual power constraints

point A′, user two is decoded first such that user one does not see any interference. To maximize his rate

$$R_1^{(\mathrm{A}')} = \log_2 \left| \mathbf{I} + \boldsymbol{H}_1 \boldsymbol{Q}_1^{(\mathrm{A}')} \boldsymbol{H}_1^{\mathrm{H}} \right|$$

the transmit covariance matrix $\boldsymbol{Q}_1^{(\mathrm{A}')}$ is chosen such that the single-user capacity is reached [52]. This is achieved when $\boldsymbol{Q}_1^{(\mathrm{A}')}$ has the same eigenbasis as $\boldsymbol{H}_1^{\mathrm{H}} \boldsymbol{H}_1$ and the power allocation follows from the water-filling algorithm [52]. Equivalently, the rate for user two seeing interference from user one reads as

$$R_2^{(\mathrm{A}')} = \log_2 \left| \mathbf{I} + \boldsymbol{H}_2^{\mathrm{H}} (\mathbf{I} + \boldsymbol{H}_1 \boldsymbol{Q}_1^{(\mathrm{A}')} \boldsymbol{H}_1^{\mathrm{H}})^{-1} \boldsymbol{H}_2 \boldsymbol{Q}_2^{(\mathrm{A}')} \right|.$$

It is maximized when $\boldsymbol{Q}_2^{(\mathrm{A}')}$ is optimized as in a single-user point-to-point system with noise-plus-interference covariance matrix $\mathbf{I} + \boldsymbol{H}_1 \boldsymbol{Q}_1^{(\mathrm{A}')} \boldsymbol{H}_1^{\mathrm{H}}$ and power limit $\mathrm{tr}(\boldsymbol{Q}_2^{(\mathrm{A}')}) \leq P_{\mathrm{max},2}$ [52]. Water-filling must again be applied for the power allocation. To achieve any point on the line segment connecting the points A and A′, user two simply uses a smaller rate than $R_2^{(\mathrm{A}')}$. Additionally, $\boldsymbol{Q}_2^{(\mathrm{A}')}$ can be scaled down accordingly to reduce the dissipated power.

When the decoding order is reversed, the points D and D′ are achieved by the same principle, i.e., user two chooses his covariance matrix $\boldsymbol{Q}_2^{(\mathrm{D}')}$ such that it commutes with $\boldsymbol{H}_2^{\mathrm{H}} \boldsymbol{H}_2$ and applies water-filling with power constraint $\mathrm{tr}(\boldsymbol{Q}_2^{(\mathrm{D}')}) \leq P_{\mathrm{max},2}$. For user one, point D′ is feasible if $\boldsymbol{Q}_1^{(\mathrm{D}')}$ has the same eigenbasis as $\boldsymbol{H}_1^{\mathrm{H}} (\mathbf{I} + \boldsymbol{H}_2 \boldsymbol{Q}_2^{(\mathrm{D}')} \boldsymbol{H}_2^{\mathrm{H}})^{-1} \boldsymbol{H}_1$ and the power allocation follows from the water-filling algorithm with power constraint $\mathrm{tr}(\boldsymbol{Q}_1^{(\mathrm{D}')}) \leq P_{\mathrm{max},1}$.

The two points B and C are achieved by the same pair of covariance matrices $\boldsymbol{Q}_1^{(\mathrm{B})} = \boldsymbol{Q}_1^{(\mathrm{C})}$ and $\boldsymbol{Q}_2^{(\mathrm{B})} = \boldsymbol{Q}_2^{(\mathrm{C})}$ but with different decoding orders. Hence, $R_1^{(\mathrm{B})} + R_2^{(\mathrm{B})} = R_1^{(\mathrm{C})} + R_2^{(\mathrm{C})}$ as can be shown with (4.5)–(4.8), and this line segment has a $-45°$ slope (it is the line segment BC in Fig. 4.1a with $\boldsymbol{Q}_1^{(\mathrm{B})}$ and $\boldsymbol{Q}_2^{(\mathrm{B})}$). Note that for point B, user two is decoded first. However, $\boldsymbol{Q}_1^{(B)}$ and $\boldsymbol{Q}_2^{(B)}$ cannot be determined by means of simply computing *two* times the capacity achieving covariance matrix for a point-to-point link with a certain noise plus interference covariance matrix. One possibility to compute them is to use the iterative algorithm proposed by Yu et al. in [71]. They found out that the KKT conditions of the sum rate maximization problem can be interpreted such that the optimum covariance matrices must be chosen to achieve the point-to-point MIMO capacity when *all* other users are regarded as interference. Since changing the covariance matrix of one users leads to a modified noise-plus-interference covariance matrix of all other users, the problem must be solved iteratively. Convergence from any starting point was proven and after only a single covariance update for every user, the sum rate is only $K-1$ nats per output dimension away from the sum capacity [71]. Any point on the line segment between B and C can be achieved either by time-sharing between the two decoding orders or by jointly decoding both users [68], or by applying rate-splitting [69]. Note that the total pentagon $\mathcal{R}(\boldsymbol{Q}_1^{(C)}, \boldsymbol{Q}_2^{(C)})$ which achieves the sum capacity is completed by the dashed lines (see Fig. 4.2 and cf. Fig. 4.1a).

Every point on the curve between A′ and B corresponds to a different pair of transmit covariance matrices $\boldsymbol{Q}_1$ and $\boldsymbol{Q}_2$ and can be achieved by successive decoding where the covariance matrices are obtained by means of a weighted sum rate maximization. On curve between A′ and B, user two has to be decoded first. This results from the fact that the slope of the tangent line of every point between A′ and B is between -45 and $-90°$. As the normal vector of the tangent line of the rate region boundary is collinear to the weight vector in the optimum of the corresponding weighted sum rate, the weight of user two is smaller than (when B is excluded) that of user one. In Appendix A.7, a rigorous proof can be found that the decoding order must be such that the weights are nondecreasing to maximize any weighted sum rate, i.e., the user with the smallest weight has to be decoded first. For the curve between A′ and B, this means that user two must be decoded before user one. We generated the curved part in Fig. 4.2 by means of a projected gradient algorithm targeted at maximizing the weighted sum rate for given weights. Apart from the individual power constraints, this algorithm is equivalent to the one described in Sect. 6.2.

An important property of the boundary of the rate region between the points A′ and B is that the curvature is such that the resulting capacity region is convex. This is a result of the following theorem:

Theorem 4.2 *The union of all pentagons $\mathcal{R}(\boldsymbol{Q}_1, \boldsymbol{Q}_2)$ emerging from all positive semi-definite covariance matrices $\boldsymbol{Q}_1$ and $\boldsymbol{Q}_2$ satisfying the individual trace constraints is convex.*

As a consequence, the convex hull operator does not have to be applied for the definition of the capacity region in (4.9). Note that this result is not obvious since the union of convex sets does not have to be convex. A proof for Theorem 5 could not be found in the literature by the author although it is stated in some publications. In [72] for example, the authors also say that the convex hull operator is not needed in the Gaussian MIMO MAC but they do not give a rationale for it.

The proof for the convexity statement goes as follows. First, we can observe that the weighted sum rate in the MIMO MAC is jointly concave with respect to the transmit covariance matrices, see Appendix A.9, and we optimize over a convex set defined by the constraints that the covariance matrices are positive semi-definite with upper bounded traces. Thus, any pair of locally optimum covariance matrices yields the unique maximum value for the weighted sum rate. However, since the weighted sum rate need not be *strictly* concave, the optimizers need not be unique in general although they achieve the unique optimum. One might thus think of a scenario where the tangent line defined by the weights touches two different rate points on the boundary. Let these two *different* rate pairs be denoted by (R_1', R_2') and (R_1'', R_2'') and let them be obtained by the covariance matrix pairs $(\boldsymbol{Q}_1', \boldsymbol{Q}_2')$ and $(\boldsymbol{Q}_1'', \boldsymbol{Q}_2'')$, respectively. Since both covariance matrix pairs are assumed to be global optimizers of the weighted sum rate maximization, the weighted sum of both rate pairs coincides. If the union of pentagons were not convex, then the boundary of the rate region between the two rate pairs (R_1', R_2') and (R_1'', R_2'') would have to be such that it lies below the two touching points with a *smaller* weighted sum rate. However, since the set of optimizers of a convex problem is also convex and thus *continuous* [54, Sect. 4.2.1, p. 137], there must be a continuum of globally optimum covariance matrix pairs connecting $(\boldsymbol{Q}_1', \boldsymbol{Q}_2')$ and $(\boldsymbol{Q}_1'', \boldsymbol{Q}_2'')$. Since the mapping from the covariance matrices to the rates is itself continuous, the two rate pairs (R_1', R_2') and (R_1'', R_2'') must be connected by a line segment on which every rate pair has the same weighted sum rate. This is a consequence of the fact that the continuum of globally optimum covariance matrices achieves the *same* globally optimum weighted sum rate metric. This excludes the possibility that the union of pentagons is nonconvex.

For the rate points C and D′ in Fig. 4.2 the same argumentation holds as for the points B and A′ except that the decoding order is now reversed, i.e., user one is decoded first. At the point D′, user two chooses his covariance matrix $\boldsymbol{Q}_2^{(\mathrm{D}')}$ such that it commutes with $\boldsymbol{H}_2^{\mathrm{H}}\boldsymbol{H}_2$ and the power policy again follows from the water-filling algorithm. For user one, the covariance matrix $\boldsymbol{Q}_1^{(\mathrm{D}')}$ is the optimizer of a point-to-point MIMO system given the noise-plus-interference covariance matrix $\mathbf{I} + \boldsymbol{H}_2\boldsymbol{Q}_2^{(\mathrm{D}')}\boldsymbol{H}_2^{\mathrm{H}}$. For orthogonal channels $\boldsymbol{H}_1$ and $\boldsymbol{H}_2$, the capacity region degenerates to a rectangle as in Fig. 4.1b, i.e., the points A′, B, C, and D′ in Fig. 4.2 merge.

4.1.1.2 Sum Power Constraint

Throughout this book, the sum power constrained MIMO MAC will be of more importance than the one with individual power constraints since the MIMO MAC with sum power constraint serves only as a construct to solve problems stated in the MIMO BC. Similar to the MSE and rate duality for linear transceivers, the dual MIMO MAC has a sum power constraint in case of nonlinear interference cancelation. Starting from the rate pentagon in (4.4) for fixed covariance matrices, the capacity region of the sum power constrained dual MIMO MAC is defined in analogy to (4.9) via

$$\mathcal{C}^{\mathrm{MAC}}(P_{\max}) := \bigcup_{\substack{\boldsymbol{Q}_i \succeq \mathbf{0}\ \forall i \in \{1,2\} \\ \sum_{i=1}^{2} \mathrm{tr}(\boldsymbol{Q}_i) \leq P_{\max}}} \mathcal{R}(\boldsymbol{Q}_1, \boldsymbol{Q}_2). \tag{4.10}$$

Using the same argumentation as for the case of individual power constraints, the union of pentagons itself defines a convex set. As the rate expressions of the two users are independent of the imposed power constraint, the nondecreasing property remains valid and the sum power constraint is active on the whole boundary of the capacity region where both rates are nonzero. A typical capacity region of a MIMO MAC with sum power constraint is depicted in Fig. 4.3a. In contrast to the case of individual power constraints shown in Fig. 4.2, there are no horizontal and no vertical line segments when only a sum power constraint is active. At point A, the whole power is allocated to user one and the covariance matrix $\boldsymbol{Q}_1^{(\mathrm{A})}$ achieves the capacity of a point-to-point MIMO link with channel $\boldsymbol{H}_1$ [52]. Since no power is left for user two, his rate is zero. In order to have a nonzero rate for user two, some amount of power has to be allocated to him such that less power is available for user one yielding a smaller rate than $R_1^{(\mathrm{A})}$ for him. Hence, a vertical line segment is impossible. Equivalently, the point D is obtained by allocating the total power $P_{\max}$ to user two and the transmit covariance matrix $\boldsymbol{Q}_2^{(\mathrm{D})}$ is the solution of the single-user point-to-point rate maximization with channel $\boldsymbol{H}_2$ [52].

The covariance matrices leading to the rate point B were obtained by the weighted sum rate maximization algorithm from Sect. 6.2 with equal weights. At point B, user two is decoded first. Interchanging the decoding order such that user one is decoded first and leaving the covariance matrices unchanged yields the rate point C. As $R_1^{(\mathrm{B})} + R_2^{(\mathrm{B})} = R_1^{(\mathrm{C})} + R_2^{(\mathrm{C})}$ [with (4.5)–(4.8)], the line segment between B and C has a slope of $-45°$ since changing the decoding order leaves the sum rate constant, cf. Appendix A.7. Any point on that line segment can be achieved either by time-sharing between the decoding orders or by jointly decoding the two users' codewords [68] or by applying rate-splitting [69]. Again, the pentagon $\mathcal{R}(\boldsymbol{Q}_1^{(\mathrm{B})}, \boldsymbol{Q}_2^{(\mathrm{B})})$ is depicted by the dashed lines. Every point on the curve between A and B corresponds to a different pair of covariance matrices $\boldsymbol{Q}_1$ and $\boldsymbol{Q}_2$ when user two is decoded before user one. For the curve between C and D, user one is decoded first.

In case of orthogonal channels where $\boldsymbol{H}_1^{\mathrm{H}}\boldsymbol{H}_2 = \mathbf{0}$, the pentagon $\mathcal{R}(\boldsymbol{Q}_1^{(\mathrm{B})}, \boldsymbol{Q}_2^{(\mathrm{B})})$ degenerates to a rectangle as shown in Fig. 4.3b. Although the boundary of the capacity region seems to hit the axes perpendicular in Fig. 4.3b, a larger zoom level would reveal that this is not the case.

4.1.2 The Multi-User MIMO MAC

In the general K-user MIMO MAC (as for any K-user MAC), there is an upper bound on the sum rate of any nonempty user subset $\mathcal{U} \subseteq \mathcal{K}$, where $\mathcal{K} := \{1, \ldots, K\}$ denotes the set of all user indices. Let $\boldsymbol{x}_{\mathcal{U}}$ denote the set of precoded symbol vectors

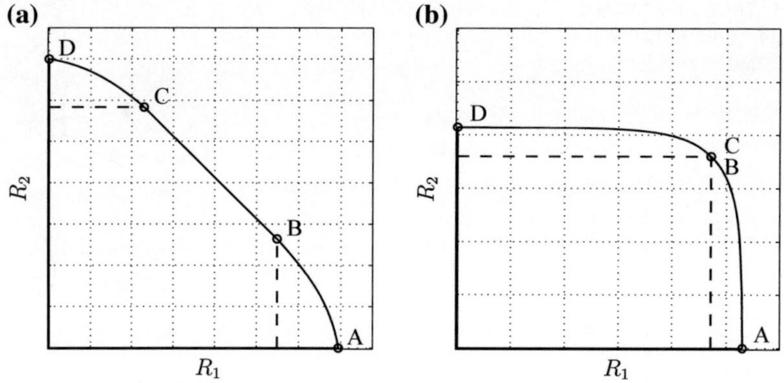

Fig. 4.3 Two user MIMO MAC capacity region with a sum power constraint: **a** typical form, **b** degenerated form (orthogonal channels)

$\{\boldsymbol{x}_i | i \in \mathcal{U}\}$ of all users out of the set $\mathcal{U}$. For a particular user subset $\mathcal{U}$, the upper bound on the sum rate of all users in $\mathcal{U}$ corresponds to the mutual information between $\boldsymbol{x}_{\mathcal{U}}$ and the received signal vector $\boldsymbol{y}$ from (2.7) given the precoded symbol vectors $\boldsymbol{x}_{\overline{\mathcal{U}}}$ of all other users belonging to the complementary set $\overline{\mathcal{U}} := \mathcal{K} \setminus \mathcal{U}$:

$$\sum_{k \in \mathcal{U}} R_k^{\text{MAC}} \leq I(\boldsymbol{x}_{\mathcal{U}}; \boldsymbol{y} | \boldsymbol{x}_{\overline{\mathcal{U}}}). \tag{4.11}$$

Given the system model in Sect. 2.2 with independent Gaussian distributed symbol vectors and independent zero mean noise $\boldsymbol{n}$ with an identity covariance matrix, the mutual information expression reads as

$$I(\boldsymbol{x}_{\mathcal{U}}; \boldsymbol{y} | \boldsymbol{x}_{\overline{\mathcal{U}}}) = \log_2 \left| \mathbf{I} + \sum_{k \in \mathcal{U}} \boldsymbol{H}_k \boldsymbol{Q}_k \boldsymbol{H}_k^{\text{H}} \right| \tag{4.12}$$

with $\boldsymbol{Q}_k := \boldsymbol{T}_k \boldsymbol{T}_k^{\text{H}} \in \mathbb{S}^{M_k}$ denoting the transmit covariance matrix of user k. Note that the presumed Gaussian symbols maximize the upper bound in (4.11) for given covariance matrices $\boldsymbol{Q}_k \; \forall k \in \mathcal{U}$. There are $2^K - 1$ different nonempty user subsets $\mathcal{U}$ and (4.11) has to hold for every $\mathcal{U}$:

$$\sum_{k \in \mathcal{U}} R_k^{\text{MAC}} \leq I(\boldsymbol{x}_{\mathcal{U}}; \boldsymbol{y} | \boldsymbol{x}_{\overline{\mathcal{U}}}) \quad \forall \mathcal{U} \subseteq \mathcal{K}, \; \mathcal{U} \neq \emptyset. \tag{4.13}$$

Obviously, there are additional K lower bounds stating that the rate of every user must be nonnegative. For fixed covariance matrices $\boldsymbol{Q}_1, \ldots, \boldsymbol{Q}_K$, the intersection of all rate pairs defined by (4.13) with the nonnegative orthant defines a polymatroid (i.e., a polyhedron where the generating set function is normalized, nondecreasing,

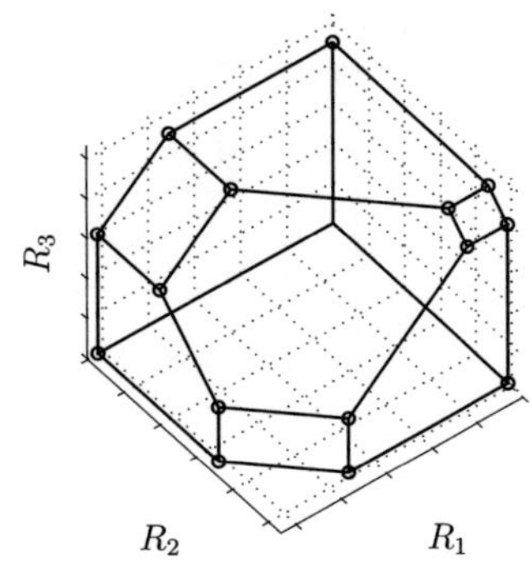

Fig. 4.4 Three user MIMO MAC polymatroid for fixed $\boldsymbol{Q}_1$, $\boldsymbol{Q}_2$, and $\boldsymbol{Q}_3$

and submodular [73]) with $2^K - 1 + K$ limiting constraints[3] and up to $K!$ vertices in the nondegenerated case [73]:

$$\mathcal{R}(\boldsymbol{Q}_1, \ldots, \boldsymbol{Q}_K) := \left\{ [R_1, \ldots, R_K]^{\mathrm{T}} \in \mathbb{R}^K_{+,0} \;\middle|\; \begin{matrix} \sum_{k\in\mathcal{U}} R_k \leq \log_2 \left| \mathbf{I} + \sum_{k\in\mathcal{U}} \boldsymbol{H}_k \boldsymbol{Q}_k \boldsymbol{H}_k^{\mathrm{H}} \right| \\ \forall \mathcal{U} \subseteq \mathcal{K},\ \mathcal{U} \neq \emptyset \end{matrix} \right\}. \tag{4.14}$$

Note that the $K!$ vertices define the corner points of the $K-1$-dimensional polyhedron where the sum rate of all users is constant, and each of these vertices corresponds to a particular decoding order π, which is a permutation of the set $\{1, \ldots, K\}$. Let $\pi[1]$ denote the user that is decoded *first* such that he sees interference from all other users $\pi[i]$ with $i > 1$, and let $\pi[K]$ denote the user that is decoded last and thus does not have to face any interference. For such a decoding order, the rate of some user $\pi[k]$ with $k \in \{1, \ldots, K\}$ reads as [cf. (4.5)–(4.8) for the two user case]

$$R^{\mathrm{MAC}}_{\pi[k]} = \log_2 \left| \mathbf{I} + \left(\mathbf{I} + \sum_{i>k} \boldsymbol{H}_{\pi[i]} \boldsymbol{Q}_{\pi[i]} \boldsymbol{H}^{\mathrm{H}}_{\pi[i]} \right)^{-1} \boldsymbol{H}_{\pi[k]} \boldsymbol{Q}_{\pi[k]} \boldsymbol{H}^{\mathrm{H}}_{\pi[k]} \right| \tag{4.15}$$

and one of the $K!$ vertices is achieved as rate tuple by the given decoding order. Other points on the $K-1$-dimensional polyhedron with constant sum rate can either be achieved via time-sharing between the $K!$ vertices, or via rate-splitting [69]. For the two user case, the polymatroid $\mathcal{R}(\boldsymbol{Q}_1, \boldsymbol{Q}_2)$ is a pentagon, see (4.4), and a three-dimensional polymatroid $\mathcal{R}(\boldsymbol{Q}_1, \boldsymbol{Q}_2, \boldsymbol{Q}_3)$ is depicted in Fig. 4.4. By definition, the polymatroid is convex.

4.1.2.1 Individual Power Constraints

In analogy to the two user case, the capacity region of a multi-user MIMO MAC with individual power constraints is given by the union of all polymatroids for which

---

[3] The K additional limiting constraints force the rates to be nonnegative. They implicitly hold due to the intersection with the nonnegative orthant.

the transmit covariance matrices are positive semi-definite and fulfill individual trace constraints:

$$\mathcal{C}^{\mathrm{MAC}}(P_{\max,1}, \ldots, P_{\max,K}) := \bigcup_{\substack{\boldsymbol{Q}_i \succeq \mathbf{0}\ \forall i \\ \mathrm{tr}(\boldsymbol{Q}_i) \leq P_{\max,i}\ \forall i}} \mathcal{R}(\boldsymbol{Q}_1, \ldots, \boldsymbol{Q}_K). \tag{4.16}$$

In the vector MAC with single-antenna users, the union of all polymatroids corresponds to a single polymatroid where all users transmit with full power. This follows from the fact that the weighted sum rate is a nondecreasing function in the powers, see Appendix A.8. Thus, Theorem 4 is also valid in the multi-user case:

Corollary 4.1 *The capacity region of a K-user vector MAC with single-antenna users and individual upper bounds $P_{\max,1}, \ldots, P_{\max,K}$ on the dissipated powers is a single polymatroid, i.e.,*

$$\mathcal{C}^{\mathrm{MAC}}(P_{\max,1}, \ldots, P_{\max,K}) = \mathcal{R}(P_{\max,1}, \ldots, P_{\max,K}).$$

For multi-antenna terminals, the union in (4.16) is necessary. Due to the individual power constraints, the capacity region in turn has a facet that is parallel to the plane defined by $R_i^{\mathrm{MAC}} = 0$ for every $i \in \{1, \ldots, K\}$, cf. the line segments A–A′ and D–D′ in Fig. 4.2. On such a facet, user i is decoded last and this covariance matrix is optimized for the single-user capacity with channel $\boldsymbol{H}_i$. The covariance matrices achieving the maximum sum rate can be determined using the algorithm in [71]. Given these optimizers, the $K!$ possible decoding orders form again a $K - 1$-dimensional polyhedron with $K!$ vertices in the nondegenerated case. All other boundary points of the capacity region can be obtained by means of a weighted sum rate maximization algorithm since the region is convex as in the two-user case (cf. Theorem 5).

4.1.2.2 Sum Power Constraint

The capacity region of the multi-user MIMO MAC can easily be extended for the case of a sum power upper bound $P_{\max}$ by changing the set of covariance matrices over which the union is taken in (4.16):

$$\mathcal{C}^{\mathrm{MAC}}(P_{\max}) := \bigcup_{\substack{\boldsymbol{Q}_i \succeq \mathbf{0}\ \forall i \\ \sum_{i=1}^{K} \mathrm{tr}(\boldsymbol{Q}_i) \leq P_{\max}}} \mathcal{R}(\boldsymbol{Q}_1, \ldots, \boldsymbol{Q}_K). \tag{4.17}$$

This artificial MAC will be of interest for determining the capacity region of the MIMO BC in Sect. 4.2. Note that the union of polymatroids is itself convex according to the same argumentation as in the two-user case in Sect. 4.1.1. Moreover, facets that are parallel to a plane $R_i^{\mathrm{MAC}} = 0$ no longer occur due to the sum power constraint.

Synopsis: In the MIMO MAC with fixed transmit covariance matrices, the set of achievable rate tuples is a polymatroid whose corner points correspond to different decoding orders. Via successive decoding and time-sharing between different decoding orders or via rate-splitting, every rate tuple of the polymatroid can be obtained. For the vector MAC with single-antenna transmitters and individual power constraints, the capacity region corresponds to a single polymatroid. In the case of a sum power constraint for the vector MAC, the capacity region is given by the union over all polymatroids where the powers of the individual users sum up to the maximum available sum power. This union is itself convex without having to apply the convex hull operator. When multi-antenna terminals are involved, the capacity region of the MIMO MAC with individual power constraints is given by the union over all polymatroids where the transmit covariance matrices are positive semi-definite and satisfy the individual power constraints. In the sum power constraint case, the union is extended such that only the sum of all covariance matrix traces is upper bounded. In both cases, the region is convex.

4.2 The Capacity Region of the MIMO BC

Due to the centralized nature of the transmitter in the broadcast channel, a sum power constraint is imposed on all data streams. Moreover, there is a fundamental difference between a single-antenna transmitter and a multi-antenna transmitter regarding the capacity achieving transmission strategy. In the former case, the broadcast channel is degraded whereas in the latter, it is not in general. Hence, these two cases are treated separately in the following two sections.

4.2.1 Single-Antenna Transmitter

4.2.1.1 Successive Decoding

The capacity region of a two-user BC with a single-antenna transmitter was first derived in 1972 by Cover in [1]. According to the system model from Sect. 2.1, the received signal of user k reads for a single-antenna transmitter as

$$y_k^{\mathrm{BC}} = H_k^{\mathrm{BC}} \sum_{i=1}^{K} P_i s_i + \eta_k. \tag{4.18}$$

Introducing the substitute

$$d_k := \frac{c_{\eta_k}}{|H_k^{\mathrm{BC}}|^2} \quad \forall k \tag{4.19}$$

as the quotient of the noise variance c_{η_k} of user k and its channel power, we can sort the users according to their scaled noise variances such that

$$d_{\pi[1]} \leq \cdots \leq d_{\pi[K]} \tag{4.20}$$

holds where $\pi[1], \ldots, \pi[K]$ denotes a permutation of $1, \ldots, K$. In the $K = 2$ user case, the 'weaker' user $\pi[2]$ sees interference from user $\pi[1]$ such that his data rate is upper bounded by

$$R_{\pi[2]}^{\mathrm{BC}} \leq \log_2 \left(1 + \frac{|P_{\pi[2]}|^2}{d_{\pi[2]} + |P_{\pi[1]}|^2}\right). \tag{4.21}$$

The property of being degraded means that in the single-antenna BC, the codeword of the 'weaker' user $\pi[2]$ can also be decoded by user $\pi[1]$ since he experiences a more favorable ratio of the noise variance and the channel power. Thus, the 'stronger' user $\pi[1]$ can subtract the impact of $s_{\pi[2]}$ on the received signal $y_{\pi[1]}^{\mathrm{BC}}$ by means of successive decoding such that his data rate is upper bounded by [1, 74][4]

$$R_{\pi[1]}^{\mathrm{BC}} \leq \log_2 \left(1 + \frac{|P_{\pi[1]}|^2}{d_{\pi[1]}}\right). \tag{4.22}$$

Let $S_k := |P_k|^2$ denote the transmit power of user k. Due to the sum power constraint in the BC, the power allocation can be parametrized via $S_k = f_k P_{\max}$ with $\sum_{k=1}^{K} f_k = 1$ and $f_k \geq 0\,\forall k$ and the transmitter applies superposition coding. Hence, the capacity region of the two user single-antenna BC is given by the union over all feasible power allocations:

$$\mathcal{C}^{\mathrm{BC}}(P_{\max}) := \bigcup_{\substack{f_1 \geq 0, f_2 \geq 0 \\ f_1 + f_2 = 1}} \left\{ \begin{bmatrix} R_1 \\ R_2 \end{bmatrix} \in \mathbb{R}_{+,0}^2 \;\middle|\; \begin{matrix} R_{\pi[1]} \leq \log_2\left(1 + \frac{f_{\pi[1]} P_{\max}}{d_{\pi[1]}}\right) \\ R_{\pi[2]} \leq \log_2\left(1 + \frac{f_{\pi[2]} P_{\max}}{d_{\pi[2]} + f_{\pi[1]} P_{\max}}\right) \end{matrix} \right\}. \tag{4.23}$$

An example of a two-user capacity region of the degraded broadcast channel is shown by the solid region in Fig. 4.5. There, user one has a stronger channel than user two and both users have the same noise variance such that $d_1 < d_2$ and $\pi[k] = k$. Thus, user one applies successive decoding and achieves a larger maximum rate than user two due to $d_1 < d_2$. The dashed triangle corresponds to the suboptimum case when the 'weaker' user shall be able to decode the data of the 'stronger' user which

---

[4] Note that in [1], the rate of the 'stronger' user additionally includes the rate of the decoded data belonging to the 'weaker' user. Moreover, the pre-log factor 0.5 in [1, 74] is due to the real-valued system model.

Fig. 4.5 Two user capacity region in a degraded broadcast channel

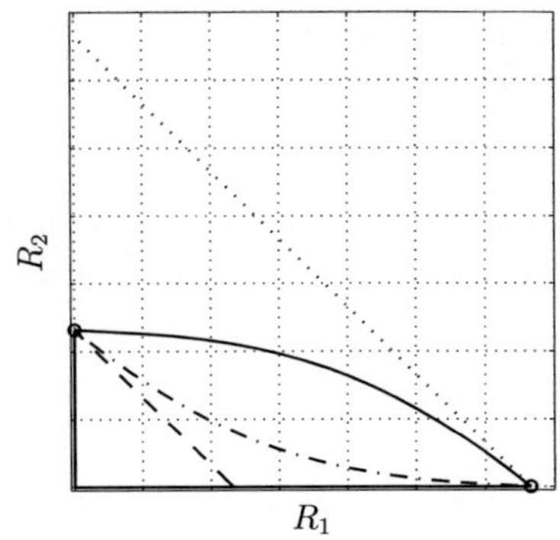

obviously limits the rate of the stronger user. Hence, the dashed line segment has a slope of $-45°$. Finally, the $-45°$ dotted line segment is meant to show that the slope of the boundary of the capacity region at the maximum rate point of user one where $f_1 = 1$ holds is lower bounded by $-45°$. For $d_1 < d_2$ and $P_{\max} < \infty$, the slope is strictly larger than $-45°$.

The extension to more than two users was published two years later by Bergmans in [75]. Therein, Bergmans showed that the capacity achieving transmit strategy is again superposition coding and every but the 'weakest' user has to apply successive decoding. In particular, user $\pi[k]$ only sees interference from users $\pi[i]$ with $i < k$ and user $\pi[k]$ is able to decode and cancel the interference from users $\pi[i]$ with $i > k$. The generalization of (4.21) and (4.22) to the K-user case hence reads as [75]

$$R^{\mathrm{BC}}_{\pi[k]} \leq \log_2\left(1 + \frac{f_{\pi[k]} P_{\max}}{d_{\pi[k]} + \sum_{i<k} f_{\pi[i]} P_{\max}}\right) \tag{4.24}$$

and the complete capacity region is given by

$$\mathcal{C}(P_{\max}) := \bigcup_{\substack{f_i \geq 0\ \forall i \\ \sum_{i=1}^{K} f_i = 1}} \left\{ [R_1, \ldots, R_K]^{\mathrm{T}} \in \mathbb{R}^K_{+,0} \,\middle|\, R_{\pi[k]} \leq \log_2\left(1 + \frac{f_{\pi[k]} P_{\max}}{d_{\pi[k]} + \sum_{i<k} f_{\pi[i]} P_{\max}}\right) \forall k \right\} \tag{4.25}$$

where $\pi[1], \ldots, \pi[K]$ follow from (4.19) and (4.20).

4.2.1.2 Scalar Dirty Paper Coding

Besides by the successive decoding strategy in the degraded single-antenna broadcast channel, the capacity region can also be achieved by means of dirty paper coding which was introduced 1983 by Costa in [76]. Therein, a point-to-point SISO system is considered where the received signal $y^{\mathrm{BC}}_{\pi[1]}$ reads as

$$y_{\pi[1]}^{\mathrm{BC}} = H_{\pi[1]}^{\mathrm{BC}} P_{\pi[1]} s_{\pi[1]} + \zeta_{\pi[1]} + \eta_{\pi[1]} \tag{4.26}$$

where the symbol $s_{\pi[1]}$, the interference $\zeta_{\pi[1]}$, and the noise $\eta_{\pi[1]}$ are all Gaussian distributed according to $s_{\pi[1]} \sim \mathcal{CN}(0, 1)$, $\zeta_{\pi[1]} \sim \mathcal{CN}(0, \sigma^2_{\zeta_{\pi[1]}})$, and $\eta_{\pi[1]} \sim \mathcal{CN}(0, c_{\eta_{\pi[1]}})$. The interference is assumed to be known noncausally at the encoder, i.e., the *complete* interference sequence is known in advance at the transmitter, but not at all at the receiver. Costa showed that the capacity for such a setup including the interference $\zeta_{\pi[1]}$ corresponds to the capacity of the same system when $\zeta_{\pi[1]} = 0$, i.e., when the interference is not present. Hence, the same capacity can be achieved without an increase of the required transmit power. Instead of subtracting the interference at the transmitter which would increase the transmit power, the encoder adapts to the interference and makes use of it by choosing codewords in the direction of the interference rather than attempting to cancel it [76]. Due to the special choice of the whole sequence $s_{\pi[1]}$, the interference $\zeta_{\pi[1]}$ and $s_{\pi[1]}$ are statistically independent.

Now going back to the two-user system from Sect. 4.2.1.1, user $\pi[1]$ sees interference from user $\pi[2]$, such that $\zeta_{\pi[1]} = H_{\pi[1]}^{\mathrm{BC}} P_{\pi[2]} s_{\pi[2]}$ holds in (4.26) due to (4.18). Applying Costa's result, the capacity of user $\pi[1]$ is exactly the same as if $\zeta_{\pi[2]} = 0$, so

$$R_{\pi[1]}^{\mathrm{BC}} \le \log_2\left(1 + \frac{|P_{\pi[1]}|^2}{d_{\pi[1]}}\right)$$

holds in analogy to (4.22). As the complete noncausal realizations of $s_{\pi[2]}$ were assumed to be known to the encoder of user $\pi[1]$ and thus fixed, user $\pi[2]$ clearly has to face interference from user $\pi[1]$ such that his maximum rate is given by (4.21). To be able to meet the assumption that the data of user $\pi[1]$ is coded without the knowledge of the interference caused by user $\pi[2]$, the data of user $\pi[2]$ was encoded first and the data of user $\pi[1]$ second.

If the 'weaker' user $\pi[2]$ shall be able to decode the codewords of the 'stronger' user $\pi[1]$ by applying successive decoding, the rate of user $\pi[1]$ is drastically limited, see the dashed line segment in Fig. 4.5. This degradation is slightly resolved when dirty paper coding is applied at the transmitter instead of successive decoding at the receiver. If the encoding order is reversed such that the 'stronger' user $\pi[1]$ is encoded before the weaker user $\pi[2]$, the rate pair has to satisfy

$$R_{\pi[2]}^{\mathrm{BC}} \le \log_2\left(1 + \frac{|P_{\pi[2]}|^2}{d_{\pi[2]}}\right)$$
$$R_{\pi[1]}^{\mathrm{BC}} \le \log_2\left(1 + \frac{|P_{\pi[1]}|^2}{d_{\pi[1]} + |P_{\pi[2]}|^2}\right).$$

The set of all feasible rate pairs for this encoding order is shown by the dash-dotted region in Fig. 4.5 which is a superset of the dashed region (successive decoding at the 'weaker' user) but a subset of the solid region, i.e., the capacity region. In the single-antenna transmitter case, the capacity region $\mathcal{C}^{\mathrm{BC}}(P_{\max})$ defined in (4.25) is achieved by only a single encoding order that satisfies (4.20). For this order, the data stream of user $\pi[K]$ is encoded first and the data stream of user $\pi[1]$ encoded last. Thus, the union over all possible encoding orders is not necessary to obtain the capacity region in the single-antenna case. In the multi-antenna transmitter case, however, the union is mandatory.

Synopsis: In a broadcast channel with a single-antenna transmitter, the capacity region can be achieved either by applying successive decoding at every but the 'weakest' user or by using Costa's dirty paper coding at the transmitter. In the former case, one exploits the property that the SISO BC is degraded, i.e., if a user can decode his own data stream then this particular data stream can also be decoded by all other users that have a smaller ratio of the noise variance and the channel power. From the user ordering according to their noise variances divided by the channel powers already follows which data stream(s) a particular user has to decode.

Another possibility to achieve the capacity is to make use of Costa's dirty paper coding. Interference that is noncausally known to the encoder but not to the decoder can be rendered harmless in terms of the achievable capacity without an increase of transmit power. Thus, the users can be encoded sequentially and for an encoding order $\pi[K], \ldots, \pi[1]$, user $\pi[k]$ sees interference only from users $\pi[i]$ with $i < k$. In the single-antenna transmitter case, the union over all encoding orders is not necessary to achieve the capacity region and the optimum encoding order is nonincreasing in the noise variance over the channel power, i.e., the user with the largest metric in (4.19) is encoded first.

4.2.2 Multi-Antenna Transmitter

If the transmitter in the BC is equipped with several antennas, the resulting BC is usually not degraded [72, 77] which means that there is no ranking of the users in terms of seeing a 'less noisy' channel than other users. Moreover, the channel matrices $\boldsymbol{H}_1^{\mathrm{BC}}, \ldots, \boldsymbol{H}_K^{\mathrm{BC}}$ do in general not have the same right singular vectors such that they cannot be decomposed into independent parallel degraded broadcast channels [77, 78]. Thus, superposition coding and the successive decoding strategy which are optimum for the single-antenna BC (see Sect. 4.2.1) are no longer capacity achieving in the multi-antenna case. Nonetheless, Costa's dirty paper coding [76] is still applicable and turned out to achieve the capacity region in the MIMO BC as we will see in the following sections. However, the generalization of scalar dirty paper coding to the vector-valued version is not straightforward. While most of the authors in the literature refer to [78, Lemma 1] and [77, Lemma 1] for the application of dirty paper coding in the vector-valued case, there is actually no rigorous proof in these two references that matches the situation that arises in a MIMO BC. An exception to this referencing is [32], where the individual data streams of a user are treated as virtual users, see Sect. 4.4.4. We give a rigorous proof how single-stream dirty paper coding must be extended to become applicable in a MIMO BC in Sect. 4.4. Interestingly, this extension exactly corresponds to the fashion our novel duality operates.

4.2.2.1 Sum Capacity

Dirty paper coding was proven to achieve the sum capacity in the MIMO BC by several authors almost at the same time. Despite of the various ways of proving this result, the use of Sato's outer bound [79] on the capacity region of a BC is common to all of them. In [79], Sato exploited the property that the capacity region of a BC only depends on the marginal transition probabilities of the channel and not on the joint distribution, i.e., only on the K marginal noise covariance matrices $\boldsymbol{C}_{\eta_1}, \ldots, \boldsymbol{C}_{\eta_K}$ and not on the cross correlations of $\boldsymbol{\eta}_1, \ldots, \boldsymbol{\eta}_K$. An upper bound on the sum capacity is thus given by the minimum capacity of the same point-to-point system with cooperating receivers where the minimization is with respect to all positive semi-definite total noise covariance matrices $\boldsymbol{C}_{\eta}$ whose main diagonal blocks are $\boldsymbol{C}_{\eta_1}, \ldots, \boldsymbol{C}_{\eta_K}$.

Early work on the sum capacity of a vector BC with multi-antenna transmitter was done by Caire and Shamai in [80, 81]. They proposed a ranked known interference transmission scheme for the vector BC utilizing dirty paper coding that asymptotically achieves the sum capacity at low SNR, and at high SNR in case of a full rank channel matrix. In the $K = 2$ user case with an $N = 2$ antenna base station, a modified, optimal scheme (including the respective precoding filter) is presented that achieves the sum capacity for any SNR under a sum power constraint.

In 2002, Vishwanath et al. derived a duality framework between the MIMO MAC with successive decoding and the MIMO BC utilizing dirty paper coding in [82], cf. Sect. 4.3. Thereby, the sum capacity of the MIMO BC could be shown to be achievable using dirty paper coding. In a first step, the sum capacity of a MIMO MAC with cooperating transmitters and worst case noise correlations was proven to be achievable in a MIMO MAC with decentralized transmitters. Using their duality framework, this rate pair with maximum sum rate can be realized in the MIMO BC as well such that Sato's upper bound on the sum capacity becomes tight. As the conversion rules how to switch from the MIMO MAC to the MIMO BC can readily be used, the sum capacity achieving transmit covariance matrices can be optimized in the dual MIMO MAC and can then be converted to the MIMO BC.

Another technique of showing that the sum capacity of the MIMO BC can be achieved by means of dirty paper coding was presented by Yu et al. in [83] and later with more details in [77]. In contrast to [82] or [32], their sum capacity result and the algorithm that delivers the sum capacity achieving transmit covariance matrix holds for any convex power constraint and thus not only under a sum power constraint. Moreover, the case is considered when the least favorable noise correlation matrix becomes singular. Yu and Cioffi give a game-theoretical interpretation for the sum capacity. Strictly speaking, the sum capacity of the MIMO BC corresponds to the

Nash equilibrium of a mutual information game, where one player optimizes the transmit covariance matrix in a MIMO BC *with cooperating receivers* to obtain the capacity for a given noise covariance matrix, and the other, fictitious player, minimizes the mutual information by varying the noise covariance matrix (the least favorable noise is indeed Gaussian distributed). As the mutual information expression of a point-to-point link is concave in the transmit covariance matrix and convex in the noise covariance matrix, convex programming algorithms are suggested to find the saddle-point, i.e., the sum capacity of the MIMO BC, with polynomial complexity. In [84], Yu relates the uplink–downlink duality for the sum capacity to the Lagrangian duality in convex optimization.

Finally, Viswanath and Tse came up with another duality between the vector MAC with successive decoding and the vector BC using dirty paper coding in 2003 [32]. Using this framework, the sum capacity of the vector BC under a sum power constraint is characterized. Moreover, it can be computed in the dual vector MAC. The authors also give an expression for the sum capacity of the MIMO BC with multi-antenna receivers and the individual data streams are treated as virtual users in their duality framework. We discuss this approach in more details in Sect. 4.4.4.

4.2.2.2 Dirty Paper Coding Region

The achievable rate region of a two-user MIMO-BC utilizing dirty-paper coding was first derived in [78] as the convex hull of all rate pairs that can be obtained where only one of the two users sees interference from the other user. These rate pairs follow from the union over all transmit covariance matrices that satisfy the sum power constraint *and* from the union over all possible encoding orders. The generalization to the multi-user case was done in [77, Theorem 1] and in [82]. For a particular encoding order $\pi[\cdot]$ in which user $\pi[1]$ is encoded *last* and user $\pi[K]$ *first* and for a particular choice of transmit covariance matrices $\boldsymbol{S}_1, \ldots, \boldsymbol{S}_K$ where $\boldsymbol{S}_i := \boldsymbol{P}_i \boldsymbol{P}_i^{\mathrm{H}}\ \forall i$, the rate of some user $\pi[k]$ with $k \in \{1, \ldots, K\}$ is upper bounded by (see Sect. 4.4.1 for a detailed derivation)

$$R_{\pi[k]}^{\mathrm{BC}} \leq \log_2 \frac{\left| \boldsymbol{C}_{\boldsymbol{\eta}_{\pi[k]}} + \boldsymbol{H}_{\pi[k]}^{\mathrm{BC}} \left(\sum_{i \leq k} \boldsymbol{S}_{\pi[i]} \right) \boldsymbol{H}_{\pi[k]}^{\mathrm{BC,H}} \right|}{\left| \boldsymbol{C}_{\boldsymbol{\eta}_{\pi[k]}} + \boldsymbol{H}_{\pi[k]}^{\mathrm{BC}} \left(\sum_{i < k} \boldsymbol{S}_{\pi[i]} \right) \boldsymbol{H}_{\pi[k]}^{\mathrm{BC,H}} \right|}. \tag{4.27}$$

Similar to the successive decoding in the MIMO MAC, user $\pi[k]$ has to face interference only from users $\pi[i]$ with $i < k$ as the interference of users $\pi[i]$ with $i > k$ is treated as side information at the encoder and thus does not have any impact on the achievable rate.

For all K users, the intersection of (4.27) and the nonnegative orthant defines a hyper-cube $\mathcal{Q}_\pi(\boldsymbol{S}_1, \ldots, \boldsymbol{S}_K)$. The set of rates that is achievable under a sum power constraint using dirty paper coding then corresponds to

$$\mathcal{C}_{\mathrm{DPC}}^{\mathrm{BC}}(P_{\max}) = \mathrm{co}\left\{ \bigcup_{\pi[\cdot]} \bigcup_{\substack{S_1,\ldots,S_K \succeq \mathbf{0} \\ \sum_{k=1}^{K} \mathrm{tr}(S_k) \le P_{\max}}} \mathcal{Q}_{\pi}(S_1,\ldots,S_K) \right\} \tag{4.28}$$

i.e., to the convex hull of unions of all hyper-cubes, where the union is taken over all permutations $\pi[\cdot]$ and all transmit covariance matrices satisfying the sum power constraint. Note that the convex hull operator is mandatory since not all rate tuples can directly be synthesized [85] such that time sharing becomes necessary. According to the duality result in [19, 82], this feasible set coincides with the capacity region of a MIMO MAC with a sum power constraint from Sect. 4.1.2.2, so $\mathcal{C}_{\mathrm{DPC}}^{\mathrm{BC}}(P_{\max}) = \mathcal{C}^{\mathrm{MAC}}(P_{\max})$. However, up to 2006, it was not clear whether $\mathcal{C}_{\mathrm{DPC}}^{\mathrm{BC}}(P_{\max})$ really corresponds to the capacity region of the MIMO BC or not.

4.2.2.3 Capacity Region

The proof for the coincidence of the MIMO BC capacity region and the dirty paper coding region is a very recent result [85, 86]. Nonetheless, many authors conjectured already years before that this would hold due to the fact that dirty paper coding is capacity achieving in case of the degraded broadcast channel (see Sect. 4.2.1.2 for the single-antenna transmitter case) and also achieves the sum capacity in the nondegraded MIMO BC (see Sect. 4.2.2.1). Another indicator mentioned in [82] was that the dirty paper coding region coincides with the capacity region of a MIMO MAC given a sum power constraint (see Sect. 4.2.2.2).

The eventual proof in [85] for the coincidence of the dirty paper coding region and the MIMO BC capacity region under any convex input constraint on the transmit covariance matrix was accomplished in several steps. First, Weingarten et al. analyzed a subclass of the MIMO BC termed *aligned degraded MIMO broadcast channel* (ADBC) which is characterized both by square channel matrices that are all the identity matrix and by noise covariance matrices that can be ordered according to the partial order on the cone of positive semi-definite matrices. In this setup, every receiver has as many antennas at the transmitter. Since the users can be ordered with respect to their noise covariance matrices for identity channel matrices, the ADBC is degraded. For the ADBC, the authors of [85] show that Gaussian inputs are optimal and that superposition coding in combination with successive decoding is capacity achieving as in the single-user case, see Sect. 4.2.1.1. However, Bergmans' proof [75] could not readily be applied to the vector-valued case since it requires some proportionality to hold between two matrices which is trivial in the scalar SISO case but does not necessarily hold in the vector-valued case. To overcome this problem, the authors showed that an enhanced channel (which is itself an ADBC) can always be created for every infeasible rate pair (R_1, R_2) lying outside the rate region which is achievable with Gaussian inputs in the original ADBC. This enhanced channel satisfies the proportionality required to apply Bergmans proof making the capacity region of it accessible. Moreover, this capacity region is a superset of the ADBC region obtained by Gaussian inputs. However, the capacity region of the enhanced channel does *not* contain the infeasible rate pair (R_1, R_2) either. By applying Bergmans' proof to the *enhanced* ADBC, the capacity region of the original ADBC is shown to be achievable by Gaussian inputs with superposition coding and successive decoding.

Next, the ADBC is generalized to an *aligned MIMO broadcast channel* (AMBC) which does not have to be degraded, i.e., the noise covariance matrices of the users need not satisfy a partial ordering, but the channel matrices are still identities. For such an AMBC, Weingarten et al. prove that dirty paper coding in combination with time sharing covers the entire capacity region. Unlike in the ADBC case, time sharing has to be considered for the AMBC, since not all rate tuples of the dirty paper coding rate region can directly be synthesized by specific covariance matrices if the noise covariance matrices cannot be ordered and thus the aligned BC is not degraded. As a consequence, parts of the boundary of the dirty paper coding region have zero curvature in some direction such that the authors of [85] use supporting hyperplanes to define the DPC region. In turn, an enhanced AMBC can be shown to exist for every original AMBC that has the same supporting hyperplane at a given rate tuple on the boundary of the original AMBC. Moreover, this enhanced AMBC is also degraded (which means that it is an ADBC) such that its capacity region is known, and the capacity region of the enhanced AMBC is again a superset of the original AMBC. The proof that DPC is capacity achieving in the AMBC, goes as follows. For a rate tuple lying outside the DPC region of an AMBC, an enhanced AMBC (which is degraded) can be constructed having the same separating hyperplane as supporting hyperplane for which the capacity region is a superset of the original DPC rate region. The originally infeasible rate tuple is also not element of the capacity region of the enhanced AMBC, from which we can conclude that DPC is capacity achieving in the AMBC.

Finally, the extension from the AMBC to the general MIMO BC with arbitrary channel matrices of arbitrary dimensions is obtained by means of a limit process exploiting the contiguity of the log-determinant function in the DPC rate expressions. In a first step, the singular value decomposition of the channel matrices is used and the receivers apply the Hermitian left singular vectors of the channel matrices as their first receiver stage. Since this operation is reversible it does neither change the capacity region nor the DPC rate region. Applying a further invertible transformation at every receiver, the filtered noise covariance matrices become block-diagonal and the effective channel matrices may have zero rows if the respective receiver has more antennas than the transmitter. These rows can be removed which corresponds to taking away the respective antenna elements. However, this removal does not have an impact on the DPC and the capacity region. For a receiver with less antenna elements than the transmitter, artificial antenna elements with uncorrelated noise can be added. If the respective channel matrix is enlarged by zero-rows to properly describe the transmission to the new antenna elements, neither the DPC nor the capacity region is influenced by the additional artificial antenna elements. In both cases the resulting channel matrices become square but need not be invertible. To achieve invertible transfer matrices a slight perturbation is added to the (so far) effective channel matrices such that the resulting DPC and capacity region is enlarged. Since these new channel matrices have full rank, they can be inverted such that an aligned MIMO BC is obtained. And for the AMBC, the capacity region was already shown to coincide with the DPC region. Due to the contiguity of the log-determinant function that arises in the DPC rate expressions, a limit process in which the slight perturbation goes to zero yields uniform convergence of the perturbed system to the unperturbed one. Hence, the capacity region of the MIMO BC is shown to be a subset of the DPC region from which follows that the DPC region coincides with the capacity region.

Note that this statement is valid for any convex constraint on the transmit covariance matrix like a sum power constraint, per-antenna power constraints, or a positive semi-definite upper bound on the transmit covariance matrix.

4.3 State-of-the-Art Rate Duality

By means of the MAC-to-BC conversion, we illustrate the state-of-the-art rate duality from [19, 82] to see the difference to our duality proposed in Sect. 4.4. This duality approach from Vishwanath et al. in [19, 82] is derived from an information-theoretic point of view and converts transmit covariance matrices $\boldsymbol{Q}_1, \ldots, \boldsymbol{Q}_K$ from the dual MIMO MAC to transmit covariance matrices $\boldsymbol{S}_1, \ldots, \boldsymbol{S}_K$ in the original MIMO BC such that the same rate tuple is obtained without an increase of the sum power. The receive filters in both systems are not considered as they are implicitly assumed to generate sufficient statistics.

Two key methods utilized are the *effective channel* and the *flipped channel* idea. The former one implies that the capacity of a point-to-point MIMO system with channel matrix $\boldsymbol{H}$ subject to additive Gaussian distortion (noise plus independent interference) with covariance matrix $\boldsymbol{X}$ equals the capacity of a point-to-point system with *effective channel* matrix $\boldsymbol{L}^{-1}\boldsymbol{H}$ subject to additive Gaussian distortion with identity covariance matrix, if $\boldsymbol{X} = \boldsymbol{L}\boldsymbol{L}^{\mathrm{H}}$. Given an arbitrary effective channel of a point-to-point system with identity noise covariance matrix, a system with reversed signal flow and Hermitian effective channel (*flipped channel*) has the same capacity, as the capacity is only determined by the nonzero singular values of the channel matrix when the noise has an identity covariance matrix [52]. Obviously, the Hermitian operator does not change the nonzero singular values.

For this state-of-the-art overview, we assume the natural decoding order since there is no gain of understanding when other permutations are considered and since this would only complicate the notation. Hence, $\pi[k] = k$ holds for all $k \in \{1, \ldots, K\}$, user 1 is decoded *first*, and user K *last*. According to (4.15), the maximum data rate of user k in the dual MIMO MAC given the natural decoding order reads as

$$R_k^{\mathrm{MAC}} = \log_2 \left| \mathbf{I}_N + \boldsymbol{X}_k^{-1} \boldsymbol{H}_k \boldsymbol{Q}_k \boldsymbol{H}_k^{\mathrm{H}} \right| \tag{4.29}$$

where the substitution $\boldsymbol{X}_k$ is defined as

$$\boldsymbol{X}_k := \mathbf{I}_N + \sum_{i>k} \boldsymbol{H}_i \boldsymbol{Q}_i \boldsymbol{H}_i^{\mathrm{H}} \in \mathbb{S}_+^N.$$

Introducing the Cholesky decomposition[5] (e.g., [45, Theorem 4.2.5, p. 41])

$$\boldsymbol{X}_k = \boldsymbol{L}_k \boldsymbol{L}_k^{\mathrm{H}} \tag{4.30}$$

and inserting two identity matrices $\mathbf{I}_{M_k} = \boldsymbol{F}_k^{-1}\boldsymbol{F}_k = \boldsymbol{F}_k^{\mathrm{H}}\boldsymbol{F}_k^{-\mathrm{H}}$, (4.29) can be expressed as

$$\begin{aligned} R_k^{\mathrm{MAC}} &= \log_2 \left| \mathbf{I}_N + \boldsymbol{L}_k^{-\mathrm{H}} \boldsymbol{L}_k^{-1} \boldsymbol{H}_k \boldsymbol{F}_k^{-1} \boldsymbol{F}_k \boldsymbol{Q}_k \boldsymbol{F}_k^{\mathrm{H}} \boldsymbol{F}_k^{-\mathrm{H}} \boldsymbol{H}_k^{\mathrm{H}} \right| \\ &= \log_2 \left| \mathbf{I}_N + \boldsymbol{L}_k^{-1} \boldsymbol{H}_k \boldsymbol{F}_k^{-1} \boldsymbol{F}_k \boldsymbol{Q}_k \boldsymbol{F}_k^{\mathrm{H}} \boldsymbol{F}_k^{-\mathrm{H}} \boldsymbol{H}_k^{\mathrm{H}} \boldsymbol{L}_k^{-\mathrm{H}} \right| \end{aligned} \tag{4.31}$$

where we used the determinant equality $\left|\mathbf{I} + \boldsymbol{C}_1\boldsymbol{C}_2\right| = \left|\mathbf{I} + \boldsymbol{C}_2\boldsymbol{C}_1\right|$ for the second equality. The matrix product $\boldsymbol{L}_k^{-1}\boldsymbol{H}_k\boldsymbol{F}_k^{-1}$ can now be regarded as the effective channel for the covariance matrix $\boldsymbol{F}_k\boldsymbol{Q}_k\boldsymbol{F}_k^{\mathrm{H}}$. How $\boldsymbol{F}_k$ must be chosen will be clarified below. For the flipped channel, we can find a covariance matrix $\boldsymbol{Z}_k \in \mathbb{S}^N$ such that

$$R_k^{\mathrm{MAC}} = \log_2 \left| \mathbf{I}_{M_k} + \boldsymbol{F}_k^{-\mathrm{H}} \boldsymbol{H}_k^{\mathrm{H}} \boldsymbol{L}_k^{-\mathrm{H}} \boldsymbol{Z}_k \boldsymbol{L}_k^{-1} \boldsymbol{H}_k \boldsymbol{F}_k^{-1} \right| \tag{4.32}$$

holds with $\operatorname{tr}(\boldsymbol{Z}_k) \le \operatorname{tr}(\boldsymbol{F}_k\boldsymbol{Q}_k\boldsymbol{F}_k^{\mathrm{H}})$. Using the reduced singular value decomposition [87, p. 27]

$$\boldsymbol{L}_k^{-1}\boldsymbol{H}_k\boldsymbol{F}_k^{-1} = \boldsymbol{U}_k\boldsymbol{D}_k\boldsymbol{V}_k^{\mathrm{H}} \in \mathbb{C}^{N\times M_k} \tag{4.33}$$

with the two (sub-)unitary matrices $\boldsymbol{U}_k \in \mathbb{C}^{N\times \min\{N,M_k\}}$ and $\boldsymbol{V}_k \in \mathbb{C}^{M_k\times \min\{N,M_k\}}$ and diagonal $\boldsymbol{D}_k \in \mathbb{R}_{+,0}^{\min\{N,M_k\}\times\min\{N,M_k\}}$, the matrix $\boldsymbol{Z}_k$ must obey [cf. (4.31) and (4.32)]

$$\left|\mathbf{I}_N + \boldsymbol{U}_k\boldsymbol{D}_k\boldsymbol{V}_k^{\mathrm{H}} \cdot \boldsymbol{F}_k\boldsymbol{Q}_k\boldsymbol{F}_k^{\mathrm{H}} \cdot \boldsymbol{V}_k\boldsymbol{D}_k\boldsymbol{U}_k^{\mathrm{H}}\right| = \left|\mathbf{I}_{M_k} + \boldsymbol{V}_k\boldsymbol{D}_k\boldsymbol{U}_k^{\mathrm{H}} \cdot \boldsymbol{Z}_k \cdot \boldsymbol{U}_k\boldsymbol{D}_k\boldsymbol{V}_k^{\mathrm{H}}\right|.$$

A possible solution for $\boldsymbol{Z}_k$ reads as

$$\boldsymbol{Z}_k = \boldsymbol{U}_k\boldsymbol{V}_k^{\mathrm{H}} \cdot \boldsymbol{F}_k\boldsymbol{Q}_k\boldsymbol{F}_k^{\mathrm{H}} \cdot \boldsymbol{V}_k\boldsymbol{U}_k^{\mathrm{H}} \tag{4.34}$$

which also satisfies $\operatorname{tr}(\boldsymbol{Z}_k) = \operatorname{tr}(\boldsymbol{V}_k^{\mathrm{H}}\boldsymbol{F}_k\boldsymbol{Q}_k\boldsymbol{F}_k^{\mathrm{H}}\boldsymbol{V}_k) \le \operatorname{tr}(\boldsymbol{F}_k\boldsymbol{Q}_k\boldsymbol{F}_k^{\mathrm{H}})$ since $\boldsymbol{V}_k$ is (sub-)unitary with $\boldsymbol{V}_k^{\mathrm{H}}\boldsymbol{V}_k = \mathbf{I}_{\min\{N,M_k\}}$.

In the original MIMO BC with identity noise covariance matrices,[6] the encoding order is the reversed decoding order of the dual MIMO MAC, such that user 1 is encoded *last* and therefore does not see any interference, whereas user K is encoded first. Then, the rate of user k in the MIMO BC reads as (see Sect. 4.4.1 for a detailed

[5] In [19, 82], the Hermitian square root $\boldsymbol{X}_k^{\frac{1}{2}}$ is used for $\boldsymbol{L}_k$.

[6] In [19, 82], only the case of identity noise covariance matrices in the MIMO BC is considered.

derivation)

$$
\begin{aligned}
R_k^{\mathrm{BC}} &= \log_2 \left| \mathbf{I}_{M_k} + \boldsymbol{Y}_k^{-1} \boldsymbol{H}_k^{\mathrm{BC}} \boldsymbol{S}_k \boldsymbol{H}_k^{\mathrm{BC,H}} \right| \\
&= \log_2 \left| \mathbf{I}_{M_k} + \boldsymbol{F}_k^{-\mathrm{H}} \boldsymbol{H}_k^{\mathrm{BC}} \boldsymbol{S}_k \boldsymbol{H}_k^{\mathrm{BC,H}} \boldsymbol{F}_k^{-1} \right|
\end{aligned} \tag{4.35}
$$

with the substitution

$$
\boldsymbol{Y}_k := \mathbf{I}_{M_k} + \boldsymbol{H}_k^{\mathrm{BC}} \sum_{i<k} \boldsymbol{S}_i \boldsymbol{H}_k^{\mathrm{BC,H}} \in \mathbb{S}_+^{M_k}
$$

and the Cholesky decomposition

$$
\boldsymbol{Y}_k = \boldsymbol{F}_k^{\mathrm{H}} \boldsymbol{F}_k. \tag{4.36}
$$

Equality between R_k^{MAC} in (4.32) and R_k^{BC} in (4.35) holds, if the dual MAC channels are chosen as the Hermitian BC channels via $\boldsymbol{H}_k = \boldsymbol{H}_k^{\mathrm{BC,H}}\ \forall k$, and if

$$
\boldsymbol{S}_k = \boldsymbol{L}_k^{-\mathrm{H}} \boldsymbol{Z}_k \boldsymbol{L}_k^{-1} \in \mathbb{S}^N. \tag{4.37}
$$

Implicitly, $\boldsymbol{Z}_k$ depends on $\boldsymbol{F}_k$ as shown in (4.34). Thus, $\boldsymbol{S}_k$ depends on $\boldsymbol{Y}_k$ which itself is a function of all $\boldsymbol{S}_i$ with $i < k$. These dependencies require that all $\boldsymbol{S}_i$ with $i < k$ have to be computed before $\boldsymbol{S}_k$, and consequently, one has to start with the computation of $\boldsymbol{S}_1$ followed by $\boldsymbol{S}_2, \ldots, \boldsymbol{S}_K$. As a consequence, the covariance matrices have to be computed serially in the duality transform. Finally, the sum power conservation $\sum_{k=1}^{K} \mathrm{tr}(\boldsymbol{S}_k) \leq \sum_{k=1}^{K} \mathrm{tr}(\boldsymbol{Q}_k)$ of the duality transformation can be proven to hold by induction, see [19, Appendix B]. Note that the conversion from BC to MAC covariance matrices can be achieved in the same fashion such that the complete duality is finally established.

4.4 Novel Rate Duality

4.4.1 Dirty Paper Coding for Vector Signals

In this section, we show how the scalar dirty paper coding technique of a SISO system (cf. Sect. 4.2.1.2) must be extended to become applicable in a MIMO BC with vector-valued signals. While most of the contributions using dirty paper coding in the MIMO BC quote [78, Lemma 1] or [77, Lemma 1] for the applicability of dirty paper coding to vector-valued signals, there is actually no rigorous proof in these two references that matches the system setup of the given MIMO BC. In [88], Jafar et al. mention reference [89] as a possible strategy how to extend scalar dirty paper coding to the vector-valued case. However, only a single data stream is considered per user in [89] as the channel matrix is approximated by its strongest rank-one transjector, and every user's precoder is chosen as the strongest left singular vector of his channel matrix which differs from our approach.

In the following, we derive how scalar dirty paper coding has to be applied to cancel a vector-valued, noncausally known interference sequence. The duality presented in the next two sections directly mirrors this behavior.

For a given encoding order $\pi[\cdot]$ which is a permutation of the natural numbers $1, \ldots, K$, user $\pi[1]$ is encoded *last* and user $\pi[K]$ is encoded *first*. As we will see, the encoder of user $\pi[1]$ will therefore treat the interference from users $\pi[i]$ with $i > 1$ as noncausally known side information that can be rendered harmless such that the receiver $\pi[1]$ will not see any interference at all.

The filtered received signal of some user $\pi[k]$ reads according to the MIMO BC system model from Sect. 2.1 as

$$\begin{aligned}\hat{\boldsymbol{s}}^{\mathrm{BC}}_{\pi[k]} &= \boldsymbol{B}_{\pi[k]}\boldsymbol{H}^{\mathrm{BC}}_{\pi[k]}\boldsymbol{P}_{\pi[k]}\boldsymbol{s}_{\pi[k]} + \underbrace{\boldsymbol{B}_{\pi[k]}\boldsymbol{H}^{\mathrm{BC}}_{\pi[k]}\sum_{i>k}\boldsymbol{P}_{\pi[i]}\boldsymbol{s}_{\pi[i]}}_{=:\boldsymbol{\zeta}_{\pi[k]}} \\ &\quad + \boldsymbol{B}_{\pi[k]}\underbrace{\Big(\boldsymbol{H}^{\mathrm{BC}}_{\pi[k]}\sum_{i<k}\boldsymbol{P}_{\pi[i]}\boldsymbol{s}_{\pi[i]} + \boldsymbol{\eta}_{\pi[k]}\Big)}_{=:\boldsymbol{\vartheta}_{\pi[k]}}. \end{aligned} \tag{4.38}$$

As the data of users $\pi[i]$ with $i > k$ has already been encoded before, $\boldsymbol{\zeta}_{\pi[k]}$ can be regarded as known side-information to the encoder of user $\pi[k]$. By means of the matched filter $\boldsymbol{B}^{\mathrm{MF}}_{\pi[k]}$ taking into account only the noise plus interference term $\boldsymbol{\vartheta}_{\pi[k]}$, a sufficient statistic of the same dimension as that of the input symbol vector $\boldsymbol{s}_{\pi[k]}$ is generated:

$$\boldsymbol{B}^{\mathrm{MF}}_{\pi[k]} = \boldsymbol{P}^{\mathrm{H}}_{\pi[k]}\boldsymbol{H}^{\mathrm{BC,H}}_{\pi[k]}\Big(\underbrace{\boldsymbol{H}^{\mathrm{BC}}_{\pi[k]}\sum_{i<k}\boldsymbol{P}_{\pi[i]}\boldsymbol{P}^{\mathrm{H}}_{\pi[i]}\boldsymbol{H}^{\mathrm{BC,H}}_{\pi[k]} + \boldsymbol{C}_{\boldsymbol{\eta}_{\pi[k]}}}_{=:\boldsymbol{C}_{\boldsymbol{\vartheta}_{\pi[k]}}}\Big)^{-1}. \tag{4.39}$$

The matrix $\boldsymbol{C}_{\boldsymbol{\vartheta}_{\pi[k]}} := \mathrm{E}[\boldsymbol{\vartheta}_{\pi[k]}\boldsymbol{\vartheta}^{\mathrm{H}}_{\pi[k]}]$ denotes the noise plus interference covariance matrix seen by user $\pi[k]$. Inserting $\boldsymbol{B}^{\mathrm{MF}}_{\pi[k]}$ from (4.39) into the expression for the filtered received signal in (4.38) yields

$$\hat{\boldsymbol{s}}^{\mathrm{BC}}_{\pi[k]} = \boldsymbol{P}^{\mathrm{H}}_{\pi[k]}\boldsymbol{H}^{\mathrm{BC,H}}_{\pi[k]}\boldsymbol{C}^{-1}_{\boldsymbol{\vartheta}_{\pi[k]}}\boldsymbol{H}^{\mathrm{BC}}_{\pi[k]}\boldsymbol{P}_{\pi[k]}\boldsymbol{s}_{\pi[k]} + \boldsymbol{\zeta}^{\mathrm{MF}}_{\pi[k]} + \boldsymbol{P}^{\mathrm{H}}_{\pi[k]}\boldsymbol{H}^{\mathrm{BC,H}}_{\pi[k]}\boldsymbol{C}^{-1}_{\boldsymbol{\vartheta}_{\pi[k]}}\boldsymbol{\vartheta}_{\pi[k]}. \tag{4.40}$$

At the transmitter, we are free to add a unitary matrix $\boldsymbol{W}_{\pi[k]}$ to the precoder which does not change the transmit covariance matrix $\boldsymbol{P}_{\pi[k]}\boldsymbol{P}^{\mathrm{H}}_{\pi[k]}$. Doing so, we define the rotated precoder via $\bar{\boldsymbol{P}}_{\pi[k]} := \boldsymbol{P}_{\pi[k]}\boldsymbol{W}_{\pi[k]}$. Obviously the MF receiver also changes to $\bar{\boldsymbol{B}}^{\mathrm{MF}}_{\pi[k]} := \boldsymbol{W}^{\mathrm{H}}_{\pi[k]}\boldsymbol{B}^{\mathrm{MF}}_{\pi[k]}$. The optimum choice for the decorrelation matrix $\boldsymbol{W}_{\pi[k]}$ is the eigenbasis of the receiver-channel-precoder chain of user $\pi[k]$ exactly as this was the case in our duality for linear filtering:

$$\boldsymbol{W}_{\pi[k]}\boldsymbol{\Lambda}_{\pi[k]}\boldsymbol{W}^{\mathrm{H}}_{\pi[k]} := \boldsymbol{B}^{\mathrm{MF}}_{\pi[k]}\boldsymbol{H}^{\mathrm{BC}}_{\pi[k]}\boldsymbol{P}_{\pi[k]}. \tag{4.41}$$

Note that the matrix $\boldsymbol{\Lambda}_{\pi[k]}$ is diagonal. Reformulating (4.40) using the *rotated* precoder and receiver yields

$$\hat{\boldsymbol{s}}^{\mathrm{BC}}_{\pi[k]} = \boldsymbol{\Lambda}_{\pi[k]}\boldsymbol{s}_{\pi[k]} + \bar{\boldsymbol{\zeta}}^{\mathrm{MF}}_{\pi[k]} + \bar{\boldsymbol{P}}^{\mathrm{H}}_{\pi[k]}\boldsymbol{H}^{\mathrm{BC,H}}_{\pi[k]}\boldsymbol{C}^{-1}_{\boldsymbol{\vartheta}_{\pi[k]}}\boldsymbol{\vartheta}_{\pi[k]}. \tag{4.42}$$

This strategy managed to diagonalize the effective channel of the desired user such that there is no interference between the individual entries of $\boldsymbol{s}_{\pi[k]}$. Moreover, the filtered noise plus interference covariance matrix is diagonalized *simultaneously* and now reads as $\boldsymbol{\Lambda}_{\pi[k]}$. As a consequence, user $\pi[k]$ has L_k independent, parallel SISO data streams and dirty paper coding can be applied to each of these streams separately. For the jth data stream, the noncausally known interfering sequence corresponding to the jth entry of $\bar{\boldsymbol{\zeta}}^{\mathrm{MF}}_{\pi[k]}$ can be rendered harmless at the transmitter using Costa's scalar dirty paper coding [76]. Correlations in the known interference $\bar{\boldsymbol{\zeta}}^{\mathrm{MF}}_{\pi[k]}$ do not matter. Hence, the mutual information between the transmitter and user $\pi[k]$ is not affected from the interference of users $\pi[i]$ with $i > k$ such that the mutual information reads as[7]

$$\begin{aligned}
I(\boldsymbol{s}_{\pi[k]}; \hat{\boldsymbol{s}}^{\mathrm{BC}}_{\pi[k]}) &= \log_2 \left| \boldsymbol{\Lambda}^2_{\pi[k]} + \boldsymbol{\Lambda}_{\pi[k]} \right| - \log_2 \left| \boldsymbol{\Lambda}_{\pi[k]} \right| \\
&= \log_2 \left| \mathbf{I} + \boldsymbol{\Lambda}_{\pi[k]} \right| \\
&= \log_2 \left| \mathbf{I} + \boldsymbol{B}^{\mathrm{MF}}_{\pi[k]}\boldsymbol{H}^{\mathrm{BC}}_{\pi[k]}\boldsymbol{P}_{\pi[k]} \right| \\
&= \log_2 \left| \mathbf{I} + \boldsymbol{P}^{\mathrm{H}}_{\pi[k]}\boldsymbol{H}^{\mathrm{BC,H}}_{\pi[k]}\boldsymbol{C}^{-1}_{\boldsymbol{\vartheta}_{\pi[k]}}\boldsymbol{H}^{\mathrm{BC}}_{\pi[k]}\boldsymbol{P}_{\pi[k]} \right| \\
&= \log_2 \left| \mathbf{I} + \boldsymbol{C}^{-1}_{\boldsymbol{\vartheta}_{\pi[k]}}\boldsymbol{H}^{\mathrm{BC}}_{\pi[k]}\boldsymbol{S}_{\pi[k]}\boldsymbol{H}^{\mathrm{BC,H}}_{\pi[k]} \right| \\
&= \log_2 \left| \boldsymbol{C}_{\boldsymbol{\vartheta}_{\pi[k]}} + \boldsymbol{H}^{\mathrm{BC}}_{\pi[k]}\boldsymbol{S}_{\pi[k]}\boldsymbol{H}^{\mathrm{BC,H}}_{\pi[k]} \right| - \log_2 \left| \boldsymbol{C}_{\boldsymbol{\vartheta}_{\pi[k]}} \right| \\
&= \log_2 \frac{\left| \boldsymbol{C}_{\boldsymbol{\eta}_{\pi[k]}} + \sum_{i\le k} \boldsymbol{H}^{\mathrm{BC}}_{\pi[k]}\boldsymbol{S}_{\pi[i]}\boldsymbol{H}^{\mathrm{BC,H}}_{\pi[k]} \right|}{\left| \boldsymbol{C}_{\boldsymbol{\eta}_{\pi[k]}} + \sum_{i< k} \boldsymbol{H}^{\mathrm{BC}}_{\pi[k]}\boldsymbol{S}_{\pi[i]}\boldsymbol{H}^{\mathrm{BC,H}}_{\pi[k]} \right|}.
\end{aligned} \tag{4.43}$$

The first equality in (4.43) directly follows from (4.42), the second line resolves zero main diagonal elements in $\boldsymbol{\Lambda}_{\pi[k]}$. For the third line, (4.41) was used with the fact that a unitary matrix has a unimodular determinant, and the fourth line is due to the definition of $\boldsymbol{B}^{\mathrm{MF}}_{\pi[k]}$ in (4.39). In the fifth line, $\left|\mathbf{I} + \boldsymbol{A}\boldsymbol{B}\right| = \left|\mathbf{I} + \boldsymbol{B}\boldsymbol{A}\right|$ was used and the transmit covariance matrix $\boldsymbol{S}_{\pi[k]} := \boldsymbol{P}_{\pi[k]}\boldsymbol{P}^{\mathrm{H}}_{\pi[k]}$ was introduced.

[7] For singular $\boldsymbol{\Lambda}_{\pi[k]}$, the first difference in (4.43) yields $-\infty + \infty$. Substreams with zero gain and zero noise could be dropped from $\boldsymbol{\Lambda}_{\pi[k]}$ to resolve this problem. However, the second line in (4.43) always yields the correct result.

Finally, the last line is due to the definition of the noise plus interference covariance matrix in (4.39). Note that in Costa's scheme [76], the decoder needs to know the interference *variance* of every SISO data stream that is used as side information at the encoder. In [90], Erez et al. introduced a high-dimensional lattice coding that asymptotically achieves the same capacity as Costa's dirty paper coding when the dimension of the lattice goes to infinity. But in contrast, the decoder does not need to know the variance of the side information at the transmitter. Instead, common randomness must be available at the encoder and decoder to perform the dithering process.

Summing up, we have proven the following theorem:

Theorem 4.3 *A vector-valued interference sequence stemming from already encoded users $\pi[i]$ with $i > k$ that is noncausally known at the encoder of user $\pi[k]$ can be rendered harmless in terms of the mutual information by applying Costa's dirty paper coding to decorrelated, parallel SISO streams.*

4.4.2 BC to MAC Conversion

Given precoding matrices $\boldsymbol{P}_1, \ldots, \boldsymbol{P}_K$ in the MIMO BC inducing transmit covariance matrices $\boldsymbol{S}_1, \ldots, \boldsymbol{S}_K$, where $\boldsymbol{S}_k := \boldsymbol{P}_k \boldsymbol{P}_k^{\mathrm{H}}\ \forall k$, we establish a duality conversion to the MIMO MAC such that the mutual information expressions in the MIMO BC are also achieved in the dual MIMO MAC consuming the same amount of total transmit power. For the sake of readability, we assume the natural encoding order $\pi[k] = k\ \forall k \in \{1, \ldots, K\}$ which can always be obtained by relabeling the users. Then, user 1 is encoded last and user K is encoded first in the MIMO BC.

We know from the previous Sect. 4.4.1, that the mutual information between the symbol vector $\boldsymbol{s}_k$ of user k and its estimate $\hat{\boldsymbol{s}}_k^{\mathrm{BC}}$ reads as [cf. (4.43)]

$$
\begin{aligned}
I_k^{\mathrm{DPC}}(\boldsymbol{P}) := I(\boldsymbol{s}_k; \hat{\boldsymbol{s}}_k^{\mathrm{BC}}) &= \log_2 \left| \mathbf{I} + \boldsymbol{B}_k^{\mathrm{MF}} \boldsymbol{H}_k^{\mathrm{BC}} \boldsymbol{P}_k \right| \\
&= \log_2 \left| \mathbf{I} + \boldsymbol{W}_k^{\mathrm{H}} \boldsymbol{B}_k^{\mathrm{MF}} \boldsymbol{H}_k^{\mathrm{BC}} \boldsymbol{P}_k \boldsymbol{W}_k \right| \\
&= \log_2 \left| \mathbf{I} + \bar{\boldsymbol{B}}_k^{\mathrm{MF}} \boldsymbol{H}_k^{\mathrm{BC}} \bar{\boldsymbol{P}}_k \right| \\
&= \log_2 \prod_{j=1}^{L_k} \left[\mathbf{I} + \bar{\boldsymbol{B}}_k^{\mathrm{MF}} \boldsymbol{H}_k^{\mathrm{BC}} \bar{\boldsymbol{P}}_k \right]_{j,j} \\
&= \sum_{j=1}^{L_k} \log_2 \left(1 + \left[\bar{\boldsymbol{B}}_k^{\mathrm{MF}} \boldsymbol{H}_k^{\mathrm{BC}} \bar{\boldsymbol{P}}_k \right]_{j,j} \right).
\end{aligned}
\tag{4.44}
$$

The first equality is taken from (4.43), and for the second, we exploited the fact that unitary matrices have unimodular determinants. The third equality is due to the definitions

$$\begin{aligned}\bar{\boldsymbol{B}}_k^{\mathrm{MF}} &:= \boldsymbol{W}_k^{\mathrm{H}} \boldsymbol{B}_k^{\mathrm{MF}} = \boldsymbol{W}_k^{\mathrm{H}} \boldsymbol{P}_k^{\mathrm{H}} \boldsymbol{H}_k^{\mathrm{BC,H}} \boldsymbol{C}_{\vartheta_k}^{-1} \\ \bar{\boldsymbol{P}}_k &:= \boldsymbol{P}_k \boldsymbol{W}_k.\end{aligned} \tag{4.45}$$

The second last equality in (4.44) holds since $\bar{\boldsymbol{B}}_k^{\mathrm{MF}} \boldsymbol{H}_k^{\mathrm{BC}} \bar{\boldsymbol{P}}_k$ is diagonal if the unitary decorrelation matrix $\boldsymbol{W}_k$ is found with the eigenvalue decomposition in (4.41). For the case of dirty paper coding, we can also define a pseudo-SINR in analogy to (3.31)

$$\begin{aligned}\mathrm{SINR}_{k,j}^{\mathrm{DPC}}(\boldsymbol{P}, \boldsymbol{b}_{k,j}) &:= \frac{|\boldsymbol{b}_{k,j}^{\mathrm{T}} \boldsymbol{H}_k^{\mathrm{BC}} \boldsymbol{p}_{k,j}|^2}{\boldsymbol{b}_{k,j}^{\mathrm{T}} (\boldsymbol{C}_{\vartheta_k} + \sum_{\ell \neq j} \boldsymbol{H}_k^{\mathrm{BC}} \boldsymbol{p}_{k,\ell} \boldsymbol{p}_{k,\ell}^{\mathrm{H}} \boldsymbol{H}_k^{\mathrm{BC,H}}) \boldsymbol{b}_{k,j}^*} \\ &= \frac{|\boldsymbol{b}_{k,j}^{\mathrm{T}} \boldsymbol{H}_k^{\mathrm{BC}} \boldsymbol{p}_{k,j}|^2}{\boldsymbol{b}_{k,j}^{\mathrm{T}} \boldsymbol{C}_{\eta_k} \boldsymbol{b}_{k,j}^* + \sum_{i<k} \sum_{\ell=1}^{L_i} |\boldsymbol{b}_{k,j}^{\mathrm{T}} \boldsymbol{H}_k^{\mathrm{BC}} \boldsymbol{p}_{i,\ell}|^2 + \sum_{\ell \neq j} |\boldsymbol{b}_{k,j}^{\mathrm{T}} \boldsymbol{H}_k^{\mathrm{BC}} \boldsymbol{p}_{k,\ell}|^2}.\end{aligned} \tag{4.46}$$

In contrast to the SINR definition from Sect. 3.2.3.3 where any kind of interference is treated as additional noise and not partially as side-information for the encoder, the pseudo-SINR definition is only meaningful when linked to the achievable rate. Other relations like to the raw bit error ratio for example are not meaningful due to the dirty paper coding component. The filtered noise component $\bar{\zeta}_k^{\mathrm{MF}}$ from (4.42) does indeed arise at receiver k, it is only rendered harmless in terms of data rate due to the utilization of DPC. Contrary, in the dual MIMO MAC with successive decoding, the impact of already decoded users i with $i < k$ on the received signal of user k can really be canceled completely.

The matched filter maximizes $\mathrm{SINR}_{k,j}^{\mathrm{DPC}}$ when neglecting the intra-user interference. The unitary rotation additionally removes the intra-user interference, i.e., the last sum in the denominator of (4.46) is zero:

$$\gamma_{k,j}(\bar{\boldsymbol{P}}) := \max_{\boldsymbol{b}_{k,j}} \mathrm{SINR}_{k,j}^{\mathrm{DPC}}(\bar{\boldsymbol{P}}, \boldsymbol{b}_{k,j}) = \mathrm{SINR}_{k,j}^{\mathrm{DPC}}(\bar{\boldsymbol{P}}, \bar{\boldsymbol{b}}_{k,j}^{\mathrm{MF}}) = \left[\bar{\boldsymbol{B}}_k^{\mathrm{MF}} \boldsymbol{H}_k^{\mathrm{BC}} \bar{\boldsymbol{P}}_k\right]_{j,j}. \tag{4.47}$$

Combining (4.44) and (4.47), the mutual information expression for the kth user regarding interference from users $i > k$ as side-information can be expressed in terms of the pseudo-SINR:

$$I_k^{\mathrm{DPC}}(\boldsymbol{P}) = \sum_{j=1}^{L_k} \log_2 \left(1 + \mathrm{SINR}_{k,j}^{\mathrm{DPC}}(\bar{\boldsymbol{P}}, \bar{\boldsymbol{b}}_{k,j}^{\mathrm{MF}})\right). \tag{4.48}$$

In the dual MIMO MAC, successive decoding is applied and the decoding order is the reversed encoding order of the BC, i.e., user 1 sees interference from all other users $2, \ldots, K$, whereas user K does not have to face any interference at all. For user k, the jth data stream then features an SINR which reads as

$$\mathrm{SINR}_{k,j}^{\mathrm{SD}}(\boldsymbol{T},\boldsymbol{g}_{k,j}) := \frac{|\boldsymbol{g}_{k,j}^{\mathrm{T}}\boldsymbol{H}_k\boldsymbol{t}_{k,j}|^2}{\|\boldsymbol{g}_{k,j}\|_2^2 + \sum_{i>k}\sum_{\ell=1}^{L_i}|\boldsymbol{g}_{k,j}^{\mathrm{T}}\boldsymbol{H}_i\boldsymbol{t}_{i,\ell}|^2 + \sum_{\ell\neq j}|\boldsymbol{g}_{k,j}^{\mathrm{T}}\boldsymbol{H}_k\boldsymbol{t}_{k,\ell}|^2}. \tag{4.49}$$

With this definition of the SINR in case of inter-user successive decoding, intra-user stream interference is not subtracted by means of successive decoding. But as we will see, the conversion from the MIMO BC to the dual MIMO MAC preserves the property that no intra-user interference is present. Under separate stream decoding, the SINRs of user k are linked to his data rate via

$$R_k^{\mathrm{SD}}(\boldsymbol{T},\boldsymbol{G}_k) = \sum_{j=1}^{L_k}\log_2\left(1+\mathrm{SINR}_{k,j}^{\mathrm{SD}}(\boldsymbol{T},\boldsymbol{g}_{k,j})\right). \tag{4.50}$$

To ensure that the dual MIMO MAC supports the same rates as the MIMO BC, $R_k^{\mathrm{SD}}(\boldsymbol{T},\boldsymbol{G}_k) = I_k^{\mathrm{DPC}}(\boldsymbol{P})$ must hold for all users k which can be accomplished by equating

$$\mathrm{SINR}_{k,j}^{\mathrm{DPC}}(\bar{\boldsymbol{P}},\bar{\boldsymbol{b}}_{k,j}^{\mathrm{MF}}) = \mathrm{SINR}_{k,j}^{\mathrm{SD}}(\boldsymbol{T},\boldsymbol{g}_{k,j}) \tag{4.51}$$

for all users k and all data streams j, see (4.48) and (4.50). For the filters in the dual MIMO MAC, we choose the same parametrization as in (3.64):

$$\begin{aligned}\boldsymbol{t}_{k,j} &= \boldsymbol{T}_k\boldsymbol{e}_j = \psi_{k,j}\boldsymbol{C}_{\boldsymbol{\eta}_k}^{\frac{1}{2},\mathrm{H}}\bar{\boldsymbol{b}}_{k,j}^{\mathrm{MF}*}\\ \boldsymbol{g}_{k,j}^{\mathrm{T}} &= \boldsymbol{e}_j^{\mathrm{T}}\boldsymbol{G}_k = \psi_{k,j}^{-1}\bar{\boldsymbol{p}}_{k,j}^{\mathrm{H}}.\end{aligned} \tag{4.52}$$

In turn, the dual MIMO MAC channels are chosen such that $\boldsymbol{H}_k\boldsymbol{C}_{\boldsymbol{\eta}_k}^{\frac{1}{2},\mathrm{H}} = \boldsymbol{H}_k^{\mathrm{BC,H}}\ \forall k$, which yields

$$\boldsymbol{G}_k\boldsymbol{H}_k\boldsymbol{T}_k = \boldsymbol{\Lambda}_k = \bar{\boldsymbol{B}}_k^{\mathrm{MF}}\boldsymbol{H}_k^{\mathrm{BC}}\bar{\boldsymbol{P}}_k.$$

Hence, no intra-user interference arises in the dual MIMO MAC such that the last sum in the denominator of (4.49) also vanishes. Moreover, the channel gain of every data stream in the dual MAC corresponds to the gain in the original BC, i.e., the two numerators of $\mathrm{SINR}_{k,j}^{\mathrm{DPC}}(\bar{\boldsymbol{P}},\bar{\boldsymbol{b}}_{k,j}^{\mathrm{MF}})$ from (4.46) and $\mathrm{SINR}_{k,j}^{\mathrm{SD}}(\boldsymbol{T},\boldsymbol{g}_{k,j})$ from (4.49) coincide. Equality in (4.51) is then achieved for

$$\begin{aligned}\psi_{k,j}^2\cdot\left[\bar{\boldsymbol{b}}_{k,j}^{\mathrm{MF,T}}\boldsymbol{C}_{\boldsymbol{\eta}_k}\bar{\boldsymbol{b}}_{k,j}^{\mathrm{MF}*} + \sum_{i<k}\sum_{\ell=1}^{L_i}|\bar{\boldsymbol{b}}_{k,j}^{\mathrm{MF,T}}\boldsymbol{H}_k^{\mathrm{BC}}\bar{\boldsymbol{p}}_{i,\ell}|^2\right]&\\ -\sum_{i>k}\sum_{\ell=1}^{L_i}\psi_{i,\ell}^2\cdot|\bar{\boldsymbol{b}}_{i,\ell}^{\mathrm{MF,T}}\boldsymbol{H}_i^{\mathrm{BC}}\bar{\boldsymbol{p}}_{k,j}|^2 &= \|\bar{\boldsymbol{p}}_{k,j}\|_2^2.\end{aligned}$$

Since above equation has to hold for all users k and all data streams j, we end up with the linear system of equations

$$\boldsymbol{Y}\cdot\left[\psi_{1,1}^2,\ldots,\psi_{K,L_K}^2\right]^{\mathrm{T}}=\left[\|\bar{\boldsymbol{p}}_{1,1}\|_2^2,\ldots,\|\bar{\boldsymbol{p}}_{K,L_K}\|_2^2\right]^{\mathrm{T}} \tag{4.53}$$

where the matrix $\boldsymbol{Y}$ is now block upper triangular in contrast to the rate duality for linear transceivers from Sect. 3.3.1 [cf. (3.67)]:

$$\boldsymbol{Y}:=\begin{bmatrix}\boldsymbol{Y}_{1,1} & \cdots & \boldsymbol{Y}_{1,K}\\ \mathbf{0} & \ddots & \vdots\\ \mathbf{0} & \mathbf{0} & \boldsymbol{Y}_{K,K}\end{bmatrix}. \tag{4.54}$$

The off-diagonal blocks with $k<i$ read as [see (3.68)]

$$\boldsymbol{Y}_{k,i}:=-\left[\left(\bar{\boldsymbol{B}}_i^{\mathrm{MF}}\boldsymbol{H}_i^{\mathrm{BC}}\bar{\boldsymbol{P}}_k\right)\odot\left(\bar{\boldsymbol{B}}_i^{\mathrm{MF}}\boldsymbol{H}_i^{\mathrm{BC}}\bar{\boldsymbol{P}}_k\right)^*\right]^{\mathrm{T}}\quad\in\mathbb{R}^{L_k\times L_i},\ \text{for } k<i$$

and $\boldsymbol{Y}_{k,k}\in\mathbb{R}_{+,0}^{L_k\times L_k}$ is diagonal with

$$\left[\boldsymbol{Y}_{k,k}\right]_{j,j}:=\bar{\boldsymbol{b}}_{k,j}^{\mathrm{MF,T}}\boldsymbol{C}_{\eta_k}\bar{\boldsymbol{b}}_{k,j}^{\mathrm{MF}*}-\sum_{i<k}\sum_{\ell=1}^{L_i}\left[\boldsymbol{Y}_{i,k}\right]_{\ell,j}\quad\forall j\in\{1,\ldots,L_k\}$$

such that $\boldsymbol{Y}$ is not only block upper triangular, but also upper triangular. Passive streams are treated in the same way as described in the rate duality for linear filtering, see the first paragraph below (3.70). As long as $\bar{\boldsymbol{B}}_k^{\mathrm{MF}}$ has no row being the zero-vector for all $k\in\{1,\ldots,K\}$, $\boldsymbol{Y}$ is a nonsingular *M-matrix*. The solution for the positive squared scaling factors $\psi_{1,1}^2,\ldots,\psi_{K,L_K}^2$ can quickly be obtained with minimum complexity via backward substitution in (4.53). To see how zero-rows in $\bar{\boldsymbol{B}}_k^{\mathrm{MF}}$ have to be handled and under which conditions they may arise, we refer to the discussion below (3.70).

To show that the dual MIMO MAC consumes the same amount of transmit power, we sum up the L rows of (4.53):

$$\sum_{k=1}^{K}\sum_{j=1}^{L_k}\psi_{k,j}^2\left[\bar{\boldsymbol{B}}_k^{\mathrm{MF}}\boldsymbol{C}_{\eta_k}\bar{\boldsymbol{B}}_k^{\mathrm{MF,H}}\right]_{j,j}=\sum_{k=1}^{K}\|\bar{\boldsymbol{P}}_k\|_{\mathrm{F}}^2=P_{\mathrm{diss}}^{\mathrm{BC}}.$$

With the precoder parametrization in the dual MIMO MAC from (4.52), the dissipated power in the MAC matches $P_{\mathrm{diss}}^{\mathrm{BC}}$ due to

$$\begin{aligned}P_{\mathrm{diss}}^{\mathrm{MAC}}&=\sum_{k=1}^{K}\|\boldsymbol{T}_k\|_{\mathrm{F}}^2=\sum_{k=1}^{K}\sum_{j=1}^{L_k}\left[\boldsymbol{T}_k^{\mathrm{H}}\boldsymbol{T}_k\right]_{j,j}\\&=\sum_{k=1}^{K}\sum_{j=1}^{L_k}\psi_{k,j}^2\left[\bar{\boldsymbol{B}}_k^{\mathrm{MF}}\boldsymbol{C}_{\eta_k}\bar{\boldsymbol{B}}_k^{\mathrm{MF,H}}\right]_{j,j}=P_{\mathrm{diss}}^{\mathrm{BC}}.\end{aligned}$$

Note that the diagonal dominance and the *M-matrix* property of $\boldsymbol{Y}$ results from the fact that the decoding order in the dual MAC is the reversed encoding order of the BC. If the decoding order had been chosen to match the encoding order, the resulting matrix $\boldsymbol{Y}$ would not have had the *M-matrix* property guaranteeing nonnegative outcomes for the squared scaling factors $\psi_{1,1}^2, \ldots, \psi_{K,L_K}^2$ and the sum power conservation.

Using the filter outcomes of the BC to MAC conversion, the dual MIMO MAC achieves the same data rates via separate stream decoding:

$$R_k^{\mathrm{SD}}(\boldsymbol{T}, \boldsymbol{G}_k) = I_k^{\mathrm{DPC}}(\boldsymbol{P}) \quad \forall k \in \{1, \ldots, K\}.$$

Although the filter outcome $\boldsymbol{G}_k$ leads to a diagonal matrix product $\boldsymbol{G}_k \boldsymbol{H}_k \boldsymbol{T}_k$, the matrix $\boldsymbol{G}_k$ is in general not the matched filter $\boldsymbol{G}_k^{\mathrm{MF}}$ for user k (see the definition of $\boldsymbol{G}_k^{\mathrm{MF}}$ in the next section). In this case, it is possible to increase the data rate in the dual MAC by applying the matched filter receiver and afterwards removing the intra-user interference via unitary decorrelation matrices. This way, the maximum mutual information $I_k^{\mathrm{SD}}(\boldsymbol{T})$ in the dual MIMO MAC is achieved even with separate stream decoding and successively removing only inter-user interference. The duality conversion thus satisfies

$$I_k^{\mathrm{SD}}(\boldsymbol{T}) \geq I_k^{\mathrm{DPC}}(\boldsymbol{P}) \quad \forall k$$

i.e., the same or a larger mutual information is possible in the dual MIMO MAC. A case in which the derived receive filter $\boldsymbol{G}_k$ already corresponds to the matched filter is for example if the filters in the original MIMO BC are chosen such that they obtain a rate pair on the boundary of the capacity region.

Synopsis: For every set of transmit covariance matrices $\boldsymbol{P}_1 \boldsymbol{P}_1^{\mathrm{H}}, \ldots, \boldsymbol{P}_K \boldsymbol{P}_K^{\mathrm{H}}$ in a MIMO BC that achieves mutual information expressions $I_1^{\mathrm{DPC}}(\boldsymbol{P}), \ldots, I_K^{\mathrm{DPC}}(\boldsymbol{P})$ using dirty paper coding, it is possible to find transmit and receive filters in a dual MIMO MAC with identity noise covariance matrix which achieve the same rate tuple. In this dual MIMO MAC, the same amount of transmit power is dissipated and successive decoding is applied. The decoding order corresponds to the reversed encoding order of the BC. Moreover, the dual MAC features at least the same mutual information expressions such that the capacity region of the dual MIMO MAC is a superset of the capacity region of the original MIMO BC.

4.4.3 MAC to BC Conversion

For some precoding matrices $\boldsymbol{T}_1, \ldots, \boldsymbol{T}_K$ that are for example outcomes of an optimization in the dual MIMO MAC, we now derive conversion rules how to convert these precoders back to the original MIMO BC where the rates are such that they correspond to the mutual information expressions of the dual MIMO MAC. Since the derivation is very similar to the BC to MAC conversion from the previous section, only the main steps are discussed.

Under the natural decoding order $\pi[k] = k\ \forall k$ where user 1 is decoded first and user K decoded last, the mutual information expression for user k can be expressed in terms of the matched filter

$$\boldsymbol{G}_k^{\mathrm{MF}} := \boldsymbol{T}_k^{\mathrm{H}}\boldsymbol{H}_k^{\mathrm{H}}\Big(\sum_{i>k}\boldsymbol{H}_i\boldsymbol{T}_i\boldsymbol{T}_i^{\mathrm{H}}\boldsymbol{H}_i^{\mathrm{H}}+\mathbf{I}\Big)^{-1}. \tag{4.55}$$

Conditioned on all $\boldsymbol{s}_i$ with $i < k$, the mutual information between the kth transmitter and the receive signal in the dual MAC with successive decoding reads as [cf. (4.15)]

$$\begin{aligned}
I_k^{\mathrm{SD}}(\boldsymbol{T}) &:= \log_2\left|\mathbf{I}+\boldsymbol{G}_k^{\mathrm{MF}}\boldsymbol{H}_k\boldsymbol{T}_k\right|\\
&= \log_2\left|\mathbf{I}+\boldsymbol{V}_k^{\mathrm{H}}\boldsymbol{G}_k^{\mathrm{MF}}\boldsymbol{H}_k\boldsymbol{T}_k\boldsymbol{V}_k\right|\\
&= \log_2\left|\mathbf{I}+\bar{\boldsymbol{G}}_k^{\mathrm{MF}}\boldsymbol{H}_k\bar{\boldsymbol{T}}_k\right|\\
&= \log_2\prod_{j=1}^{L_k}\left[\mathbf{I}+\bar{\boldsymbol{G}}_k^{\mathrm{MF}}\boldsymbol{H}_k\bar{\boldsymbol{T}}_k\right]_{j,j}\\
&= \sum_{j=1}^{L_k}\log_2\left(1+\left[\bar{\boldsymbol{G}}_k^{\mathrm{MF}}\boldsymbol{H}_k\bar{\boldsymbol{T}}_k\right]_{j,j}\right)
\end{aligned} \tag{4.56}$$

where $\bar{\boldsymbol{T}}_k := \boldsymbol{T}_k\boldsymbol{V}_k$ and $\bar{\boldsymbol{G}}_k^{\mathrm{MF}} := \boldsymbol{V}_k^{\mathrm{H}}\boldsymbol{G}_k^{\mathrm{MF}}$. To let the fourth line of (4.56) hold, the unitary matrix $\boldsymbol{V}_k$ must be chosen such that

$$\boldsymbol{V}_k\boldsymbol{\Xi}_k\boldsymbol{V}_k^{\mathrm{H}} := \boldsymbol{G}_k^{\mathrm{MF}}\boldsymbol{H}_k\boldsymbol{T}_k \tag{4.57}$$

holds with unitary $\boldsymbol{V}_k$ and diagonal $\boldsymbol{\Xi}_k$. From (4.49), it can be observed that

$$\mathrm{SINR}_{k,j}^{\mathrm{SD}}(\bar{\boldsymbol{T}},\bar{\boldsymbol{g}}_{k,j}^{\mathrm{MF}}) = \left[\bar{\boldsymbol{G}}_k^{\mathrm{MF}}\boldsymbol{H}_k\bar{\boldsymbol{T}}_k\right]_{j,j} \tag{4.58}$$

i.e., the mutual information $I_k^{\mathrm{SD}}(\boldsymbol{T})$ can be achieved with separate stream decoding without intra-user interference and only inter-user successive decoding must be applied. For the filter parametrization in (3.77), i.e.,

$$\begin{aligned}
\boldsymbol{p}_{k,j} &= \boldsymbol{P}_k\boldsymbol{e}_j = \varphi_{k,j}\boldsymbol{G}_k^{\mathrm{MF,H}}\boldsymbol{V}_k\boldsymbol{e}_j\\
\boldsymbol{b}_{k,j}^{\mathrm{T}} &= \boldsymbol{e}_j^{\mathrm{T}}\boldsymbol{B}_k = \varphi_{k,j}^{-1}\boldsymbol{e}_j^{\mathrm{T}}\boldsymbol{V}_k^{\mathrm{H}}\boldsymbol{T}_k^{\mathrm{H}}\boldsymbol{C}_{\eta_k}^{-\frac{1}{2}}
\end{aligned} \tag{4.59}$$

we equate all pseudo-SINRs of the original MIMO BC in (4.46) with the SINRs in the dual MIMO MAC defined in (4.49) for the particular filters in (4.59). In turn,

$\boldsymbol{B}_k \boldsymbol{H}_k^{\mathrm{BC}} \boldsymbol{P}_k = \bar{\boldsymbol{G}}_k^{\mathrm{MF}} \boldsymbol{H}_k \bar{\boldsymbol{T}}_k$ holds for the choice of the channels in the dual MAC in Theorem 1 such that only the denominators of the two SINR expressions have to be equated and no intra-user interference arises in the original BC:

$$\varphi_{k,j}^2 \cdot \left[\|\bar{\boldsymbol{g}}_{k,j}^{\mathrm{MF}}\|_2^2 + \sum_{i>k} \sum_{\ell=1}^{L_i} |\bar{\boldsymbol{g}}_{k,j}^{\mathrm{MF,T}} \boldsymbol{H}_i \bar{\boldsymbol{t}}_{i,\ell}|^2 \right] \\ - \sum_{i<k} \sum_{\ell=1}^{L_i} \varphi_{i,\ell}^2 \cdot |\bar{\boldsymbol{g}}_{i,\ell}^{\mathrm{MF,T}} \boldsymbol{H}_k \bar{\boldsymbol{t}}_{k,j}|^2 = \|\bar{\boldsymbol{t}}_{k,j}\|_2^2$$

Since above equation has to hold for all users k and all data streams j, the following linear system of equations is obtained:

$$\boldsymbol{Z} \cdot \left[\varphi_{1,1}^2, \ldots, \varphi_{K,L_K}^2\right]^{\mathrm{T}} = \left[\|\bar{\boldsymbol{t}}_{1,1}\|_2^2, \ldots, \|\bar{\boldsymbol{t}}_{K,L_K}\|_2^2\right]^{\mathrm{T}}. \tag{4.60}$$

For the MAC to BC conversion, the matrix $\boldsymbol{Z}$ is block lower triangular:

$$\boldsymbol{Z} := \begin{bmatrix} \boldsymbol{Z}_{1,1} & \boldsymbol{0} & \boldsymbol{0} \\ \vdots & \ddots & \boldsymbol{0} \\ \boldsymbol{Z}_{K,1} & \cdots & \boldsymbol{Z}_{K,K} \end{bmatrix}.$$

The off-diagonal blocks $\boldsymbol{Z}_{k,j}$ with $k > j$ read as

$$\boldsymbol{Z}_{k,j} := -\left[\left(\bar{\boldsymbol{G}}_j^{\mathrm{MF}} \boldsymbol{H}_k \bar{\boldsymbol{T}}_k\right) \odot \left(\bar{\boldsymbol{G}}_j^{\mathrm{MF}} \boldsymbol{H}_k \bar{\boldsymbol{T}}_k\right)^* \right]^{\mathrm{T}} \in \mathbb{R}^{L_k \times L_j} \quad \forall k > j$$

and the main diagonal blocks $\boldsymbol{Z}_{k,k} \in \mathbb{R}_{+,0}^{L_k \times L_k}$ are diagonal matrices with

$$\left[\boldsymbol{Z}_{k,k}\right]_{j,j} := \|\bar{\boldsymbol{g}}_{k,j}^{\mathrm{MF}}\|_2^2 - \sum_{i>k} \sum_{\ell=1}^{L_i} \left[\boldsymbol{Z}_{i,k}\right]_{\ell,j} \quad \forall j \in \{1, \ldots, L_k\}.$$

Since $\boldsymbol{Z}$ is not only block lower triangular but also lower triangular, the squared scaling factors $\varphi_{1,1}^2, \ldots, \varphi_{K,L_K}^2$ can be obtained by means of forward substitution with very low complexity. Zero-rows and zero-columns in $\boldsymbol{Z}$ belonging to passive data streams have to be removed before the forward substitution is applied, see the discussion below (4.54). Moreover, the sum power conservation can again be shown by summing up all rows of (4.60) and using (4.59) to see that $P_{\mathrm{diss}}^{\mathrm{MAC}} = P_{\mathrm{diss}}^{\mathrm{BC}}$ holds.

Due to the conservation of the individual SINRs, the data rates in the MIMO BC correspond to the mutual information expressions in the dual MIMO MAC:

$$R_k^{\mathrm{DPC}}(\boldsymbol{P}, \boldsymbol{B}_k) := \sum_{j=1}^{L_k} \log_2\left(1 + \mathrm{SINR}_{k,j}^{\mathrm{DPC}}(\boldsymbol{P}, \boldsymbol{b}_{k,j})\right) = I_k^{\mathrm{SD}}(\boldsymbol{T}).$$

If the dual MAC precoders $\boldsymbol{T}$ are not chosen to lead to a rate tuple on the boundary of the rate region, the data rate in the original MIMO BC may be increased by applying matched filter receivers and afterwards decorrelate all matrix products $\boldsymbol{B}_k^{\mathrm{MF}} \boldsymbol{H}_k^{\mathrm{BC}} \boldsymbol{P}_k$ via unitary decorrelation matrices for all $k \in \{1, \ldots, K\}$. Hence, the duality conversion satisfies

$$I_k^{\mathrm{DPC}}(\boldsymbol{P}) \geq I_k^{\mathrm{SD}}(\boldsymbol{T}) \quad \forall k$$

i.e., the same or a larger mutual information is possible in the original MIMO BC.

Synopsis: Precoding matrices $\boldsymbol{T}_1, \ldots, \boldsymbol{T}_K$ in the dual MIMO MAC which induce mutual information expressions $I_1^{\mathrm{SD}}(\boldsymbol{T}), \ldots, I_K^{\mathrm{SD}}(\boldsymbol{T})$ can be converted to transmit and receive filters in the original MIMO BC such that these filters achieve the same data rates by means of dirty paper coding and the same amount of dissipated power. Moreover, at least the same mutual information expressions are feasible in the original MIMO BC which means that the capacity region of the MIMO BC is a superset of the capacity region of the dual MIMO MAC where successive decoding is applied. In combination with the result that the capacity region of the MIMO BC is a subset of the capacity of the dual MIMO MAC, we conclude that both regions coincide.

4.4.4 Advantages of the Novel Rate Duality

The rate duality of Vishwanath et al. in [19, 82] has been summarized in Sect. 4.3 for the case of the MAC to BC conversion. There, the transmit covariance matrices $\boldsymbol{S}_1, \ldots, \boldsymbol{S}_K$ of the MIMO BC have to be computed successively, since $\boldsymbol{S}_k$ depends on all $\boldsymbol{S}_i$ with $i < k$ [see (4.37)]. Moreover, K Cholesky decompositions of $N \times N$ matrices have to be computed [see (4.30)], and K further Cholesky decompositions must be computed for matrices of dimensions $M_k \times M_k$ [see (4.36)], where $k \in \{1, \ldots, K\}$. In addition, many matrix products occur and K reduced singular value decompositions have to be computed [see (4.33)], one for each $N \times M_k$ matrix with $k \in \{1, \ldots, K\}$. Additionally, the authors in [19, 82] do not explicitly mention how dirty paper coding is implemented in the vector-valued case. It is not clear, whether the individual streams of a particular user are also treated as noncausally known side-information as this is the case in [32].

Our proposed duality is derived rather from a signal processing point of view than from an information-theoretic perspective. Hence, implementation issues are already addressed. Having computed the matched filter receivers [see (4.55)], K eigenvalue decompositions of positive semi-definite matrices with small dimensions $M_k \times M_k$, $k \in \{1, \ldots, K\}$ have to be computed [see (4.57)]. The linear system of equations (4.60) from which the scaling factors are computed can quickly be solved

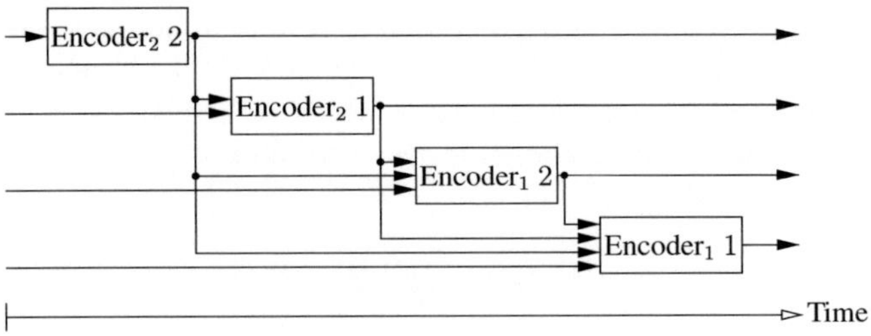

Fig. 4.6 Schematic view of the encoding delay for the duality of Viswanath and Tse in [32]

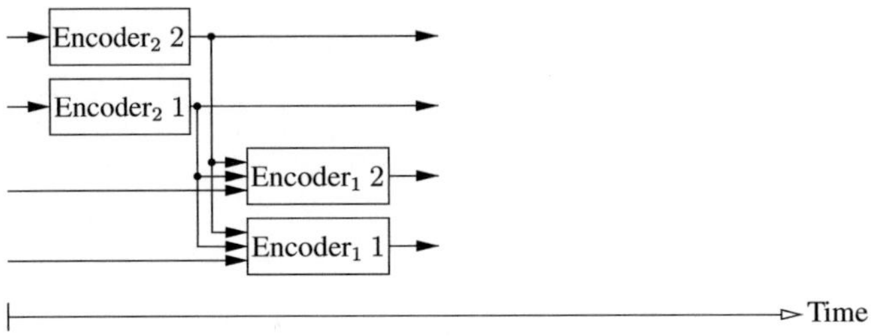

Fig. 4.7 Schematic view of the encoding delay for the proposed duality

by means of forward or backward substitution and there are no dependencies during the computation. Due to the decorrelation filters, no intra-user interference is present which reduces the encoding delay induced by dirty paper coding.

To see the benefit of these decorrelation filters, the duality of Viswanath and Tse in [32] is now investigated. Therein, every data stream of a user is treated as a virtual user. Hence, the already encoded data streams of a user are passed to the encoder of the current data stream of the same user where they are treated as known side-information such that their interference can be canceled by coding. Since the side-information has to be known noncausally for dirty paper coding, the total encoding delay is L times the time required to encode a single data stream, where L is the total number of streams transmitted. This is depicted in Fig. 4.6 for the example of two transmitters with two data streams each, where 'Encoder$_k$ j' denotes the encoder for the jth data stream of user k. Since the second data stream of user 2 acts as self-interference for the first data stream, it is removed by the encoder for the first data stream of user 2. To this end, the complete sequence must be known such that the first stream of user 2 cannot be encoded before the encoding of the second data stream has been finished. The encoder for the second stream of user 1 treats interference from user 1 as known side-information and so on, such that the total encoding delay is four times the encoding time of a single data stream.

In contrast, the intra-user interference of our proposed duality is removed by means of filtering and not by means of coding. Due to the lack of self-interference between the data streams of a particular user, all data streams of a single user can be encoded simultaneously, see Fig. 4.7 for the two users with two data streams example.

Dirty paper coding is only used to remove the inter-user interference of the already encoded users. Hence, the total encoding delay is only K times the encoding delay of a single data stream. In the chosen example, the encoding delay is thus half as long as the one resulting from Viswanath's and Tse's duality.

Chapter 5
Matrix-Based Gradient-Projection Algorithm

Having established the duality relationship between the MIMO BC and the MIMO MAC in Chaps. 3 and 4, we can handle optimizations originally arising in the BC in its dual MIMO MAC where they have more favorable properties like being convex, for example, as in the weighted sum rate maximization problem. Due to the limited resource power, we have to deal with constrained optimizations. In particular, an upper bound on the dissipated sum power is imposed for the transceiver design in the following two chapters. An attractive iterative scheme that is targeted at solving constrained optimization problems is the gradient-projection algorithm which can be regarded as an extension of the steepest ascent method derived by Cauchy to the case of a constrained optimization. In this chapter, we extensively derive the matrix-based gradient projection algorithm such that it can readily be applied in Chaps. 6 and 7 for the transceiver design with and without interference cancelation, respectively.

In Sect. 5.1, the standard gradient-projection algorithm is derived when applied to the maximization of a utility over a general closed convex set. Different step-size strategies are analyzed in terms of applicability and convergence. The extension to the *relaxed* gradient-projection algorithm is explained in Sect. 5.2, where an additional outer-step size mechanism comes into play which is intended to speed up convergence. Finally, the orthogonal projections onto the constraint set (this projection is obligatory for the gradient-projection algorithm) and onto the tangent cone (this projection is optional) are explicitly derived in Sect. 5.3 for two specific constraint sets. First, when the gradient-projection algorithm is based on the precoders, the upper bound on the sum power leads to a ball-shaped constraint set. Second, if based on the transmit covariance matrices, the shape of the constraint set not only follows from the upper bound on the sum power but also from the positive semi-definiteness structure of the transmit covariance matrices.

R. Hunger, *Analysis and Transceiver Design for the MIMO Broadcast Channel*, Foundations in Signal Processing, Communications and Networking 8,
DOI: 10.1007/978-3-642-31692-0_5,

5.1 Standard Gradient-Projection Algorithm

The standard gradient-projection algorithm was first proposed in 1964 by Goldstein [91] and independently in 1965 by Levitin and Polyak [92, Sect. 5]. In its original form, a vector-valued, continuously differentiable function f is minimized over a nonempty closed convex set $\mathcal{C}$ under a Lipschitz assumption on the gradient ∇f:

$$\underset{\boldsymbol{x}}{\text{minimize}}\, f(\boldsymbol{x}) \quad \text{s.t.: } \boldsymbol{x} \in \mathcal{C}.$$

Goldstein, Levitin, and Polyak showed that under certain conditions on the step-size $s^{(n)}$, the recursion

$$\boldsymbol{x}^{(n+1)} = \mathcal{P}_{\mathcal{C}}\big(\boldsymbol{x}^{(n)} - s^{(n)}\nabla f(\boldsymbol{x}^{(n)})\big) \tag{5.1}$$

converges to a stationary point. In particular, if f is convex, the global minimum of f on $\mathcal{C}$ is obtained. In (5.1), the operator $\mathcal{P}_{\mathcal{C}}(\cdot)$ denotes the orthogonal projection of its argument onto the set $\mathcal{C}$ which is unique due to the convexity of $\mathcal{C}$. The simple update rule makes it attractive even for problems with high dimensionality since only the gradient ∇f has to be computed. Moreover, $\boldsymbol{x}^{(n)} \in \mathcal{C}$ is ensured for every n due to the projection, i.e., the algorithm moves along the boundary of $\mathcal{C}$.[1] If $\mathcal{C}$ is constituted by the intersection or the Cartesian product of several constraints, the orthogonal projection is not influenced by a newly encountered active constraint as this would be the case in Rosen's projection for example [93]. However, it is only applicable if the orthogonal projection can be carried out with reasonable complexity which limits the constraint sets that the gradient-projection algorithm can handle. In [93, 91], Bertsekas and Goldstein name spheres, boxes, orthants, and Cartesian products of spheres as adequate candidates for the constraint set $\mathcal{C}$. As we will see in Sect. 5.3, the sum power constraint of the MIMO BC either leads to a ball-shaped constraint set (when working on the precoders) or a Cartesian product of cones with a sum trace constraint as constraint set (when working on the covariance matrices). In both cases, the orthogonal projection can be computed efficiently.

In the following, we extend the vector-valued gradient-projection algorithm to the matrix-valued case for complex matrices. An easy to follow step-by-step derivation will finally lead to an update rule in the style of (5.1) for the matrix-valued case. Moreover, we show that moving along the direction of the negative gradient of the unconstrained function f does not yield the steepest descent direction when the projection is taken into account. The direction of the steepest descent of the composition of the function f *and* the projection will be derived as well in the following. As the gradient-projection algorithm will serve as the core of the (weighted) sum rate maximization in the next two chapters, we investigate the *maximization* of a real-valued function $f : \mathbf{Z} \mapsto f(\mathbf{Z})$ with matrix-valued, block-diagonal argument $\mathbf{Z} := \text{blockdiag}\{\mathbf{Z}_k\}_{k=1}^K$ over a nonempty, closed convex set $\mathcal{C}$:

[1] The unconstrained infimum of f is assumed not to be attained in $\mathcal{C}$ and every unconstrained gradient step is assumed to leave $\mathcal{C}$ to let this statement hold.

$$\underset{Z}{\text{maximize}}\ f(Z) \quad \text{s.t.:}\ Z \in \mathcal{C}. \tag{5.2}$$

Since there is a small, but subtle difference in the total differential $\mathrm{d}f$ whether Z is Hermitian (covariance matrix approach) or not (precoder approach), we differentiate these two cases and treat them separately in the next two subsections.

5.1.1 Precoder-Based Gradient-Projection Algorithm

In the precoder-based approach, the utility f in (5.2) is maximized with respect to the block-diagonal, but otherwise unstructured precoders $T := \text{blockdiag}\{T_k\}_{k=1}^K \in \mathbb{C}^{M \times L}$ of the dual MIMO MAC. Hence, $Z_k = T_k \in \mathbb{C}^{M_k \times L_k}$ is an unstructured $M_k \times L_k$ matrix with M_k denoting the number of antennas at terminal k and L_k denoting the number of data streams associated to user k. An incremental increase $\mathrm{d}T := \text{blockdiag}\{\mathrm{d}T_k\}_{k=1}^K$ of T leads to the differential [see (A24) in Appendix A11]

$$\begin{aligned}\mathrm{d}f(T, \mathrm{d}T) &= 2\sum_{k=1}^{K} \mathrm{Re}\left[\mathrm{tr}\left(\frac{\partial_{\mathrm{w}} f(T)}{\partial_{\mathrm{w}} T_k^{\mathrm{T}}} \cdot \mathrm{d}T_k\right)\right] \\ &= 2\mathrm{Re}\left[\mathrm{tr}\left(\frac{\partial_{\mathrm{w}} f(T)}{\partial_{\mathrm{w}} T^{\mathrm{T}}} \cdot \mathrm{d}T\right)\right]\end{aligned} \tag{5.3}$$

where $\frac{\partial_{\mathrm{w}}}{\partial_{\mathrm{w}}(\cdot)}$ denotes the Wirtinger derivative,[2] see Appendix A10 for a brief introduction, and (A25) in Appendix A11 for the definition of the derivative of a scalar function with respect to a matrix. A differential increment $\mathrm{d}T$ is an ascent direction at T and thus leads to an incremental increase of the function f, if $\mathrm{d}f(T, \mathrm{d}T) > 0$. Under a fixed Frobenius norm constraint on $\mathrm{d}T$, i.e., for fixed $\|\mathrm{d}T\|_{\mathrm{F}}^2 = \sum_{k=1}^K \|\mathrm{d}T_k\|_{\mathrm{F}}^2$, the steepest ascent direction of $f(T)$ which obtains the largest increment $\mathrm{d}f$ ignoring the constraint $T \in \mathcal{C}$ is given by

$$\mathrm{d}T = \left(\frac{\partial_{\mathrm{w}} f(T)}{\partial_{\mathrm{w}} T^{\mathrm{T}}}\right)^{\mathrm{H}} \mathrm{d}\alpha = \frac{\partial_{\mathrm{w}} f(T)}{\partial_{\mathrm{w}} T^{*}} \cdot \mathrm{d}\alpha \tag{5.4}$$

with some differential $\mathrm{d}\alpha > 0$ that sets the squared Frobenius norm $\|\mathrm{d}T\|_{\mathrm{F}}^2$ to the fixed value. The statement that $\mathrm{d}T$ is the steepest ascent direction follows from the Cauchy–Schwarz inequality for example. For $T \in \mathcal{C}$, even a differential update might leave the feasible set such that $T + \mathrm{d}T \notin \mathcal{C}$. A simple remedy to overcome this problem is to project the updated variable back onto $\mathcal{C}$. However, two aspects have to be taken into consideration. First, the projection must be chosen appropriately. We

[2] For a convenient notation, it is necessary to define the (Wirtinger) derivative with respect to a block-diagonal matrix as a block-diagonal matrix itself, whose main diagonal blocks correspond to the (Wirtinger) derivative of the function with respect to the main diagonal blocks of the argument.

choose the *orthogonal* projection $\mathcal{P}_{\mathcal{C}}(\cdot)$ which yields the point on the boundary[3] $\partial\mathcal{C}$ with minimum distance. Since $\mathcal{C}$ is closed and convex and since the Frobenius norm is *strictly* convex, this point is unique which is known as the closest point theorem [94, Theorem 2.4.1, p. 50]. Second, the update direction $\mathrm{d}\boldsymbol{T}$ must be chosen appropriately. It turns out that a general ascent direction $\mathrm{d}\boldsymbol{T}$ satisfying $\mathrm{d}f(\boldsymbol{T}, \mathrm{d}\boldsymbol{T}) > 0$ may lead to

$$\mathrm{d}f\big(\mathcal{P}_{\mathcal{C}}(\boldsymbol{T}), \mathrm{d}\mathcal{P}_{\mathcal{C}}(\boldsymbol{T}, \mathrm{d}\boldsymbol{T})\big) < 0$$

which means that the directional update $\mathrm{d}\boldsymbol{T}$ is indeed an ascent direction in the unconstrained case, but may be a descent direction *after* the projection [95, Sect. 4.9, p. 95]. Note that $\mathrm{d}\mathcal{P}_{\mathcal{C}}(\boldsymbol{T}, \mathrm{d}\boldsymbol{T})$ is the differential after the projection that is induced by a differential change $\mathrm{d}\boldsymbol{T}$ of the precoders. In contrast, if $\mathrm{d}\boldsymbol{T}$ is chosen like in (5.4) as the direction of the steepest ascent of the unconstrained f, this problem does not arise as we show next. To this end, we need the following proposition, which is proven in [96, Proposition 2]:

Proposition 5.1 *Let $\mathcal{C}$ be a closed, convex set and let $\boldsymbol{T} \in \mathcal{C}$. Then, the directional derivative of the orthogonal projection $\mathcal{P}_{\mathcal{C}}(\boldsymbol{T} + \gamma\mathrm{d}\boldsymbol{T})$ onto $\mathcal{C}$ in the direction $\mathrm{d}\boldsymbol{T}$ corresponds to the orthogonal projection of $\mathrm{d}\boldsymbol{T}$ onto the tangent cone $\mathcal{T}_{\mathcal{C}}(\boldsymbol{T})$ of $\mathcal{C}$ at $\boldsymbol{T}$:*

$$\mathrm{d}\mathcal{P}_{\mathcal{C}}(\boldsymbol{T}, \mathrm{d}\boldsymbol{T}) = \frac{\mathrm{d}}{\mathrm{d}\gamma}\mathcal{P}_{\mathcal{C}}(\boldsymbol{T} + \gamma\mathrm{d}\boldsymbol{T})\Big|_{\gamma=0} = \mathcal{P}_{\mathcal{T}_{\mathcal{C}}(\boldsymbol{T})}(\mathrm{d}\boldsymbol{T}).$$

In other words, the first-order Taylor expansion of the orthogonal projection of $\boldsymbol{T} + \mathrm{d}\boldsymbol{T}$ with $\boldsymbol{T} \in \mathcal{C}$ onto the set $\mathcal{C}$ corresponds to $\boldsymbol{T}$ plus the orthogonal projection of $\mathrm{d}\boldsymbol{T}$ onto the tangent cone at $\boldsymbol{T}$. Using Proposition 1 and (5.3), the differential $\mathrm{d}f\big(\mathcal{P}_{\mathcal{C}}(\boldsymbol{T}), \mathrm{d}\mathcal{P}_{\mathcal{C}}(\boldsymbol{T}, \mathrm{d}\boldsymbol{T})\big)$ can be written as

$$\begin{aligned} \mathrm{d}f\big(\mathcal{P}_{\mathcal{C}}(\boldsymbol{T}), \mathrm{d}\mathcal{P}_{\mathcal{C}}(\boldsymbol{T}, \mathrm{d}\boldsymbol{T})\big) &= 2\mathrm{Re}\left[\mathrm{tr}\left(\frac{\partial_{\mathrm{W}} f(\boldsymbol{T}')}{\partial_{\mathrm{W}} \boldsymbol{T}'^{\mathrm{T}}}\Big|_{\boldsymbol{T}'=\mathcal{P}_{\mathcal{C}}(\boldsymbol{T})} \cdot \mathrm{d}\mathcal{P}_{\mathcal{C}}(\boldsymbol{T}, \mathrm{d}\boldsymbol{T})\right)\right] \\ &= 2\mathrm{Re}\left[\mathrm{tr}\left(\frac{\partial_{\mathrm{W}} f(\boldsymbol{T})}{\partial_{\mathrm{W}} \boldsymbol{T}^{\mathrm{T}}} \cdot \mathcal{P}_{\mathcal{T}_{\mathcal{C}}(\boldsymbol{T})}(\mathrm{d}\boldsymbol{T})\right)\right] \end{aligned} \tag{5.5}$$

where $\boldsymbol{T} \in \mathcal{C}$ is assumed for the second equality to hold. Now if $\mathrm{d}\boldsymbol{T}$ is chosen as the direction of the steepest ascent of the unconstrained f *without* projection as in (5.4), the differential in (5.5) becomes

$$\mathrm{d}f\big(\mathcal{P}_{\mathcal{C}}(\boldsymbol{T}), \mathrm{d}\mathcal{P}_{\mathcal{C}}(\boldsymbol{T}, \mathrm{d}\boldsymbol{T})\big)\Big|_{\mathrm{d}\boldsymbol{T}=\mathrm{d}\alpha\frac{\partial_{\mathrm{W}} f(\boldsymbol{T})}{\partial_{\mathrm{W}} \boldsymbol{T}^*}} = 2\left\|\mathcal{P}_{\mathcal{T}_{\mathcal{C}}(\boldsymbol{T})}\left(\frac{\partial_{\mathrm{W}} f(\boldsymbol{T})}{\partial_{\mathrm{W}} \boldsymbol{T}^*}\right)\right\|_{\mathrm{F}}^2 \mathrm{d}\alpha \geq 0 \tag{5.6}$$

where equality only holds if the projection onto the tangent cone vanishes, i.e., at a stationary point. Moving along the direction of the unconstrained gradient with a

[3] Clearly, the point that has to be projected has to lie outside $\mathcal{C}$. Otherwise, the projection is superfluous.

step-size s, the first-order Taylor expansion of the composition of the function f and the orthogonal projection $\mathcal{P}_{\mathcal{C}}(\cdot)$ reads for $\boldsymbol{T} \in \mathcal{C}$ as [see (5.6)]

$$f\left(\mathcal{P}_{\mathcal{C}}\left(\boldsymbol{T} + s \cdot \frac{\partial_{\mathrm{W}} f(\boldsymbol{T})}{\partial_{\mathrm{W}} \boldsymbol{T}^*}\right)\right) = f(\boldsymbol{T}) + 2\left\|\mathcal{P}_{\mathcal{T}_{\mathcal{C}}(\boldsymbol{T})}\left(\frac{\partial_{\mathrm{W}} f(\boldsymbol{T})}{\partial_{\mathrm{W}} \boldsymbol{T}^*}\right)\right\|_{\mathrm{F}}^2 \cdot s + e_1(s) \quad (5.7)$$

with some error term $e_1(s)$ that satisfies $e_1(s)/s \to 0$ as $s \to 0$.

Although the steepest ascent direction of the unconstrained $f(\boldsymbol{T})$ in (5.4) remains an ascent direction according to (5.6) when the orthogonal projection is taken into consideration, it is *not* the direction of the steepest ascent of the *composition* of f and the orthogonal projection $\mathcal{P}_{\mathcal{C}}(\cdot)$. The steepest ascent of the composition $f(\mathcal{P}_{\mathcal{C}}(\cdot))$ is given by

$$\mathrm{d}\boldsymbol{T} = \mathcal{P}_{\mathcal{T}_{\mathcal{C}}(\boldsymbol{T})}\left(\frac{\partial_{\mathrm{W}} f(\boldsymbol{T})}{\partial_{\mathrm{W}} \boldsymbol{T}^*}\right)\mathrm{d}\beta \quad (5.8)$$

with $\mathrm{d}\beta > 0$, which means that the direction of the steepest ascent of $f(\mathcal{P}_{\mathcal{C}}(\cdot))$ is along the steepest ascent of the unconstrained f *projected* onto the tangent cone. If the two differentials in (5.4) and (5.8) shall have the same Frobenius norm [for example to show that (5.8) leads to a larger differential $\mathrm{d}f$ than (5.4)], $\mathrm{d}\beta \geq \mathrm{d}\alpha$ holds and equality is only achieved if the Wirtinger derivative of $f(\boldsymbol{T})$ with respect to $\boldsymbol{T}^*$ already lies in the tangent cone.

The proof that (5.8) is the direction of the steepest ascent of the composition follows from the definition of the orthogonal projection onto the tangent cone $\mathcal{T}_{\mathcal{C}}(\boldsymbol{T})$. For every $\boldsymbol{V}$ having the same block-diagonal structure as $\boldsymbol{T}$, its projection $\mathcal{P}_{\mathcal{T}_{\mathcal{C}}(\boldsymbol{T})}(\boldsymbol{V})$ satisfies for every $\boldsymbol{Y} \in \mathcal{T}_{\mathcal{C}}(\boldsymbol{T})$ by definition the inequality

$$\begin{aligned}
\left\|(\boldsymbol{T} + \boldsymbol{V}) - \left(\boldsymbol{T} + \mathcal{P}_{\mathcal{T}_{\mathcal{C}}(\boldsymbol{T})}(\boldsymbol{V})\right)\right\|_{\mathrm{F}}^2 &\leq \|(\boldsymbol{T} + \boldsymbol{V}) - (\boldsymbol{T} + \boldsymbol{Y})\|_{\mathrm{F}}^2 \\
\Leftrightarrow \left\|\boldsymbol{V} - \mathcal{P}_{\mathcal{T}_{\mathcal{C}}(\boldsymbol{T})}(\boldsymbol{V})\right\|_{\mathrm{F}}^2 &\leq \|\boldsymbol{V} - \boldsymbol{Y}\|_{\mathrm{F}}^2 \\
\Leftrightarrow -2\mathrm{Re}\left[\mathrm{tr}\left(\boldsymbol{V}^{\mathrm{H}} \mathcal{P}_{\mathcal{T}_{\mathcal{C}}(\boldsymbol{T})}(\boldsymbol{V})\right)\right] + \|\mathcal{P}_{\mathcal{T}_{\mathcal{C}}(\boldsymbol{T})}(\boldsymbol{V})\|_{\mathrm{F}}^2 &\leq -2\mathrm{Re}\left[\mathrm{tr}(\boldsymbol{V}^{\mathrm{H}}\boldsymbol{Y})\right] + \|\boldsymbol{Y}\|_{\mathrm{F}}^2
\end{aligned}$$

from which we conclude that

$$2\mathrm{Re}\left[\mathrm{tr}(\boldsymbol{V}^{\mathrm{H}}\boldsymbol{Y})\right] \leq \|\boldsymbol{Y}\|_{\mathrm{F}}^2 - \|\mathcal{P}_{\mathcal{T}_{\mathcal{C}}(\boldsymbol{T})}(\boldsymbol{V})\|_{\mathrm{F}}^2 + 2\mathrm{Re}\left[\mathrm{tr}\left(\boldsymbol{V}^{\mathrm{H}} \mathcal{P}_{\mathcal{T}_{\mathcal{C}}(\boldsymbol{T})}(\boldsymbol{V})\right)\right]. \quad (5.9)$$

Note that equality only holds if $\boldsymbol{Y} = \mathcal{P}_{\mathcal{T}_{\mathcal{C}}(\boldsymbol{T})}(\boldsymbol{V})$ since the outcome of the orthogonal projection is unique. Setting

$$\begin{aligned}
\boldsymbol{V} &= \frac{\partial_{\mathrm{W}} f(\boldsymbol{T})}{\partial_{\mathrm{W}} \boldsymbol{T}^*}\mathrm{d}\beta \\
\boldsymbol{Y} &= \mathcal{P}_{\mathcal{T}_{\mathcal{C}}(\boldsymbol{T})}(\mathrm{d}\boldsymbol{T})
\end{aligned} \quad (5.10)$$

with $\mathrm{d}\beta > 0$, the combination of (5.5) and (5.9) yields

$$\begin{aligned}\mathrm{d}f\big(\mathcal{P}_{\mathcal{C}}(\boldsymbol{T}),\mathrm{d}\mathcal{P}_{\mathcal{C}}(\boldsymbol{T},\mathrm{d}\boldsymbol{T})\big) &\leq \frac{1}{\mathrm{d}\beta}\Big(\|\mathcal{P}_{\mathcal{T}_{\mathcal{C}}(\boldsymbol{T})}(\mathrm{d}\boldsymbol{T})\|_{\mathrm{F}}^2-\|\mathcal{P}_{\mathcal{T}_{\mathcal{C}}(\boldsymbol{T})}(\boldsymbol{V})\|_{\mathrm{F}}^2+2\mathrm{Re}\big[\mathrm{tr}\big(\boldsymbol{V}^{\mathrm{H}}\mathcal{P}_{\mathcal{T}_{\mathcal{C}}(\boldsymbol{T})}(\boldsymbol{V})\big)\big]\Big)\\ &\leq \frac{1}{\mathrm{d}\beta}\Big(\|\mathrm{d}\boldsymbol{T}\|_{\mathrm{F}}^2-\|\mathcal{P}_{\mathcal{T}_{\mathcal{C}}(\boldsymbol{T})}(\boldsymbol{V})\|_{\mathrm{F}}^2+2\mathrm{Re}\big[\mathrm{tr}\big(\boldsymbol{V}^{\mathrm{H}}\mathcal{P}_{\mathcal{T}_{\mathcal{C}}(\boldsymbol{T})}(\boldsymbol{V})\big)\big]\Big).\end{aligned} \tag{5.11}$$

The second inequality in (5.11) follows from $\|\mathcal{P}_{\mathcal{T}_{\mathcal{C}}(\boldsymbol{T})}(\mathrm{d}\boldsymbol{T})\|_{\mathrm{F}}^2 \leq \|\mathrm{d}\boldsymbol{T}\|_{\mathrm{F}}^2$ due to the projection. Note that (5.11) is an upper bound for the differential $\mathrm{d}f\big(\mathcal{P}_{\mathcal{C}}(\boldsymbol{T}), \mathrm{d}\mathcal{P}_{\mathcal{C}}(\boldsymbol{T},\mathrm{d}\boldsymbol{T})\big)$ for any choice of $\mathrm{d}\boldsymbol{T}$. To determine the direction of the steepest ascent of the composition with maximum differential $\mathrm{d}f(\mathcal{P}_{\mathcal{C}}(\boldsymbol{T}), \mathrm{d}\mathcal{P}_{\mathcal{C}}(\boldsymbol{T},\mathrm{d}\boldsymbol{T}))$, one has to fix $\|\mathrm{d}\boldsymbol{T}\|_{\mathrm{F}}^2$ such that the right-hand side of (5.11) is constant since $\boldsymbol{V}$ is constant as well [see (5.10)]. Without loss of generality, we set

$$\|\mathrm{d}\boldsymbol{T}\|_{\mathrm{F}}^2 = \|\mathcal{P}_{\mathcal{T}_{\mathcal{C}}(\boldsymbol{T})}(\boldsymbol{V})\|_{\mathrm{F}}^2 \tag{5.12}$$

and get

$$\mathrm{d}f(\mathcal{P}_{\mathcal{C}}(\boldsymbol{T}),\mathrm{d}\mathcal{P}_{\mathcal{C}}(\boldsymbol{T},\mathrm{d}\boldsymbol{T})) \leq 2\mathrm{Re}\left[\mathrm{tr}\left(\frac{\partial_{\mathrm{w}}f(\boldsymbol{T})}{\partial_{\mathrm{w}}\boldsymbol{T}^{\mathrm{T}}}\cdot\mathcal{P}_{\mathcal{T}_{\mathcal{C}}(\boldsymbol{T})}\left(\frac{\partial_{\mathrm{w}}f(\boldsymbol{T})}{\partial_{\mathrm{w}}\boldsymbol{T}^*}\right)\right)\right]\mathrm{d}\beta. \tag{5.13}$$

With the choice

$$\mathrm{d}\boldsymbol{T} = \mathcal{P}_{\mathcal{T}_{\mathcal{C}}(\boldsymbol{T})}\left(\frac{\partial_{\mathrm{w}}f(\boldsymbol{T})}{\partial_{\mathrm{w}}\boldsymbol{T}^*}\right)\mathrm{d}\beta \tag{5.14}$$

the condition (5.12) is fulfilled and the inequality (5.13) becomes an equality [see (5.5)] since the projector is idempotent. Thus, $\mathrm{d}\boldsymbol{T}$ from (5.8) is the steepest ascent including the projection and finally leads to the differential [cf. (5.5)]

$$\mathrm{d}f\big(\mathcal{P}_{\mathcal{C}}(\boldsymbol{T}),\mathrm{d}\mathcal{P}_{\mathcal{C}}(\boldsymbol{T},\mathrm{d}\boldsymbol{T})\big) = 2\left\|\mathcal{P}_{\mathcal{T}_{\mathcal{C}}(\boldsymbol{T})}\left(\frac{\partial_{\mathrm{w}}f(\boldsymbol{T})}{\partial_{\mathrm{w}}\boldsymbol{T}^*}\right)\right\|_{\mathrm{F}}^2\mathrm{d}\beta.$$

Hence, moving along the direction of the projected gradient from a $\boldsymbol{T}\in\mathcal{C}$ with step-size s, the projection onto the constraint set has the first order Taylor expansion

$$f\left(\mathcal{P}_{\mathcal{C}}\left(\boldsymbol{T}+s\cdot\mathcal{P}_{\mathcal{T}_{\mathcal{C}}(\boldsymbol{T})}\left(\frac{\partial_{\mathrm{w}}f(\boldsymbol{T})}{\partial_{\mathrm{w}}\boldsymbol{T}^*}\right)\right)\right) = f(\boldsymbol{T})+2\left\|\mathcal{P}_{\mathcal{T}_{\mathcal{C}}(\boldsymbol{T})}\left(\frac{\partial_{\mathrm{w}}f(\boldsymbol{T})}{\partial_{\mathrm{w}}\boldsymbol{T}^*}\right)\right\|_{\mathrm{F}}^2\cdot s+e_2(s) \tag{5.15}$$

with some error term $e_2(s)$ that satisfies $e_2(s)/s \to 0$ as $s \to 0$. Note that (5.15) coincides with (5.7) until the linear term, which means that the first order change of the utility is the same irrespective of whether moving along the direction of the unconstrained gradient or along its projection onto the tangent cone. The steepest ascent property of the projected version results from the fact that the same first order increase of the utility is achieved by a smaller norm of the direction matrix, i.e., $\mathrm{d}\beta \geq \mathrm{d}\alpha$.

To summarize, the direction of the steepest ascent of the utility *including* the projection corresponds to the Wirtinger derivative of $f(\boldsymbol{T})$ with respect to $\boldsymbol{T}^*$ projected onto the tangent cone. Thus, locally irrelevant information of the gradient of the unconstrained utility $f(\boldsymbol{T})$ is removed. Moreover, the unconstrained gradient need not vanish at a stationary point,[4] whereas the gradient of the utility *including* the orthogonal projection goes to zero when approaching a stationary point.

To solve the constrained optimization problems which arise in the next two chapters, we will investigate the following two recursions:

$$\boldsymbol{T}^{(n+1)} = \mathcal{P}_{\mathcal{C}}\left(\boldsymbol{T}^{(n)} + s^{(n)} \cdot \frac{\partial_{\mathrm{W}} f(\boldsymbol{T})}{\partial_{\mathrm{W}} \boldsymbol{T}^*}\bigg|_{\boldsymbol{T}=\boldsymbol{T}^{(n)}}\right) \tag{5.16}$$

$$\boldsymbol{T}^{(n+1)} = \mathcal{P}_{\mathcal{C}}\left(\boldsymbol{T}^{(n)} + s^{(n)} \cdot \mathcal{P}_{\mathcal{T}_{\mathcal{C}}(\boldsymbol{T}^{(n)})}\left(\frac{\partial_{\mathrm{W}} f(\boldsymbol{T})}{\partial_{\mathrm{W}} \boldsymbol{T}^*}\bigg|_{\boldsymbol{T}=\boldsymbol{T}^{(n)}}\right)\right). \tag{5.17}$$

Different strategies how to choose the iteration-dependent step-size $s^{(n)} > 0$ to ensure convergence to a stationary point of optimization (5.2) are discussed in Sect. 5.1.3. The orthogonal projections onto the constraint set $\mathcal{C}$ and the tangent cone $\mathcal{T}_{\mathcal{C}}(\boldsymbol{T})$ are derived in Sect. 5.3.1.

Synopsis: The direction of the steepest ascent of a utility f is given by the Wirtinger derivative of f with respect to the conjugate block-diagonal precoder $\boldsymbol{T}^*$. Since moving a step towards the direction of the steepest ascent of $f(\boldsymbol{T})$ may leave the constraint set, the so chosen precoder update is projected back onto the constraint set $\mathcal{C}$. When this orthogonal projection is taken into consideration, the Wirtinger derivative remains an ascent direction for the composition of the function f and the orthogonal projection. To show this, we exploited the fact that the first order Taylor expansion of the orthogonal projection onto the convex set $\mathcal{C}$ corresponds to an orthogonal projection onto the tangent cone. While the Wirtinger derivative of the unconstrained function f remains an ascent direction for the composition of the projection and the function f, it is *not* the direction of the steepest ascent, which is achieved when moving along the direction of the Wirtinger derivative projected onto the tangent cone.

Two recursions for the block-diagonal precoder will be investigated: in the first one, we move along the direction of the Wirtinger derivative of $f(\boldsymbol{T})$ with respect to $\boldsymbol{T}^*$ (steepest ascent of the unconstrained function $f(\boldsymbol{T})$) with an appropriate step-size starting from the precoder of the previous step. The sum is then projected onto the constraint set $\mathcal{C}$, see (5.16). In the second recursion, we move along the projection of the Wirtinger derivative, where the projection is onto the tangent cone. The sum is afterwards again projected back onto the constraint set $\mathcal{C}$, see (5.17).

---

[4] The supremum of f is assumed to lie outside $\mathcal{C}$.

5.1.2 Covariance-Based Gradient-Projection Algorithm

In the covariance-based approach, the utility f in (5.2) is maximized with respect to the block-diagonal matrix $\boldsymbol{Q} := \text{blockdiag}\{\boldsymbol{Q}_k\}_{k=1}^K \in \mathbb{S}^M$ of the dual MIMO MAC, where $\boldsymbol{Q}_k \in \mathbb{S}^{M_k}$ is positive semi-definite for all k.

A differential change $\mathrm{d}\boldsymbol{Q} := \text{blockdiag}\{\mathrm{d}\boldsymbol{Q}_k\}_{k=1}^K$ with Hermitian, but not necessarily positive semi-definite $\mathrm{d}\boldsymbol{Q}_k = (\mathrm{d}\boldsymbol{Q}_k)^{\mathrm{H}}\ \forall k$ which satisfies $\boldsymbol{Q} + \mathrm{d}\boldsymbol{Q} \succcurlyeq \boldsymbol{0}$ leads to the differential change of the function f which reads as [see (A26) and (A27) in Appendix A12]

$$\mathrm{d}f(\boldsymbol{Q}, \mathrm{d}\boldsymbol{Q}) = \sum_{k=1}^K \mathrm{tr}\left(\frac{\partial_{\mathrm{W}} f(\boldsymbol{Q})}{\partial_{\mathrm{W}} \boldsymbol{Q}_k^{\mathrm{T}}} \cdot \mathrm{d}\boldsymbol{Q}_k\right) = \mathrm{tr}\left(\frac{\partial_{\mathrm{W}} f(\boldsymbol{Q})}{\partial_{\mathrm{W}} \boldsymbol{Q}^{\mathrm{T}}} \cdot \mathrm{d}\boldsymbol{Q}\right). \tag{5.18}$$

In the Euclidean space, the direction of the steepest ascent of $f(\boldsymbol{Q})$ is given by the differential $\mathrm{d}\boldsymbol{Q}$ that maximizes $\mathrm{d}f(\boldsymbol{Q}, \mathrm{d}\boldsymbol{Q})$ in (5.18) given a fixed Frobenius norm of $\mathrm{d}\boldsymbol{Q}$. Moreover, $\boldsymbol{Q} + \mathrm{d}\boldsymbol{Q} \succcurlyeq \boldsymbol{0}$ must be fulfilled. For the weighted sum rate maximization with interference cancelation treated in Chap. 6, $\frac{\partial_{\mathrm{W}} f(\boldsymbol{Q})}{\boldsymbol{Q}_k^{\mathrm{T}}} \succcurlyeq \boldsymbol{0}$ holds for all $\boldsymbol{Q}_k \succcurlyeq \boldsymbol{0}$ and for all k such that the direction of the steepest ascent of $f(\boldsymbol{Q})$ in the (unconstrained) Euclidean space is given by [cf. (5.4)]

$$\mathrm{d}\boldsymbol{Q} = \left(\frac{\partial_{\mathrm{W}} f(\boldsymbol{Q})}{\partial_{\mathrm{W}} \boldsymbol{Q}^{\mathrm{T}}}\right)^{\mathrm{H}} \mathrm{d}\alpha = \frac{\partial_{\mathrm{W}} f(\boldsymbol{Q})}{\partial_{\mathrm{W}} \boldsymbol{Q}^{\mathrm{T}}} \cdot \mathrm{d}\alpha \tag{5.19}$$

with some differential $\mathrm{d}\alpha > 0$. According to Proposition 1 and (5.18), the first-order Taylor expansion of the utility when moving along the direction of the unconstrained gradient of $f(\boldsymbol{Q})$ with step-size s from $\boldsymbol{Q} \in \mathcal{C}$ reads as [cf. (5.7)]

$$f\left(\mathcal{P}_{\mathcal{C}}\left(\boldsymbol{Q} + s \cdot \frac{\partial_{\mathrm{W}} f(\boldsymbol{Q})}{\partial_{\mathrm{W}} \boldsymbol{Q}^{\mathrm{T}}}\right)\right) = f(\boldsymbol{Q}) + \left\|\mathcal{P}_{\mathcal{T}_{\mathcal{C}}(\boldsymbol{Q})}\left(\frac{\partial_{\mathrm{W}} f(\boldsymbol{Q})}{\partial_{\mathrm{W}} \boldsymbol{Q}^{\mathrm{T}}}\right)\right\|_{\mathrm{F}}^2 \cdot s + e_3(s) \tag{5.20}$$

where the error term $e_3(s)$ satisfies $e_3(s)/s \to 0$ as $s \to 0$.

As we can see from (5.20), moving along the direction of the unconstrained gradient $\mathrm{d}\boldsymbol{Q}$ of $f(\boldsymbol{Q})$ is an ascent direction also when the projection onto the constraint set $\mathcal{C}$ is taken into account. However, it is not the direction of the steepest ascent of the *composition* of the function f *and* the projection onto the constraint set. According to the derivation in Sect. 5.1.1, this direction is given by the projection of the Wirtinger derivative of $f(\boldsymbol{Q})$ onto the tangent cone at $\boldsymbol{Q}$:

$$\mathrm{d}\boldsymbol{Q} = \mathcal{P}_{\mathcal{T}_{\mathcal{C}}(\boldsymbol{Q})}\left(\frac{\partial_{\mathrm{W}} f(\boldsymbol{Q})}{\partial_{\mathrm{W}} \boldsymbol{Q}^{\mathrm{T}}}\right) \mathrm{d}\beta \tag{5.21}$$

with some differential $\mathrm{d}\beta > 0$. Equivalent to the precoder-based constrained utility maximization, the first-order Taylor expansion when moving along the projected gradient with step-size s yields the same linear term as in (5.20):

$$f\left(\mathcal{P}_{\mathcal{C}}\left(\boldsymbol{Q}+s\cdot\mathcal{P}_{\mathcal{T}_{\mathcal{C}}(\boldsymbol{Q})}\left(\frac{\partial_{\mathrm{W}} f(\boldsymbol{Q})}{\partial_{\mathrm{W}} \boldsymbol{Q}_k^{\mathrm{T}}}\right)\right)\right)=f(\boldsymbol{Q})+\left\|\mathcal{P}_{\mathcal{T}_{\mathcal{C}}(\boldsymbol{Q})}\left(\frac{\partial_{\mathrm{W}} f(\boldsymbol{Q})}{\partial_{\mathrm{W}} \boldsymbol{Q}^{\mathrm{T}}}\right)\right\|_{\mathrm{F}}^2\cdot s+e_4(s) \tag{5.22}$$

In turn, the error term $e_4(s)$ satisfies $e_4(s)/s \to 0$ as $s \to 0$.

To solve the constrained optimization problems which arise in the next two chapters, we investigate the following two recursions [cf. (5.16) and (5.17)]:

$$\boldsymbol{Q}^{(n+1)}=\mathcal{P}_{\mathcal{C}}\left(\boldsymbol{Q}^{(n)}+s^{(n)}\cdot\left.\frac{\partial_{\mathrm{W}} f(\boldsymbol{Q})}{\partial_{\mathrm{W}} \boldsymbol{Q}^{\mathrm{T}}}\right|_{\boldsymbol{Q}=\boldsymbol{Q}^{(n)}}\right) \tag{5.23}$$

$$\boldsymbol{Q}^{(n+1)}=\mathcal{P}_{\mathcal{C}}\left(\boldsymbol{Q}^{(n)}+s^{(n)}\cdot\mathcal{P}_{\mathcal{T}_{\mathcal{C}}(\boldsymbol{Q}^{(n)})}\left(\left.\frac{\partial_{\mathrm{W}} f(\boldsymbol{Q})}{\partial_{\mathrm{W}} \boldsymbol{Q}^{\mathrm{T}}}\right|_{\boldsymbol{Q}=\boldsymbol{Q}^{(n)}}\right)\right) \tag{5.24}$$

Step-size strategies on how to determine $s^{(n)} > 0$ to ensure convergence to a stationary point of (5.2) are discussed in Sect. 5.1.3 whereas the projections onto the constraint set $\mathcal{C}$ and the tangent cone are derived in Sect. 5.3.2.

5.1.3 Step-Size Strategies

Having determined the directions of the constrained or unconstrained steepest ascent, the iteration-dependent step-size $s^{(n)}$ in (5.16), (5.17), (5.23), and (5.24) has to be chosen appropriately in order to ensure convergence to a stationary point of the optimization in (5.2). In the following, different strategies for this choice are introduced and evaluated in terms of applicability.

5.1.3.1 Lipschitz Constant Based Step-Size Rule

In his 1964 paper, Goldstein derived a sufficient condition on the step-size $s^{(n)}$ to let the projected gradient algorithm converge to a stationary point for the case when the second derivative (in the sense of Gâteaux) of the function f is uniformly bounded in $\mathcal{C}$ [91]. For $g(t) := f(\boldsymbol{Z}+t\boldsymbol{V})$, Goldstein requires the existence of a constant L_0 such that the absolute value of the second derivative of g evaluated at $t = 0$ satisfies

$$\left|\left.\frac{\partial^2 g(t)}{\partial t^2}\right|_{t=0}\right| \leq L_0 \cdot \|\boldsymbol{V}\|_{\mathrm{F}}^2 \tag{5.25}$$

for all $\boldsymbol{Z} \in \mathcal{C}$. For any L_0 fulfilling (5.25), convergence of the gradient-projection algorithm to a limit point is achieved for any step-size sequence satisfying (see Appendix A13 for the derivation)

$$\varepsilon \leq s^{(n)} \leq \frac{2}{L_0} - \varepsilon \quad \forall n \geq n_0 \tag{5.26}$$

where $0 < \varepsilon \leq \frac{1}{L_0}$ and $n_0 < \infty$. More precisely, $\lim_{n\to\infty}(\mathbf{Z}^{(n+1)} - \mathbf{Z}^{(n)}) = \mathbf{0}$, the sequence $f(\mathbf{Z}^{(n)})$ is nondecreasing for $n \geq n_0$, and for compact $\mathcal{C}$ and continuously differentiable f, convergence to a stationary point of the optimization in (5.2) is ensured. For concave f, the global maximum of f in $\mathcal{C}$ is achieved. Note that (5.26) is only a sufficient condition that is not necessary. Due to the conservative bounds that are applied during its derivation (see Appendix A13) and since L_0 is chosen with respect to the complete constraint set $\mathcal{C}$ and may thus strongly deviate from the local situation, convergence is extremely slow. Additionally, note that condition (5.26) is sufficient for convergence only if $\boldsymbol{V}$ is chosen as the Wirtinger derivative of f. For the case that the direction $\boldsymbol{V}$ is chosen as the projection of the Wirtinger derivative onto the tangent cone at $\mathbf{Z}$, we were not able to find a simple rule as (5.26). Instead, the Armijo rule that is described in Sect. 5.1.3.3 should be used in this case.

Levitin and Polyak essentially come to the same result under slightly milder conditions [92]. Instead of being twice differentiable, they require the first derivative of f to be uniformly Lipschitz continuous in $\mathcal{C}$ which is less restrictive.

The problem for this Lipschitz constant based step-size rule is that some L_0 must be known to find an $s^{(n)}$ which satisfies (5.26), that may be hard to obtain. An example for a function f where such an L_0 can be derived rather easily is the sum rate in the MIMO MAC, see (4.12) with $\mathcal{U} = \{1, \ldots, K\}$. For $\boldsymbol{H} := [\boldsymbol{H}_1, \ldots, \boldsymbol{H}_K]$, block-diagonal $\boldsymbol{Q} = \text{blockdiag}\{\boldsymbol{Q}_k\}_{k=1}^K$, and for $\mathbf{Z} = \boldsymbol{Q}$, the function $g(t)$ reads for Hermitian $\boldsymbol{V}$ as

$$g(t) := \log_2 \left|\mathbf{I} + \boldsymbol{H}(\boldsymbol{Q} + t\boldsymbol{V})\boldsymbol{H}^{\mathrm{H}}\right|.$$

The first derivative of $g(t)$ with respect to t is given by [see Appendix A6]

$$\frac{\partial g(t)}{\partial t} = \frac{1}{\ln 2} \operatorname{tr}\left[\left(\mathbf{I} + \boldsymbol{H}(\boldsymbol{Q} + t\boldsymbol{V})\boldsymbol{H}^{\mathrm{H}}\right)^{-1} \boldsymbol{H}\boldsymbol{V}\boldsymbol{H}^{\mathrm{H}}\right]$$

and the absolute value of the second derivative evaluated at $t = 0$ reads as (see Appendix A6)

$$\begin{aligned}\left|\left.\frac{\partial^2 g(t)}{\partial t^2}\right|_{t=0}\right| &= \frac{1}{\ln 2} \operatorname{tr}\left[(\mathbf{I} + \boldsymbol{H}\boldsymbol{Q}\boldsymbol{H}^{\mathrm{H}})^{-1}\boldsymbol{H}\boldsymbol{V}\boldsymbol{H}^{\mathrm{H}}(\mathbf{I} + \boldsymbol{H}\boldsymbol{Q}\boldsymbol{H}^{\mathrm{H}})^{-1}\boldsymbol{H}\boldsymbol{V}\boldsymbol{H}^{\mathrm{H}}\right] \\ &\leq \frac{1}{\ln 2}\|\boldsymbol{H}\boldsymbol{V}\boldsymbol{H}^{\mathrm{H}}\|_{\mathrm{F}}^2 \leq \|\boldsymbol{V}\|_{\mathrm{F}}^2 \cdot \frac{1}{\ln 2}\|\boldsymbol{H}\|_{\mathrm{F}}^4.\end{aligned}$$

The second inequality results from the fact that the Frobenius norm is submultiplicative. In combination with (5.25), a possible value for L_0 is $L_0 = \frac{1}{\ln 2}\|\boldsymbol{H}\|_{\mathrm{F}}^4$.

5.1.3.2 Exact Line Search Based Step-Size Rule

McCormick and Tapia propose to choose the step-size $s^{(n)}$ in iteration n such that $f(\boldsymbol{Z}^{(n+1)})$ is maximized over all $s^{(n)} \geq 0$ which corresponds to a one-dimensional line search [96]. In order to achieve this, many orthogonal projections and function evaluations would have to be executed. Moreover, the maximum need not exist, i.e., the supremum need not be attained for finite $s^{(n)} < \infty$,which can be visualized easily for a ball-shaped constraint set for example. Hence, we do not make use of this kind of step-size rule.

5.1.3.3 Generalized Armijo Step-Size Rule

An inexact line search technique for unconstrained optimizations that is known as Armijo rule has first been extended to the case of constrained optimizations by Bertsekas [93]. We offer a more general transition from the unconstrained to the constrained optimization in this subsection. This transition must be done carefully to ensure convergence also in the constrained case. For the generalized Armijo rule, we parameterize the step-size $s^{(n)}$ in iteration n via

$$s^{(n)} = s_0^{(n)} \cdot \beta^{m^{(n)}} \tag{5.27}$$

where $0 < \beta < 1$ and $s_0^{(n)}$ must be bounded above and below by iteration-independent positive (nonzero) numbers. We will choose $s_0^{(n)}$ such that the Frobenius norm/the trace of the unconstrained gradient scaled by $s_0^{(n)}$ has the same order of magnitude as the Frobenius norm/the trace of $\boldsymbol{Z}^{(n)}$. This will reduce the dependency of the speed of convergence on the sum power $P_{\max}$ when $m^{(n)} \in \mathbb{N}_0$ and reduce the number of step-size corrections when $m^{(n)} \in \mathbb{Z}$. The choice of the exponent $m^{(n)}$ will be treated separately for the precoder-based and the covariance-based approach.

The basic idea of the Armijo rule is that the increase of a utility must be in relation to the step-size. More precisely, the utility increase must be at least as large as the step-size times the slope of the utility multiplied by some iteration independent constant σ that satisfies $0 < \sigma < 1$. This excludes the possibility of a marginal increase of the utility for large step-sizes as they may occur in the on-demand step-size reduction in the next subsection. The maximization of the *composition* of the utility f and the orthogonal projection can be regarded as an unconstrained optimization problem for which the Armijo rule, when used for the covariance matrices, leads to the exponent

$$m^{(n)} := \min_{m \in \mathbb{N}_0} m \quad \text{s.t.:}$$
$$f\left(\mathcal{P}_{\mathcal{C}}(\boldsymbol{Q}^{(n)} + s_0^{(n)} \beta^m \boldsymbol{V}^{(n)})\right) - f\left(\boldsymbol{Q}^{(n)}\right) \geq \sigma \cdot \left\| \mathcal{P}_{\mathcal{T}_{\mathcal{C}}(\boldsymbol{Q})}\left(\boldsymbol{V}^{(n)}\right)\right\|_{\mathrm{F}}^2 \cdot s_0^{(n)} \beta^m \tag{5.28}$$

where the slope (the squared Frobenius norm) follows from (5.20) and (5.22). Note that $\boldsymbol{V}^{(n)}$ must either be chosen as the gradient $\frac{\partial_{\mathrm{W}} f(\boldsymbol{Q})}{\partial_{\mathrm{W}} \boldsymbol{Q}^{\mathrm{T}}}|_{\boldsymbol{Q}=\boldsymbol{Q}^{(n)}}$ or as its projection onto the tangent cone to let (5.28) be valid. If $\boldsymbol{V}^{(n)}$ is the gradient of the unconstrained f, the Armijo rule in (5.28) would require to compute its projection onto the tangent cone. To overcome this additional computational burden, (5.28) may be modified by *under*-estimating the slope[5] of the constrained f which basically corresponds to reducing σ as shown in the following.

According to Proposition 5.1, the projection onto the tangent cone corresponds to the limit process

$$\mathcal{P}_{\mathcal{T}_{\mathcal{C}}(\boldsymbol{Q})}(\boldsymbol{V}) = \lim_{s \to 0} \frac{\mathcal{P}_{\mathcal{C}}(\boldsymbol{Q} + s\boldsymbol{V}) - \boldsymbol{Q}}{s}. \tag{5.29}$$

For finite s, the quotient $\frac{1}{s}\|\mathcal{P}_{\mathcal{C}}(\boldsymbol{Q} + s\boldsymbol{V}) - \boldsymbol{Q}\|_{\mathrm{F}}$ is nonincreasing in s [93, Lemma 2.2]. Thus, we can under-estimate the squared norm of the projected gradient on the right-hand side of the constraint in (5.28) via

$$\left\|\mathcal{P}_{\mathcal{T}_{\mathcal{C}}(\boldsymbol{Q})}\left(\boldsymbol{V}^{(n)}\right)\right\|_{\mathrm{F}}^2 \geq \frac{\|\mathcal{P}_{\mathcal{C}}(\boldsymbol{Q}^{(n)} + s_0^{(n)}\beta^m \boldsymbol{V}^{(n)}) - \boldsymbol{Q}^{(n)}\|_{\mathrm{F}}^2}{\left(s_0^{(n)}\beta^m\right)^2}$$

and come to a modified version of the Armijo rule in (5.28):

$$m^{(n)} := \min_{m \in \mathbb{N}_0} m \quad \text{s.t.:}$$

$$f\left(\mathcal{P}_{\mathcal{C}}(\boldsymbol{Q}^{(n)} + s_0^{(n)}\beta^m \boldsymbol{V}^{(n)})\right) - f\left(\boldsymbol{Q}^{(n)}\right) \geq \sigma \cdot \frac{\|\mathcal{P}_{\mathcal{C}}(\boldsymbol{Q}^{(n)} + s_0^{(n)}\beta^m \boldsymbol{V}^{(n)}) - \boldsymbol{Q}^{(n)}\|_{\mathrm{F}}^2}{s_0^{(n)}\beta^m}. \tag{5.30}$$

Note that (5.30) does not require the projection of the gradient $\boldsymbol{V}^{(n)}$ onto the tangent cone. Besides choosing $m^{(n)}$ only from $\mathbb{N}_0$ as proposed by Bertsekas in [93], it is possible to allow for larger step-sizes by choosing $m^{(n)}$ out of $\mathbb{Z}$. This further speeds up convergence but leads to additional projections and function evaluations.[6]

If the utility maximization is based on the precoders instead of the covariance matrices, the basic structure of (5.28) and (5.30) remains valid, only the expressions for the slope of the constrained f change, see (5.7) and (5.15). In this case, we may either use

[5] Clearly, the under-estimated slope must remain positive.

[6] Choosing $m^{(n)} \in \mathbb{Z}$ works fine if the exact slope (squared Frobenius norm of the tangent cone projection) as in (5.28) and (5.31) is used. However, if the under-estimated slope from (5.30) and (5.32) shall be used, this estimation should not depend on m. As a consequence, β^m should be replaced by $\beta^0 = 1$ in the right hand side of (5.30) and (5.32). Otherwise, a smallest integer m might not exist since the slope estimate would go to zero for $m \to -\infty$.

$$m^{(n)} := \min_{m \in \mathbb{N}_0} m \quad \text{s.t.:}$$

$$f\left(\mathcal{P}_{\mathcal{C}}(\boldsymbol{T}^{(n)} + s_0^{(n)} \beta^m \boldsymbol{V}^{(n)})\right) - f\left(\boldsymbol{T}^{(n)}\right) \geq \sigma \cdot 2 \left\| \mathcal{P}_{\mathcal{T}_{\mathcal{C}}(\boldsymbol{T})} \left(\boldsymbol{V}^{(n)}\right) \right\|_{\mathrm{F}}^2 \cdot s_0^{(n)} \beta^m \tag{5.31}$$

or the variant that does not require the projection onto the tangent cone

$$m^{(n)} := \min_{m \in \mathbb{N}_0} m \quad \text{s.t.:}$$

$$f\left(\mathcal{P}_{\mathcal{C}}(\boldsymbol{T}^{(n)} + s_0^{(n)} \beta^m \boldsymbol{V}^{(n)})\right) - f\left(\boldsymbol{T}^{(n)}\right) \geq \sigma \cdot 2 \, \frac{\left\| \mathcal{P}_{\mathcal{C}}(\boldsymbol{T}^{(n)} + s_0^{(n)} \beta^m \boldsymbol{V}^{(n)}) - \boldsymbol{T}^{(n)} \right\|_{\mathrm{F}}^2}{s_0^{(n)} \beta^m} \tag{5.32}$$

where $\boldsymbol{V}^{(n)}$ must either be the gradient $\frac{\partial_{\mathrm{w}} f(\boldsymbol{T})}{\partial_{\mathrm{w}} \boldsymbol{T}^*}|_{\boldsymbol{T}=\boldsymbol{T}^{(n)}}$ [in this case, (5.32) becomes attractive] or its projection onto the tangent cone. Again, $m^{(n)} \in \mathbb{Z}$ can be chosen to speed up convergence.

5.1.3.4 Open Loop Step-Size Rule

An exogenous choice for the step-size satisfying

$$\lim_{n \to \infty} s^{(n)} = 0 \quad \text{with} \quad \sum_{n=0}^{\infty} s^{(n)} = \infty$$

does not require knowledge of the Lipschitz constant of the Wirtinger derivative of f but still leads to a convergent sequence of the utility [98]. However, the speed of convergence is very low.

5.1.3.5 On-Demand Step-Size Reduction

In this approach, the step-size $s^{(n)}$ is given by the product of two scalars via

$$s^{(n)} = s_0^{(n)} \cdot r^{(n)}$$

where $s_0^{(n)}$ is bounded above and below by some positive constants. In addition, $r^{(n)}$ is nonincreasing in n and must be reduced if required until

$$f\left(\mathcal{P}_{\mathcal{C}}(\boldsymbol{Z}^{(n)} + s_0^{(n)} r^{(n)} \boldsymbol{V}^{(n)})\right) - f\left(\boldsymbol{Z}^{(n)}\right) > 0 \tag{5.33}$$

holds. This reduction can for example be obtained if $r^{(n)}$ follows from $r^{(n-1)}$ by multiplying $r^{(n-1)}$ by a positive constant that is smaller than one, or by increasing the inverse step-size $1/r^{(n-1)}$ by one until (5.33) holds. Note that, despite its appealing simplicity, this choice for the step-size may converge to a limit point that is *not*

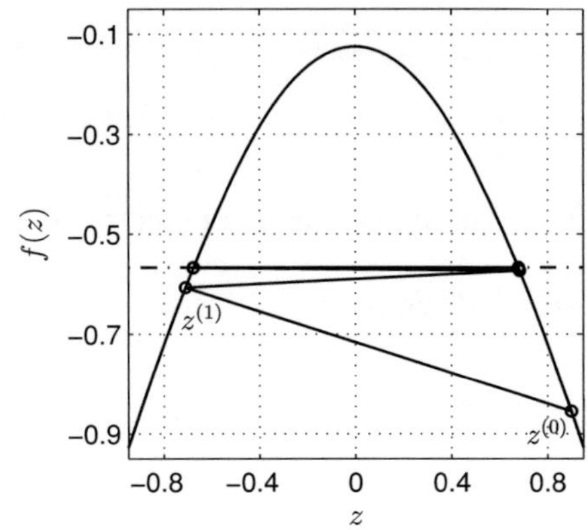

Fig. 5.1 Example for a concave function f where the on-demand step-size reduction fails

stationary! This behavior can occur in a one-dimensional, unconstrained problem already.

To visualize this, let f be piece-wise defined via [99] (see Fig. 5.1)

$$f(z) = \begin{cases} -\frac{1}{2}z^2 + \frac{1}{2}z & \text{for } z \leq -\frac{1}{2} \\ -z^2 - \frac{1}{8} & \text{for } |z| \leq \frac{1}{2} \\ -\frac{1}{2}z^2 - \frac{1}{2}z & \text{for } z \geq \frac{1}{2}. \end{cases}$$

Note that f is concave and continuously differentiable. Furthermore, $z = 0$ is the maximizer of f. Starting from any $z^{(0)} > 0.5$, the recurrence relation $z^{(n+1)} = z^{(n)} + s \cdot \frac{\partial f(z)}{\partial z}|_{z=z^{(n)}}$ leads for fixed s satisfying[7]

$$1 < s < \frac{4z^{(0)}}{1 + 2z^{(0)}}$$

to an increasing sequence $f(z^{(n)})$ that converges to the limit[8]

$$\lim_{n\to\infty} f(z^{(n)}) = f\left(\frac{s}{4 - 2s}\right) = -\frac{s(4 - s)}{2(4 - 2s)^2} \tag{5.34}$$

which is *not* a stationary point (i.e., not the global optimum in this case) of the maximization of f. However, convergence is only achieved with respect to the utility but not with respect to its argument:

---

[7] The lower bound $1 < s$ ensures that $z^{(1)} < -0.5$ in the first step and that the sign of $z^{(n)}$ is alternating in n. Due to the symmetry, $1 < s$ leads to $|z^{(n)}| > 0.5$ for all n. The upper bound enforces an increase of the utility. Due to the symmetry and concavity of f, an increase of the utility corresponds to the necessity that $|z^{(n+1)}| < |z^{(n)}|$ which leads to the upper bound for s.

[8] According to Footnote 7, the sign of $z^{(n)}$ is alternating and $|z^{(n)}| > 0.5\ \forall n$. Hence, the limit in (5.34) can be obtained by solving $-z = z + s \cdot \frac{\partial f(z)}{\partial z}$ for z.

$$\limsup_{n\to\infty} z^{(n)} = +\frac{s}{4-2s}$$
$$\liminf_{n\to\infty} z^{(n)} = -\frac{s}{4-2s}.$$

This example illustrates that an increase of the utility alone does not suffice to ensure convergence to a stationary point of the maximization of f. In principle, one could detect that $z^{(n)}$ does not converge and hence that the limit is not a stationary point. As a remedy, the step-size could be reduced to overcome this problem. Note that the Lipschitz constant of the derivative of f is 2. Hence, any step-size $s \leq 1-\varepsilon$ with $\varepsilon > 0$ would ensure convergence, see Sect. 5.1.3.1.

Although convergence to a nonstationary point can be observed in theory in the case of above piece-wise defined function f, it is unclear whether this will happen in the case of the (weighted) sum rate maximization which is twice continuously differentiable along any ray in contrast to the constructed example above. The *relaxed* gradient-projection algorithm described in the next subsection helps to avoid convergence to such limits that are not stationary if the on-demand step-size reduction strategy is chosen.

5.2 Relaxed Gradient-Projection Algorithm

The recursion formulas for the standard gradient-projection algorithm in (5.16)–(5.17) and (5.23)–(5.24) can be extended to the *relaxed* gradient-projection algorithm, which was first investigated by Demyanov and Rubinov in [100, p. 148] and was later applied for example in [101–103]. For reasons that will become clear later in this section, we apply this extension only to the covariance-based approach.

The two recursions in (5.23)–(5.24) can be stated in an abstract way as

$$\boldsymbol{Q}^{(n+1)} = \mathcal{P}_{\mathcal{C}}\left(\boldsymbol{Q}^{(n)} + s^{(n)}\boldsymbol{V}^{(n)}\right) \tag{5.35}$$

where the direction $\boldsymbol{V}^{(n)}$ either corresponds to the Wirtinger derivative of the unconstrained function $f(\boldsymbol{Q})$ or to its projection onto the tangent cone $\mathcal{T}_{\mathcal{C}}(\boldsymbol{Q}^{(n)})$. For the *relaxed* gradient-projection algorithm, the fixed-point iteration in (5.35) which is also called *Picard* iteration [104] is embedded into an additional outer line search and can thus be interpreted as a *Mann* iteration [104]:

$$\begin{aligned}\boldsymbol{Q}^{(n+1)} &= \boldsymbol{Q}^{(n)} + \omega^{(n)}\left[\mathcal{P}_{\mathcal{C}}\left(\boldsymbol{Q}^{(n)} + s^{(n)}\boldsymbol{V}^{(n)}\right) - \boldsymbol{Q}^{(n)}\right]\\ &= \left(1-\omega^{(n)}\right)\boldsymbol{Q}^{(n)} + \omega^{(n)}\mathcal{P}_{\mathcal{C}}\left(\boldsymbol{Q}^{(n)} + s^{(n)}\boldsymbol{V}^{(n)}\right).\end{aligned} \tag{5.36}$$

The iteration-dependent step-size $\omega^{(n)}$ satisfies $\omega^{(n)} \in [0,1]\ \forall n$. Since $\mathcal{P}_{\mathcal{C}}(\boldsymbol{Q}^{(n)} + s^{(n)}\boldsymbol{V}^{(n)}) \in \mathcal{C}$, it is easy to see that the convex combination $\boldsymbol{Q}^{(n+1)}$ is also element of $\mathcal{C}$

for any $\omega^{(n)} \in [0, 1]$. Moreover, if both $\boldsymbol{Q}^{(n)}$ and the projection $\mathcal{P}_{\mathcal{C}}(\boldsymbol{Q}^{(n)} + s^{(n)}\boldsymbol{V}^{(n)})$ satisfy the sum power constraint with equality and thus lie on the boundary of $\mathcal{C}$, then $\boldsymbol{Q}^{(n+1)}$ also consumes the full amount of transmit power which is a desired property since the weighted sum rate maximization is increasing in the sum power (see Appendix A8). This property of preserving the dissipation of the maximum available transmit power does not hold for $\omega^{(n)} \in (0, 1)$ when working on the precoders instead of the transmit covariance matrices[9] so that we apply the relaxed gradient-projection algorithm only for the covariance matrices.

Note that (5.2) corresponds to (5.35) for $\omega^{(n)} = 1$, whereas for $\omega^{(n)} = 0$, the result of the previous iteration is obtained. By varying $\omega^{(n)} \in [0, 1]$, every point on the convex combination of $\boldsymbol{Q}^{(n)}$ and $\mathcal{P}_{\mathcal{C}}(\boldsymbol{Q}^{(n)} + s^{(n)}\boldsymbol{V}^{(n)})$ can be reached. Since $\omega^{(n)}$ is bounded, the supremum of $f(\boldsymbol{Q}^{(n+1)})$ with respect to $\omega^{(n)}$ is attained for some $\omega^{(n)} \in [0, 1]$.

Several choices for the outer step-size $\omega^{(n)}$ are available in the literature, see for example [100–103]. However, we will implement a heuristic that was applied in a similar context for the sum power iterative water-filling algorithm in [105]. Let $\varphi_{n+1}(\omega)$ denote the utility f evaluated at $\boldsymbol{Q}^{(n+1)}$ from (5.2) when $\omega^{(n)}$ is replaced by ω, i.e.,

$$\varphi_{n+1}(\omega) := f\Big(\boldsymbol{Q}^{(n)} + \omega\big[\mathcal{P}_{\mathcal{C}}\big(\boldsymbol{Q}^{(n)} + s^{(n)}\boldsymbol{V}^{(n)}\big) - \boldsymbol{Q}^{(n)}\big]\Big). \tag{5.37}$$

We assume that the step-size $s^{(n)}$ has been chosen such that the standard gradient-projection algorithm (5.35) increases the utility from step n to $n + 1$, i.e., $\varphi_{n+1}(1) > \varphi_{n+1}(0)$. Clearly, this requires that $\boldsymbol{Q}^{(n)}$ is not a stationary point such that an increase of the utility is possible. The heuristic results from approximating $\varphi_{n+1}(\omega)$ in the closed interval $\omega \in [0, 1]$ by the parabola

$$\tilde{\varphi}_{n+1}(\omega) := c_1\omega^2 + c_2\omega + c_3. \tag{5.38}$$

If the apex ω_0 of the parabola $\tilde{\varphi}_{n+1}$ satisfies $\omega_0 \in (0, 1]$, and if, moreover, the exact function $\varphi_{n+1}(\omega_0)$ evaluated at the apex ω_0 of the parabola satisfies $\varphi_{n+1}(\omega_0) > \varphi_{n+1}(1)$, the relaxed gradient-projection algorithm achieves a larger increase of the utility from step n to step $n + 1$ than the standard gradient-projection algorithm with only two additional evaluations of the function f as we will see in the following.

The three coefficients c_1, c_2, and c_3 that govern the parabola $\tilde{\varphi}_{n+1}(\omega)$ in (5.38) are determined by means of three sampling points $\omega = 0$, $\omega = 1$, and $\omega = 0.5$. For $\omega = 0$, the value $\varphi_{n+1}(0)$ corresponds to evaluating the function f at the covariance matrix $\boldsymbol{Q}^{(n)}$ of iteration n. This result has already been computed in the previous iteration. As mentioned before, $\varphi_{n+1}(1)$ is the utility that is achieved by the standard gradient-projection update (5.35). Using these three sampling points, we obtain

[9] An exception to this is when the update direction is orthogonal to the precoder.

$$\begin{bmatrix} c_1 \\ c_2 \\ c_3 \end{bmatrix} = \begin{bmatrix} +2 & -4 & +2 \\ -1 & +4 & -3 \\ 0 & 0 & +1 \end{bmatrix} \cdot \begin{bmatrix} \varphi_{n+1}(1) \\ \varphi_{n+1}(0.5) \\ \varphi_{n+1}(0) \end{bmatrix}. \tag{5.39}$$

If the parabola $\tilde{\varphi}_{n+1}$ is supposed to have a unique maximum, it must be strictly concave, which means that

$$c_1 < 0 \quad \Leftrightarrow \quad \frac{\varphi_{n+1}(1) + \varphi_{n+1}(0)}{2} < \varphi_{n+1}(0.5)$$

needs to hold. The apex ω_0 of the parabola is given by

$$\omega_0 = -\frac{c_2}{2c_1} = \frac{\varphi_{n+1}(1) - 4\varphi_{n+1}(0.5) + 3\varphi_{n+1}(0)}{4\varphi_{n+1}(1) - 8\varphi_{n+1}(0.5) + 4\varphi_{n+1}(0)}. \tag{5.40}$$

Note that ω_0 is only relevant for $0 < \omega_0 \le 1$. For $c_1 < 0$ and $\varphi_{n+1}(1) > \varphi_{n+1}(0)$, the inequality $0 < \omega_0$ is automatically fulfilled since $c_2 > 0$ holds in this case. If $0 < \omega_0 < 1$, the maximum $\tilde{\varphi}_{n+1}(\omega_0)$ of the interpolated parabola is larger than $\varphi_{n+1}(1)$. However, we also need to verify that the true utility $\varphi_{n+1}(\omega_0)$ is larger than $\varphi_{n+1}(1)$. If so, we set $\omega^{(n)} = \omega_0$. Otherwise, we set $\omega^{(n)} = 1$ which corresponds to the standard gradient-projection algorithm. In this case, the relaxed version does not yield any gain in this step n. Algorithm 1 summarizes how the step-size $\omega^{(n)}$ of the outer line search is computed.

The convergence to a nonstationary point of the on-demand step-size reduction for the constructed example in Sect. 5.1.3.5 resulted from only a very small increase of the utility despite the huge variation of the optimization variable within one iteration. In contrast, the relaxed gradient projection algorithm performs an additional (inexact) line-search to overcome this problem.

Algorithm 5.1 Outer step-size computation for the relaxed gradient-projection algorithm

Require: Standard gradient-projection leads to increase of utility $\Leftrightarrow \varphi_{n+1}(1) > \varphi_{n+1}(0)$

$\omega^{(n)} \leftarrow 1$ — initialize outer step-size under the assumption that standard PG will be used
Compute $\varphi_{n+1}(0)$, $\varphi_{n+1}(0.5)$, and $\varphi_{n+1}(1)$ via (5.37) — evaluate function at the sampling points
$c_1 \leftarrow 2\varphi_{n+1}(1) - 4\varphi_{n+1}(0.5) + 2\varphi_{n+1}(0)$ — compute first coefficient of the interpolating parabola
$c_2 \leftarrow -\varphi_{n+1}(1) + 4\varphi_{n+1}(0.5) - 3\varphi_{n+1}(0)$ — compute second coefficient
if $c_1 \ge 0$ **then**
    exit — Exit if parabola is not strictly concave
end if
$\omega_0 \leftarrow -\frac{c_2}{2c_1}$ — compute apex of the parabola
if $\omega_0 \ge 1$ **then**
    exit — exit, if apex outside interval (0, 1)
end if
if $\varphi_{n+1}(\omega_0) > \varphi_{n+1}(1)$ **then**
    $\omega^{(n)} \leftarrow \omega_0$ — set outer step-size if true utility at ω_0 is better than that of standard PG
end if

5.3 Orthogonal Projection of a Block-Diagonal Matrix

The orthogonal projection of a matrix onto a set is one task of a more generally formulated matrix nearness problem [106]. There, the minimum distance of a matrix $\boldsymbol{Z}$ to a set $\mathcal{C}$ defining a class of matrices with specific properties has to be computed. Moreover, efficient algorithms to determine the element in $\mathcal{C}$ which achieves the minimum distance are of interest. Such matrix nearness problems arise for example in the context of rounding errors due to the finite word length precision [106]. A trivial example is a usually real-valued Hermitian form which has some very small, but nonzero imaginary part due to finite word length precision. Since the property of being real-valued is a priori known, the outcome of the Hermitian form computation should be improved by mapping it to that element of the real numbers with minimum distance (by simply discarding the imaginary part). Another example is when one wants to determine the minimum distance of a matrix to a class of matrices having the property of being rank-deficient or ill-conditioned.

In our context, the set $\mathcal{C}$ defines the class of matrices obeying the sum power constraint imposed in the MIMO BC and also in the MIMO MAC due to the duality. When operating on the transmit covariance matrices, the elements of $\mathcal{C}$ also have to be positive semi-definite. In Sect. 5.3.1, the orthogonal projection onto the set $\mathcal{C}$ is computed when working on the precoding matrices $\boldsymbol{T}_1, \ldots, \boldsymbol{T}_K$ of the dual MIMO MAC, whereas Sect. 5.3.2 derives the projection when working on the transmit covariance matrices $\boldsymbol{Q}_1, \ldots, \boldsymbol{Q}_K$.

5.3.1 *Orthogonal Projection of Precoding Matrices*

For the block-diagonal precoding matrix $\boldsymbol{T} := \mathrm{blockdiag}\{\boldsymbol{T}_k\}_{k=1}^K$, the sum power constraint $P_{\mathrm{diss}}^{\mathrm{MAC}} \leq P_{\max}$ defines the constraint set $\mathcal{C}$ via [see (2.6)]

$$\mathcal{C} := \left\{\boldsymbol{T} \in \mathcal{B}_{\boldsymbol{T}} \middle| \|\boldsymbol{T}\|_{\mathrm{F}}^2 \leq P_{\max}\right\} \subset \mathbb{C}^{M \times L}$$

which means that $\mathcal{C}$ is a ball. For a compact notation, we define the set of block-diagonal matrices whose main diagonal blocks have appropriate dimensions via

$$\mathcal{B}_{\boldsymbol{T}} := \left\{\boldsymbol{T} \middle| \boldsymbol{T} = \mathrm{blockdiag}\{\boldsymbol{T}_k\}_{k=1}^K \wedge \boldsymbol{T}_k \in \mathbb{C}^{M_k \times L_k}\ \forall k\right\}.$$

5.3.1.1 Constraint Set Projection

Let $\boldsymbol{T}' := \mathrm{blockdiag}\{\boldsymbol{T}'_k\}_{k=1}^K \in \mathcal{B}_{\boldsymbol{T}}$ be a block-diagonal matrix out of $\mathcal{B}_{\boldsymbol{T}}$. The projection of $\boldsymbol{T}'$ onto the constraint set $\mathcal{C}$ is given by

$$\mathcal{P}_{\mathcal{C}}(\boldsymbol{T}') = \boldsymbol{T}' + \check{\boldsymbol{E}} \tag{5.41}$$

where $\check{\boldsymbol{E}}$ is the minimizer of the following optimization:

$$\check{\boldsymbol{E}} := \underset{\boldsymbol{E}}{\operatorname{argmin}} \, \|\boldsymbol{E}\|_{\mathrm{F}}^2 \quad \text{s.t.: } \boldsymbol{T}' + \boldsymbol{E} \in \mathcal{C} \tag{5.42}$$

Since $\mathcal{C}$ is convex and since the Frobenius norm is used as distance measure, the optimizer is unique, cf. the closest point theorem [94, Theorem 2.4.1, p. 50]. The Lagrangian function associated to (5.42) reads as

$$\mathcal{L}(\boldsymbol{E}, \nu) = \operatorname{tr}(\boldsymbol{E}^{\mathrm{H}}\boldsymbol{E}) + \nu\left[\operatorname{tr}\left((\boldsymbol{T}' + \boldsymbol{E})^{\mathrm{H}}(\boldsymbol{T}' + \boldsymbol{E})\right) - P_{\max}\right] \tag{5.43}$$

where $\nu \geq 0$ is the Lagrangian multiplier corresponding to the sum power constraint that defines $\mathcal{C}$. From setting the Wirtinger derivative of $\mathcal{L}(\boldsymbol{E}, \nu)$ with respect to $\boldsymbol{E}^*$ to zero and evaluating the result at the optimum variables which are denoted via $\check{(\cdot)}$, we find

$$\check{\boldsymbol{E}} = -\frac{\check{\nu}}{1+\check{\nu}}\boldsymbol{T}'.$$

Obviously, $\check{\nu} = 0 \Leftrightarrow \check{\boldsymbol{E}} = \boldsymbol{0}$ holds if and only if $\|\boldsymbol{T}'\|_{\mathrm{F}}^2 \leq P_{\max}$ making the projection superfluous. Otherwise, the constraint in (5.42) is active, i.e., $\check{\nu} > 0$, yielding

$$\boldsymbol{T}' + \check{\boldsymbol{E}} = \frac{1}{1+\check{\nu}}\boldsymbol{T}' = \frac{\sqrt{P_{\max}}}{\|\boldsymbol{T}'\|_{\mathrm{F}}} \cdot \boldsymbol{T}'$$

which means that the orthogonal projection simply rescales $\boldsymbol{T}'$ such that $\boldsymbol{T}' + \check{\boldsymbol{E}} \in \partial\mathcal{C}$, i.e., the projection is onto the sphere $\partial\mathcal{C}$. Summing up, the orthogonal projection of the block-diagonal matrix $\boldsymbol{T}'$ onto $\mathcal{C}$ reads as

$$\mathcal{P}_{\mathcal{C}}(\boldsymbol{T}') = \begin{cases} \boldsymbol{T}' & \text{if } \|\boldsymbol{T}'\|_{\mathrm{F}}^2 \leq P_{\max} \Leftrightarrow \boldsymbol{T}' \in \mathcal{C} \\ \frac{\sqrt{P_{\max}}}{\|\boldsymbol{T}'\|_{\mathrm{F}}} \cdot \boldsymbol{T}' & \text{if } \|\boldsymbol{T}'\|_{\mathrm{F}}^2 \geq P_{\max}. \end{cases} \tag{5.44}$$

5.3.1.2 Tangent Cone Projection

Besides the projection onto the set $\mathcal{C}$, the recursion formula (5.17) for updating the precoders also entails an orthogonal projection onto the tangent cone. Given a block-diagonal matrix $\boldsymbol{T} \in \mathcal{C}$, the precoder update $s\boldsymbol{V}$ in the direction $\boldsymbol{V}$, which is also block-diagonal, is projected onto the tangent cone $\mathcal{T}_{\mathcal{C}}(\boldsymbol{T})$ at $\boldsymbol{T}$. Since we deal only with convex sets $\mathcal{C}$, the tangent cone corresponds to the closure of the set of feasible directions [107, Proposition 4.6.2, p. 250]. Thus, we define the tangent cone $\mathcal{T}_{\mathcal{C}}(\boldsymbol{T})$ for convex $\mathcal{C}$ as follows [107, Definition 4.6.1, p. 248]:

Definition 5.1 Let $\mathcal{C}$ be a closed, convex set and let $\boldsymbol{Z} \in \mathcal{C}$. A matrix $\boldsymbol{V}$ is said to be a feasible direction of $\mathcal{C}$ at $\boldsymbol{Z}$ if there exists an $\bar{\alpha} > 0$ such that $\boldsymbol{Z} + \alpha \boldsymbol{V} \in \mathcal{C}$ for all $\alpha \in [0, \bar{\alpha}]$. The closure of the set of all feasible directions $\mathcal{F}_{\mathcal{C}}(\boldsymbol{Z})$ of $\mathcal{C}$ at $\boldsymbol{Z}$ is called the tangent cone $\mathcal{T}_{\mathcal{C}}(\boldsymbol{Z}) = \mathrm{cl}\big(\mathcal{F}_{\mathcal{C}}(\boldsymbol{Z})\big)$.

From the definition directly follows, that if $\boldsymbol{T} \in \mathrm{int}(\mathcal{C})$, then the tangent cone $\mathcal{T}_{\mathcal{C}}(\boldsymbol{T})$ corresponds to the set of all block-diagonal matrices whose main diagonal blocks have the same dimension as those of $\boldsymbol{T}$ and the projection onto the tangent cone is given by the identity mapping, i.e., $\mathcal{P}_{\mathcal{T}_{\mathcal{C}}(\boldsymbol{T})}(s\boldsymbol{V}) = s\boldsymbol{V}$. Thus, the projection onto the tangent cone is only interesting for $\boldsymbol{T} \in \partial\mathcal{C}$, i.e., when $\boldsymbol{T}$ is on the boundary of $\mathcal{C}$, which in our case means that $\|\boldsymbol{T}\|_{\mathrm{F}}^2 = P_{\max}$.

For $\boldsymbol{T} \in \partial\mathcal{C}$, the condition $\boldsymbol{T} + \alpha\boldsymbol{V} \in \mathcal{C}$ requires the direction $\boldsymbol{V}$ to be block-diagonal. For such $\boldsymbol{V}$,

$$\begin{aligned}\boldsymbol{T} + \alpha\boldsymbol{V} \in \mathcal{C} &\Leftrightarrow \|\boldsymbol{T} + \alpha\boldsymbol{V}\|_{\mathrm{F}}^2 \leq P_{\max}\\ &\Leftrightarrow P_{\max} + \alpha^2\|\boldsymbol{V}\|_{\mathrm{F}}^2 + 2\alpha\mathrm{Re}\big[\mathrm{tr}(\boldsymbol{T}^{\mathrm{H}}\boldsymbol{V})\big] \leq P_{\max}\\ &\Leftrightarrow \alpha^2\|\boldsymbol{V}\|_{\mathrm{F}}^2 + 2\alpha\mathrm{Re}\big[\mathrm{tr}(\boldsymbol{T}^{\mathrm{H}}\boldsymbol{V})\big] \leq 0.\end{aligned}$$

Since $\alpha > 0$ can be made arbitrarily small, the open set of feasible directions is given by

$$\mathcal{F}_{\mathcal{C}}(\boldsymbol{T}) = \{\boldsymbol{0}\} \cup \left\{\boldsymbol{V} \in \mathcal{B}_{\boldsymbol{T}} \middle| \mathrm{Re}\big[\mathrm{tr}(\boldsymbol{T}^{\mathrm{H}}\boldsymbol{V})\big] < 0\right\}.$$

The tangent cone $\mathcal{T}_{\mathcal{C}}(\boldsymbol{T})$ is then given by the closure of $\mathcal{F}_{\mathcal{C}}(\boldsymbol{T})$:

$$\mathcal{T}_{\mathcal{C}}(\boldsymbol{T}) = \left\{\boldsymbol{V} \in \mathcal{B}_{\boldsymbol{T}} \middle| \mathrm{Re}\big[\mathrm{tr}(\boldsymbol{T}^{\mathrm{H}}\boldsymbol{V})\big] \leq 0\right\}. \tag{5.45}$$

At the point $\boldsymbol{T} \in \partial\mathcal{C}$, the projection of a direction matrix $\boldsymbol{V}$ onto the tangent cone $\mathcal{T}_{\mathcal{C}}(\boldsymbol{T})$ can be expressed as

$$\mathcal{P}_{\mathcal{T}_{\mathcal{C}}(\boldsymbol{T})}(\boldsymbol{V}) = \boldsymbol{V} + \check{\boldsymbol{E}}$$

where $\check{\boldsymbol{E}}$ is defined as follows:

$$\check{\boldsymbol{E}} := \underset{\boldsymbol{E}}{\mathrm{argmin}}\ \|\boldsymbol{E}\|_{\mathrm{F}}^2 \quad \text{s.t.:}\ \boldsymbol{V} + \boldsymbol{E} \in \mathcal{T}_{\mathcal{C}}(\boldsymbol{T}). \tag{5.46}$$

The Lagrangian function associated to (5.46) reads by means of (5.45) as[10]

$$\mathcal{L}(\boldsymbol{E}, \nu) = \|\boldsymbol{E}\|_{\mathrm{F}}^2 + \nu\mathrm{Re}\big(\mathrm{tr}\left[\boldsymbol{T}^{\mathrm{H}}(\boldsymbol{V} + \boldsymbol{E})\right]\big)$$

with $\nu \geq 0$. Setting the Wirtinger derivative of $\mathcal{L}(\boldsymbol{E}, \nu)$ with respect to $\boldsymbol{E}^*$ to zero yields

[10] The block-diagonal structure constraint for $\boldsymbol{E}$ is automatically fulfilled which can be seen from the solution (5.47).

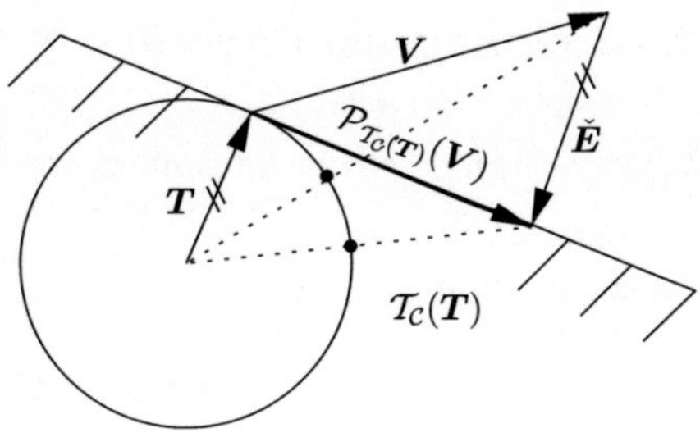

Fig. 5.2 Projection onto the tangent cone

$$\check{\boldsymbol{E}} = -\frac{\check{\nu}}{2}\boldsymbol{T}.$$

In turn, $\check{\nu} = 0 \Leftrightarrow \check{\boldsymbol{E}} = \boldsymbol{0}$ holds if and only if $\mathrm{Re}[\mathrm{tr}(\boldsymbol{T}^{\mathrm{H}}\boldsymbol{V})] \leq 0$, which for block-diagonal $\boldsymbol{V}$ means that $\boldsymbol{V} \in \mathcal{T}_\mathcal{C}(\boldsymbol{T})$. Otherwise, the trace constraint $\mathrm{Re}(\mathrm{tr}[\boldsymbol{T}^{\mathrm{H}}(\boldsymbol{V} - \check{\nu}/2 \cdot \boldsymbol{T})]) \leq 0$ is active and equality holds, from which we obtain

$$\check{\boldsymbol{E}} = -\frac{\mathrm{Re}\big[\,\mathrm{tr}(\boldsymbol{T}^{\mathrm{H}}\boldsymbol{V})\big]}{P_{\max}} \cdot \boldsymbol{T} \tag{5.47}$$

i.e., the projection is onto the tangent plane at $\boldsymbol{T}$. Summing up, the orthogonal projection of a block-diagonal $\boldsymbol{V}$ onto the tangent cone $\mathcal{T}_\mathcal{C}(\boldsymbol{T})$ is given by

$$\mathcal{P}_{\mathcal{T}_\mathcal{C}(\boldsymbol{T})}(\boldsymbol{V}) = \begin{cases} \boldsymbol{V} & \text{for } \boldsymbol{T} \in \mathrm{int}(\mathcal{C}) \text{ or } \boldsymbol{T} \in \partial\mathcal{C} \wedge \mathrm{Re}[\mathrm{tr}(\boldsymbol{T}^{\mathrm{H}}\boldsymbol{V})] \leq 0 \\ \boldsymbol{V} - \frac{\mathrm{Re}[\mathrm{tr}(\boldsymbol{T}^{\mathrm{H}}\boldsymbol{V})]}{P_{\max}} \cdot \boldsymbol{T} & \text{for } \boldsymbol{T} \in \partial\mathcal{C} \wedge \mathrm{Re}[\mathrm{tr}(\boldsymbol{T}^{\mathrm{H}}\boldsymbol{V})] \geq 0. \end{cases} \tag{5.48}$$

A graphical visualization of the tangent-cone projection is shown in Fig. 5.2 which also reveals that larger steps on the boundary of $\mathcal{C}$ are possible when the tangent cone search direction is used instead of the standard Wirtinger derivative, i.e., the steepest ascent property of the projected Wirtinger derivative is illustrated.

Synopsis: For the precoder-based gradient-projection algorithm, the sum power limitation in the dual MIMO MAC leads to a ball-shaped constraint set $\mathcal{C}$. The orthogonal projection of a block-diagonal matrix $\boldsymbol{T}' \notin \mathcal{C} \Leftrightarrow \|\boldsymbol{T}'\|_{\mathrm{F}}^2 > P_{\max}$ onto the set $\mathcal{C}$ is obtained by rescaling $\boldsymbol{T}'$ such that its squared Frobenius norm matches $P_{\max}$. For $\boldsymbol{T}' \in \mathcal{C} \Leftrightarrow \|\boldsymbol{T}'\|_{\mathrm{F}}^2 \leq P_{\max}$, the orthogonal projection onto $\mathcal{C}$ is the identity mapping.

The tangent cone of a block-diagonal matrix $\boldsymbol{T}$ with constraint set $\mathcal{C}$ is the closure of the set of feasible directions for convex $\mathcal{C}$. For $\boldsymbol{T} \in \mathrm{int}(\mathcal{C}) \Leftrightarrow \|\boldsymbol{T}\|_{\mathrm{F}}^2 < P_{\max}$, the tangent cone corresponds to the set of all block-diagonal matrices having the same structure as $\boldsymbol{T}$, and the projection of a block-diagonal matrix $\boldsymbol{V}$ onto the tangent cone simply yields $\boldsymbol{V}$. For $\boldsymbol{T} \in \partial\mathcal{C} \Leftrightarrow \|\boldsymbol{T}\|_{\mathrm{F}}^2 = P_{\max}$, the tangent cone is a half-space. In this case, the projection of $\boldsymbol{V}$ onto the tangent cone either corresponds to the projection onto the tangent plane at $\boldsymbol{T}$ when $\boldsymbol{V}$ is not element of the half-space, or to the identity mapping, when $\boldsymbol{V}$ was already element of the half-space.

5.3.2 Orthogonal Projection of Covariance Matrices

When working on the transmit covariance matrices that are stored in the block-diagonal matrix $\boldsymbol{Q} := \mathrm{blockdiag}\{\boldsymbol{Q}_k\}_{k=1}^K \in \mathbb{C}^{M\times M}$, the sum power constraint $P_{\mathrm{diss}}^{\mathrm{MAC}} \leq P_{\max}$ and the positive semi-definiteness constraints define the constraint set $\mathcal{C}$ via

$$\mathcal{C} := \left\{\boldsymbol{Q} \in \mathcal{B}_{\boldsymbol{Q}} \middle| \mathrm{tr}(\boldsymbol{Q}) \leq P_{\max} \wedge \boldsymbol{Q} \succcurlyeq \boldsymbol{0}\right\} \subset \mathbb{S}^M \tag{5.49}$$

where the set $\mathcal{B}_{\boldsymbol{Q}}$ of block-diagonal matrices of appropriate dimensions is defined via

$$\mathcal{B}_{\boldsymbol{Q}} := \left\{\boldsymbol{Q} \middle| \boldsymbol{Q} = \mathrm{blockdiag}\{\boldsymbol{Q}_k\}_{k=1}^K \wedge \boldsymbol{Q}_k \in \mathbb{C}^{M_k\times M_k}\ \forall k\right\}$$

for a compact notation.

5.3.2.1 Constraint Set Projection

Let $\boldsymbol{Q}' \in \mathcal{B}_{\boldsymbol{Q}}$ be a block-diagonal matrix which we assume to be Hermitian. Its projection onto the constraint set $\mathcal{C}$ in (5.49) is given by

$$\mathcal{P}_{\mathcal{C}}(\boldsymbol{Q}') = \boldsymbol{Q}' + \check{\boldsymbol{E}}$$

where $\check{\boldsymbol{E}} = \check{\boldsymbol{E}}^{\mathrm{H}}$ is of course Hermitian and is given by[11]

$$\check{\boldsymbol{E}} := \underset{\boldsymbol{E}}{\mathrm{argmin}}\ \frac{1}{2}\|\boldsymbol{E}\|_{\mathrm{F}}^2 \quad \text{s.t.: } \boldsymbol{Q}' + \boldsymbol{E} \succcurlyeq \boldsymbol{0} \ \text{ and } \ \mathrm{tr}(\boldsymbol{Q}' + \boldsymbol{E}) \leq P_{\max}. \tag{5.50}$$

We published the solution to above optimization for the special case $\boldsymbol{Q}' \succcurlyeq \boldsymbol{0}$ and $\mathrm{tr}(\boldsymbol{Q}') > P_{\max}$ in [108]. The same result was obtained independently by Liu et al. in [109]. However, since we also investigate the case of moving along the direction of the gradient *with* projection onto the tangent cone [see (5.24)], we have to solve (5.50) also for the case when $\boldsymbol{Q}'$ is not positive semi-definite. The solution to (5.50) for general $\boldsymbol{Q}' = \boldsymbol{Q}'^{\mathrm{H}} \in \mathcal{B}_{\boldsymbol{Q}}$ will be derived in the following.

To this end, we define the eigenvalue decomposition $\boldsymbol{Q}' = \boldsymbol{U}\boldsymbol{\Lambda}\boldsymbol{U}^{\mathrm{H}}$ with unitary $\boldsymbol{U}$ and diagonal $\boldsymbol{\Lambda} = \mathrm{diag}\{\lambda_m\}_{m=1}^M$, where $\lambda_m \in \mathbb{R}\ \forall m$ may also be negative since $\boldsymbol{Q}'$ only need to be Hermitian. The Lagrangian function associated to (5.50) reads as

$$\mathcal{L}(\boldsymbol{E}, \nu, \boldsymbol{\Xi}) = \frac{1}{2}\|\boldsymbol{E}\|_{\mathrm{F}}^2 + \nu\left[\mathrm{tr}(\boldsymbol{Q}' + \boldsymbol{E}) - P_{\max}\right] - \mathrm{tr}\left[(\boldsymbol{Q}' + \boldsymbol{E})\boldsymbol{\Xi}\right]$$

---

[11] The factor $\frac{1}{2}$ in front of the distance $\|\boldsymbol{E}\|_{\mathrm{F}}^2$ in (5.50) does not change the minimizer $\check{\boldsymbol{E}}$. Its only purpose is to get rid of the factor $\frac{1}{2}$ in front of the Lagrangian multiplier ν when solving for $\check{\boldsymbol{E}}$ after setting the Wirtinger derivative with respect to $\boldsymbol{E}^{\mathrm{T}}$ to zero. This is due to the fact that $\frac{\partial_{\mathrm{w}}}{\partial_{\mathrm{w}}\boldsymbol{E}^{\mathrm{T}}}\|\boldsymbol{E}\|_{\mathrm{F}}^2 = 2\boldsymbol{E}$ for $\boldsymbol{E} = \boldsymbol{E}^{\mathrm{H}}$ with a pre-factor of 2, see (A28) in Appendix A12.

with $\nu \geq 0$ and $\boldsymbol{\Xi} \succcurlyeq \mathbf{0}$. The first KKT condition follows from setting the Wirtinger derivative of $\mathcal{L}(\boldsymbol{E}, \nu, \boldsymbol{\Xi})$ with respect to $\boldsymbol{E}^{\mathrm{T}}$ to zero and evaluating at the optimum variables. This yields

$$\check{\boldsymbol{E}} = -\check{\nu}\mathbf{I}_M + \check{\boldsymbol{\Xi}}. \tag{5.51}$$

The remaining KKT conditions read as

$$\boldsymbol{Q}' + \check{\boldsymbol{E}} \succcurlyeq \mathbf{0} \tag{5.52}$$

$$\check{\boldsymbol{\Xi}}(\boldsymbol{Q}' + \check{\boldsymbol{E}}) = \mathbf{0} \tag{5.53}$$

$$\mathrm{tr}(\boldsymbol{Q}' + \check{\boldsymbol{E}}) \leq P_{\mathrm{max}} \tag{5.54}$$

$$\check{\nu}\big[\mathrm{tr}(\boldsymbol{Q} + \check{\boldsymbol{E}}) - P_{\mathrm{max}}\big] = 0. \tag{5.55}$$

From (5.51) and (5.53), we find

$$\check{\boldsymbol{\Xi}}(\boldsymbol{Q}' - \check{\nu}\mathbf{I}_M + \check{\boldsymbol{\Xi}}) = \mathbf{0} \quad \Rightarrow \quad \check{\boldsymbol{\Xi}}^2 - \check{\nu}\check{\boldsymbol{\Xi}} + \check{\boldsymbol{\Xi}}\boldsymbol{Q}' = \mathbf{0} \tag{5.56}$$

from which we conclude that $\check{\boldsymbol{\Xi}}\boldsymbol{Q}'$ must be Hermitian. This implies that $\check{\boldsymbol{\Xi}}$ and $\boldsymbol{Q}'$ commute since both are Hermitian, which itself means that $\check{\boldsymbol{\Xi}}$ has the same eigenbasis $\boldsymbol{U}$ as $\boldsymbol{Q}'$ has. Using the eigenvalue decomposition

$$\check{\boldsymbol{\Xi}} = \boldsymbol{U}\check{\boldsymbol{\Sigma}}\boldsymbol{U}^{\mathrm{H}} \quad \text{with} \quad \check{\boldsymbol{\Sigma}} = \mathrm{diag}\{\check{\sigma}_m\}_{m=1}^{M} \tag{5.57}$$

the condition $\check{\boldsymbol{\Xi}} \succcurlyeq \mathbf{0}$ implies $\check{\sigma}_m \geq 0\ \forall m$. Due to the common eigenbasis, (5.56) can be reformulated element-wise as

$$\check{\sigma}_m(\check{\sigma}_m - \check{\nu} + \lambda_m) = 0 \quad \forall m$$

with the two possible solutions $\check{\sigma}_m = 0$ and $\check{\sigma}_m = \check{\nu} - \lambda_m$, where the second one is only possible, if $\check{\nu} - \lambda_m \geq 0$ since $\check{\sigma}_m \geq 0$ has to hold. To see that $\check{\sigma}_m = \check{\nu} - \lambda_m$ is the desired solution when $\check{\nu} - \lambda_m > 0$, we rewrite (5.52) by means of (5.51) and (5.57) as

$$\lambda_m - \check{\nu} + \check{\sigma}_m \geq 0 \quad \forall m.$$

Now if $\lambda_m - \check{\nu} < 0$, we find $\check{\sigma}_m > 0$ in order to satisfy above inequality, and hence, the second expression $\check{\sigma}_m = \check{\nu} - \lambda_m$ is the correct one. Instead, if $\lambda_m - \check{\nu} > 0$, the second expression would be negative and thus, $\check{\sigma}_m = 0$. Summing up, we have

$$\check{\sigma}_m = [\check{\nu} - \lambda_m]_+ \quad \forall m \tag{5.58}$$

where the operator $[\cdot]_+ := \max\{0, \cdot\}$ sets negative values to zero. Using (5.51), (5.57), and (5.58), we obtain

$$\begin{aligned}\mathcal{P}_{\mathcal{C}}(\boldsymbol{Q}') = \boldsymbol{Q}' + \check{\boldsymbol{E}} &= \boldsymbol{U}\big(\boldsymbol{\Lambda} - \check{\nu}\mathbf{I}_M + \check{\boldsymbol{\Sigma}}\big)\boldsymbol{U}^{\mathrm{H}} \\ &= \boldsymbol{U}\big(\boldsymbol{\Lambda} - \check{\nu}\mathbf{I}_M + [\check{\nu}\mathbf{I}_M - \boldsymbol{\Lambda}]_+\big)\boldsymbol{U}^{\mathrm{H}} \\ &= \boldsymbol{U}[\boldsymbol{\Lambda} - \check{\nu}\mathbf{I}_M]_+\boldsymbol{U}^{\mathrm{H}}.\end{aligned} \tag{5.59}$$

The sum power constraint in (5.54) can be reformulated using (5.59) as

$$\mathrm{tr}(\boldsymbol{Q}' + \check{\boldsymbol{E}}) = \sum_{m=1}^{M} [\lambda_m - \check{\nu}]_+ \leq P_{\max}. \tag{5.60}$$

The Lagrangian multiplier $\check{\nu}$ can only be zero if $\sum_{m=1}^{M}[\lambda_m]_+ \leq P_{\max}$. In this case, (5.51) and (5.59) lead to

$$\check{\boldsymbol{E}} = \boldsymbol{U}[-\boldsymbol{\Lambda}]_+\boldsymbol{U}^{\mathrm{H}} \Rightarrow \mathcal{P}_{\mathcal{C}}(\boldsymbol{Q}') = \boldsymbol{U}[\boldsymbol{\Lambda}]_+\boldsymbol{U}^{\mathrm{H}} \quad \text{for} \sum_{m=1}^{M}[\lambda_m]_+ \leq P_{\max} \tag{5.61}$$

which means that all negative eigenvalues of $\boldsymbol{Q}'$ are simply set to zero.

If the sum of all positive eigenvalues of $\boldsymbol{Q}'$ is larger than $P_{\max}$, the Lagrangian multiplier $\check{\nu}$ is positive, i.e., $\check{\nu} > 0$. As a consequence, (5.60) must hold with equality. The reformulation of (5.59) to

$$\mathcal{P}_{\mathcal{C}}(\boldsymbol{Q}') = \boldsymbol{U}[\boldsymbol{\Lambda} - \check{\nu}\mathbf{I}_M]_+\boldsymbol{U}^{\mathrm{H}} = \boldsymbol{U}\big[[\boldsymbol{\Lambda}]_+ - \check{\nu}\mathbf{I}_M\big]_+\boldsymbol{U}^{\mathrm{H}} \quad \text{for} \sum_{m=1}^{M}[\lambda_m]_+ > P_{\max} \tag{5.62}$$

allows for a two-step interpretation: first, all negative eigenvalues of $\boldsymbol{Q}'$ are set to zero, and second, all eigenvalues are reduced by the same value $\check{\nu}$ as long as they are nonnegative until

$$\mathrm{tr}\left[\mathcal{P}_{\mathcal{C}}(\boldsymbol{Q}')\right] = \sum_{m=1}^{M}\big[[\lambda_m]_+ - \check{\nu}\big]_+ = P_{\max} \quad \text{for} \sum_{m=1}^{M}[\lambda_m]_+ > P_{\max}. \tag{5.63}$$

The second step of the projection has a nice water-spilling interpretation which is closely related to the well-known water-filling algorithm and the optimum Lagrangian multiplier can also be found in at most M steps.

Let $s_1, \ldots, s_M$ be a permutation of the indices $1, \ldots, M$ such that

$$[\lambda_{s_1}]_+ \geq [\lambda_{s_2}]_+ \geq \ldots \geq [\lambda_{s_M}]_+ \geq 0$$

and let $\bar{M}$ denote the largest integer such that $\lambda_{s_{\bar{M}}} > 0$. Since $\sum_{m=1}^{M}[\lambda_m]_+ > P_{\max}$, such an $\bar{M}$ exists. The active sum power constraint (5.63) can then be rewritten to

Algorithm 5.2 Projection of $\boldsymbol{Q}'$ onto the constraint set $\mathcal{C}$

Require: Block-diagonal, Hermitian matrix $\boldsymbol{Q}'$	
$\boldsymbol{U} \cdot \mathrm{diag}\{\lambda_m\}_{m=1}^{M} \cdot \boldsymbol{U}^{\mathrm{H}} \leftarrow \boldsymbol{Q}'$	compute eigenvalue decomposition of $\boldsymbol{Q}'$
$\lambda_m \leftarrow [\lambda_m]_+ \ \forall m \in \{1, \ldots, M\}$	set all negative eigenvalues to zero
if $\sum_{m=1}^{M} \lambda_m \leq P_{\max}$ **then**	
$\quad \mathcal{P}_{\mathcal{C}}(\boldsymbol{Q}') \leftarrow \boldsymbol{U} \cdot \mathrm{diag}\{\lambda_m\}_{m=1}^{M} \boldsymbol{U}^{\mathrm{H}}$	projection is given by setting all negative eigenvalues to zero
$\quad$ **exit**	exit the program
end if	
$s_1, \ldots, s_M : \lambda_{s_1} \geq \lambda_{s_2} \geq \ldots \geq \lambda_{s_M}$	find a permutation that sorts $\lambda_1, \ldots, \lambda_M$
$\bar{M} \leftarrow \max_{m \in \{1,\ldots,M\}} m \ \text{ s.t.: } \lambda_{s_m} > 0$	find largest $\bar{M}$ such that $\lambda_{s_{\bar{M}}} > 0$
for $a = \bar{M}$ downto 1 **do**	
$\quad \nu(a) \leftarrow \dfrac{\sum_{m=1}^{a} \lambda_{s_m} - P_{\max}}{a}$	compute Lagrangian multiplier under the hypothesis $\lambda_{s_a} > \nu(a)$
$\quad$ **if** $\lambda_{s_a} > \nu(a)$ **then**	
$\quad\quad$ **break**	exit the loop since hypothesis is true
$\quad$ **end if**	
end for	
$\check{\nu} \leftarrow \nu(a)$	set optimum Lagrangian multiplier
$\mathcal{P}_{\mathcal{C}}(\boldsymbol{Q}') \leftarrow \boldsymbol{U} \cdot \mathrm{diag}\{[\lambda_m - \check{\nu}]_+\}_{m=1}^{M} \cdot \boldsymbol{U}^{\mathrm{H}}$	compute the projection according to (5.62)

$$P_{\max} = \sum_{m=1}^{\bar{M}} [\lambda_{s_m} - \check{\nu}]_+. \tag{5.64}$$

The optimum Lagrangian multiplier $\check{\nu}$ is found successively by first assuming that $\lambda_{s_a} > \nu(a)$ holds for $a = \bar{M}$, where $\nu(a)$ is the hypothetical optimum Lagrangian multiplier. This implies that $\lambda_{s_m} > \nu(a)$ for all $m \in \{1, \ldots, a\}$. Under this assumption, (5.64) yields

$$\nu(a) = \frac{\sum_{m=1}^{a} \lambda_{s_m} - P_{\max}}{a}.$$

If the hypothesis $\lambda_{s_m} > \nu(a)$ shall be true for all $m \in \{1, \ldots, a\}$, we only have to check whether $\lambda_{s_a} > \nu(a)$. If so, the optimum $\check{\nu}$ has been found and is given by $\check{\nu} = \nu(a)$. Otherwise, a is reduced by one, $\nu(a)$ is recomputed for the new a, and the test $\lambda_{s_a} > \nu(a)$ is performed again. A complete pseudo-code algorithm for the projection of a block-diagonal, Hermitian matrix $\boldsymbol{Q}'$ is shown in Algorithm 5.2. Note that the block-diagonal structure of $\boldsymbol{Q}'$ can be exploited during the computation of its eigenvalue decomposition to reduce the computational complexity. Only the eigenvalue decompositions of the main diagonal blocks of $\boldsymbol{Q}'$ have to be computed that have smaller dimensions.

A graphical visualization of the water-spilling interpretation of the second step of the orthogonal projection is shown in Fig. 5.3 for the case $M = 4$. The largest three eigenvalues λ_{s_1}, λ_{s_2}, and λ_{s_3} of $\boldsymbol{Q}'$ are all positive, whereas $\lambda_{s_4} = 0$.

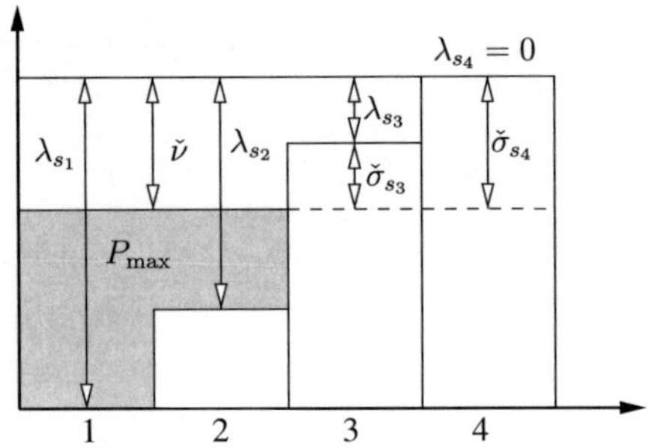

Fig. 5.3 Water-spilling interpretation of the second step of the orthogonal projection

We can interpret $\mathrm{tr}(\boldsymbol{Q}') = \sum_{m=1}^{4} \lambda_{s_m}$ as the amount of water in the box which is larger than the maximum allowed dissipated power $P_{\max}$ such that the second step of the orthogonal projection really has to be applied. The orthogonal projection onto the constraint set is achieved by spilling some water until the remaining amount of water (the shaded area) corresponds to $P_{\max}$. Doing so, the water level is reduced by the optimum Lagrangian multiplier $\check{\nu}$. Since this level is larger than λ_{s_3} and λ_{s_4}, the nonnegativity constraints of these two eigenvalues become active and the corresponding Lagrangian multipliers $\check{\sigma}_{s_3}$ and $\check{\sigma}_{s_4}$ are chosen such that $\check{\nu} = \lambda_{s_3} + \check{\sigma}_{s_3}$ and $\check{\nu} = \lambda_{s_4} + \check{\sigma}_{s_4}$.

Synopsis: The projection of a block-diagonal, Hermitian matrix $\boldsymbol{Q}'$ onto the constraint set $\mathcal{C}$ consists of two steps. First, all negative eigenvalues are set to zero. If the trace of the resulting matrix is smaller than or equal to the maximum allowed dissipated power $P_{\max}$, the second step is superfluous. Otherwise, all eigenvalues are reduced by a common value as long as they are nonnegative until the trace has been reduced to the desired value $P_{\max}$. This procedure has a nice water-spilling interpretation in analogy to the well-known water-filling principle.

5.3.2.2 Tangent Cone Projection

Since $\mathcal{C}$ is a closed, convex set, the tangent cone $\mathcal{T}_{\mathcal{C}}(\boldsymbol{Q})$ is again defined as the closure of the set of feasible directions, see Definition 1. For its computation, we need to distinguish four cases that emerge from the two constraints $\boldsymbol{Q} \succeq \boldsymbol{0}$ and $\mathrm{tr}(\boldsymbol{Q}) \leq P_{\max}$ which are either active or not:

1. $\boldsymbol{Q} \succ \boldsymbol{0}$ and $\mathrm{tr}(\boldsymbol{Q}) < P_{\max}$:
   Since $\boldsymbol{Q}$ is positive definite and the trace constraint is not active, the set of feasible directions is given by all block-diagonal and Hermitian $\boldsymbol{V}$. This results from the fact that for $\boldsymbol{Q} \in \mathrm{int}(\mathcal{C})$, we have that $\boldsymbol{Q} + \alpha \boldsymbol{V} \in \mathcal{C}$ for all block-diagonal

$\boldsymbol{V} = \boldsymbol{V}^{\mathrm{H}}$ and sufficiently small α. Thus, we obtain

$$\mathcal{T}_{\mathcal{C}}(\boldsymbol{Q}) = \left\{\boldsymbol{V} \in \mathcal{B}_{\boldsymbol{Q}} \middle| \boldsymbol{V} = \boldsymbol{V}^{\mathrm{H}}\right\} \\ \text{for } \boldsymbol{Q} \succ \boldsymbol{0} \wedge \operatorname{tr}(\boldsymbol{Q}) < P_{\max}. \tag{5.65}$$

2. $\boldsymbol{Q} \succ \boldsymbol{0}$ and $\operatorname{tr}(\boldsymbol{Q}) = P_{\max}$:
In this case, the sum power constraint is active. To ensure that the update $\boldsymbol{Q}+\alpha\boldsymbol{V}$ remains in $\mathcal{C}$ for sufficiently small α, $\boldsymbol{V}$ need not only be block-diagonal and Hermitian, but also needs to satisfy $\operatorname{tr}(\boldsymbol{V}) \leq 0$. Hence, the tangent cone can be expressed as

$$\mathcal{T}_{\mathcal{C}}(\boldsymbol{Q}) = \left\{\boldsymbol{V} \in \mathcal{B}_{\boldsymbol{Q}} \middle| \operatorname{tr}(\boldsymbol{V}) \leq 0 \wedge \boldsymbol{V} = \boldsymbol{V}^{\mathrm{H}}\right\} \\ \text{for } \boldsymbol{Q} \succ \boldsymbol{0} \wedge \operatorname{tr}(\boldsymbol{Q}) = P_{\max}. \tag{5.66}$$

Before we handle cases 3 and 4, we need to clarify which bearing a small perturbation $\alpha\boldsymbol{V}$ of a positive semi-definite matrix $\boldsymbol{Q} \succeq \boldsymbol{0}$ has on the eigenvalues $\lambda_1, \ldots, \lambda_M$ of $\boldsymbol{Q}$. In particular, we want to know how those eigenvalues of $\boldsymbol{Q}$ that are zero change in their first-order Taylor approximation when $\boldsymbol{Q}$ is perturbed by $\alpha\boldsymbol{V}$. It is shown in Appendix A14 that the first-order derivative (with respect to α evaluated at $\alpha = 0$) of an eigenvalue λ_m with algebraic multiplicity one reads as $\boldsymbol{u}_m^{\mathrm{H}}\boldsymbol{V}\boldsymbol{u}_m$, where $\boldsymbol{u}_m$ is the eigenvector corresponding to the eigenvalue λ_m. If λ_m has an algebraic multiplicity that is larger than one, then the derivatives of these common eigenvalues correspond to the eigenvalues of the matrix $\boldsymbol{U}_0^{\mathrm{H}}\boldsymbol{V}\boldsymbol{U}_0$, where $\boldsymbol{U}_0$ is any subunitary (minimal) basis that spans the eigen-space which corresponds to the eigenvalue. With this knowledge, we can handle cases 3 and 4.

3. $\boldsymbol{Q} \succeq \boldsymbol{0}$, $\operatorname{tr}(\boldsymbol{Q}) < P_{\max}$, and $\mathcal{A} := \{m|\lambda_m = 0\} \neq \emptyset$:
In this case, $\boldsymbol{Q}$ is not positive definite but positive semi-definite with $|\mathcal{A}| > 0$ eigenvalues being zero. Since we are looking for the *closure* of the set of feasible directions and not for the set itself, a first order Taylor expansion of the eigenvalues of $\boldsymbol{Q} + \alpha\boldsymbol{V}$ suffices to determine the closure.[12] Let $\lambda_1, \ldots, \lambda_M$ denote the eigenvalues of $\boldsymbol{Q}$. Clearly, the update $\boldsymbol{Q} + \alpha\boldsymbol{V}$ may have different eigenvalues,

[12] To illustrate this circumstance, assume that $\boldsymbol{Q} = \begin{bmatrix} 1 & 0 \\ 0 & 0 \end{bmatrix}$ and $\boldsymbol{V} = \begin{bmatrix} v_1 & v_2 \\ v_2 & v_3 \end{bmatrix} \in \mathbb{R}^{2\times 2}$. To let $\boldsymbol{V}$ be a feasible direction, $\boldsymbol{Q} + \alpha\boldsymbol{V} \succeq \boldsymbol{0}$ must hold for sufficiently small $\alpha > 0$. For a real-valued, symmetric 2×2 matrix, positive semi-definiteness means that both the trace and the determinant are nonnegative. This implies first, that $1+\alpha(v_1+v_3) \geq 0$, which is always fulfilled for sufficiently small α, and second, that $(1+\alpha v_1)\alpha v_3 - \alpha^2 v_2^2 = \alpha v_3 + \alpha^2(v_1 v_3 - v_2^2) \geq 0$. The first condition does not entail any limitations on v_1 and v_3. For $v_3 > 0$, the second condition is always satisfied for any v_1 and v_2. However, to let $\boldsymbol{V}$ be a feasible direction, $v_3 = 0$ implies $v_2 = 0$. The (open) set of feasible directions is thus given by $\mathcal{F}_{\mathcal{C}}(\boldsymbol{Q}) = \{\boldsymbol{V}|v_3 > 0\} \cup \{\boldsymbol{V}|v_2 = v_3 = 0\}$. If we neglect the quadratic term in the condition $\alpha v_3 + \alpha^2(v_1 v_3 - v_2^2) \geq 0$ and thus replace the second condition by $\alpha v_3 \geq 0$, we end up with $v_3 \geq 0$, which defines the tangent cone to $\mathcal{T}_{\mathcal{C}}(\boldsymbol{Q}) = \{\boldsymbol{V}|v_3 \geq 0\} = \operatorname{cl}(\mathcal{F}_{\mathcal{C}}(\boldsymbol{Q}))$ which clearly coincides with the closure of the set of feasible directions. This example explains the fact that a first-order analysis of the eigenvalues is sufficient to determine the *closure* of the set of feasible directions.

and we need to ensure that all of them remain nonnegative in the first order Taylor approximation. Since $\alpha > 0$ can be made arbitrarily small and since the eigenvalues vary continuously, only the zero eigenvalues lead to additional constraints in the tangent cone definition. Let $\mathcal{A} = \{m|\lambda_m = 0\}$ define the set of indices for which the corresponding eigenvalues are zero, i.e., the eigenvalue zero has an algebraic (and geometric) multiplicity of $|\mathcal{A}|$. According to (A40), the differential change of those $|\mathcal{A}|$ vanishing eigenvalues is given by the eigenvalues of $\boldsymbol{U}_0^{\mathrm{H}} \cdot \boldsymbol{V} \cdot \boldsymbol{U}_0 \mathrm{d}\alpha$ and must remain nonnegative to ensure that $\boldsymbol{Q} + \alpha \boldsymbol{V} \succcurlyeq \boldsymbol{0}$ holds in the first order Taylor approximation. Thus, we have to ensure that $\boldsymbol{U}_0^{\mathrm{H}} \boldsymbol{V} \boldsymbol{U}_0$ is positive semi-definite. The matrix $\boldsymbol{U}_0 \in \mathbb{C}^{M \times |\mathcal{A}|}$ denotes any subunitary basis that spans the eigen-space which corresponds to the vanishing eigenvalue. This leads to the tangent cone

$$\mathcal{T}_{\mathcal{C}}(\boldsymbol{Q}) = \left\{ \boldsymbol{V} \in \mathcal{B}_{\boldsymbol{Q}} \middle| \boldsymbol{U}_0^{\mathrm{H}} \boldsymbol{V} \boldsymbol{U}_0 \succcurlyeq \boldsymbol{0} \wedge \boldsymbol{V} = \boldsymbol{V}^{\mathrm{H}} \right\}$$
$$\text{for } \boldsymbol{Q} \succcurlyeq \boldsymbol{0} \wedge \mathcal{A} = \{m|\lambda_m = 0\} \neq \emptyset \wedge \mathrm{tr}(\boldsymbol{Q}) < P_{\max}. \qquad (5.67)$$

4. $\boldsymbol{Q} \succcurlyeq \boldsymbol{0}$, $\mathrm{tr}(\boldsymbol{Q}) = P_{\max}$, and $\mathcal{A} := \{m|\lambda_m = 0\} \neq \emptyset$:
   In the fourth configuration, not only the positive semi-definiteness constraint is active for some user(s), but also the sum power constraint. Consequently, the tangent cone reads as

$$\mathcal{T}_{\mathcal{C}}(\boldsymbol{Q}) = \left\{ \boldsymbol{V} \in \mathcal{B}_{\boldsymbol{Q}} \middle| \boldsymbol{U}_0^{\mathrm{H}} \boldsymbol{V} \boldsymbol{U}_0 \succcurlyeq \boldsymbol{0} \wedge \mathrm{tr}(\boldsymbol{V}) \leq 0 \wedge \boldsymbol{V} = \boldsymbol{V}^{\mathrm{H}} \right\}$$
$$\text{for } \boldsymbol{Q} \succcurlyeq \boldsymbol{0} \wedge \mathcal{A} = \{m|\lambda_m = 0\} \neq \emptyset \wedge \mathrm{tr}(\boldsymbol{Q}) = P_{\max}. \qquad (5.68)$$

Depending on $\boldsymbol{Q}$, (5.65), (5.66), (5.67), or (5.68) define the tangent cone for the four different classes of transmit covariance matrices $\boldsymbol{Q}$.

In the following, we compute the projection of a block-diagonal, Hermitian matrix $\boldsymbol{V} = \boldsymbol{V}^{\mathrm{H}}$ onto the tangent cone $\mathcal{T}_{\mathcal{C}}(\boldsymbol{Q})$ for above four classes of $\boldsymbol{Q}$. Hence, we already assume that $\boldsymbol{V} \in \mathcal{B}_{\boldsymbol{Q}}$.

1. $\boldsymbol{Q} \succ \boldsymbol{0}$ and $\mathrm{tr}(\boldsymbol{Q}) < P_{\max}$:
   For $\mathcal{T}_{\mathcal{C}}(\boldsymbol{Q})$ in (5.65), $\boldsymbol{V} \in \mathcal{T}_{\mathcal{C}}(\boldsymbol{Q})$ already holds such that

$$\mathcal{P}_{\mathcal{T}_{\mathcal{C}}(\boldsymbol{Q})}(\boldsymbol{V}) = \boldsymbol{V} \quad \text{for } \boldsymbol{Q} \succ \boldsymbol{0} \ \wedge \ \mathrm{tr}(\boldsymbol{Q}) < P_{\max}. \qquad (5.69)$$

2. $\boldsymbol{Q} \succ \boldsymbol{0}$ and $\mathrm{tr}(\boldsymbol{Q}) = P_{\max}$:
   If only the sum power constraint is active for $\boldsymbol{Q}$, the projection of $\boldsymbol{V}$ onto the set (5.66) must yield a nonpositive trace and is given by

$$\mathcal{P}_{\mathcal{T}_{\mathcal{C}}(\boldsymbol{Q})}(\boldsymbol{V}) = \boldsymbol{V} + \check{\boldsymbol{E}} \qquad (5.70)$$

   where $\check{\boldsymbol{E}}$ is the minimizer of the optimization

$$\check{\boldsymbol{E}} := \underset{\boldsymbol{E}}{\operatorname{argmin}} \frac{1}{2}\|\boldsymbol{E}\|_{\mathrm{F}}^2 \quad \text{s.t.: } \operatorname{tr}(\boldsymbol{V}+\boldsymbol{E}) \leq 0.$$

Obviously, $\boldsymbol{E}$ must be Hermitian. The according Lagrangian function reads as

$$\mathcal{L}(\boldsymbol{E}, \nu) = \frac{1}{2}\|\boldsymbol{E}\|_{\mathrm{F}}^2 + \nu \operatorname{tr}(\boldsymbol{V}+\boldsymbol{E})$$

with the nonnegative Lagrangian multiplier $\nu \geq 0$. Setting the Wirtinger derivative of $\mathcal{L}(\boldsymbol{E}, \nu)$ with respect to $\boldsymbol{E}^{\mathrm{T}}$ to zero, we obtain

$$\check{\boldsymbol{E}} = -\check{\nu}\mathbf{I}_M$$

with the optimum Lagrangian multiplier

$$\check{\nu} = \begin{cases} 0 & \text{for } \operatorname{tr}(\boldsymbol{V}) \leq 0 \\ \frac{\operatorname{tr}(\boldsymbol{V})}{M} & \text{for } \operatorname{tr}(\boldsymbol{V}) \geq 0. \end{cases}$$

Thus, the projection onto the tangent cone can be summarized to be

$$\mathcal{P}_{\mathcal{T}_{\mathcal{C}}(\boldsymbol{Q})}(\boldsymbol{V}) = \begin{cases} \boldsymbol{V} & \text{for } \operatorname{tr}(\boldsymbol{V}) \leq 0 \\ \boldsymbol{V} - \frac{\operatorname{tr}(\boldsymbol{V})}{M} \cdot \mathbf{I}_M & \text{for } \operatorname{tr}(\boldsymbol{V}) \geq 0 \end{cases} \quad \text{for } \boldsymbol{Q} \succ \mathbf{0} \text{ and } \operatorname{tr}(\boldsymbol{Q}) = P_{\max}. \tag{5.71}$$

3. $\boldsymbol{Q} \succeq \mathbf{0}$, $\operatorname{tr}(\boldsymbol{Q}) < P_{\max}$, and $\mathcal{A} := \{m | \lambda_m = 0\} \neq \emptyset$:
For an inactive sum power constraint but rank-deficient $\boldsymbol{Q}$, the tangent cone is given by (5.67). The projection of $\boldsymbol{V} \in \mathcal{B}_{\boldsymbol{Q}}$ onto $\mathcal{T}_{\mathcal{C}}(\boldsymbol{Q})$ follows from (5.70) where $\check{\boldsymbol{E}}$ is the minimizer of

$$\check{\boldsymbol{E}} := \underset{\boldsymbol{E}}{\operatorname{argmin}} \frac{1}{2}\|\boldsymbol{E}\|_{\mathrm{F}}^2 \quad \text{s.t.: } \boldsymbol{U}_0^{\mathrm{H}}(\boldsymbol{V}+\boldsymbol{E})\boldsymbol{U}_0 \succeq \mathbf{0}.$$

Note that for $\boldsymbol{V} \succeq \mathbf{0}$, we have $\check{\boldsymbol{E}} = \mathbf{0}$ and thus $\mathcal{P}_{\mathcal{T}_{\mathcal{C}}(\boldsymbol{Q})}(\boldsymbol{V}) = \boldsymbol{V}$. Since the derivation is only a special variant of that for case 4, we skip the solution for the case when $\boldsymbol{V}$ is not positive semi-definite [set $\check{\nu} = 0$ in (5.79)].
4. $\boldsymbol{Q} \succeq \mathbf{0}$, $\operatorname{tr}(\boldsymbol{Q}) = P_{\max}$, and $\mathcal{A} := \{m | \lambda_m = 0\} \neq \emptyset$:
This configuration leads to the most complex projection since not only the sum power constraint is active for $\boldsymbol{Q}$, but also a positive semi-definiteness constraint. For this setup, the projection onto the tangent cone is similar to the projection onto the constraint set $\mathcal{C}$ which was discussed in Sect. 5.3.2.1. The Hermitian matrix $\check{\boldsymbol{E}}$ in (5.70) is the minimizer of the optimization [see (5.68)]

$$\check{\boldsymbol{E}} := \underset{\boldsymbol{E}}{\operatorname{argmin}} \frac{1}{2}\|\boldsymbol{E}\|_{\mathrm{F}}^2 \quad \text{s.t.: } \operatorname{tr}(\boldsymbol{V}+\boldsymbol{E}) \leq 0 \text{ and } \boldsymbol{U}_0^{\mathrm{H}}(\boldsymbol{V}+\boldsymbol{E})\boldsymbol{U}_0 \succeq \mathbf{0}. \tag{5.72}$$

Note that $\boldsymbol{U}_0 \in \mathbb{C}^{M\times|\mathcal{A}|}$ satisfies $\boldsymbol{U}_0^{\mathrm{H}}\boldsymbol{U}_0 = \mathbf{I}_{|\mathcal{A}|}$ and can be chosen as any basis of the eigen-space that is spanned by the vanishing eigenvalue. The associated Lagrangian function reads as

$$\mathcal{L}(\boldsymbol{E}, \nu, \boldsymbol{\Xi}) = \frac{1}{2}\|\boldsymbol{E}\|_{\mathrm{F}}^2 + \nu\big[\operatorname{tr}(\boldsymbol{V}+\boldsymbol{E})\big] - \operatorname{tr}\big[\boldsymbol{U}_0^{\mathrm{H}}(\boldsymbol{V}+\boldsymbol{E})\boldsymbol{U}_0\cdot\boldsymbol{\Xi}\big]$$

where $\nu \geq 0$ and $\boldsymbol{\Xi} \in \mathbb{S}^{|\mathcal{A}|}$. Setting the derivative of $\mathcal{L}(\boldsymbol{E}, \nu, \boldsymbol{\Xi})$ with respect to $\boldsymbol{E}^{\mathrm{T}}$ to zero and evaluating at the optimum variables, we obtain

$$\check{\boldsymbol{E}} = -\check{\nu}\mathbf{I}_M + \boldsymbol{U}_0\check{\boldsymbol{\Xi}}\boldsymbol{U}_0^{\mathrm{H}}. \tag{5.73}$$

The remaining KKT conditions read as

$$\boldsymbol{U}_0^{\mathrm{H}}(\boldsymbol{V}+\check{\boldsymbol{E}})\boldsymbol{U}_0 \succcurlyeq \mathbf{0} \quad\Rightarrow\quad \boldsymbol{U}_0^{\mathrm{H}}\boldsymbol{V}\boldsymbol{U}_0 - \check{\nu}\mathbf{I}_{|\mathcal{A}|} + \check{\boldsymbol{\Xi}} \succcurlyeq \mathbf{0} \tag{5.74}$$

$$\check{\boldsymbol{\Xi}}\cdot\boldsymbol{U}_0^{\mathrm{H}}(\boldsymbol{V}+\check{\boldsymbol{E}})\boldsymbol{U}_0 = \mathbf{0} \quad\Rightarrow\quad \check{\boldsymbol{\Xi}}\cdot\boldsymbol{U}_0^{\mathrm{H}}\boldsymbol{V}\boldsymbol{U}_0 - \check{\nu}\check{\boldsymbol{\Xi}} + \check{\boldsymbol{\Xi}}^2 = \mathbf{0} \tag{5.75}$$

$$\operatorname{tr}(\boldsymbol{V}+\check{\boldsymbol{E}}) \leq 0 \quad\Rightarrow\quad \operatorname{tr}(\boldsymbol{V}) - M\check{\nu} + \operatorname{tr}(\check{\boldsymbol{\Xi}}) \leq 0 \tag{5.76}$$

$$\check{\nu}\operatorname{tr}(\boldsymbol{V}+\check{\boldsymbol{E}}) = 0 \quad\Rightarrow\quad \check{\nu}\big(\operatorname{tr}(\boldsymbol{V}) - M\check{\nu} + \operatorname{tr}(\check{\boldsymbol{\Xi}})\big) = 0. \tag{5.77}$$

From (5.75), we find that $\check{\boldsymbol{\Xi}}\cdot\boldsymbol{U}_0^{\mathrm{H}}\boldsymbol{V}\boldsymbol{U}_0$ must be Hermitian. Hence, $\check{\boldsymbol{\Xi}}$ must have the same eigenbasis as $\boldsymbol{U}_0^{\mathrm{H}}\boldsymbol{V}\boldsymbol{U}_0$. Using the eigenvalue decomposition $\boldsymbol{U}_0^{\mathrm{H}}\boldsymbol{V}\boldsymbol{U}_0 = \boldsymbol{W}\boldsymbol{L}\boldsymbol{W}^{\mathrm{H}}$ with unitary $\boldsymbol{W}$ and diagonal $\boldsymbol{L} := \operatorname{diag}\{\ell_m\}_{m=1}^{|\mathcal{A}|}$, we can parameterize $\check{\boldsymbol{\Xi}}$ via

$$\check{\boldsymbol{\Xi}} = \boldsymbol{W}\check{\boldsymbol{\Sigma}}\boldsymbol{W}^{\mathrm{H}}$$

where $\check{\boldsymbol{\Sigma}} := \operatorname{diag}\{\check{\sigma}_m\}_{m=1}^{|\mathcal{A}|} \succcurlyeq \mathbf{0}$ is diagonal and positive semi-definite. (5.75) can then be rewritten as

$$\check{\boldsymbol{\Sigma}}^2 - \check{\nu}\check{\boldsymbol{\Sigma}} + \check{\boldsymbol{\Sigma}}\boldsymbol{L} = \mathbf{0}$$

which in combination with (5.74) leads to

$$\check{\sigma}_m = [\check{\nu} - \ell_m]_+ \quad \forall m \in \{1, \ldots, |\mathcal{A}|\}. \tag{5.78}$$

This allows to express the matrix $\check{\boldsymbol{E}}$ in (5.73) as

$$\check{\boldsymbol{E}} = -\check{\nu}\mathbf{I}_M + \boldsymbol{U}_0\boldsymbol{W}\big[\check{\nu}\mathbf{I}_{|\mathcal{A}|} - \boldsymbol{L}\big]_+\boldsymbol{W}^{\mathrm{H}}\boldsymbol{U}_0^{\mathrm{H}}. \tag{5.79}$$

Setting $\check{\nu}$ to zero solves case 3. However, in case 4 we also have to meet the trace constraint in (5.76):

$$\operatorname{tr}(\boldsymbol{V} + \check{\boldsymbol{E}}) = \operatorname{tr}(\boldsymbol{V}) - M\check{\nu} + \sum_{m=1}^{|\mathcal{A}|} [\check{\nu} - \ell_m]_+ \leq 0. \tag{5.80}$$

The Lagrangian multiplier $\check{\nu}$ can only be zero if

$$\operatorname{tr}(\boldsymbol{V}) + \sum_{m=1}^{|\mathcal{A}|} [-\ell_m]_+ \leq 0 \quad \Rightarrow \quad \check{\nu} = 0 \tag{5.81}$$

which for positive semi-definite $\boldsymbol{V} \succcurlyeq \boldsymbol{0}$ is only possible when $\boldsymbol{V} = \boldsymbol{0}$ and thus also $\boldsymbol{L} = \boldsymbol{0}$. For indefinite or negative (semi-)definite $\boldsymbol{V}$, however, (5.81) might hold. If (5.81) holds, we have $\boldsymbol{U}_0 \boldsymbol{W}[-\boldsymbol{L}]_+ \boldsymbol{W}^{\mathrm{H}} \boldsymbol{U}_0^{\mathrm{H}}$. Otherwise, $\check{\nu} > 0$ and (5.80) must hold with equality which follows from (5.77). The implicit equation

$$\operatorname{tr}(\boldsymbol{V}) - M\check{\nu} + \sum_{m=1}^{|\mathcal{A}|} [\check{\nu} - \ell_m]_+ = 0 \tag{5.82}$$

can be solved for $\check{\nu}$ in at most $|\mathcal{A}| + 1$ steps by probing the $|\mathcal{A}| + 1$ relevant cases for which the operator $[\cdot]_+$ is active or not.
To this end, let $s_1, \ldots, s_{|\mathcal{A}|}$ denote a permutation of $1, \ldots, |\mathcal{A}|$ such that

$$\ell_{s_1} \leq \ell_{s_2} \leq \cdots \leq \ell_{s_{|\mathcal{A}|}}. \tag{5.83}$$

Then, we can rewrite (5.82) to

$$\operatorname{tr}(\boldsymbol{V}) - M\check{\nu} + \sum_{m=1}^{|\mathcal{A}|} [\check{\nu} - \ell_{s_m}]_+ = 0. \tag{5.84}$$

Starting with $a = |\mathcal{A}|$ and the hypothesis that

$$\nu(a) - \ell_{s_m} > 0 \, \forall m \in \{1, \ldots, a\} \Leftrightarrow \nu(a) - \ell_{s_a} > 0$$

with the hypothetical optimum Lagrangian multiplier $\nu(a)$, the implicit equation in (5.84) yields

$$\begin{aligned} &\operatorname{tr}(\boldsymbol{V}) - M\nu(a) + \sum_{m=1}^{a} \left(\nu(a) - \ell_{s_m}\right) = 0 \\ \Leftrightarrow \quad &\nu(a) = \frac{\operatorname{tr}(\boldsymbol{V}) - \sum_{m=1}^{a} \ell_{s_m}}{M - a}. \end{aligned} \tag{5.85}$$

Note that $|\mathcal{A}| = M$ is not possible since this would imply that $\boldsymbol{Q} = \boldsymbol{0}$, which contradicts $\operatorname{tr}(\boldsymbol{Q}) = P_{\max} > 0$. Hence, the denominator in (5.85) is always nonzero. To see whether the hypothesis is true, we have to verify that $\nu(a) - \ell_{s_a} > 0$. If this inequality holds, the optimum $\check{\nu}$ has been found and is given by $\check{\nu} = \nu(a)$. Otherwise, a is reduced by one, $\nu(a)$ is recomputed via (5.85) and $\nu(a) - \ell_{s_a} > 0$ is checked again and so on. The last test is for $a = 0$. Note that for $a = 0$, the condition $\nu(a) - \ell_{s_a} > 0$ need not be tested since s_0 is not defined. If $\nu(a) - \ell_{s_a} \leq 0$ is true for all $a \in \{1, \ldots, |\mathcal{A}|\}$, i.e., if all tests have failed so far, then $a = 0$ is the solution [cf. (5.71)] that corresponds to the case, where $\boldsymbol{U}_0 \boldsymbol{V} \boldsymbol{U}_0$ is positive definite and the ν necessary to fulfill $\operatorname{tr}(\boldsymbol{V} - \nu \mathbf{I}) \leq 0$ is small enough such that $\boldsymbol{U}_0(\boldsymbol{V} - \nu\mathbf{I})\boldsymbol{U}_0 \succcurlyeq \boldsymbol{0}$ holds.

A pseudo-code implementation of this procedure is shown in Algorithm 5.3. In the first line, a threshold is defined that determines which eigenvalues are treated as being 'numerically' zero. This value clearly depends on the word length of the machine. Next, an eigenvalue decomposition of $\boldsymbol{Q}$ is performed and the set of eigenvalues which are (numerically) zero is determined. In Line 4, the eigenvectors of the numerically zero eigenvalues are collected in the subunitary matrix $\boldsymbol{U}_0$ and another eigenvalue decomposition is computed in Line 5. A permutation of the indices in $\mathcal{A}$ that sorts $\ell_1, \ldots, \ell_{|\mathcal{A}|}$ in a nondecreasing manner is found in Line 6 [cf. (5.83)]. Starting in Line 7, we probe the assumption of a inactive $\max\{0, \cdot\}$ operators with a ranging from $|\mathcal{A}|$ down to 0. Under this assumption, the Lagrangian multiplier $\nu(a)$ is computed according to (5.85). If the assumption of a inactive constraints shall be true, $\nu(a) - \ell_{s_a} > 0$ must hold, which we test in Line 10. However, for $a = 0$, this test is superfluous and s_0 is not defined. Hence, the if-condition in Line 9 prevents the execution of Line 10. If $\nu(a) - \ell_{s_a} > 0$, the loop is left in Line 11, and we directly jump to Line 15. If not, the number of inactive operators $[\cdot]_+$ is decreased by one, and we go back to Line 8. After the optimum a has been found, the optimum Lagrangian multiplier $\check{\nu}$ which handles the trace constraint is set accordingly in Line 15. In Line 16, the Lagrangian multiplier associated to the positive semi-definiteness constraint is computed. Both are used to compute the correcting matrix $\check{\boldsymbol{E}}$ in Line 17 and finally, the projection onto the tangent cone is computed in Line 18.

Algorithm 5.3 Tangent cone projection of $\boldsymbol{V}$ for rank-deficient $\boldsymbol{Q}$ with $\mathrm{tr}(\boldsymbol{Q}) = P_{\max}$.

Require: Rank-deficient $\boldsymbol{Q} \succcurlyeq \mathbf{0}$ with $\mathrm{tr}(\boldsymbol{Q}) = P_{\max}$ and direction $\boldsymbol{V}$
1: $\varepsilon_{\text{threshold}} \leftarrow 10^{-12}$ — set tolerance below which an eigenvalue is regarded as zero
2: $\boldsymbol{U} \cdot \mathrm{diag}\{\lambda_m\}_{m=1}^{M} \cdot \boldsymbol{U}^{\mathrm{H}} \leftarrow \boldsymbol{Q}$ — perform eigenvalue decomposition of $\boldsymbol{Q}$
3: $\mathcal{A} \leftarrow \{m|\lambda_m \leq \varepsilon_{\text{threshold}}\}$ — find indices of eigenvalues that are numerically zero
4: Set $\boldsymbol{U}_0$ as columns of $\boldsymbol{U}$ with indices in $\mathcal{A}$ — extract eigenvectors of numerically zero eigenvalues
5: $\boldsymbol{W} \cdot \mathrm{diag}\{\ell_m\}_{m=1}^{|\mathcal{A}|} \cdot \boldsymbol{W}^{\mathrm{H}} \leftarrow \boldsymbol{U}_0^{\mathrm{H}} \boldsymbol{V} \boldsymbol{U}_0$ — perform eigenvalue decomposition
6: $s_1, \ldots, s_{|\mathcal{A}|} : \ell_{s_1} \leq \ell_{s_2} \leq \ldots \leq \ell_{s_{|\mathcal{A}|}}$ — find permutation of elements in $\mathcal{A}$ that sorts $\ell_1, \ldots, \ell_{|\mathcal{A}|}$
7: **for** $a = |\mathcal{A}|$ downto 0 **do**
8:   $\nu(a) \leftarrow \dfrac{\mathrm{tr}(\boldsymbol{V}) - \sum_{m=1}^{a} \ell_{s_m}}{M - a}$ — compute Lagrangian multiplier assuming a inactive constraints
9:   **if** $a > 0$ **then**
10:     **if** $\nu(a) - \ell_{s_a} > 0$ **then**
11:       **break** — exit loop since optimum a has been found
12:     **end if**
13:   **end if**
14: **end for**
15: $\check{\nu} \leftarrow \nu(a)$ — set optimum Lagrangian multiplier of trace constraint
16: $\check{\boldsymbol{\Xi}} \leftarrow \boldsymbol{W}\big[\check{\nu}\mathbf{I} - \mathrm{diag}\{\ell_m\}_{m=1}^{M}\big]_{+} \boldsymbol{W}^{\mathrm{H}}$ — set optimum Lagrangian multiplier of pos. semdef. constraint
17: $\check{\boldsymbol{E}} \leftarrow -\check{\nu}\mathbf{I}_M + \boldsymbol{U}_0 \check{\boldsymbol{\Xi}} \boldsymbol{U}_0^{\mathrm{H}}$ — compute the update matrix $\check{\boldsymbol{E}}$ that maps $\boldsymbol{V}$ to the tangent cone
18: $\mathcal{P}_{\mathcal{T}_{\mathcal{C}}(\boldsymbol{Q})}(\boldsymbol{V}) \leftarrow \boldsymbol{V} + \check{\boldsymbol{E}}$ — compute the projection

Synopsis: For the projection of a block-diagonal, Hermitian matrix $\boldsymbol{V}$ onto the tangent cone $\mathcal{T}_{\mathcal{C}}(\boldsymbol{Q})$, we have to distinguish four different cases depending on whether the two constraints $\boldsymbol{Q} \succcurlyeq \mathbf{0}$ and $\mathrm{tr}(\boldsymbol{Q}) \leq P_{\max}$ are active in each case or not. If both constraints are inactive, the projection onto the tangent cone is the identity mapping since the tangent cone is given by the set of all block-diagonal, Hermitian matrices with appropriate dimension. If only the constraint $\boldsymbol{Q} \succcurlyeq \mathbf{0}$ is active and if in addition $\boldsymbol{V} \succcurlyeq \mathbf{0}$ as will be the case in the weighted sum rate maximization with interference cancelation, the projection is again the identity mapping.

If only the sum power constraint $\mathrm{tr}(\boldsymbol{Q}) \leq P_{\max}$ is binding, the projection of $\boldsymbol{V}$ with $\mathrm{tr}(\boldsymbol{V}) > 0$ onto the tangent cone is given by subtracting a scaled identity matrix from $\boldsymbol{V}$, where the scaling must be chosen such that the result has a vanishing trace. Finally, when both constraints are active, the projection onto the tangent cone can be computed in a finite number of steps similar to the water-filling algorithm. In turn, a scaled identity matrix is subtracted from $\boldsymbol{V}$ but a certain matrix-valued Hermitian form needs to remain positive semi-definite. Therefore, in addition to subtracting the scaled identity matrix, the projection also consists of adding a positive semi-definite matrix which ensures the positive semi-definiteness of the matrix-valued Hermitian form.

5.3.2.3 Orthogonal Projection Applied to the Tangent Cone Search Direction

In the covariance matrix based approach, we may further analyze the constraint set projection

$$\mathcal{P}_{\mathcal{C}}\left(\boldsymbol{Q}+s\cdot\mathcal{P}_{\mathcal{T}_{\mathcal{C}}(\boldsymbol{Q})}\left(\frac{\partial_{\mathrm{w}} f(\boldsymbol{Q})}{\partial_{\mathrm{w}}\boldsymbol{Q}^{\mathrm{T}}}\right)\right)$$

that occurs in the recurrence relation (5.24) with the projection of the gradient onto the tangent cone. First, recall that for the two less important cases 1 and 3 in Sect. 5.3.2.2 where the trace of $\boldsymbol{Q}$ is strictly smaller than $P_{\max}$, the projection of the gradient onto the tangent cone is simply the identity mapping (for case 3, this requires the gradient to be positive semi-definite). The case $\mathrm{tr}(\boldsymbol{Q}) = P_{\max}$ is more important since the (weighted) sum rate utility is increasing in the sum power such that the optimum covariance matrix will always make use of the whole power budget. For positive definite $\boldsymbol{Q} \succ \mathbf{0}$ (case 2) and a positive semi-definite gradient, the tangent cone projection is given by [see (5.71)]

$$\mathcal{P}_{\mathcal{T}_{\mathcal{C}}(\boldsymbol{Q})}\left(\frac{\partial_{\mathrm{w}} f(\boldsymbol{Q})}{\partial_{\mathrm{w}}\boldsymbol{Q}^{\mathrm{T}}}\right)=\frac{\partial_{\mathrm{w}} f(\boldsymbol{Q})}{\partial_{\mathrm{w}}\boldsymbol{Q}^{\mathrm{T}}}-\underbrace{\frac{\mathrm{tr}\left(\frac{\partial_{\mathrm{w}} f(\boldsymbol{Q})}{\partial_{\mathrm{w}}\boldsymbol{Q}^{\mathrm{T}}}\right)}{M}}_{=:\alpha}\cdot\mathbf{I}_M.$$

When used as search direction with step-size s, the projection onto the constraint set $\mathcal{C}$ can be expressed as

$$\mathcal{P}_{\mathcal{C}}\left(\boldsymbol{Q}+s\cdot\frac{\partial_{\mathrm{w}} f(\boldsymbol{Q})}{\partial_{\mathrm{w}}\boldsymbol{Q}^{\mathrm{T}}}-s\alpha\mathbf{I}_M\right)=\mathcal{P}_{\mathcal{C}}\left(\boldsymbol{Q}+s\cdot\frac{\partial_{\mathrm{w}} f(\boldsymbol{Q})}{\partial_{\mathrm{w}}\boldsymbol{Q}^{\mathrm{T}}}\right) \tag{5.86}$$

which means that it does not make any difference whether the unconstrained gradient or its projection onto the tangent cone is used as search direction when $\boldsymbol{Q} \succ \mathbf{0}$! To understand the equality in (5.86), we refer to (5.62): the projection onto the constraint set corresponds to subtracting the scaled identity matrix $\check{\nu}\mathbf{I}_M$ from the argument and afterwards setting the negative eigenvalues to zero. Proving (5.86) is thus equivalent to showing that $\check{\nu} \geq s\alpha$, i.e., that the water level must be decreased by at least $s\alpha$. Assume that we reduce the water-level by $s\alpha$. Then, we have

$$\mathrm{tr}\left[\boldsymbol{Q}+s\cdot\frac{\partial_{\mathrm{w}} f(\boldsymbol{Q})}{\partial_{\mathrm{w}}\boldsymbol{Q}^{\mathrm{T}}}-s\alpha\mathbf{I}_M\right]=\mathrm{tr}(\boldsymbol{Q})=P_{\max}. \tag{5.87}$$

However, subtracting the scaled identity matrix $s\alpha\mathbf{I}_M$ may have generated negative eigenvalues which are set to zero during the projection onto the constraint set, see (5.62). If negative eigenvalues of the argument of the trace operator in (5.87) are set to zero, the trace is increased. But it must be kept at $P_{\max}$ such that the water-level must be further decreased. Hence, we have $\check{\nu} \geq s\alpha$ which proves (5.86).

Synopsis Choosing the search direction as the projection of the gradient onto the tangent cone leads to the steepest ascent direction. However, when applied to the constraint set projection, the steepest ascent direction yields the same result as the conventional gradient (that is not projected onto the tangent cone) in most cases. More precisely, the constraint set projection of the two search directions can only differ when $\boldsymbol{Q}$ is rank deficient and fulfills $\mathrm{tr}(\boldsymbol{Q}) = P_{\mathrm{max}}$. Even in this case, the two search directions share the same first-order Taylor approximation of the utility, see (5.20) and (5.22). Hence, the performance difference of the two search directions is expected to be small and can only be nonzero when the covariance matrix is rank-deficient in some intermediate step of the recurrence relation (5.24).

Chapter 6
MIMO BC Transceiver Design with Interference Cancelation

For the sum power constrained MIMO broadcast channel with interference cancelation, two interesting optimizations will be investigated and solved in this chapter. The first one corresponds to the computation of the sum capacity whereas in the second, the weighted sum rate is maximized where arbitrary, nonnegative weights are assigned to the K users. Making use of the duality presented in Chap. 4, both problems are converted to the dual MIMO MAC where they turn out to have a concave utility that must be maximized over a convex set. For their solution, we apply the gradient-projection algorithms described in Chap. 5. Both the covariance-based and the precoder-based approach are investigated. In addition, we present an overview of the state-of-the-art and discuss applications that require the solution of a weighted sum rate maximization problem.

6.1 Sum Capacity Computation

As mentioned in Sect. 4.2.2.1, the sum capacity of the MIMO BC can be achieved by means of dirty paper coding. In this section, we present an algorithm that computes the sum capacity and the corresponding optimum transmit covariance matrices. To begin with, an overview of hitherto existing approaches is given.

6.1.1 Literature Overview

From the various techniques to compute the sum capacity in the MIMO BC and the respective transmit covariance matrices, only few authors tackle the problem directly in the BC. Two examples are the ranked known interference scheme of Caire and Shamai in [80, 81] and the alternating optimization approach of Yu and Cioffi in [83, 77]. The first scheme is only applicable in the two-user vector BC with

R. Hunger, *Analysis and Transceiver Design for the MIMO Broadcast Channel*, Foundations in Signal Processing, Communications and Networking 8, DOI: 10.1007/978-3-642-31692-0_6, © Springer-Verlag Berlin Heidelberg 2013

a two-antenna base station. In the second approach, the sum rate of a virtual point-to-point MIMO system where the K users cooperate is maximized with respect to the transmit covariance matrix and minimized with respect to the (constrained) total noise covariance matrix in an alternating fashion, see Sect. 4.2.2.1 for more details. Basically all other algorithms operate in the dual MIMO MAC.

The very first approach in the dual MIMO MAC of Jindal et al. in 2002 applies iterative water-filling with a sum power constraint [110] and can be regarded as a modification of the iterative water-filling algorithm of Yu and Cioffi [71]. As the MIMO MAC is considered in [71], *individual* power constraints and thus individual water levels are given. The idea of [110] is to treat the links from the K users to the base station as K parallel, noninterfering point-to-point links and to maximize the sum rate under the assumption of *fixed* effective channels and a common water level. The KKT conditions of the problem in the dual MIMO MAC give rise to this interpretation of parallel point-to-point links whose effective channels depend on the transmit covariance matrices of all other users. Convergence is proven in [110] only for the two user case. Since the sum power constraint couples all covariance matrices via the common water level, the algorithm is prone to oscillations for more than two users and usually fails to converge to the global optimum. Two possibilities to overcome this problem were included in the corresponding journal [111] that appeared two years later, see below.

An algorithm for the sum rate maximization which operates in the Lagrange-dual domain was presented by Yu in 2003 [112]. In the original formulation of the sum rate maximization problem, all transmit covariance matrices are coupled by the sum power constraint. To overcome the coupling of the *matrices*, Yu introduced K scalar individual power constraints for the K covariance matrices and a further constraint on the sum of these *scalar* powers. Thus, the coupling is only due to the *scalar* sum power constraint and the Lagrange dualization is only with respect to this scalar constraint. For the solution of the Lagrange dual problem, a bisection method is proposed for this one-dimensional problem. Note that the Lagrange dual function corresponds to applying iterative water-filling with a fixed water-level. In contrast to [110], the algorithm converges to the global optimum for an arbitrary number of users. However, the speed of convergence is very low and it is not clear how to choose the interval for the bisection in which the Lagrangian multiplier has to lie. In addition, the algorithm consists of two nested loops and one has to decide for the number of iterations of the inner loop in which the iterative water-filling is applied.

Two convergent extensions of the sum power iterative water-filling algorithm [110] for more than $K = 2$ users were proposed by Jindal himself in [111]. In the first one, the effective channel of a user is constituted by covariance matrices not only of the last iteration, but the $K - 1$ last iterations. Hence, all covariance matrices of the $K - 1$ last iterations need to be stored in the memory. In the second one, the covariance matrices of the current iteration are computed only on the basis of those from the last iteration. However, the outcome of the iterative water-filling is not directly used for the covariance matrices of the current iteration. Instead, the outcome contributes only by a factor of $1/K$ to them whereas the remaining part is given by the covariance matrices of the last iteration, which are scaled by $(K-1)/K$.

Hence, the Mann iteration [104] is used instead of the Picard iteration to solve the fixed point equation that results from the KKT conditions. Another hybrid algorithm that mixes the original iterative water-filling from [110] with the damped one is also proposed to overcome the slow convergence introduced by the damping factor $1/K$ without having to renounce convergence.

While the iteration-independent factor $1/K$ of the Mann iteration is sufficient for convergence, the resulting low speed of convergence can be improved when variable damping factors are allowed. Using a parabolically interpolated dependency of the sum rate on the damping factor, Böhnke and Kammeyer choose this factor to be the apex of the parabola [105] as in our relaxed gradient-projection algorithm from Sect. 5.2.

Another way to make the original iterative water-filling algorithm [110] convergent for more than two users is given by Codreanu et al. in [113]. In their algorithm, two out of the K users are randomly picked and their covariance matrices are optimized by means of the two-user iterative water-filling algorithm subject to no increase in the sum power of those two users. Convergence to the global optimum is claimed to be achieved with probability one.

A covariance-based gradient-projection algorithm that is similar to the one described in the next subsection was proposed by Liu et al. in [109]. However, conjugate search directions are used where the search direction of the previous iteration contributes to the current one. Due to the gradient approach, the orthogonal projection onto the constraint set was used as well.

The precoder-based (standard) gradient-projection algorithm from Sect. 6.1.3 which does not require any eigenvalue decompositions was published by the author in [114]. Clearly, the more general algorithms from Sect. 6.2.2 that maximize any *weighted* sum rate may also be used for the standard sum rate maximization with equal weights.

6.1.2 Covariance-Based Sum Capacity Computation

In the MIMO BC with transmit covariance matrices $\boldsymbol{S}_1, \ldots, \boldsymbol{S}_K$, the maximum possible data rate at which user $\pi[k]$ can transmit with arbitrarily small error probability is given by [see (4.43)]

$$R_{\pi[k]}^{\mathrm{BC}}(\boldsymbol{S}) = \log_2 \frac{\left|\boldsymbol{C}_{\eta_{\pi[k]}} + \sum_{i \le k} \boldsymbol{H}_{\pi[k]}^{\mathrm{BC}} \boldsymbol{S}_{\pi[i]} \boldsymbol{H}_{\pi[k]}^{\mathrm{BC,H}}\right|}{\left|\boldsymbol{C}_{\eta_{\pi[k]}} + \sum_{i < k} \boldsymbol{H}_{\pi[k]}^{\mathrm{BC}} \boldsymbol{S}_{\pi[i]} \boldsymbol{H}_{\pi[k]}^{\mathrm{BC,H}}\right|} \tag{6.1}$$

where $\pi[\cdot]$ denotes a permutation of the indices $1, \ldots, K$. In (6.1), $\boldsymbol{S}$ is a block-diagonal matrix that contains the individual covariance matrices $\boldsymbol{S}_1, \ldots, \boldsymbol{S}_K$ as diagonal blocks. Moreover, (6.1) assumes that user $\pi[K]$ is encoded first whereas user $\pi[1]$ is encoded last and thus does not see any kind of interference. Due to the indices mismatch of the channels and the covariance matrices in the BC, the individual determinants of the sum rate expression

$$R^{\mathrm{BC}}(\boldsymbol{S}) := \sum_{k=1}^{K} R^{\mathrm{BC}}_{\pi[k]}(\boldsymbol{S}) \tag{6.2}$$

cannot be combined. Another aspect that speaks in favor of tackling the sum rate maximization problem in the dual MIMO MAC results from the fact that $R^{\mathrm{BC}}(\boldsymbol{S})$ is in general not concave in $\boldsymbol{S}$ for $K > 1$. Finally, it is unclear from (6.2) in which order the users have to be encoded. We conclude, that the sum power constrained sum capacity computation

$$\begin{aligned} \underset{\boldsymbol{S}_1,\ldots,\boldsymbol{S}_K,\pi[\cdot]}{\text{maximize}} \ \sum_{k=1}^{K} R^{\mathrm{BC}}_{\pi[k]}(\boldsymbol{S}) \quad \text{s.t.: } & \boldsymbol{S}_k \succeq \mathbf{0} \ \forall k \in \{1,\ldots,K\} \\ & \sum_{k=1}^{K} \operatorname{tr}(\boldsymbol{S}_k) \leq P_{\max} \\ & \pi[\cdot] \text{ is a permutation of } 1,\ldots,K \end{aligned} \tag{6.3}$$

is hard to solve directly in the BC. Since the capacity regions of the MIMO BC and the MIMO MAC coincide under a sum power constraint, we can convert (6.3) to a problem in the dual MAC as described in Sect. 4.4.2 and solve the optimization in this dual domain where it turns out to be a convex problem. Afterwards, the solution is converted back to the BC using the MAC to BC conversion procedure from Sect. 4.4.3.

As shown in Sect. 4.4.2, the dual MAC can be chosen to have an identity noise covariance matrix if the MAC channels $\boldsymbol{H}_1,\ldots,\boldsymbol{H}_K$ follow from the BC channels and the noise covariance matrices in the BC via

$$\boldsymbol{H}_k = \boldsymbol{H}_k^{\mathrm{BC,H}} \boldsymbol{C}_{\eta_k}^{-\frac{1}{2},\mathrm{H}} \quad \forall k.$$

The MAC sum rate $R^{\mathrm{MAC}}(\boldsymbol{Q})$ that is achieved by means of successive decoding with transmit covariance matrices $\boldsymbol{Q}_1,\ldots,\boldsymbol{Q}_K$ reads as [see (4.12)]

$$R^{\mathrm{MAC}}(\boldsymbol{Q}) := \log_2 \left| \mathbf{I} + \sum_{k=1}^{K} \boldsymbol{H}_k \boldsymbol{Q}_k \boldsymbol{H}_k^{\mathrm{H}} \right| = \log_2 \left| \mathbf{I} + \boldsymbol{H}\boldsymbol{Q}\boldsymbol{H}^{\mathrm{H}} \right| \tag{6.4}$$

where $\boldsymbol{H} := [\boldsymbol{H}_1,\ldots,\boldsymbol{H}_K] \in \mathbb{C}^{N\times M}$ denotes the composite channel matrix and $\boldsymbol{Q} = \operatorname{blockdiag}\{\boldsymbol{Q}_k\}_{k=1}^{K} \in \mathbb{S}^M$ is again block-diagonal. Note that (6.4) holds for any of the $K!$ decoding orders, i.e., the sum rate is invariant with respect to the order. However, different individual rates are achieved for different decoding orders, see (4.15) for the individual rate expressions. Every rate tuple then corresponds to one vertex of the $K-1$-dimensional polyhedron in which the sum rate is constant, see Sect. 4.1.2.

In the dual MAC, the sum capacity computation under a sum power constraint can be expressed as

$$\underset{\boldsymbol{Q}_1,\ldots,\boldsymbol{Q}_K}{\text{maximize}} \ \log_2\left|\mathbf{I}+\sum_{k=1}^{K}\boldsymbol{H}_k\boldsymbol{Q}_k\boldsymbol{H}_k^{\mathrm{H}}\right| \ \text{s.t.:}\ \boldsymbol{Q}_k \succeq \mathbf{0}\ \forall k\in\{1,\ldots,K\}$$

$$\sum_{k=1}^{K}\operatorname{tr}(\boldsymbol{Q}_k)\leq P_{\max}. \tag{6.5}$$

Clearly, the constraint set in (6.5) is convex since the set of positive semi-definite matrices is a convex cone, the trace constraint is affine, and the intersection of convex sets is again convex. In contrast to the BC utility in (6.3), the MAC utility in (6.5) is concave. We prove this by showing that the sum rate $R^{\mathrm{MAC}}(\boldsymbol{Q})$ is concave when restricted to a line segment $\boldsymbol{Q}+t\boldsymbol{V}\succeq\mathbf{0}$ that intersects its domain [54 Definition 3.1.1, p. 67].

Let $g(t) := \log_2\left|\mathbf{I}+\boldsymbol{H}(\boldsymbol{Q}+t\boldsymbol{V})\boldsymbol{H}^{\mathrm{H}}\right|$ with Hermitian $\boldsymbol{V}\in\mathbb{C}^{M\times M}$. Using Appendix A6, the derivative of $g(t)$ with respect to t read as

$$\frac{\partial g(t)}{\partial t}=\frac{1}{\ln 2}\cdot\operatorname{tr}\left[\left(\mathbf{I}+\boldsymbol{H}(\boldsymbol{Q}+t\boldsymbol{V})\boldsymbol{H}^{\mathrm{H}}\right)^{-1}\boldsymbol{H}\boldsymbol{V}\boldsymbol{H}^{\mathrm{H}}\right]$$

and the second derivative can be computed by means of Appendix A5 as

$$\frac{\partial^2 g(t)}{\partial t^2}=-\frac{1}{\ln 2}\cdot\operatorname{tr}\left[\left(\mathbf{I}+\boldsymbol{H}(\boldsymbol{Q}+t\boldsymbol{V})\boldsymbol{H}^{\mathrm{H}}\right)^{-1}\boldsymbol{H}\boldsymbol{V}\boldsymbol{H}^{\mathrm{H}}\left(\mathbf{I}+\boldsymbol{H}(\boldsymbol{Q}+t\boldsymbol{V})\boldsymbol{H}^{\mathrm{H}}\right)^{-1}\boldsymbol{H}\boldsymbol{V}\boldsymbol{H}^{\mathrm{H}}\right].$$

For $\boldsymbol{Q}+t\boldsymbol{V}\succeq\mathbf{0}$, the positive semi-definite square root $(\mathbf{I}+\boldsymbol{H}(\boldsymbol{Q}+t\boldsymbol{V})\boldsymbol{H}^{\mathrm{H}})^{-\frac{1}{2}}$ exists, and we can rewrite the second derivative as

$$\frac{\partial^2 g(t)}{\partial t^2}=-\frac{1}{\ln 2}\cdot\left\|\left(\mathbf{I}+\boldsymbol{H}(\boldsymbol{Q}+t\boldsymbol{V})\boldsymbol{H}^{\mathrm{H}}\right)^{-\frac{1}{2}}\boldsymbol{H}\boldsymbol{V}\boldsymbol{H}^{\mathrm{H}}\left(\mathbf{I}+\boldsymbol{H}(\boldsymbol{Q}+t\boldsymbol{V})\boldsymbol{H}^{\mathrm{H}}\right)^{-\frac{1}{2}}\right\|_{\mathrm{F}}^{2}\leq 0$$

from which we conclude that g is concave[1] and thus also the sum rate $R^{\mathrm{MAC}}(\boldsymbol{Q})$. As a consequence, any stationary point of (6.5) is a global optimizer. However, for $\boldsymbol{H}\boldsymbol{V}\boldsymbol{H}^{\mathrm{H}}=\mathbf{0}$, the second derivative of $g(t)$ vanishes which means that $R^{\mathrm{MAC}}(\boldsymbol{Q})$ need not be *strictly* concave in $\boldsymbol{Q}$. Hence, the optimizing $\boldsymbol{Q}$ need not be unique. To visualize this, assume that $\boldsymbol{H}_1=\boldsymbol{H}_2$. In this case, a modification of $\boldsymbol{Q}_1$ leaves the sum rate unchanged as long as $\boldsymbol{Q}_1+\boldsymbol{Q}_2$ remains constant—only the sum $\boldsymbol{Q}_1+\boldsymbol{Q}_2$ comes into play. However, if the optimizer of (6.5) is not unique, the set containing

[1] To show the concavity of $R^{\mathrm{MAC}}(\boldsymbol{Q})$, the block-diagonal structure of $\boldsymbol{Q}$ and $\boldsymbol{V}$ need not be taken into consideration explicitly.

all optimizers is convex and thus continuous [54 Section 4.2.1, p. 137] and every element of the set achieves the sum capacity.

In order to be able to apply the gradient-projection algorithm from Chap. 5, the Wirtinger derivative of $R^{\mathrm{MAC}}(\boldsymbol{Q})$ with respect to $\boldsymbol{Q}_k^{\mathrm{T}}$ has to be computed for every k, see Sect. 5.1.2. Since $\boldsymbol{Q}_k$ is Hermitian, the Wirtinger derivative must be computed as derived in Appendix A12, and the individual entries of the derivative are given by the derivative with respect to the individual entries, see (A25). Using Appendix A6, we obtain

$$\frac{\partial_{\mathrm{W}} R^{\mathrm{MAC}}(\boldsymbol{Q})}{\partial_{\mathrm{W}}[\boldsymbol{Q}_k]_{i,j}} = \frac{1}{\ln 2} \cdot \mathrm{tr}\left[\left(\mathbf{I} + \boldsymbol{H}\boldsymbol{Q}\boldsymbol{H}^{\mathrm{H}}\right)^{-1} \boldsymbol{H}_k \boldsymbol{e}_i \boldsymbol{e}_j^{\mathrm{T}} \boldsymbol{H}_k^{\mathrm{H}}\right]$$
$$= \frac{1}{\ln 2} \cdot \boldsymbol{e}_j^{\mathrm{T}} \boldsymbol{H}_k^{\mathrm{H}} \left(\mathbf{I} + \boldsymbol{H}\boldsymbol{Q}\boldsymbol{H}^{\mathrm{H}}\right)^{-1} \boldsymbol{H}_k \boldsymbol{e}_i$$

with $\boldsymbol{e}_j$ denoting the jth canonical unit vector of appropriate dimension. Hence, the Wirtinger derivative with respect to the transposed covariance matrix $\boldsymbol{Q}_k^{\mathrm{T}}$ of user k reads as

$$\boldsymbol{V}_k := \frac{\partial_{\mathrm{W}} R^{\mathrm{MAC}}(\boldsymbol{Q})}{\partial_{\mathrm{W}} \boldsymbol{Q}_k^{\mathrm{T}}} = \frac{1}{\ln 2} \cdot \boldsymbol{H}_k^{\mathrm{H}} \left(\mathbf{I} + \boldsymbol{H}\boldsymbol{Q}\boldsymbol{H}^{\mathrm{H}}\right)^{-1} \boldsymbol{H}_k \succcurlyeq \mathbf{0} \tag{6.6}$$

and is itself positive semi-definite. Using the function $R^{\mathrm{MAC}}(\boldsymbol{Q})$ in (6.4) and the Wirtinger derivatives in (6.6), we can apply the covariance-based gradient-projection algorithm from Sect. 5.1.2.

If the generalized Armijo rule in (5.27) is chosen for the step-size control mechanism, we set the scalar $s_0^{(n)}$ to

$$s_0^{(n)} = \frac{P_{\max}}{\mathrm{tr}(\boldsymbol{V}^{(n)})} = P_{\max} \cdot \mathrm{tr}^{-1}\left(\left. \frac{\partial_{\mathrm{W}} R^{\mathrm{MAC}}(\boldsymbol{Q})}{\partial_{\mathrm{W}} \boldsymbol{Q}^{\mathrm{T}}} \right|_{\boldsymbol{Q}=\boldsymbol{Q}^{(n)}} \right)$$
$$= \frac{P_{\max} \cdot \ln 2}{\sum_{k=1}^{K} \mathrm{tr}\left[\boldsymbol{H}_k^{\mathrm{H}} \left(\mathbf{I} + \boldsymbol{H}\boldsymbol{Q}^{(n)}\boldsymbol{H}^{\mathrm{H}}\right)^{-1} \boldsymbol{H}_k\right]} = \frac{P_{\max} \cdot \ln 2}{\mathrm{tr}\left[\left(\mathbf{I} + \boldsymbol{H}\boldsymbol{Q}^{(n)}\boldsymbol{H}^{\mathrm{H}}\right)^{-1} \boldsymbol{H}\boldsymbol{H}^{\mathrm{H}}\right]}. \tag{6.7}$$

To show that $s_0^{(n)}$ is a valid scaling of the step-size $s^{(n)}$, a positive lower and a finite upper bound for (6.7) must exist, see Sect. 5.1.3.3 or [93]. A lower bound follows from

$$\mathrm{tr}(\boldsymbol{V}^{(n)}) \leq \frac{1}{\ln 2} \cdot \|\boldsymbol{H}\|_{\mathrm{F}}^2$$

which results from the fact that all eigenvalues of $(\mathbf{I} + \boldsymbol{H}\boldsymbol{Q}^{(n)}\boldsymbol{H}^{\mathrm{H}})^{-1}$ are smaller than or equal to one. For the upper bound, note that $\bar{\boldsymbol{Q}} \succcurlyeq \boldsymbol{Q}$ for $\bar{\boldsymbol{Q}} = P_{\max} \cdot \mathbf{I}$ due to $\mathrm{tr}(\boldsymbol{Q}) \leq P_{\max}$. Therefore,

$$\mathrm{tr}(\boldsymbol{V}^{(n)}) = \frac{1}{\ln 2} \cdot \mathrm{tr}\left[\left(\mathbf{I} + \boldsymbol{H}\boldsymbol{Q}^{(n)}\boldsymbol{H}^{\mathrm{H}}\right)^{-1} \boldsymbol{H}\boldsymbol{H}^{\mathrm{H}}\right] \geq \frac{1}{\ln 2} \cdot \mathrm{tr}\left[\left(\mathbf{I} + P_{\max} \cdot \boldsymbol{H}\boldsymbol{H}^{\mathrm{H}}\right)^{-1} \boldsymbol{H}\boldsymbol{H}^{\mathrm{H}}\right].$$

Since a positive lower bound and a finite upper bound exist for the trace of the gradient, this choice for the step-size scaling $s_0^{(n)}$ is valid.

Synopsis: The covariance-based sum capacity computation with given sum power constraint is a nonconvex problem in the MIMO BC. In the dual domain, the sum rate is concave with respect to the covariance matrices and can be maximized efficiently by means of the projected gradient algorithm from Chap. 5. Having obtained the optimizing covariance matrices in the dual MIMO MAC, these can be converted back to the MIMO BC by means of the rate duality from Chap. 4.

6.1.3 Precoder-Based Sum Capacity Computation

The sum rate can not only be maximized by optimizing the covariance matrices, but also by optimizing the precoders. Doing so, the projections onto the constraint set and onto the tangent cone are much easier and do not require any eigenvalue decomposition, see Sect. 5.3.1. Another advantage results from the possibility to limit the maximum number of data streams per user by choosing the precoders to be tall matrices.[2] If the number of columns L_k of the precoder $\boldsymbol{T}_k \in \mathbb{C}^{M_k \times L_k}$ of user k is smaller than his number of antennas M_k, then user k will multiplex at most L_k data streams.

However, these benefits are achieved at the expense of the loss of concavity of the utility. This can already be observed in the single-user single-antenna case. While $\log_2(1 + \|\boldsymbol{h}_1\|_2^2 \cdot Q_1)$ is concave in Q_1, the expression $\log_2(1 + \|\boldsymbol{h}_1\|_2^2 \cdot |T_1|^2)$ is not concave in T_1. When switching from the covariance matrices to the precoders, additional stationary points of the sum rate maximization will therefore arise and these can have different utilities. For example, $\boldsymbol{T} = \boldsymbol{0}$ is a stationary point that achieves the global minimum, but there can also be stationary points that are suboptimum although they make use of the whole power budget $P_{\max}$. In the single-user single-stream case where $L_1 = 1$ and $\boldsymbol{T}_1$ is a column vector, every eigenvector of $\boldsymbol{H}_1^{\mathrm{H}} \boldsymbol{H}_1$ with squared norm $P_{\max}$ is a stationary point. Clearly, only the one belonging to the largest eigenvalue achieves the largest sum rate. Fortunately, the projected gradient algorithm will not converge to such a suboptimal stationary point unless it has already been initialized with a scaled version of a suboptimal eigenvector. However, it is not clear whether this will also hold in the multi-user case with multiple streams per user allowed.

In analogy to (6.4), the sum rate in the dual MIMO MAC reads for a block-diagonal composite precoder $\boldsymbol{T} := \mathrm{blockdiag}\{\boldsymbol{T}_k\}_{k=1}^K \in \mathbb{C}^{M \times L}$ as

---

[2] In this case, the sum capacity need not necessarily be achievable.

$$R^{\mathrm{MAC}}(\boldsymbol{T}) := \log_2\left|\mathbf{I} + \sum_{k=1}^{K} \boldsymbol{H}_k\boldsymbol{T}_k\boldsymbol{T}_k^{\mathrm{H}}\boldsymbol{H}_k^{\mathrm{H}}\right| = \log_2\left|\mathbf{I} + \boldsymbol{H}\boldsymbol{T}\boldsymbol{T}^{\mathrm{H}}\boldsymbol{H}^{\mathrm{H}}\right|. \tag{6.8}$$

For the two recursions (5.16) and (5.17), the Wirtinger derivative of $R^{\mathrm{MAC}}(\boldsymbol{T})$ with respect to the conjugate precoders is needed. To this end, we first compute the Wirtinger derivative with respect to the entry $[\boldsymbol{T}_k^*]_{i,j}$ (see Appendix A6):

$$\begin{aligned}\frac{\partial_{\mathrm{W}} R^{\mathrm{MAC}}(\boldsymbol{T})}{\partial_{\mathrm{W}}[\boldsymbol{T}_k^*]_{i,j}} &= \frac{1}{\ln 2}\cdot\operatorname{tr}\left[\left(\mathbf{I}+\boldsymbol{H}\boldsymbol{T}\boldsymbol{T}^{\mathrm{H}}\boldsymbol{H}^{\mathrm{H}}\right)^{-1}\boldsymbol{H}_k\boldsymbol{T}_k\boldsymbol{e}_j\boldsymbol{e}_i^{\mathrm{T}}\boldsymbol{H}_k^{\mathrm{H}}\right]\\ &= \frac{1}{\ln 2}\cdot\boldsymbol{e}_i^{\mathrm{T}}\boldsymbol{H}_k^{\mathrm{H}}\left(\mathbf{I}+\boldsymbol{H}\boldsymbol{T}\boldsymbol{T}^{\mathrm{H}}\boldsymbol{H}^{\mathrm{H}}\right)^{-1}\boldsymbol{H}_k\boldsymbol{T}_k\boldsymbol{e}_j.\end{aligned}$$

Then, the complete Wirtinger derivative with respect to $\boldsymbol{T}_k^*$ reads as [see (A25) in Appendix A11]

$$\boldsymbol{V}_k := \frac{\partial_{\mathrm{W}} R^{\mathrm{MAC}}(\boldsymbol{T})}{\partial_{\mathrm{W}}\boldsymbol{T}_k^*} = \frac{1}{\ln 2}\cdot\boldsymbol{H}_k^{\mathrm{H}}\left(\mathbf{I}+\boldsymbol{H}\boldsymbol{T}\boldsymbol{T}^{\mathrm{H}}\boldsymbol{H}^{\mathrm{H}}\right)^{-1}\boldsymbol{H}_k\boldsymbol{T}_k \quad \forall k. \tag{6.9}$$

Above equation reveals that any block-diagonal precoder $\boldsymbol{T}$ that satisfies $\boldsymbol{H}\boldsymbol{T} = \mathbf{0}$ is also a stationary point. Obviously, it depends on the number of antennas and data streams per user whether a nonzero $\boldsymbol{T} \in \mathcal{B}_{\boldsymbol{T}}$ can exist such that $\boldsymbol{H}\boldsymbol{T} = \mathbf{0}$.

If the unconstrained gradient is used for the search direction as in (5.16), the update rule reads as

$$\boldsymbol{T}_k^{(n+1)} = \kappa^{(n)}\cdot\left(\mathbf{I} + \frac{s^{(n)}}{\ln 2}\cdot\boldsymbol{H}_k^{\mathrm{H}}\left(\mathbf{I}+\boldsymbol{H}\boldsymbol{T}^{(n)}\boldsymbol{T}^{(n),\mathrm{H}}\boldsymbol{H}^{\mathrm{H}}\right)^{-1}\boldsymbol{H}_k\right)\boldsymbol{T}_k^{(n)} \quad \forall k \tag{6.10}$$

where $\kappa^{(n)}$ is either equal to one if the resulting *total* squared Frobenius norm $\|\boldsymbol{T}^{(n+1)}\|_{\mathrm{F}}^2$ is at most $P_{\max}$ or $\kappa^{(n)} < 1$ such that $\|\boldsymbol{T}^{(n+1)}\|_{\mathrm{F}}^2 = P_{\max}$ with equality. Thus, $\kappa^{(n)}$ performs the orthogonal projection onto the constraint set, see (5.44). If $\boldsymbol{T}^{(0)} \in \mathcal{B}_{\boldsymbol{T}}$ is initialized such that $\|\boldsymbol{T}^{(0)}\|_{\mathrm{F}}^2 = P_{\max}$, moving along the direction of the unconstrained gradient will always leave the constraint set $\mathcal{C}$ ($\boldsymbol{H}\boldsymbol{T} \neq \mathbf{0}$ assumed) and thus consume more power such that the projection onto the constraint set is necessary and $\kappa^{(n)} < 1$ will hold for every n.

For the steepest ascent direction, the total gradient $\boldsymbol{V} := \operatorname{blockdiag}\{\boldsymbol{V}_k\}_{k=1}^K$ with $\boldsymbol{V}_k$ from (6.9) must be projected onto the tangent cone. For $\|\boldsymbol{T}\|_{\mathrm{F}}^2 = P_{\max}$ and since $\operatorname{tr}(\boldsymbol{T}^{\mathrm{H}}\boldsymbol{V}) > 0$ [see (6.9)], the projection is given by [see (5.48)]

$$\mathcal{P}_{\mathcal{T}_{\mathcal{C}}(\boldsymbol{T})}(\boldsymbol{V}) = \left[\boldsymbol{V} - \frac{\mathrm{Re}[\mathrm{tr}(\boldsymbol{T}^{\mathrm{H}}\boldsymbol{V})]}{P_{\max}}\boldsymbol{T}\right]$$
$$= \frac{1}{\ln 2} \cdot \mathrm{blockdiag}\left\{\left[\boldsymbol{H}_k^{\mathrm{H}}(\mathbf{I} + \boldsymbol{H}\boldsymbol{T}\boldsymbol{T}^{\mathrm{H}}\boldsymbol{H}^{\mathrm{H}})^{-1}\boldsymbol{H}_k - \alpha\mathbf{I}\right] \cdot \boldsymbol{T}_k\right\}_{k=1}^{K} \tag{6.11}$$

where the scalar α reads as

$$\alpha = \frac{\mathrm{Re}[\mathrm{tr}(\boldsymbol{T}^{\mathrm{H}}\boldsymbol{V})] \cdot \ln 2}{P_{\max}}$$
$$= \frac{1}{P_{\max}} \cdot \sum_{k=1}^{K} \mathrm{tr}\left[\boldsymbol{T}_k^{\mathrm{H}}\boldsymbol{H}_k^{\mathrm{H}}(\mathbf{I} + \boldsymbol{H}\boldsymbol{T}\boldsymbol{T}^{\mathrm{H}}\boldsymbol{H}^{\mathrm{H}})^{-1}\boldsymbol{H}_k\boldsymbol{T}_k\right]$$
$$= \frac{1}{P_{\max}} \cdot \mathrm{tr}\left[(\mathbf{I} + \boldsymbol{H}\boldsymbol{T}\boldsymbol{T}^{\mathrm{H}}\boldsymbol{H}^{\mathrm{H}})^{-1}\boldsymbol{H}\boldsymbol{T}\boldsymbol{T}^{\mathrm{H}}\boldsymbol{H}^{\mathrm{H}}\right]$$
$$= \frac{N - \mathrm{tr}\left[(\mathbf{I} + \boldsymbol{H}\boldsymbol{T}\boldsymbol{T}^{\mathrm{H}}\boldsymbol{H}^{\mathrm{H}})^{-1}\right]}{P_{\max}} \geq 0. \tag{6.12}$$

Hence, the precoder update rule in (5.17) can be expressed user-wise by

$$\boldsymbol{T}_k^{(n+1)} = \kappa'^{(n)} \cdot \left(\mathbf{I} + \frac{s^{(n)}}{\ln 2}\left[\boldsymbol{H}_k^{\mathrm{H}}(\mathbf{I} + \boldsymbol{H}\boldsymbol{T}^{(n)}\boldsymbol{T}^{(n)\,\mathrm{H}}\boldsymbol{H}^{\mathrm{H}})^{-1}\boldsymbol{H}_k - \alpha^{(n)}\mathbf{I}\right]\right)\boldsymbol{T}_k^{(n)} \quad \forall k \tag{6.13}$$

where $\kappa'^{(n)} \leq 1$ performs the simultaneous projection onto the constraint set and $\alpha^{(n)}$ follows from (6.12) with $\boldsymbol{T}$ replaced by $\boldsymbol{T}^{(n)}$. Clearly, $\boldsymbol{T}^{(0)}$ should be initialized with $\|\boldsymbol{T}^{(0)}\|_{\mathrm{F}}^2 = P_{\max}$, otherwise the tangent-cone projection would be the identity mapping, see (5.48). For such an initialization, $\boldsymbol{T}^{(n+1)} \in \partial\mathcal{C}$ holds for all n, i.e., $\kappa'^{(n)} > 1$ and $\|\boldsymbol{T}^{(n+1)}\|_{\mathrm{F}}^2 = P_{\max}$ $\forall n$ and the constraint set projection is mandatory ($\boldsymbol{H}\boldsymbol{T} \neq \mathbf{0}$ assumed).

The update rules (6.10) and (6.13) show that the precoders of iteration $n+1$ follow from the precoders of iteration n by a left-hand side multiplication with a square matrix. Hence, the rank of the precoders cannot increase during the iterations which has to be considered for the initialization of the precoders. If $\boldsymbol{T}_k \in \mathbb{C}^{M_k \times L_k}$ shall in principle be able to multiplex up to $L_k \leq M_k$ data streams, then the respective precoder $\boldsymbol{T}_k^{(0)}$ must be initialized with rank L_k.

For the Armijo step-size control from Sect. 5.1.3.3, we choose the common step-size factor $s_0^{(n)}$ in (5.27) such that the unconstrained total gradient scaled by $s_0^{(n)}$ has the same Frobenius norm as the total precoder:

$$s_0^{(n)} = \frac{\|\boldsymbol{T}^{(n)}\|_{\mathrm{F}}}{\|\boldsymbol{V}^{(n)}\|_{\mathrm{F}}} = \frac{\|\boldsymbol{T}^{(n)}\|_{\mathrm{F}} \cdot \ln 2}{\sqrt{\sum_{k=1}^{K} \|\boldsymbol{H}_k^{\mathrm{H}}(\mathbf{I} + \boldsymbol{H}\boldsymbol{T}^{(n)}\boldsymbol{T}^{(n)\,\mathrm{H}}\boldsymbol{H}^{\mathrm{H}})^{-1}\boldsymbol{H}_k\boldsymbol{T}_k^{(n)}\|_{\mathrm{F}}^2}}$$

The initial precoder $\boldsymbol{T}^{(0)}$ must be chosen such that $\boldsymbol{H}\boldsymbol{T}^{(0)} \neq \mathbf{0}$ to let $s_0^{(n)}$ have a positive lower bound.

Synopsis: The sum rate of the dual MIMO MAC can also be maximized by optimizing the precoders instead of the transmit covariance matrices. In this case, the projections onto the constraint set and onto the tangent cone are very simple. Moreover, the maximum number of data streams per user can be controlled by limiting the number of columns of the precoders. For square precoders, the sum capacity may be achieved if they are initialized with full rank. Unfortunately, the precoder-based optimization has additional stationary points that are not globally optimum. It is unclear, whether the gradient-projection algorithm will converge to such stationary points or not.

6.1.4 Simulation Results

In this subsection, we compare the presented algorithms to existing ones with respect to the speed of convergence. In particular, we analyze the precoder- and covariance-based gradient-projection algorithms which either use the unconstrained gradient as search direction or its projection onto the tangent cone. In addition, we investigate the performance of the standard iterative water-filling algorithm of Jindal et al. in [111] and its extension to an iteration-dependent memory factor of Böhnke et al. in [105], see Sect. 6.1.1 for more details.

For our simulations, $K = 3$ users are served by an N-antenna base station in the MIMO BC, where either $N = 3$ or $N = 6$ holds. Each of the three users is equipped with two antennas, i.e., $M_1 = M_2 = M_3 = 2$. The covariance-based algorithms are initialized by scaled identity matrices consuming the total amount of available power, i.e.,

$$\boldsymbol{Q}_k^{(0)} = \frac{P_{\max}}{6} \cdot \mathbf{I}_2 \quad \forall k \in \{1, 2, 3\}$$

and the precoder-based ones are equivalently initialized by square matrices via

$$\boldsymbol{T}_k^{(0)} = \sqrt{\frac{P_{\max}}{6}} \cdot \mathbf{I}_2 \quad \forall k \in \{1, 2, 3\}.$$

The noise covariance matrices in the original BC are also assumed to be identity matrices for all three users. Figures 6.1 and 6.2 show the relative distance to the sum capacity versus the number of iterations for $10 \log_{10} P_{\max} = -5\,\text{dB}$ and $10 \log_{10} P_{\max} = +5\,\text{dB}$, respectively, for a randomly chosen channel realization. This relative distance to capacity is defined as one minus the quotient of the achieved sum rate and the sum capacity. The left figures correspond to the $N = 6$ antenna base

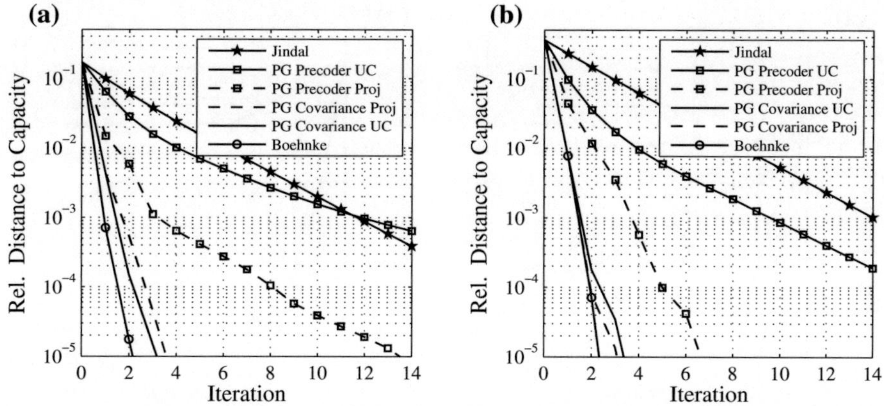

Fig. 6.1 Relative distance to sum capacity for $K = 3$ users and $10 \log_{10} P_{\max} = -5\,\mathrm{dB}$. **a** $N = 6$ antennas at the base station. **b** $N = 3$ antennas at the base station

station, whereas the right ones show the case with $N = 3$. We used the Armijo line search from Sect. 5.1.3.3 for all gradient-projection-based algorithms where we also allowed for an increase of the step-size, i.e., the exponent $m^{(n)}$ in (5.28) and (5.31) was taken from $\mathbb{Z}$. Note that the parameter β of the Armijo rule was set to $\beta = 0.5$ in all cases.

In Fig. 6.1a, we see that for $10 \log_{10} P_{\max} = -5\,\mathrm{dB}$ and $N = 6$, the scaled identity initialization achieves about 80 % of the sum capacity. Given this starting point, the iterative water-filling algorithm of Jindal [111] (`Jindal`–star marker) shows the smallest initial speed of convergence which follows from the conservatively chosen (convergence ensuring) fraction that the water-filling solution contributes to the covariance matrix update. Up to a relative distance to capacity of about 10^{-3}, the precoder-based projected gradient algorithm using the unconstrained gradient from (6.9) as search direction (`PG Precoder UC`–solid curve, square marker) outperforms the iterative water-filling algorithm. Note that $\sigma = 0.1$ was used for the Armijo line search. Since the optimum step-size often turned out to be infinite when the unconstrained gradient of the sum rate utility is used as search direction for the precoder-based approach, σ was chosen that small. The performance is drastically increased if the projection of the gradient onto the tangent cone is used as search direction (`PG Precoder Proj`–dashed curve, square marker). In this case as well as for the covariance-based algorithms, $\sigma = 0.3$ was used. Since the projection onto the tangent cone is very simple in case of the precoders, the larger speed of convergence is achieved by a negligible complexity increase. An even faster convergence is achieved by the covariance-based gradient projection algorithm both for the unconstrained gradient as search direction (`PG Covariance UC`–solid curve, no marker) and for its projection onto the tangent cone (`PG Covariance Proj`–dashed curve, no marker). One reason for the quick convergence is the fact that the projection onto the constraint set can switch off individual data streams completely, see the

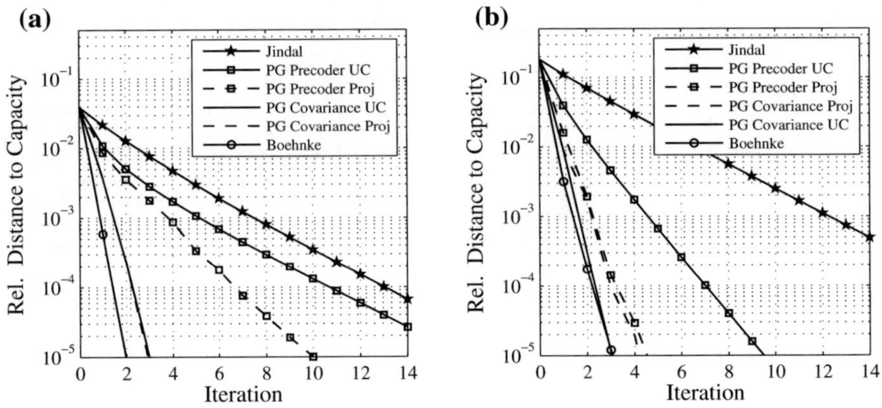

Fig. 6.2 Relative distance to sum capacity for $K = 3$ users and $10 \log_{10} P_{\max} = +5$ dB. **a** $N = 6$ antennas at the base station. **b** $N = 3$ antennas at the base station

water-spilling interpretation in Sect. 5.3.2. For the precoder-based algorithms, this is impossible within one step. There, the streams are switched off asymptotically over several iterations. The fastest convergence is obtained by the modified sum power iterative water-filling algorithm of Böhnke [105] (`Boehnke`–circle marker). Since the iterative water-filling partially exploits the structure of KKT conditions of the problem by rewriting it as parallel, noninterfering channels and since the memory-factor is chosen nearly optimal, this problem-tailored algorithm is slightly faster than the gradient-projection-based ones.

In Fig. 6.1b, the base station has only $N = 3$ antennas. Due to the reduced number of degrees of freedom, the full-rank initialization of all covariance matrices and precoders achieves only 60 % of the sum capacity. The convergence behavior basically remains unchanged except that the precoder-based algorithm utilizing the projected gradient as search direction (`PG Precoder Proj`–dashed curve, square marker) now features an iteration-independent slope.

Raising the transmit power by 10 dB leads to the results in Fig. 6.2. If enough degrees of freedom are available at the base station to let all three users apply full multiplexing, the asymptotically optimum transmit covariance matrices are scaled identities as we will see in Chap. 8. Thus, this initialization already achieves 96 % of the sum capacity for an $N = 6$ antenna base station with $10 \log_{10} P_{\max} = +5$ dB, see Fig. 6.2a. As we mentioned in Sect. 5.3.2.3, the constraint set projection leads to the same result for full rank covariance matrices irrespective of whether the unconstrained gradient is used (`PG Covariance UC`–solid curve, no marker) or its projection onto the tangent cone (`PG Covariance Proj`–dashed curve, no marker). Hence, these two algorithms feature the same performance for the given setup.

For only $N = 3$ antennas at the base station, the full rank initialization of all precoders and covariance matrices yields only 80 percent of the sum capacity, see Fig. 6.2b. In this case, both the two covariance-based gradient-projection approaches

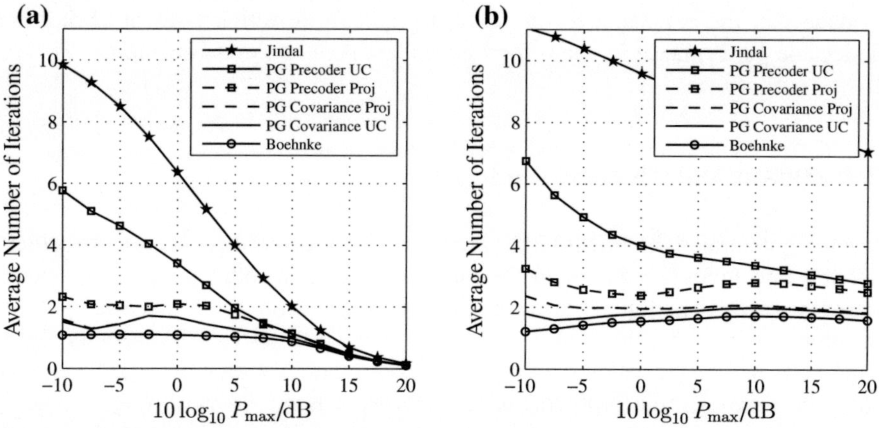

Fig. 6.3 Average number of iterations to achieve 0.995·sum capacity for $K = 3$ users. **a** $N = 6$ antennas at the base station. **b** $N = 3$ antennas at the base station

and the precoder-based approach with the projection of the gradient onto the tangent cone have excellent convergence properties.

While Figs. 6.1 and 6.2 show the convergence behavior for a randomly chosen channel realization, Fig. 6.3 shows channel-averaged results. More precisely, the average number of iterations that are required to achieve 0.995 times the sum capacity is plotted versus the logarithmic transmit power $P_{\max}$ for $N = 6$ (Fig. 6.3a) and $N = 3$ (Fig. 6.3b) antennas at the base station. The entries of the composite channel matrix $\boldsymbol{H}$ were chosen as i.i.d. zero-mean complex-valued Gaussian random variables with variance one. In Fig. 6.3a, we can observe that the average number of iterations decays to zero as the transmit power increases. Again, this results from the fact that for high power, the initialization with scaled identity matrices is asymptotically optimum since enough degrees of freedom are available. For small transmit powers, the standard sum power iterative water-filling of Jindal (`Jindal`–star marker) requires 10 iterations on average. Apart from the precoder-based gradient-projection algorithm with the unconstrained gradient as search direction (`PG Precoder UC`–solid curve, square marker), all other algorithms require at most only two iterations to achieve 0.995 times the sum capacity independent of the given transmit power.

If the base station is equipped with only $N = 3$ antennas, scaled identity matrices are no longer asymptotically optimum. Hence, all algorithms require at least two iterations at large transmit powers. In turn, the two covariance-based gradient projection algorithms (without markers) and the algorithm of Böhnke (`Boehnke`–circle marker) need about two iterations for convergence independent of the transmit power. The two precoder-based algorithms (with square markers) need slightly more, whereas the standard sum power iterative water-filling algorithm of Jindal (`Jindal`–star marker) takes much more.

Note that the extension to the relaxed gradient-projection does not bring any noticeable performance increase for the standard sum rate maximization.

6.2 Weighted Sum Rate Maximization

The full potential of the MIMO BC can only be unleashed if there is a way to achieve not only rate tuples that maximize the sum rate but also any other point on the boundary of the capacity region. By means of the *weighted* sum rate maximization (possibly in combination with time-sharing) any point on the boundary of the capacity region can be attained. In the following, we present a short motivation where such weighted sum rate maximization problems arise and why it is necessary to have a quickly converging algorithm that solves this optimization. After a literature overview, the gradient-projection-based algorithm is derived which is targeted at maximizing any weighted sum rate with nonnegative weights.

6.2.1 Motivation

The weighted sum rate (WSR) maximization plays an important role when it comes to utility maximization, scheduling, and Quality-of-Service optimizations. In [115] and [116], Palomar et al. give an overview of decomposition methods for the network utility maximization. The presented decomposition technique can be used to decompose a big problem into smaller subproblems that can be handled and solved efficiently. Often, the Lagrange dual problem is decomposed instead of the primal one which is then termed *dual decomposition*. Thus it makes sense only if strong duality holds with vanishing duality gap. A frequently arising decomposition is into a subproblem where the weighted sum rate has to be maximized where the weights are Lagrangian multipliers, and into a master problem at the higher level which is in charge of updating the Lagrangian multipliers. An example for such a network utility maximization can be found in the cross-layer optimization from Brehmer et al. in [117], where a utility function is used to model the properties of the upper OSI layers. In the presented two-step approach, a rate vector that maximizes the utility in the rate space has to be found first. Then, the physical layer parameters like encoding order, covariance matrices, and time-sharing fractions that achieve the respective rate tuple must be determined. During the first step (maximization in the rate space), a WSR problem arises when solving the Lagrange dual function during a correction step which can be seen as the projection onto the boundary of the capacity region. The weights of this WSR problem are given by the (nonnegative) Lagrangian multipliers which correspond to the inequalities that the rate tuple is dominated by a vector which belongs to the capacity region.

In the scheduling context, the WSR problem is part of the proportional fair scheduler where the weights are chosen as the inverse average throughputs that have so far been achieved, see [118 Section V.A]. Instead, if the weights of the WSR are chosen as the queue system states, one ends up with a stability-optimal transmission policy [119].

Finally, the weighted sum rate maximization problem often makes up the core of Quality-of-Service optimizations like the power minimization with given rate requirements from Wunder et al. in [120] and [121].

These examples show that the WSR problem arises in many contexts which motivates the demand for efficient techniques to solve this problem.

6.2.2 Literature Overview

The very first algorithm for the maximization of any weighted sum rate in the MIMO BC utilizing dirty paper coding was introduced in 2003 by Viswanathan et al. see [118]. Like all other WSR maximizing approaches, it operates in the dual MIMO MAC. In every step of the iterative algorithm, the covariance matrix of only a single user experiences a structural update. More precisely, a rank-one covariance matrix update in the direction of the strongest eigenmode of all gradients is applied to that user, whose gradient matrix has the largest eigenvalue. The optimum step-size is obtained by means of a bisection which involves many function evaluations. Since a structural update is applied to the covariance matrix of only a single user, and since this update is only a rank-one update, the speed of convergence is very low. Although claimed to be optimal by Viswanathan, the authors of [122] disagree on the global optimality of the presented approach. Besides their theoretical argumentation that the Zoutendijk method used in [118] is not closed and thus need not necessarily converge, they also find a channel configuration for which the algorithm in [118] in fact fails to achieve the maximum weighted sum rate.

Three years later, Kobayashi and Caire came up with an extension of the sum power iterative water-filling algorithm [111] from Jindal to the case of different weights [123]. Following the idea of Yu and Cioffi [71], the multi-point-to-point communication in the dual MAC is considered as K parallel single-input multiple-output links with independent effective channels. However, their contribution is only applicable for single-antenna terminals. During the power allocation step, an exact line-search must be done to find the Lagrangian multiplier that ensures the sum power limitation.

A precoder-based conjugate gradient-projection algorithm was proposed by Böhnke et al. in [124] and [62]. In fact, the authors use the projection of the gradient onto the tangent cone as search direction but without actually knowing that this is the direction of the steepest ascent of the composition of the utility function and the projection. In their approach, the projection onto the tangent cone is just a vehicle to reduce the impact of the old search direction of the previous iteration onto the current one. For the step-size control mechanism, they choose the on-demand step-size reduction described in Sect. 5.1.3.5 which may in principle fail to converge to a stationary point due to a too small increase of the utility compared to the step-size.

In 2008, Liu et al. also published a gradient-projection-based variant for the maximization of the weighted sum rate [122] which can be regarded as an extension of their previous sum rate maximization algorithm in [109]. It operates on the covariance matrices and uses the standard (unconstrained) gradient as search direction. Equivalent to the approach from Böhnke in [124], conjugate search directions are used but the unconstrained gradient is neither normalized nor projected onto the tangent cone. Due to the additional outer line-search, their approach belongs to the class of relaxed gradient projection algorithms.

In the same year, we proposed a very similar approach using the standard unconstrained gradient as search direction and not the conjugate ones in [108]. In addition, we introduced the water-spilling interpretation for the orthogonal projection of the covariance matrices onto the constraint set, see Sect. 5.3.2.

6.2.3 Covariance-Based Weighted Sum Rate Maximization

The so far only known way to optimally solve the sum power constrained weighted sum rate maximization problem in the MIMO BC for more than one user is to tackle the optimization in the dual MIMO MAC and then convert the solution back to the original BC via the rate duality with interference cancelation. For a particular decoding order in which user $\pi[1]$ is decoded first and user $\pi[k]$ last, the maximum error-free data rate of some user $\pi[k]$ with $k \in \{1, \ldots K\}$ reads as [see (4.15)]

$$R_{\pi[k]}^{\mathrm{MAC}}(\boldsymbol{Q}) = \log_2 \left| \mathbf{I} + \Big(\mathbf{I} + \sum_{i>k} \boldsymbol{H}_{\pi[i]} \boldsymbol{Q}_{\pi[i]} \boldsymbol{H}_{\pi[i]}^{\mathrm{H}} \Big)^{-1} \boldsymbol{H}_{\pi[k]} \boldsymbol{Q}_{\pi[k]} \boldsymbol{H}_{\pi[k]}^{\mathrm{H}} \right|. \tag{6.14}$$

For the weighted sum rate maximization, nonnegative weights $w_1, \ldots, w_K \in \mathbb{R}_{+,0}$ are associated to the K users. As derived in Appendix A7, the optimum decoding order directly follows from these weights and an optimum decoding order[3] is achieved when the users are decoded in nondecreasing priorities, i.e., the user with the smallest weight has to be decoded first whereas the user with the largest weight must be decoded last. To improve readability, we assume that $w_1 \leq w_2 \leq \cdots \leq w_K$ holds such that the optimum decoding order is the natural one in which user one is decoded first and user K last, i.e., $\pi[k] = k \; \forall k$. This setup can always be achieved by relabeling all users. Then, (6.14) can be written as

$$R_k^{\mathrm{MAC}}(\boldsymbol{Q}) = \log_2 \left| \mathbf{I} + \sum_{i \geq k} \boldsymbol{H}_i \boldsymbol{Q}_i \boldsymbol{H}_i^{\mathrm{H}} \right| - \log_2 \left| \mathbf{I} + \sum_{i>k} \boldsymbol{H}_i \boldsymbol{Q}_i \boldsymbol{H}_i^{\mathrm{H}} \right|. \tag{6.15}$$

[3] For some degenerated cases, the optimum decoding order is not unique. Think of a two user scenario where the two channel matrices are orthogonal such that the two users do not interfere with each other. In this case, the decoding order is arbitrary. However, choosing the decoding order in nondecreasing priorities is always optimal.

Using this difference representation for the rate of user k, the weighted sum rate $f(\boldsymbol{Q}, \boldsymbol{w})$ that depends on the transmit covariance matrices in $\boldsymbol{Q} = \text{blockdiag}\{\boldsymbol{Q}_k\}_{k=1}^K$ and on the user weights $\boldsymbol{w} := [w_1, \ldots, w_K]^{\mathrm{T}}$ reads as

$$
\begin{aligned}
f(\boldsymbol{Q}, \boldsymbol{w}) &:= \sum_{k=1}^{K} w_k R_k^{\mathrm{MAC}}(\boldsymbol{Q}) \\
&= \sum_{k=1}^{K} w_k \log_2 \left| \mathbf{I} + \sum_{i \geq k} \boldsymbol{H}_i \boldsymbol{Q}_i \boldsymbol{H}_i^{\mathrm{H}} \right| - \sum_{k=1}^{K} w_k \log_2 \left| \mathbf{I} + \sum_{i > k} \boldsymbol{H}_i \boldsymbol{Q}_i \boldsymbol{H}_i^{\mathrm{H}} \right| \\
&= \sum_{k=1}^{K} v_k \log_2 \left| \mathbf{I} + \sum_{i \geq k} \boldsymbol{H}_i \boldsymbol{Q}_i \boldsymbol{H}_i^{\mathrm{H}} \right|
\end{aligned}
\tag{6.16}
$$

where the nonnegative scalars $v_1, \ldots, v_K$ are defined via

$$
v_k := \begin{cases} w_k - w_{k-1} & \text{for } k > 1 \\ w_1 & \text{for } k = 1. \end{cases} \tag{6.17}
$$

In Appendix A9, it is shown that $f(\boldsymbol{Q}, \boldsymbol{w})$ is concave in $\boldsymbol{Q}$. Hence, any stationary point of the sum power constrained weighted sum rate maximization is a global optimizer and the covariance-based gradient-projection algorithm utilizing the Armijo rule for the step-size control will converge to such a stationary point. Since the utility again need not be *strictly* concave, the set containing all optimizers need not be singleton.

For the gradient-projection algorithm in Chap. 5, the Wirtinger derivative of $f(\boldsymbol{Q}, \boldsymbol{w})$ with respect to $\boldsymbol{Q}_j^{\mathrm{T}}$ has to be computed for every user $j \in \{1, \ldots, K\}$, see (5.23) and (5.24). Using Appendix A6, the derivative of (6.16) with respect to the entry in the mth row and nth column of $\boldsymbol{Q}_j$ reads as

$$
\begin{aligned}
\frac{\partial_{\mathrm{W}} f(\boldsymbol{Q}, \boldsymbol{w})}{\partial_{\mathrm{W}} [\boldsymbol{Q}_j]_{m,n}} &= \sum_{k=1}^{j} \frac{v_k}{\ln 2} \cdot \operatorname{tr}\left[\left(\mathbf{I} + \sum_{i \geq k} \boldsymbol{H}_i \boldsymbol{Q}_i \boldsymbol{H}_i^{\mathrm{H}} \right)^{-1} \boldsymbol{H}_j \boldsymbol{e}_m \boldsymbol{e}_n^{\mathrm{T}} \boldsymbol{H}_j^{\mathrm{H}} \right] \\
&= \sum_{k=1}^{j} \frac{v_k}{\ln 2} \cdot \boldsymbol{e}_n^{\mathrm{T}} \boldsymbol{H}_j^{\mathrm{H}} \left(\mathbf{I} + \sum_{i \geq k} \boldsymbol{H}_i \boldsymbol{Q}_i \boldsymbol{H}_i^{\mathrm{H}} \right)^{-1} \boldsymbol{H}_j \boldsymbol{e}_m.
\end{aligned}
$$

The complete Wirtinger derivative with respect to $\boldsymbol{Q}_j^{\mathrm{T}}$ can therefore be expressed as

$$
\frac{\partial_{\mathrm{W}} f(\boldsymbol{Q}, \boldsymbol{w})}{\partial_{\mathrm{W}} \boldsymbol{Q}_j^{\mathrm{T}}} = \sum_{k=1}^{j} \frac{v_k}{\ln 2} \cdot \boldsymbol{H}_j^{\mathrm{H}} \left(\mathbf{I} + \sum_{i \geq k} \boldsymbol{H}_i \boldsymbol{Q}_i \boldsymbol{H}_i^{\mathrm{H}} \right)^{-1} \boldsymbol{H}_j \succeq \mathbf{0} \tag{6.18}
$$

and is again positive semi-definite. For the generalized Armijo rule in (5.27), the scalar $s_0^{(n)}$ is in turn chosen as the quotient of $P_{\max}$ and the trace of the total gradient $\mathrm{tr}(\boldsymbol{V}^{(n)})$ to keep the number of step-size corrections small and almost independent of the available transmit power:

$$s^{(n)} = \frac{P_{\max}}{\mathrm{tr}(\boldsymbol{V}^{(n)})}.$$

This time, the trace of the gradient reads as

$$\begin{aligned}\mathrm{tr}\,(\boldsymbol{V}^{(n)}) :&= \mathrm{tr}\left(\left.\frac{\partial_{\mathrm{W}} f(\boldsymbol{Q},\boldsymbol{w})}{\partial_{\mathrm{W}} \boldsymbol{Q}^{\mathrm{T}}}\right|_{\boldsymbol{Q}=\boldsymbol{Q}^{(n)}} \right) \\ &= \sum_{j=1}^{K}\sum_{k=1}^{j} \frac{v_k}{\ln 2} \cdot \mathrm{tr}\left[\boldsymbol{H}_j^{\mathrm{H}} \Big(\mathbf{I} + \sum_{i\geq k} \boldsymbol{H}_i \boldsymbol{Q}_i^{(n)} \boldsymbol{H}_i^{\mathrm{H}} \Big)^{-1} \boldsymbol{H}_j \right].\end{aligned}$$

A positive lower and a finite upper bound on $s_0^{(n)}$ can be shown to exist in the same way as for the sum capacity computation in Sect. 6.1.2.

6.2.4 Precoder-Based Weighted Sum Rate Maximization

In the precoder parametrization, the weighted sum rate in the dual MIMO MAC reads as [cf. (6.16)]

$$f(\boldsymbol{T},\boldsymbol{w}) := \sum_{k=1}^{K} v_k \log_2 \Big| \mathbf{I} + \sum_{i\geq k} \boldsymbol{H}_i \boldsymbol{T}_i \boldsymbol{T}_i^{\mathrm{H}} \boldsymbol{H}_i^{\mathrm{H}} \Big|$$

where the difference of the weights v_k, $k \in \{1,\ldots,K\}$, is defined as in (6.17) and the block-diagonal precoder is defined via $\boldsymbol{T} = \mathrm{blockdiag}\{\boldsymbol{T}_k\}_{k=1}^{K} \in \mathbb{C}^{M\times L}$. The Wirtinger derivative of $f(\boldsymbol{T},\boldsymbol{w})$ with respect to the conjugate precoder $\boldsymbol{T}_j^* \; \forall j$, which is needed for the two recursions of the gradient-projection algorithm in (5.16) and (5.17), follows from the derivative with respect to the individual entries of $\boldsymbol{T}_j^*$ according to Appendix A6:

$$\begin{aligned}\frac{\partial_{\mathrm{W}} f(\boldsymbol{T},\boldsymbol{w})}{\partial_{\mathrm{W}} [\boldsymbol{T}_j^*]_{m,n}} &= \sum_{k=1}^{j} \frac{v_k}{\ln 2} \cdot \mathrm{tr}\left[\Big(\mathbf{I} + \sum_{i\geq k} \boldsymbol{H}_i \boldsymbol{T}_i \boldsymbol{T}_i^{\mathrm{H}} \boldsymbol{H}_i^{\mathrm{H}} \Big)^{-1} \boldsymbol{H}_j \boldsymbol{T}_j \boldsymbol{e}_n \boldsymbol{e}_m^{\mathrm{T}} \boldsymbol{H}_j^{\mathrm{H}} \right] \\ &= \sum_{k=1}^{j} \frac{v_k}{\ln 2} \cdot \boldsymbol{e}_m^{\mathrm{T}} \boldsymbol{H}_j^{\mathrm{H}} \Big(\mathbf{I} + \sum_{i\geq k} \boldsymbol{H}_i \boldsymbol{T}_i \boldsymbol{T}_i^{\mathrm{H}} \boldsymbol{H}_i^{\mathrm{H}} \Big)^{-1} \boldsymbol{H}_j \boldsymbol{T}_j \boldsymbol{e}_n.\end{aligned}$$

Therefore, the complete Wirtinger derivative with respect to $\boldsymbol{T}_j^*$ reads as

$$\frac{\partial_{\mathrm{w}} f(\boldsymbol{T}, \boldsymbol{w})}{\partial_{\mathrm{w}} \boldsymbol{T}_j^*} = \sum_{k=1}^{j} \frac{v_k}{\ln 2} \cdot \boldsymbol{H}_j^{\mathrm{H}} \Big(\mathbf{I} + \sum_{i \geq k} \boldsymbol{H}_i \boldsymbol{T}_i \boldsymbol{T}_i^{\mathrm{H}} \boldsymbol{H}_i^{\mathrm{H}}\Big)^{-1} \boldsymbol{H}_j \boldsymbol{T}_j. \tag{6.19}$$

Switching from the covariance-based to the precoder-based approach again leads to the loss of concavity and generates additional stationary points. Moreover, $\boldsymbol{HT} \neq \mathbf{0}$ must be ensured for the initialization. Although we can not prove that the maximum weighted sum rate can be achieved by the gradient-projection algorithm if $\boldsymbol{T}$ is not initialized as a (suboptimum) stationary point, all precoders $\boldsymbol{T}_1, \ldots, \boldsymbol{T}_K$ should at least be initialized as square, full-rank matrices in order to have the theoretical possibility to achieve the maximum. Reducing the number of columns of a precoder allows to limit the maximum number of data streams of this user but of course may crop the achievable rate region. When used in conjunction with the Armijo step-size rule, we choose the scaling $s_0^{(n)}$ in (5.27) as the quotient of the Frobenius norms of $\boldsymbol{T}$ and the total Wirtinger derivative:

$$s_0^{(n)} = \frac{\|\boldsymbol{T}^{(n)}\|_{\mathrm{F}}}{\sqrt{\sum_{j=1}^{K} \left\| \frac{\partial_{\mathrm{w}} f(\boldsymbol{T}, \boldsymbol{w})}{\partial_{\mathrm{w}} \boldsymbol{T}_j^*} \Big|_{\boldsymbol{T}=\boldsymbol{T}^{(n)}} \right\|_{\mathrm{F}}^2}}.$$

The initial $\boldsymbol{T}^{(0)}$ must satisfy $\boldsymbol{HT}^{(0)} \neq \mathbf{0}$ to let $s_0^{(n)}$ have a positive lower bound.

6.2.5 Simulation Results

In this subsection, we analyze the convergence behavior of the precoder- and covariance-based gradient projection algorithms when applied to the weighted sum rate maximization. Again, $K = 3$ two-antennas users are served by the base station in the MIMO BC. The user weights are set to $[w_1, w_2, w_3] = [\frac{1}{2}, \frac{2}{2}, \frac{3}{2}]$ and the base station is either equipped with $N = 6$ or with $N = 3$ antennas. In addition to the algorithms derived in Sects. 6.2.3 and 6.2.4, the performance of the approaches from Böhnke in [124, 62] and from Viswanathan in [118] is investigated. We did not implement the algorithm from Liu in [122] due to the lack of information about how the 'appropriate' step-sizes of the inner and outer loop have to be chosen.

The composite covariance matrix $\boldsymbol{Q}^{(0)}$ and the composite precoder $\boldsymbol{T}^{(0)}$ are initialized in the same way as for the sum rate maximization with equal weights, i.e.,

$$\boldsymbol{Q}_k^{(0)} = \frac{P_{\max}}{6} \cdot \mathbf{I}_2 \quad \text{and} \quad \boldsymbol{T}_k^{(0)} = \sqrt{\frac{P_{\max}}{6}} \cdot \mathbf{I}_2 \quad \forall k \in \{1, 2, 3\}.$$

For unequal weights, however, this initialization is no longer asymptotically optimum in the high SNR regime for $N = 6$ antennas at the base station, see Chap. 8. In turn, the Armijo rule was used for the step-size control mechanism of the proposed gradient-projection algorithms with $\beta = 0.5$ and $m^{(n)}$ taken from $\mathbb{Z}$, see (5.27). The slope reduction parameter σ was set to $\sigma = 0.3$ for all gradient-projection based algorithms except for the precoder-based one with the standard Wirtinger derivative as search direction. There, $\sigma = 0.1$ was chosen to allow for large steps. Figures 6.4 and 6.5 show the relative distance to the maximum weighted sum rate versus the iteration number for the two sum powers $10 \log_{10} P_{\max} = -5$ dB and $10 \log_{10} P_{\max} = +5$ dB, respectively. The channel matrix $\boldsymbol{H} = [\boldsymbol{H}_1, \boldsymbol{H}_2, \boldsymbol{H}_3]$ for the $N = 6$ antenna case was randomly set to

$$\boldsymbol{H} = \frac{1}{10} \cdot \begin{bmatrix} +7-\mathrm{j}4 & +10-\mathrm{j}5 & +5+\mathrm{j}7 & -8+\mathrm{j}9 & -6+\mathrm{j}5 & +5-\mathrm{j}5 \\ -1+\mathrm{j}7 & +11-\mathrm{j}7 & +12-\mathrm{j}7 & -2-\mathrm{j}2 & -5+\mathrm{j}9 & +8-\mathrm{j}4 \\ -8-\mathrm{j}10 & -5+\mathrm{j}8 & +5-\mathrm{j}5 & -11-\mathrm{j}1 & -9-\mathrm{j}7 & -5-\mathrm{j}2 \\ -1+\mathrm{j}18 & -3 & +5+\mathrm{j}1 & +4+\mathrm{j}8 & -1+\mathrm{j}3 & +2-\mathrm{j}6 \\ -6-\mathrm{j}3 & +3+\mathrm{j}2 & +9-\mathrm{j}1 & +13+\mathrm{j}3 & +1-\mathrm{j}6 & +12-\mathrm{j}4 \\ -2+\mathrm{j}3 & +1+\mathrm{j}1 & +6+\mathrm{j}6 & +11-\mathrm{j}12 & +10-\mathrm{j}3 & -5+\mathrm{j}4 \end{bmatrix}. \tag{6.20}$$

For the $N = 3$ antenna case, the last three rows of $\boldsymbol{H}$ in (6.20) were removed.

In Fig. 6.1a, the base station is equipped with $N = 6$ antennas. For the initializations $\boldsymbol{Q}^{(0)}$ and $\boldsymbol{T}^{(0)}$, respectively, about 75 % of the maximum WSR are achieved. The lowest speed of convergence is obtained by the algorithm from Viswanathan [118] (`Viswanathan`–star marker) followed by the precoder-based projected gradient algorithm with the unconstrained Wirtinger derivative as search direction (`PG Precoder UC`–solid curve, square marker). Böhnke's approach [124, 62] (`Boehnke`–circle marker) using conjugate search directions features a faster decay of the relative distance to the maximum WSR but it is outperformed by the precoder-based gradient-projection algorithm when the tangent cone projection of the Wirtinger derivative is chosen as search direction (`PG Precoder Proj`–dashed curve, square marker). This results from the circumstance that in contrast to Böhnke, we choose the maximum possible step-size that fulfills Armijo's criterion. If our projected gradient algorithm is based on the covariances instead of the precoders, the speed of convergence is even higher and only 2 iterations are required to reduce the relative distance to the maximum WSR below 10^{-3}. This results from the fact that the orthogonal projection onto the constraint set can completely disable individual data streams within one step whereas the precoder-based variants can switch off data steams only asymptotically over several iterations. Especially at small transmit powers where only few data streams are active in general, this turns out to be advantageous. For the given channel realization at $10 \log_{10} P_{\max} = -5$ dB, three streams are active at the optimum. The relaxed gradient-projection algorithms using either the unconstrained gradient (`Relaxed PG UC`–solid curve, triangle down marker) or its projection onto the tangent cone (`Relaxed PG Proj`–dashed curve, triangle down marker) as search direction outperform the standard gradient projection algorithm working on the covariances (`PG Covariance UC`–solid curve, no marker). For the latter one, the curve associated to the tangent cone projection of the search direction coincides with the solid curve without marker and is hence not plotted in the figure.

Figure 6.4b shows the situation for a base station with only $N = 3$ antennas, where the channel matrix in (6.20) is cropped. There, the convergence behavior

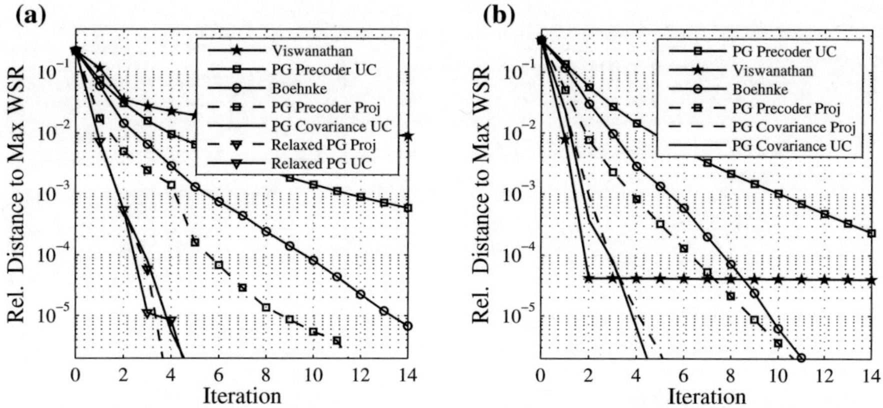

Fig. 6.4 Relative distance to maximum WSR for $K = 3$ users and $10 \log_{10} P_{\max} = -5\,\text{dB}$. **a** $N = 6$ antennas at the base station. **b** $N = 3$ antennas at the base station

remains qualitatively the same for most of the algorithms. However, two changes can be observed. First, the relaxed gradient-projection algorithms now yield the same results as their covariance-based counterparts (`PG Covariance Proj`–dashed curve, no marker) and (`PG Covariance UC`–solid curve, no marker) without the additional outer line search and are thus not plotted. Second, the approach from Viswanathan (`Viswanathan`–star marker) shows a very quick reduction of the relative distance to the maximum WSR during the first two iterations followed by an extremely slow further reduction during the next iterations. This can be explained as follows. For the given setup, the maximum WSR is achieved when user one does not send any data in the dual MAC, i.e. $\check{\boldsymbol{Q}}_1 = \mathbf{0}$. User two and user three have one active data stream each at the optimum such that $\operatorname{rank}(\check{\boldsymbol{Q}}_2) = 1$ and $\operatorname{rank}(\check{\boldsymbol{Q}}_3) = 1$. During one iteration of Viswanathan's algorithm, the covariance matrices of all users are scaled by a factor t that satisfies $0 \leq t \leq 1$, and a rank-one update $P_{\max}(1-t)\boldsymbol{u}_k\boldsymbol{u}_k^{\mathrm{H}}$ with $\|\boldsymbol{u}_k\|_2 = 1$ is added to that user k whose Wirtinger derivative has the largest eigenvalue. During one iteration, this algorithms can either keep the number of active data streams unchanged, reduce it to one, or increase it by one.

The optimum step-size during the first iteration is $t^{(1)} = 0$ and user three has the largest eigenvalue of the Wirtinger derivative. Hence, $\boldsymbol{Q}_1^{(1)} = \boldsymbol{Q}_2^{(1)} = \mathbf{0}$ and $\boldsymbol{Q}_3^{(1)} = P_{\max}\boldsymbol{u}_3\boldsymbol{u}_3^{\mathrm{H}}$ holds after the first iteration. During the second iteration, user two features the largest eigenvalue of the Wirtinger derivatives and consequently, $\boldsymbol{Q}_2^{(2)} = p_2\boldsymbol{u}_2\boldsymbol{u}_2^{\mathrm{H}}$ and $\boldsymbol{Q}_3^{(2)} = p_3\boldsymbol{u}_3\boldsymbol{u}_3^{\mathrm{H}}$ holds whereas $\boldsymbol{Q}_1^{(2)}$ remains the zero matrix. The optimum stream-configuration has thus been found after two iterations. But now, the adaptation of the two powers p_2 and p_3 to their optimum values is extremely slow which results in an almost horizontal part of the curve in Fig. 6.4b for the remaining iterations.

Raising the transmit power to $10 \log_{10} P_{\max} = +5\,\text{dB}$ leads to the results shown in Fig. 6.5. The left plot (Fig. 6.5a) shows the $N = 6$ antenna case. As one can see,

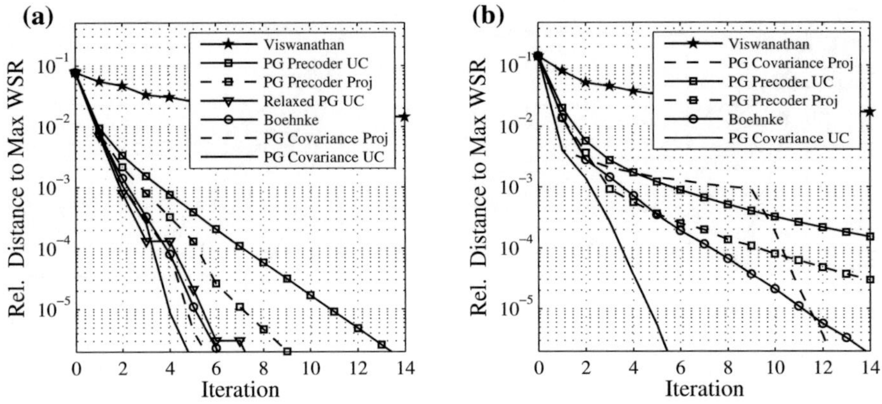

Fig. 6.5 Relative distance to maximum WSR for $K = 3$ users and $10 \log_{10} P_{\max} = +5\,\text{dB}$. **a** $N = 6$ antennas at the base station. **b** $N = 3$ antennas at the base station

the conjugate search directions approach from Böhnke (`Boehnke`–circle marker) performs better for the increased transmit power $P_{\max}$. The relaxed gradient-projection algorithm (`Relaxed PG UC`–triangle down marker) seems to leave the relative distance to the maximum WSR from steps three to four and six to seven almost constant. Indeed, the WSR is increased only by a very small portion due to very small steps in the direction of the Wirtinger derivative. This results from the fact that in this case, the slope of the first order Taylor expansion of the weighted sum rate utility matches the squared Frobenius norm of the tangent cone projection [see the right hand side of (5.28)] only in a very small neighborhood around zero and quickly flattens beyond. A very small (but nonzero) eigenvalue that is treated as nonzero by the tangent cone projection leads to this behavior. In this case, it would be better to switch from the exact slope in (5.28) (squared Frobenius norm of the Wirtinger derivative projected onto the tangent cone) to its underestimated version in (5.30) to allow for a larger step-size.

Figure 6.5b shows the simulation results for an $N = 3$ antenna base station. In this case, the covariance-based gradient-projection algorithm utilizing the Wirtinger derivative as search direction (`PG Covariance UC`–solid curve, no marker) shows the best performance followed by the conjugate search directions approach from Böhnke (`Boehnke`–circle marker).

Figures 6.4 and 6.5 show the convergence behavior for a particular channel realization. Channel-averaged results are plotted in Fig. 6.6, where the average number of iterations needed to achieve 0.998 times the maximum WSR is plotted versus the logarithmic transmit power $P_{\max}$. In turn, the entries of the channel matrix were chosen as i.i.d. zero-mean Gaussian random variables with variance one and we averaged over $1,000$ realizations. Although the algorithm from Viswanathan is claimed not to achieve the maximum WSR in every case by Liu in [122], we did not encounter this phenomenon during the 1,000 channels that we simulated.

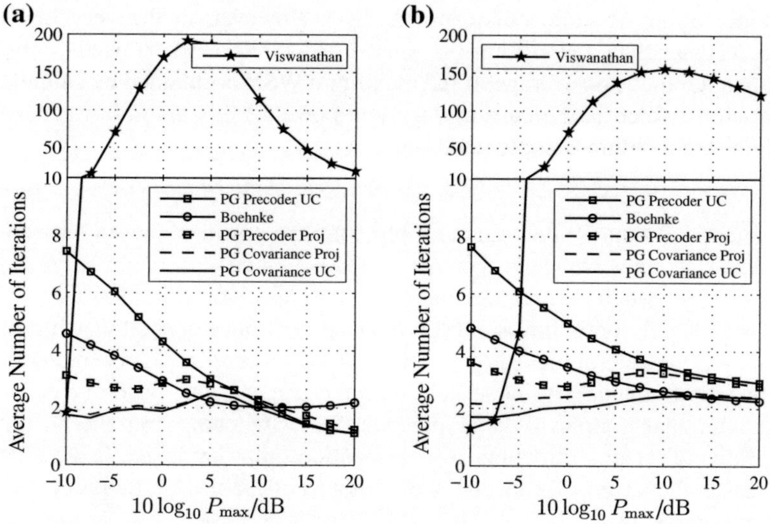

Fig. 6.6 Average number of iterations to achieve 0.998·maximum WSR for $K = 3$ users. **a** $N = 6$ antennas at the base station. **b** $N = 3$ antennas at the base station

The performance of Viswanathan's algorithm (`Viswanathan`–star marker) strongly depends on the available transmit power. As mentioned before, this approach is effective in the low power regime where only one or two data streams are active. In the $N = 6$ antenna case (Fig. 6.6a) at $10\log_{10} P_{\max} = -10\,\mathrm{dB}$, only two iterations are needed on average. However, at 5 dB, almost 200 iterations are necessary to achieve the desired accuracy! At higher transmit powers, this large amount reduces again since the initialization already yields the correct structure of the asymptotically optimum solution, only the power that is allocated to the users has to be adapted, see Chap. 8. The precoder-based gradient-projection algorithms feature a sane performance. In the low power regime, using the tangent cone projection of the Wirtinger derivative as search direction (`PG Precoder Proj`–dashed curve, square marker) performs much better than the standard Wirtinger derivative (`PG Precoder UC`–solid curve, square marker). Between $+5\,\mathrm{dB}$ and $+10\,\mathrm{dB}$, Böhnke's conjugate search directions approach (`Boehnke`–circle marker) performs best. Outside, the covariance-based gradient-projection algorithms feature the overall best performance that is almost independent of the available transmit power $P_{\max}$. In the low power regime, the tangent cone search direction (`PG Covariance Proj`–dashed curve, no marker) yields slightly more iterations on average than the unconstrained search direction (`PG Covariance UC`–solid curve, no marker). For the given metric, the relaxed gradient-projection algorithms need exactly the same number of iterations on average as their standard covariance-based counterparts.

Reducing the number of antennas at the base station to $N = 3$ leads to a similar result. The smallest number of iterations is obtained by the covariance-based gradient-projection algorithm using the Wirtinger derivative as search direction

(`PG Covariance UC`–solid curve, no marker). However, in the very low power regime, Viswanathan's algorithm (`Viswanathan`–circle marker) needs only about one or two iterations on average. If the maximum WSR is obtained by single-stream transmission (which is the case at very low powers), this algorithm achieves the global optimum within a single iteration.

Synopsis: The presented gradient-projection algorithms haven proven to feature nice convergence properties for the weighted sum rate maximization problem. Although the precoder-based variants may have stationary points that are not globally optimum, we did not encounter convergence to such points. However, they are attractive due to the circumstance that the orthogonal projections onto the constraint set and onto the tangent cone have a very small computational complexity in the precoder case. Switching from the Wirtinger derivative as search direction to its projection onto the tangent cone clearly increases the speed of convergence without noticeable additional computational complexity.

If the weighted sum rate maximization is based on the covariance matrices, an even faster convergence is obtained. However, this comes at the price of a higher computation complexity of the orthogonal projections. Only two iterations are required on average to reach the maximum weighted sum rate up to two per mill independent of the available transmit power. Using the tangent cone projection of the Wirtinger derivative as search direction usually does not bring any gain in the covariance-based case. In many cases, exactly the same performance is achieved due to the reasons mentioned in Sect. 5.3.2.3. However, it is important for a deeper theoretical understanding of the gradient-projection algorithm and indispensable for finding the correct slope of the first order Taylor approximation of the weighted sum rate utility when moving along the direction of the (possibly projected) Wirtinger derivative. Since we allowed the Armijo rule to select the maximum possible step-size such that the inexact line search is very accurate, the extension towards the relaxed gradient-projection algorithm did not bring substantial performance gains.

Chapter 7
Linear Transceiver Design for the MIMO BC

In the preceding chapter, a globally optimal way to achieve rate tuples lying on the boundary of the capacity region was presented using the weighted sum rate maximization approach. In order to realize such a Pareto optimal rate tuple, dirty paper coding has to be applied which is attended by a very high computational complexity as we will see in Sect. 7.1. Practical schemes for the implementation of dirty paper coding are therefore still very challenging and under ongoing research. A remedy to this circumstance is to apply a *linear* transceiver design where the total inter-user interference is treated as additional noise instead of partially known side-information. In particular, this strategy drastically simplifies the decoding procedure which essentially boils down to the decoding of parallel AWGN channels due to the decorrelating property of our rate duality from Sect. 3.3. Besides some loss of performance, however, this tremendous complexity reduction comes at the price of the loss of concavity of the rate expressions making rate-based optimizations more difficult. In Sect. 7.2, we present two algorithms for the maximization of the sum rate when inter-user interference is treated as additional noise. The extension to the weighted sum rate maximization with different weights follows in Sect. 7.3. Finally, a simple transceiver design that is based on the sum MSE minimization in an uncoded system is presented in Sect. 7.4.

7.1 Motivation for Linear Transceivers

In order to fully exploit the potential of the MIMO broadcast channel with more than one user, dirty paper coding has to be applied in general unless the channel is degraded by chance which usually is not the case. Thanks to the unitary decorrelation matrices of our rate duality in Chap. 4, scalar dirty paper coding is sufficient for the cancelation of the vector-valued interference terms, see Sect. 4.4.1. In this section, we shed some light on how the noncausally known side-information is actually taken into consideration in order to render it harmless at the receiver in terms of capacity. To this

R. Hunger, *Analysis and Transceiver Design for the MIMO Broadcast Channel*, Foundations in Signal Processing, Communications and Networking 8,
DOI: 10.1007/978-3-642-31692-0_7, © Springer-Verlag Berlin Heidelberg 2013

end, we review two approaches to the practical implementation of dirty paper coding. The first one in Sect. 7.1.1 is based on typical sequences, whereas the second one transforms a power-constrained AWGN channel into a modulo-lattice additive noise (MLAN) channel, see Sect. 7.1.2. Both procedures turn out to feature an extremely high computational complexity making the suboptimal linear transceiver designs an attractive alternative.

From (4.42), we observe that the filtered receive signal of a particular user in the MIMO BC can be interpreted as parallel, independent AWGN channels. This results from the decorrelation matrices. Hence, we may concentrate on a single data stream and analyze the implementation of *scalar* dirty paper encoding and decoding. In analogy to (4.42), the end-to-end relationship between the transmitted symbol $s[t] \in \mathbb{R}$ and the received symbol $\hat{s}[t] \in \mathbb{R}$ can be expressed in an abstract way for a single time instant $t \in \{1, \ldots, n\}$ as[1]

$$\hat{s}[t] = \lambda s[t] + \zeta[t] + z[t] \tag{7.1}$$

where $\zeta[t]$ denotes the side-information, $\lambda \in \mathbb{R}_+$, and $z[t]$ is the effective additive noise at time instant t. For the encoding process, however, the complete noncausally known interference sequence $\zeta[1], \ldots, \zeta[n]$ must be known in advance. Thus, we group n time instances and rewrite (7.1) to its vector-valued version

$$\hat{\boldsymbol{s}} = \lambda \boldsymbol{s} + \boldsymbol{\zeta} + \boldsymbol{z} \tag{7.2}$$

where $\hat{\boldsymbol{s}}$, $\boldsymbol{s}$, $\boldsymbol{\zeta}$, and $\boldsymbol{z}$ are elements of $\mathbb{R}^n$.

7.1.1 Typical Sequences Approach to Dirty Paper Coding

In his 1983 correspondence, Costa showed the principle existence of a capacity-achieving code by extending the capacity results from Gel'fand and Pinsker in [125] to the case of continuous alphabets. According to their contribution in [125], the capacity of the channel

$$\hat{s} = \lambda s + \zeta + z$$

with $s \sim \mathcal{N}(0, 1)$, $\zeta \sim \mathcal{N}(0, \sigma_\zeta^2)$, and $z \sim \mathcal{N}(0, \sigma_z^2)$ corresponds to the supremum of the difference $I(u(\alpha); \hat{s}) - I(u(\alpha); \zeta)$, where $u(\alpha)$ denotes an auxiliary random variable that depends on the parameter α. The maximization is done with respect to the joint distribution of $u(\alpha)$ and s given ζ. According to Costa, the parametrization of the random variable $u(\alpha)$ via

---

[1] We restrict ourselves to the real-valued implementation of dirty paper coding for the complexity analysis. Since the inphase and quadrature component are independent, the complex-valued case corresponds to applying real-valued dirty paper coding two times. Due to the fact that the scalar λ in (7.1) is equal to one in Costa's correspondence, the following expressions look slightly different to those in [76].

$$u(\alpha) = s + \frac{\alpha}{\lambda} \cdot \zeta \tag{7.3}$$

maximizes the mutual information difference $I(u(\alpha); \hat{s}) - I(u(\alpha); \zeta)$ if α is set to

$$\check{\alpha} = \frac{\lambda^2}{\lambda^2 + \sigma_z^2}. \tag{7.4}$$

Note that $\check{\alpha}$ does not depend on the variance σ_ζ^2 of the interference. For this choice, the capacity

$$C := I\big(u(\check{\alpha}); \hat{s}\big) - I\big(u(\check{\alpha}); \zeta\big) = \frac{1}{2}\log_2\left(1 + \frac{\lambda^2}{\sigma_z^2}\right) \tag{7.5}$$

is achieved using the typical sequences approach.

If no side-information was available at the transmitter that has to be rendered harmless, a possible encoding-decoding scheme would be the following [66, p. 244]. The transmitter would generate a codebook consisting of $\lceil 2^{nC} \rceil$ sequences/codewords $\boldsymbol{u}^{(1)}, \ldots, \boldsymbol{u}^{(\lceil 2^{nC} \rceil)}$ whose i.i.d. entries are taken from a Gaussian distribution with variance $1 - \varepsilon$ for some arbitrarily small $\varepsilon > 0$. For large n, the power constraint $\frac{1}{n} \cdot \|\boldsymbol{u}^{(w)}\|_2^2 \leq 1$ is fulfilled with high probability for all $w \in \{1, \ldots, \lceil 2^{nC} \rceil\}$. The receiver has to know this codebook as well. If the data w shall be conveyed, the sender transmits the sequence $\boldsymbol{s} = \boldsymbol{u}^{(w)}$. The receiver has to test all $\lceil 2^{nC} \rceil$ codewords of the codebook for joint typicality (see [66, Sect. 8.6, p. 194] for the definition) with respect to the received signal $\hat{\boldsymbol{s}} = \lambda \boldsymbol{s} + \boldsymbol{z}$. For large n, the probability that there is exactly one such codeword goes to one. Even for the case without known side-information at the transmitter, the complexity of the decoding process is exponential in n since $\lceil 2^{nC} \rceil$ sequences have to be tested for joint typicality with respect to the received signal.

Now if noncausally known side-information shall be rendered harmless in terms of capacity, the codebook contains $\lceil 2^{nI(u(\check{\alpha});\hat{s})} \rceil$ codewords instead of 'only' $\lceil 2^{nC} \rceil$. The entries of the codewords are again drawn from a Gaussian distribution, but this time with variance $1 + \check{\alpha}^2/\lambda^2 \cdot \sigma_\zeta^2$, see (7.3). These $\lceil 2^{nI(u(\check{\alpha});\hat{s})} \rceil$ codewords are placed into $\lceil 2^{nC} \rceil$ bins such that every bin contains approximately the same number of $2^{nI(u(\check{\alpha});\zeta)}$ codewords, see (7.5). In turn, the receiver has to know which codewords are associated to which bin. The message w that the sender wants to convey corresponds to the index of the bin. From the approximately $2^{nI(u(\check{\alpha});\zeta)}$ codewords in bin w, the encoder picks the one that is jointly typical with the interference sequence $\boldsymbol{\zeta}$. This is equivalent to finding a sequence $\boldsymbol{u}$ in bin w such that [76]

$$\left(\boldsymbol{u} - \frac{\check{\alpha}}{\lambda} \cdot \boldsymbol{\zeta}\right)^{\mathrm{T}} \boldsymbol{\zeta} \approx 0. \tag{7.6}$$

The so found sequence $\boldsymbol{u}$ is used to generate the transmit sequence $\boldsymbol{s}$ via

$$s = u - \frac{\check{\alpha}}{\lambda} \cdot \zeta \tag{7.7}$$

whose average power then reads as

$$\begin{aligned}\frac{1}{n} \cdot \|s\|_2^2 &= \frac{1}{n} \cdot \left[\|u\|_2^2 - 2 \cdot \frac{\check{\alpha}}{\lambda} \cdot u^{\mathrm{T}} \zeta + \frac{\check{\alpha}^2}{\lambda^2} \cdot \|\zeta\|_2^2\right] \\ &\approx \frac{1}{n} \cdot \left[\|u\|_2^2 - \frac{\check{\alpha}^2}{\lambda^2} \cdot \|\zeta\|_2^2\right]\end{aligned}$$

where the approximation follows from (7.6). For large n, we find

$$\begin{aligned}&\frac{1}{n} \cdot \|\zeta\|_2^2 \to \sigma_\zeta^2 \\ &\frac{1}{n} \cdot \|u\|_2^2 \to 1 + \frac{\check{\alpha}^2}{\lambda^2} \cdot \sigma_\zeta^2\end{aligned}$$

due to the Gaussian distributions from which the entries of ζ and u are drawn. For large n,

$$\frac{1}{n} \cdot \|s\|_2^2 \approx 1$$

therefore holds with high probability such that the transmit power constraint is fulfilled. If side-information did not have to be suppressed, the encoder would only have to pick the bin that is associated to message w. Otherwise, the encoder becomes more complex since it has to test the $2^{nI(u(\check{\alpha});\zeta)}$ codewords in bin w for joint typicality with respect to the interference ζ. Obviously, the number of codewords per bin again scales exponentially in n. In addition, the situation at the *decoder* is even worse! Instead of having to probe $\lceil 2^{nC} \rceil$ codewords for joint typicality with respect to the received sequence $\hat{s}$, the decoder has to test $\lceil 2^{nI(u(\check{\alpha});\hat{s})} \rceil$ codewords to find the correct bin. This corresponds to an increase of complexity by a factor of approximately $2^{nI(u(\check{\alpha});\zeta)}$, i.e., the originally exponential complexity for the case without side-information is multiplied by a term which is itself exponential in n. More precisely, we find from (7.3) and (7.4) that

$$\begin{aligned}I(u(\check{\alpha}); \zeta) &= \mathcal{H}\big(u(\check{\alpha})\big) - \mathcal{H}\big(u(\check{\alpha})|\zeta\big) = \mathcal{H}\big(u(\check{\alpha})\big) - \mathcal{H}(s) \\ &= \log_2 \sqrt{2\pi \mathrm{e}} + \frac{1}{2} \cdot \log_2 \left(1 + \frac{\check{\alpha}^2}{\lambda^2} \cdot \sigma_\zeta^2\right) - \left(\log_2 \sqrt{2\pi \mathrm{e}} + \frac{1}{2} \cdot \log_2(1)\right) \\ &= \frac{1}{2} \cdot \log_2 \left(1 + \frac{\check{\alpha}^2}{\lambda^2} \cdot \sigma_\zeta^2\right) \\ &= \frac{1}{2} \cdot \log_2 \left(1 + \frac{\lambda^2 \sigma_\zeta^2}{(\lambda^2 + \sigma_z^2)^2}\right).\end{aligned}$$

Hence, the computational complexity at the receiver is increased by a factor of approximately

$$2^{nI(u(\breve{\alpha});\zeta)} = \left[1 + \frac{\lambda^2 \sigma_\zeta^2}{(\lambda^2 + \sigma_z^2)^2}\right]^{\frac{n}{2}}$$

which is even increasing in the variance of the interference. This huge complexity makes the typical sequences approach to dirty paper coding very difficult to implement.

7.1.2 Modulo-Lattice Additive Noise Channel Approach

Besides the typical sequences approach, the capacity of the channel with side-information at the transmitter may also be achieved by means of multidimensional lattice quantization and MMSE scaling [90, 126]. To this end, a lattice

$$\Lambda := \{\boldsymbol{l} | \boldsymbol{l} = \boldsymbol{G}\boldsymbol{p} \wedge \boldsymbol{p} \in \mathbb{Z}^n\}$$

is created using a generator matrix $\boldsymbol{G} \in \mathbb{R}^{n \times n}$. This lattice defines the fundamental Voronoi region $\mathcal{V}$ as the set of points in $\mathbb{R}^n$ whose distance to the origin is smaller than the distance to any other point of the lattice Λ [127]. The transmitter picks an element $\boldsymbol{v}$ of the modulation alphabet (which corresponds to a set of discrete points in $\mathcal{V}$) and subtracts the scaled side-information $\breve{\alpha}/\lambda \cdot \boldsymbol{\zeta}$ as well as a dithering signal $\boldsymbol{d}$ which is uniformly distributed over $\mathcal{V}$ and known both to the transmitter and the receiver. This difference is then quantized with respect to the lattice Λ, and the quantization error is chosen as the transmit signal $\boldsymbol{s}$:

$$\boldsymbol{s} = \left[\boldsymbol{v} - \frac{\breve{\alpha}}{\lambda} \cdot \boldsymbol{\zeta} - \boldsymbol{d}\right] \bmod \Lambda \tag{7.8}$$

Due to the dithering $\boldsymbol{d}$, the transmit signal $\boldsymbol{s}$ looks as if it is taken from a uniform distribution over $\mathcal{V}$. Hence, the generator matrix $\boldsymbol{G}$ of the lattice Λ must be chosen such that its fundamental Voronoi region $\mathcal{V}$ has a second moment that is equal to one in order to satisfy the power constraint. Upon reception of $\hat{\boldsymbol{s}}$ in (7.2), the receiver computes

$$\tilde{\boldsymbol{s}} = \left[\frac{\breve{\alpha}}{\lambda} \cdot \hat{\boldsymbol{s}} + \boldsymbol{d}\right] \bmod \Lambda. \tag{7.9}$$

The resulting channel that relates $\tilde{\boldsymbol{s}}$ and $\boldsymbol{v}$ is equivalent in distribution to the modulo-lattice additive noise (MLAN) channel [90, 126]

$$\tilde{\boldsymbol{s}} = \left[\boldsymbol{v} + \boldsymbol{z}'\right] \bmod \Lambda \quad \text{with} \quad \boldsymbol{z}' = [(1 - \breve{\alpha})\boldsymbol{d} + \breve{\alpha}/\lambda \cdot \boldsymbol{z}] \bmod \Lambda. \tag{7.10}$$

Note that (7.10) follows from (7.9) due to [90, Eqs. (68)–(73)]

$$
\begin{aligned}
\tilde{s} &= \left[\frac{\check{\alpha}}{\lambda} \cdot (\lambda s + \zeta + z) + d\right] \bmod \Lambda \\
&= \left[\check{\alpha} s + v + \left(-v + \frac{\check{\alpha}}{\lambda} \cdot \zeta + d\right) + \frac{\check{\alpha}}{\lambda} \cdot z\right] \bmod \Lambda \\
&= \left[\check{\alpha} s + v - s + \frac{\check{\alpha}}{\lambda} \cdot z\right] \bmod \Lambda \\
&= \left[v - (1 - \check{\alpha}) s + \frac{\check{\alpha}}{\lambda} \cdot z\right] \bmod \Lambda .
\end{aligned}
$$

As mentioned before, the dithering d leads to the fact that s can be interpreted to stem from a uniform distribution over $\mathcal{V}$ such that (7.10) holds since d and $-d$ result from the same distribution.

Such an MLAN channel has been investigated by Erez et al. in [128], and it was shown that it has the same capacity as a standard AWGN channel up to a penalty term that decreases to zero when the dimension n of the lattice goes to infinity and the lattice is chosen appropriately (its normalized second moment must converge to $\frac{1}{2\pi e}$ as $n \to \infty$). Thus, the noncausally known side-information can be rendered harmless in terms of capacity using the modulo-lattice approach. Instead of drawing a random code according to a uniform distribution over the fundamental Voronoi region and then applying maximum likelihood decoding at the receiver, a nested lattice code can be used [128]. The coarse lattice is then used for shaping and the fine lattice defines the codewords such that it is a good channel code. For this kind of encoding, lattice decoding is optimal.

Unfortunately, the dimension n of the lattice has to be large to achieve a capacity penalty term that is close to zero. This means that the quantization/modulo operations at the transmitter and at the receiver must be done in high dimensions such that the lattice encoding/decoding process (finding the closest point in a lattice) involves an extremely high computational complexity making the implementation of dirty paper coding difficult.

Synopsis: Despite being capacity achieving, the implementation of dirty paper coding is a challenging task that is accompanied by a huge computational complexity. Two possible ways to implement DPC are the typical sequences approach and the modulo-lattice additive noise channel approach. In the first one, the complexity of the *decoder* is drastically increased compared to a system without side-information. Instead of having to test $\lceil 2^{nC} \rceil$ codewords at rate C for joint typicality with respect to the received signal as this would be the case if no side-information had to be suppressed, the receiver now has to test the m-fold amount of codewords, where m is again exponential in the codeword length n and is moreover increasing in the variance of the known side-information. The second approach is based on the conversion of an AWGN channel into a modulo-lattice additive noise channel. Using high-dimensional modulo operators at the transmitter and the receivers, the capacity of a system without noncausally known side-information can be achieved by means of high-dimensional lattices. However, the lattice encoding/decoding again features an extremely high computational complexity due to the search for the closest point in an infinite lattice.

The tremendous complexity of the dirty paper coding implementation motivates the design and analysis of linear transceivers that treat the complete inter-user interference as additional noise and not partially as known side-information that has to be canceled.

7.2 Sum Rate Maximization

For the linear transceiver design, the total inter-user interference is treated as additional noise and not partially as noncausally known side-information. As a consequence, the (weighted) sum rate expression is neither concave in the original MIMO BC nor in its dual MIMO MAC as we will see later on. Working in the dual domain does therefore *not* reveal a hidden convexity of the problem as this is the case when interference is partly treated as side-information, see Chap. 6. Nonetheless, we derive a gradient-projection based algorithm for the (weighted) sum rate maximization again in the dual MAC instead of the BC. The reason for doing so results from the fact that both the rate expressions and the derivatives have simpler expressions in the dual MAC that can be computed more efficiently than in the BC. Due to the iterative nature of the algorithm, these complexity savings sum up in every iteration and hence over-compensate the additional effort which results from the MAC to BC conversion.

This section begins with some basic properties of the sum rate expression, which are highlighted in Sect. 7.2.1. Afterwards, a combinatorial algorithm probing all possible stream permutations is derived in Sect. 7.2.2. In order to reduce the computational complexity, a greedy variant that keeps the stream allocations of the previous steps fixed is deduced in Sect. 7.2.3. Simulation results follow in Sect. 7.2.4. An overview of algorithms in the literature which are also targeted at maximizing the sum rate for linear transceivers with multi-antenna terminals is not explicitly revisited again. Instead, we refer to Sect. 3.3.3 where several algorithms have already been introduced.

7.2.1 Basic Properties of the Sum Rate Expression

For the linear transceiver case, the maximum possible rate at which user k can convey data in the dual MAC with arbitrarily small error probability is given by [cf. (3.74)]

$$\begin{aligned} R_k^{\mathrm{MAC}}(\boldsymbol{Q}) &:= \log_2 \Big| \mathbf{I} + \sum_{i=1}^{K} \boldsymbol{H}_i \boldsymbol{Q}_i \boldsymbol{H}_i^{\mathrm{H}} \Big| - \log_2 \Big| \mathbf{I} + \sum_{i \neq k} \boldsymbol{H}_i \boldsymbol{Q}_i \boldsymbol{H}_i^{\mathrm{H}} \Big| \\ &= \log_2 \Big| \mathbf{I} + \Big(\mathbf{I} + \sum_{i \neq k} \boldsymbol{H}_i \boldsymbol{Q}_i \boldsymbol{H}_i^{\mathrm{H}} \Big)^{-1} \boldsymbol{H}_k \boldsymbol{Q}_k \boldsymbol{H}_k^{\mathrm{H}} \Big| \end{aligned} \tag{7.11}$$

where $\boldsymbol{Q} := \mathrm{blockdiag}\{\boldsymbol{Q}_k\}_{k=1}^{K}$ contains the K transmit covariance matrices. The rate expression in (7.11) induces a sum rate that can be written as

$$R^{\mathrm{MAC}}(\boldsymbol{Q}) := K \log_2 \Big| \mathbf{I} + \sum_{i=1}^{K} \boldsymbol{H}_i \boldsymbol{Q}_i \boldsymbol{H}_i^{\mathrm{H}} \Big| - \sum_{k=1}^{K} \log_2 \Big| \mathbf{I} + \sum_{i \neq k} \boldsymbol{H}_i \boldsymbol{Q}_i \boldsymbol{H}_i^{\mathrm{H}} \Big|. \tag{7.12}$$

In contrast to the MAC sum rate expression with successive decoding in (6.4), above term is in general not concave in $\boldsymbol{Q}$ for $K > 1$. Moreover, it even need not be increasing in $\boldsymbol{Q}$, which means that there may exist some $\boldsymbol{Q}' \succ \boldsymbol{Q}$ for which $R^{\mathrm{MAC}}(\boldsymbol{Q}') < R^{\mathrm{MAC}}(\boldsymbol{Q})$. To see this, we compute the derivative of $R^{\mathrm{MAC}}(\boldsymbol{Q} + \alpha \boldsymbol{D})$ with respect to α, evaluate it at $\alpha = 0$, and show that depending on $\boldsymbol{Q}$ and $\boldsymbol{H}$, there may exist a $\boldsymbol{D} = \mathrm{blockdiag}\{\boldsymbol{D}_k\}_{k=1}^{K} \succ \mathbf{0}$ for which the derivative is negative. Using the substitutions

$$\begin{aligned} \boldsymbol{X} &:= \mathbf{I} + \sum_{i=1}^{K} \boldsymbol{H}_i \boldsymbol{Q}_i \boldsymbol{H}_i^{\mathrm{H}} \\ \boldsymbol{X}_k &:= \mathbf{I} + \sum_{i \neq k} \boldsymbol{H}_i \boldsymbol{Q}_i \boldsymbol{H}_i^{\mathrm{H}} \end{aligned} \tag{7.13}$$

the derivative of $R^{\mathrm{MAC}}(\boldsymbol{Q} + \alpha \boldsymbol{D})$ with respect to α reads according to Appendix A.6 as

$$\begin{aligned} \frac{\partial R^{\mathrm{MAC}}(\boldsymbol{Q} + \alpha \boldsymbol{D})}{\partial \alpha} \Big|_{\alpha=0} &= \frac{K}{\ln 2} \sum_{i=1}^{K} \mathrm{tr}\left[\boldsymbol{X}^{-1} \boldsymbol{H}_i \boldsymbol{D}_i \boldsymbol{H}_i^{\mathrm{H}} \right] \\ &\quad - \frac{1}{\ln 2} \sum_{k=1}^{K} \sum_{i \neq k} \mathrm{tr}\left[\boldsymbol{X}_k^{-1} \boldsymbol{H}_i \boldsymbol{D}_i \boldsymbol{H}_i^{\mathrm{H}} \right]. \end{aligned}$$

For the special choice

$$\boldsymbol{D}_k = \begin{cases} \mathbf{0} & \text{for } k \neq m \\ \boldsymbol{d}_m \boldsymbol{d}_m^{\mathrm{H}} & \text{for } k = m \end{cases}$$

the derivative simplifies to

$$\left.\frac{\partial R^{\mathrm{MAC}}(\boldsymbol{Q} + \alpha \boldsymbol{D})}{\partial \alpha}\right|_{\alpha=0} = \frac{1}{\ln 2} \cdot \boldsymbol{d}_m^{\mathrm{H}} \underbrace{\Big[\boldsymbol{H}_m^{\mathrm{H}} \Big(K \boldsymbol{X}^{-1} - \sum_{k \neq m} \boldsymbol{X}_k^{-1} \Big) \boldsymbol{H}_m \Big]}_{=:\boldsymbol{Z}_m} \boldsymbol{d}_m.$$

Depending on $\boldsymbol{H}$ and $\boldsymbol{Q}$, there may exist an $m \in \{1, \ldots, K\}$ for which $\boldsymbol{Z}_m$ has a negative eigenvalue. Choosing m as the index of such a user and $\boldsymbol{d}_m$ as the unit-norm eigenvector that corresponds to the smallest (negative) eigenvalue of $\boldsymbol{Z}_m$ leads to

$$R^{\mathrm{MAC}}(\boldsymbol{Q} + \alpha \boldsymbol{D}) < R^{\mathrm{MAC}}(\boldsymbol{Q})$$

for sufficiently small $\alpha > 0$ although $\boldsymbol{Q} + \alpha \boldsymbol{D} \succcurlyeq \boldsymbol{Q}$. Since $R^{\mathrm{MAC}}(\boldsymbol{Q} + \alpha \boldsymbol{D})$ varies continuously in $\boldsymbol{D} \succcurlyeq \mathbf{0}$, we may set

$$\boldsymbol{D}_k = \begin{cases} \varepsilon \mathbf{I}_{M_k} & \text{for } k \neq m \\ \boldsymbol{d}_m \boldsymbol{d}_m^{\mathrm{H}} + \varepsilon \mathbf{I}_{M_m} & \text{for } k = m \end{cases}$$

and find a sufficiently small $\varepsilon > 0$ such that

$$R^{\mathrm{MAC}}(\boldsymbol{Q} + \alpha \boldsymbol{D}) < R^{\mathrm{MAC}}(\boldsymbol{Q})$$

still holds although $\boldsymbol{Q} + \alpha \boldsymbol{D} \succ \boldsymbol{Q}$ even holds with strict inequality since $\boldsymbol{D}$ is full rank. Even though the sum rate need not be increasing in the transmit covariance matrices, the following, slightly weaker theorem holds:

Theorem 7 *For linear transceivers, the sum rate expression $R^{\mathrm{MAC}}(\alpha \boldsymbol{Q})$ in the dual MIMO MAC is increasing in $\alpha \geq 0$ for every $\boldsymbol{Q} \succcurlyeq \mathbf{0}$ that satisfies $\boldsymbol{H}\boldsymbol{Q} \neq \mathbf{0}$. For $\boldsymbol{H} \neq \mathbf{0}$, the sum power constrained sum rate maximization therefore makes use of the total transmit power $P_{\max}$ at the optimum.*

For the proof, we compute the derivative of $R_k^{\mathrm{MAC}}(\alpha \boldsymbol{Q})$ with respect to α using the second row in (7.11), Appendices A.6 and A.5, and the identity $\mathrm{tr}(\boldsymbol{A}\boldsymbol{B}) = \mathrm{tr}(\boldsymbol{B}\boldsymbol{A})$:

$$
\begin{aligned}
\frac{\partial R_k^{\mathrm{MAC}}(\alpha \boldsymbol{Q})}{\partial \alpha} &= \frac{\partial}{\partial \alpha} \log_2 \Big| \mathbf{I} + \underbrace{\Big(\frac{1}{\alpha} \cdot \mathbf{I} + \sum_{i \neq k} \boldsymbol{H}_i \boldsymbol{Q}_i \boldsymbol{H}_i^{\mathrm{H}} \Big)}_{=: \boldsymbol{X}'_k}{}^{-1} \boldsymbol{H}_k \boldsymbol{Q}_k \boldsymbol{H}_k^{\mathrm{H}} \Big| \\
&= \frac{1}{\ln 2} \cdot \operatorname{tr}\Big[\big(\mathbf{I} + \boldsymbol{X}_k'^{-1} \boldsymbol{H}_k \boldsymbol{Q}_k \boldsymbol{H}_k^{\mathrm{H}} \big)^{-1} \boldsymbol{X}_k'^{-1} \cdot \alpha^{-2} \cdot \boldsymbol{X}_k'^{-1} \boldsymbol{H}_k \boldsymbol{Q}_k \boldsymbol{H}_k^{\mathrm{H}} \Big] \\
&= \frac{1}{\alpha^2 \ln 2} \cdot \operatorname{tr}\Big[\big(\boldsymbol{X}'_k + \boldsymbol{H}_k \boldsymbol{Q}_k \boldsymbol{H}_k^{\mathrm{H}} \big)^{-1} \boldsymbol{X}_k'^{-1} \boldsymbol{H}_k \boldsymbol{Q}_k \boldsymbol{H}_k^{\mathrm{H}} \Big] \\
&= \frac{1}{\alpha^2 \ln 2} \cdot \operatorname{tr}\Big[\big(\boldsymbol{X}'_k + \boldsymbol{H}_k \boldsymbol{Q}_k \boldsymbol{H}_k^{\mathrm{H}} \big)^{-1} \boldsymbol{X}_k'^{-1} \cdot \Big(\big(\boldsymbol{X}'_k + \boldsymbol{H}_k \boldsymbol{Q}_k \boldsymbol{H}_k^{\mathrm{H}} \big) - \boldsymbol{X}'_k \Big) \Big] \\
&= \frac{1}{\alpha^2 \ln 2} \cdot \operatorname{tr}\Big[\boldsymbol{X}_k'^{-1} - \big(\boldsymbol{X}'_k + \boldsymbol{H}_k \boldsymbol{Q}_k \boldsymbol{H}_k^{\mathrm{H}} \big)^{-1} \Big] \geq 0
\end{aligned}
$$

Above expression is nonnegative since $\boldsymbol{X}'_k + \boldsymbol{H}_k \boldsymbol{Q}_k \boldsymbol{H}_k^{\mathrm{H}} \succcurlyeq \boldsymbol{X}'_k$. Furthermore, equality only holds, if $\boldsymbol{H}_k \boldsymbol{Q}_k = \mathbf{0}$. Since $R^{\mathrm{MAC}}(\alpha \boldsymbol{Q})$ is the sum of the individual rates, Theorem 7 is proven.

7.2.2 Combinatorial Sum Rate Maximization

For more than one user, the nonconcavity of the sum rate utility with respect to the transmit covariance matrices is to such an extent that the sum power constrained sum rate maximization may feature several local maxima that are not globally optimal. On account of this, it is impossible to find an iterative algorithm with polynomial complexity that achieves the global optimum from any starting point. Nonetheless, we would like to have an algorithm for the maximization of the sum rate under linear filtering that comes as close to the global optimum as possible. This objective is pursued by trying to cover as many different local optima as possible by utilizing a gradient-projection based local optimizer which is initialized by different starting points that correspond to different stream configurations. In order to ensure that the local optimum features a stream configuration that is close to the one it has been initialized with, we choose the precoder-based gradient-projection variant since the covariance-based utility is not concave either. As mentioned in Sect. 6.1.3, the precoder-based variant cannot increase the rank of the initial precoding matrices during the iterations. Hence, we can control the maximum number of active streams per user at the local optimum by the choice of the precoder initialization.

In the following, we first derive the relevant Wirtinger derivatives and the precoder-based rate expressions for a *single* stream configuration such that the gradient-projection algorithm from Sect. 5.1.1 becomes applicable. Afterwards, we discuss which stream configurations should be probed to cover as many different local optima as possible. These two parts finally make up our combinatorial algorithm for the sum rate maximization that has partly been published in [65].

Given a particular stream configuration in which L_k data streams are associated to user k for every $k \in \{1, \ldots, K\}$, the dimensions of the precoders are readily defined via $\boldsymbol{T}_k \in \mathbb{C}^{M_k \times L_k} \forall k$. How $L_1, \ldots, L_K$ should be chosen will be clarified later. When operating on the precoders, the rate of user k can be expressed as [cf. (7.11)]

$$\begin{aligned} R_k^{\mathrm{MAC}}(\boldsymbol{T}) &:= \log_2 \Big| \underbrace{\mathbf{I} + \sum_{i=1}^{K} \boldsymbol{H}_i \boldsymbol{T}_i \boldsymbol{T}_i^{\mathrm{H}} \boldsymbol{H}_i^{\mathrm{H}}}_{=:\boldsymbol{X}} \Big| - \log_2 \Big| \underbrace{\mathbf{I} + \sum_{i \neq k} \boldsymbol{H}_i \boldsymbol{T}_i \boldsymbol{T}_i^{\mathrm{H}} \boldsymbol{H}_i^{\mathrm{H}}}_{=:\boldsymbol{X}_k} \Big| \\ &= \log_2 \Big| \mathbf{I} + \Big(\mathbf{I} + \sum_{i \neq k} \boldsymbol{H}_i \boldsymbol{T}_i \boldsymbol{T}_i^{\mathrm{H}} \boldsymbol{H}_i^{\mathrm{H}} \Big)^{-1} \boldsymbol{H}_k \boldsymbol{T}_k \boldsymbol{T}_k^{\mathrm{H}} \boldsymbol{H}_k^{\mathrm{H}} \Big| \end{aligned} \tag{7.14}$$

and leads to the sum rate $R^{\mathrm{MAC}}(\boldsymbol{T}) := \sum_{k=1}^{K} R_k^{\mathrm{MAC}}(\boldsymbol{T})$. For the precoder-based gradient-projection recursions in (5.16) and (5.17), the Wirtinger derivatives of the sum rate utility with respect to every conjugate precoder $\boldsymbol{T}_j^* \ \forall j$ must be computed. Using the substitutions $\boldsymbol{X}$ and $\boldsymbol{X}_j$, the Wirtinger derivative of $R_j^{\mathrm{MAC}}(\boldsymbol{T})$ in (7.14) with respect to the entry $[\boldsymbol{T}_j^*]_{m,n}$ reads as (see Appendix A.6)

$$\begin{aligned} \frac{\partial_{\mathrm{W}} R_j^{\mathrm{MAC}}(\boldsymbol{T})}{\partial_{\mathrm{W}} [\boldsymbol{T}_j^*]_{m,n}} &= \frac{1}{\ln 2} \cdot \mathrm{tr} \Big[\Big(\mathbf{I} + \boldsymbol{X}_j^{-1} \boldsymbol{H}_j \boldsymbol{T}_j \boldsymbol{T}_j^{\mathrm{H}} \boldsymbol{H}_j^{\mathrm{H}} \Big)^{-1} \boldsymbol{X}_j^{-1} \boldsymbol{H}_j \boldsymbol{T}_j \boldsymbol{e}_n \boldsymbol{e}_m^{\mathrm{T}} \boldsymbol{H}_j^{\mathrm{H}} \Big] \\ &= \frac{1}{\ln 2} \cdot \boldsymbol{e}_m^{\mathrm{T}} \boldsymbol{H}_j^{\mathrm{H}} \boldsymbol{X}^{-1} \boldsymbol{H}_j \boldsymbol{T}_j \boldsymbol{e}_n \end{aligned}$$

such that the derivative with respect to the total conjugate precoder $\boldsymbol{T}_j^*$ yields

$$\frac{\partial_{\mathrm{W}} R_j^{\mathrm{MAC}}(\boldsymbol{T})}{\partial_{\mathrm{W}} \boldsymbol{T}_j^*} = \frac{1}{\ln 2} \cdot \boldsymbol{H}_j^{\mathrm{H}} \boldsymbol{X}^{-1} \boldsymbol{H}_j \boldsymbol{T}_j. \tag{7.15}$$

For $k \neq j$, the derivative of $R_k^{\mathrm{MAC}}(\boldsymbol{T})$ with respect to $\boldsymbol{T}_j^*$ looks slightly different. By means of Appendices A.6 and A.5, we find

$$\begin{aligned} \frac{\partial_{\mathrm{W}} R_k^{\mathrm{MAC}}(\boldsymbol{T})}{\partial_{\mathrm{W}} [\boldsymbol{T}_j^*]_{m,n}} \Big|_{k \neq j} &= -\frac{1}{\ln 2} \cdot \mathrm{tr} \Big[\Big(\mathbf{I} + \boldsymbol{X}_k^{-1} \boldsymbol{H}_k \boldsymbol{T}_k \boldsymbol{T}_k^{\mathrm{H}} \boldsymbol{H}_k \Big)^{-1} \boldsymbol{X}_k^{-1} \boldsymbol{H}_j \boldsymbol{T}_j \boldsymbol{e}_n \boldsymbol{e}_m^{\mathrm{T}} \boldsymbol{H}_j^{\mathrm{H}} \boldsymbol{X}_k^{-1} \\ &\qquad \boldsymbol{H}_k \boldsymbol{T}_k \boldsymbol{T}_k^{\mathrm{H}} \boldsymbol{H}_k^{\mathrm{H}} \Big] \\ &= -\frac{1}{\ln 2} \cdot \boldsymbol{e}_m^{\mathrm{T}} \boldsymbol{H}_j^{\mathrm{H}} \boldsymbol{X}_k^{-1} \boldsymbol{H}_k \boldsymbol{T}_k \boldsymbol{T}_k^{\mathrm{H}} \boldsymbol{H}_k^{\mathrm{H}} \boldsymbol{X}^{-1} \boldsymbol{H}_j \boldsymbol{T}_j \boldsymbol{e}_n \\ &= -\frac{1}{\ln 2} \cdot \boldsymbol{e}_m^{\mathrm{T}} \boldsymbol{H}_j^{\mathrm{H}} \boldsymbol{X}_k^{-1} (\boldsymbol{X} - \boldsymbol{X}_k) \boldsymbol{X}^{-1} \boldsymbol{H}_j \boldsymbol{T}_j \boldsymbol{e}_n \\ &= +\frac{1}{\ln 2} \cdot \boldsymbol{e}_m^{\mathrm{T}} \boldsymbol{H}_j^{\mathrm{H}} (\boldsymbol{X}^{-1} - \boldsymbol{X}_k^{-1}) \boldsymbol{H}_j \boldsymbol{T}_j \boldsymbol{e}_n. \end{aligned}$$

Thus, the Wirtinger derivative with respect to the complete conjugate precoder $\boldsymbol{T}_j^*$ is given by

$$\left.\frac{\partial_{\mathrm{w}} R_k^{\mathrm{MAC}}(\boldsymbol{T})}{\partial_{\mathrm{w}} \boldsymbol{T}_j^*}\right|_{k\neq j} = \frac{1}{\ln 2}\cdot \boldsymbol{H}_j^{\mathrm{H}}\left(\boldsymbol{X}^{-1} - \boldsymbol{X}_k^{-1}\right)\boldsymbol{H}_j\boldsymbol{T}_j. \tag{7.16}$$

Combining (7.15) and (7.16), the derivative of the total sum rate with respect to the conjugate precoder of user j can be written as

$$\frac{\partial_{\mathrm{w}} R^{\mathrm{MAC}}(\boldsymbol{T})}{\partial_{\mathrm{w}} \boldsymbol{T}_j^*} = \frac{1}{\ln 2}\cdot \boldsymbol{H}_j^{\mathrm{H}}\Big(K\boldsymbol{X}^{-1} - \sum_{k\neq j}\boldsymbol{X}_k^{-1}\Big)\boldsymbol{H}_j\boldsymbol{T}_j. \tag{7.17}$$

This allows us to apply the recursion formula (5.16) for the iterative gradient-projection algorithm with standard Wirtinger derivative as search direction. Equivalently, the projection of the Wirtinger derivative onto the tangent cone may be used as search direction which yields the recursion formula (5.17). Moving along the direction of the (projected) gradient leaves the constraint set in both cases such that the re-projection onto the constraint set is mandatory, see Appendix A.16. For the Armijo step-size rule, the scaling $s_0^{(n)}$ in (5.27) is again chosen as the quotient of the Frobenius norm of $\boldsymbol{T}$ and the (possibly projected) Wirtinger derivative:

$$s_0^{(n)} = \frac{\|\boldsymbol{T}^{(n)}\|_{\mathrm{F}}}{\left\|\left.\frac{\partial_{\mathrm{w}} R^{\mathrm{MAC}}(\boldsymbol{T})}{\partial_{\mathrm{w}}\boldsymbol{T}^*}\right|_{\boldsymbol{T}=\boldsymbol{T}^{(n)}}\right\|_{\mathrm{F}}} \quad \text{or} \quad s_0^{(n)} = \frac{\|\boldsymbol{T}^{(n)}\|_{\mathrm{F}}}{\left\|\mathcal{P}_{\mathcal{T}_{\mathcal{C}}(\boldsymbol{T}^{(n)})}\left(\left.\frac{\partial_{\mathrm{w}} R^{\mathrm{MAC}}(\boldsymbol{T})}{\partial_{\mathrm{w}}\boldsymbol{T}^*}\right|_{\boldsymbol{T}=\boldsymbol{T}^{(n)}}\right)\right\|_{\mathrm{F}}} \tag{7.18}$$

Given a particular stream configuration with L_k data streams associated to user k, we need to specify the precoder initialization. As we will see below, there is no rate gain dedicating more data streams to a particular user than his terminal has antennas. Thus, $L_k \leq M_k$ holds for all users $k \in \{1,\ldots,K\}$. Similar to the sum rate maximization for systems with interference cancelation, the precoders are initialized as scaled identity matrices such that the columns of all precoders have the same norm. For $0 < L_k < M_k$, only the first L_k columns of the scaled identity matrix are chosen, and $L_k = 0$ leads to a zero-vector of dimension M_k.[2] For example if $M_k = 3$, the four possible initializations for the precoder $\boldsymbol{T}_k^{(0)}$ which correspond to $L_k \in \{0, 1, 2, 3\}$ would read as

$$\boldsymbol{T}_k^{(0)} = \begin{bmatrix}0\\0\\0\end{bmatrix}, \qquad \boldsymbol{T}_k^{(0)} = \sqrt{\frac{P_{\max}}{L}}\cdot\begin{bmatrix}1\\0\\0\end{bmatrix}$$

$$\boldsymbol{T}_k^{(0)} = \sqrt{\frac{P_{\max}}{L}}\cdot\begin{bmatrix}1&0\\0&1\\0&0\end{bmatrix}, \quad \boldsymbol{T}_k^{(0)} = \sqrt{\frac{P_{\max}}{L}}\cdot\begin{bmatrix}1&0&0\\0&1&0\\0&0&1\end{bmatrix}$$

[2] Equivalently, the user could simply be dropped from the system model.

Table 7.1 Reasonable stream configurations for $K = 3$, $N = 6$, and $M_1 = M_2 = M_3 = 2$

Number of active streams (L)	Stream configurations
1	100, 010, 001
2	200, 020, 002, 110, 101, 011
3	210, 120, 102, 201, 012, 021, 111
4	220, 202, 022, 112, 121, 211
5	221, 212, 122
6	222

where $L := \sum_{k=1}^{K} L_k$ again denotes the total number of streams of the current configuration.

Our combinatorial sum rate maximization approach consists of carrying out such a locally optimal gradient-projection-based algorithm for every reasonable stream configuration.[3] It remains to determine those candidates which potentially lead to the global maximum. As one can see from (7.11), the rate expressions only depend on the transmit covariance matrices. Hence, it suffices to dedicate at most as many data streams to a user as his terminal has antennas because the outer product $\boldsymbol{T}_k\boldsymbol{T}_k^{\mathrm{H}}$ may achieve the maximum possible rank M_k when $\boldsymbol{T}_k$ has M_k columns, i.e., when $L_k = M_k$. Therefore, we know that $L_k \leq M_k$ holds for every user $k \in \{1, \ldots, K\}$. This inequality can be generalized to rank-deficient $\boldsymbol{H}_k$ by restricting the number of active streams that are associated to user k by $L_k \leq \operatorname{rank}(\boldsymbol{H}_k)$ since only the product $\boldsymbol{H}_k\boldsymbol{Q}_k\boldsymbol{H}_k^{\mathrm{H}}$ plays a role in the rate expressions, see (7.11).

In the low power regime with $P_{\max} \rightarrow 0$, single stream transmission is asymptotically optimum, see Appendix A.15 for the proof. In the high power regime with $P_{\max} \rightarrow \infty$, exactly $\operatorname{rank}(\boldsymbol{H})$ data streams must be active in order to achieve the maximum multiplexing gain. A rigorous proof for this will be derived in Chap. 8. We conjecture that at most $\operatorname{rank}(\boldsymbol{H})$ active data streams suffice to achieve the maximum sum rate for any transmit power $P_{\max}$. This is a commonly used assumption in the literature, but to the best of the author's knowledge, there is no proof available.

Summing up, the combinatorial sum rate maximization algorithm probes all stream configurations in which $\operatorname{rank}(\boldsymbol{H}) \leq \min\{M, N\}$ data streams are allocated to the K users such that $0 \leq L_k \leq \operatorname{rank}(\boldsymbol{H}_k)$ holds for every $k \in \{1, \ldots, K\}$. Table 7.1 shows all reasonable stream configurations for a $K = 3$ user setup in which every user has $M_1 = M_2 = M_3 = 2$ antennas and the base station is equipped with $N = 6$ antennas. The channel matrix $\boldsymbol{H}$ is assumed to have full rank in this setup.

[3] For $L = 1$ when only a single data stream is active, the gradient-projection algorithm need not be run actually. In this case, the optimum precoder of user k is given by the dominant eigenvector of the channel Gram $\boldsymbol{H}_k^{\mathrm{H}}\boldsymbol{H}_k$.

Synopsis: Due to the nonconcavity of the sum rate utility in a multi-user system with linear transceivers, the sum rate maximization is a nonconvex problem. In addition, several local maxima may exist which prevents from finding the global optimum by means of an iterative algorithm with polynomial complexity. In order to cover many different local maxima to which hopefully the global optimum belongs to, we propose a combinatorial algorithm that probes all reasonable stream configurations as initializations. For every such stream configuration, a precoder-based gradient-projection algorithm finds a local optimum of the sum rate.

By utilizing the precoder-based variant instead of the covariance-based, we can control the maximum number of active streams and thus try to find many different local optima. A reasonable stream configuration must obey the inequality $L_k \leq \text{rank}(\boldsymbol{H}_k) \leq M_k$ for every k, i.e., at most $\text{rank}(\boldsymbol{H}_k)$ data streams need to be allocated to user k. Moreover, the total number of active data streams L may not exceed $\text{rank}(\boldsymbol{H}) \leq \min\{M, N\}$. Although we cannot prove this conjecture, it is a common assumption which is intuitively derived from the circumstance that in the low power regime $P_{\max} \to 0$, only a single data stream is active in the optimum. In the high power regime $P_{\max} \to \infty$, exactly $\text{rank}(\boldsymbol{H})$ streams must be active in order to achieve the maximum multiplexing gain. For transmit powers in between, we conjecture that $1 \leq L \leq \text{rank}(\boldsymbol{H})$ streams are active at the optimum.

7.2.3 Successive Sum Rate Maximization

If the base station has at least as many antennas as the user terminals have in sum ($N \geq M$), the number of stream allocations that are probed by the combinatorial approach is given by

$$\prod_{k=1}^{K}(M_k + 1) - 1$$

for full rank $\boldsymbol{H}$ and hence grows exponentially in the number of users K. For example, 26 different initializations are probed for the setup in Table 7.1. One possibility to overcome this huge number of permutations is to switch from the combinatorial approach to a greedy one. Let

$$[\check{L}_1^{(i)}, \check{L}_2^{(i)}, \ldots, \check{L}_K^{(i)}]$$

be the stream configuration for $L = i$ active streams (at the initialization) that achieves the maximum sum rate. For $L = i + 1$ active streams, we test only those stream configurations for which

Table 7.2 Stream configurations tested by the greedy algorithm for $K = 3$, $N = 6$, $M_1 = M_2 = M_3 = 2$

Number of active streams (L)	Stream configurations
1	100, **010**, 001
2	020, 110, **011**
3	012, **021**, 111
4	022, **121**
5	221, **122**
6	**222**

$$\exists j: \; L_j^{(i+1)} = \check{L}_j^{(i)} + 1$$
$$\forall k \neq j: L_k^{(i+1)} = \check{L}_k^{(i)}.$$

Obviously, $\check{L}_j^{(i)} + 1 \leq \operatorname{rank}(\boldsymbol{H}_j)$ and $i + 1 \leq \operatorname{rank}(\boldsymbol{H})$ must hold, i.e., it must be possible to allocate a further stream to user j. This way, it is ensured that at most K permutations have to be probed for every i which results in an upper bound of $K \cdot \min\{M, N\}$ tests for full rank $\boldsymbol{H}$. Table 7.2 shows a possible evolution of stream configurations that are tested by the greedy algorithm. Note that the bold-faced configurations are assumed to achieve the maximum sum rate for a fixed number of active streams L. For $L = 1$, the maximum sum rate is achieved by the permutation

$$[\check{L}_1^{(1)}, \check{L}_2^{(1)}, \check{L}_3^{(1)}] = [0, 1, 0].$$

Hence, we do not have to check the permutations [2, 0, 0], [0, 0, 2], and [1, 0, 1] for $L = 2$ since no stream is associated to user 2 for those three. Assuming that the initialization [0, 1, 1] achieves the maximum sum rate for $L = 2$ which is larger than the sum rate of the configuration [0, 1, 0], only those permutations have to be probed for $L = 3$ for which user two and three have at least one active data stream at the initialization. Increasing L is stopped either if $L = \operatorname{rank}(\boldsymbol{H})$ or if the maximum sum rate did not increase from step $L = i$ to $L = i + 1$. From that point of view, Table 7.2 shows the worst case in which the maximum sum rate for a *fixed* number of active streams is increasing in L, i.e., the overall maximum sum rate is achieved for $L = 6$.

7.2.4 Simulation Results

To begin with, we first analyze the convergence behavior of the precoder-based gradient-projection algorithm for a fixed channel realization and a fixed initial stream configuration. For that purpose, we choose the channel matrix $\boldsymbol{H}$ from (6.20) where $K = 3$ two-antenna users ($M_1 = M_2 = M_3 = 2$) communicate with an $N = 6$ antenna base station. Given the particular stream configuration $[L_1, L_2, L_3] =$

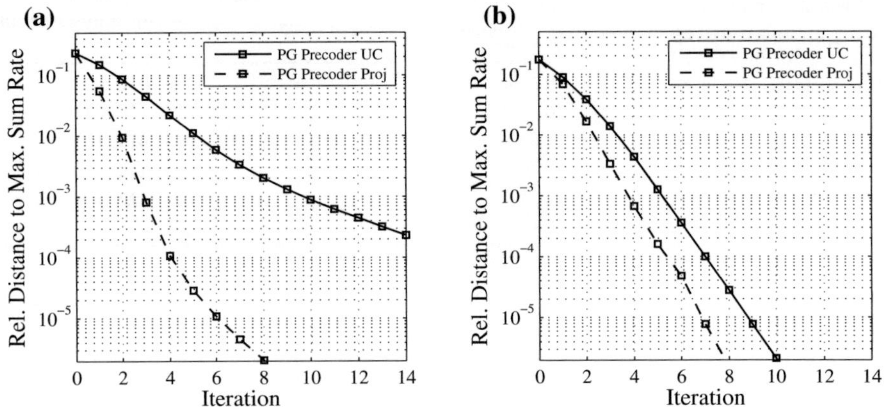

Fig. 7.1 Relative distance to maximally achieved sum rate for $K = 3$ users and $N = 6$ antennas. **a** $10\log_{10} P_{\max} = -5\,\text{dB}$, **b** $10\log_{10} P_{\max} = +5\,\text{dB}$

[2, 1, 1] the relative distance to the maximally achieved[4] sum rate versus the iteration number is plotted in Fig. 7.1 for the standard Wirtinger derivative as search direction (`PG Precoder UC`—solid curve, square marker) and its projection onto the tangent cone as search direction (`PG Precoder Proj`—dashed curve, square marker). For both variants, the Armijo rule was chosen as step-size control mechanism with $\beta = 0.5$ and $\sigma = 0.1$ for the standard Wirtinger derivative and $\sigma = 0.3$ for its projection onto the tangent cone as search direction, respectively, see Sect. 5.1.3.3. The left Fig. 7.1a shows the situation for $10\log_{10} P_{\max} = -5\,\text{dB}$ whereas $10\log_{10} P_{\max} = +5\,\text{dB}$ holds in the right Fig. 7.1b. As one can see in Fig. 7.1a, the standard gradient search direction needs about 10 iterations to reach the maximally achieved sum rate up to one per mill. If the Wirtinger derivative is projected onto the tangent cone, this search direction requires only 3 iterations to reach the same target and thus features a much faster decay of the distance to the achieved maximum. For $10\log_{10} P_{\max} = +5\,\text{dB}$ (Fig. 7.1b), the standard Wirtinger derivative (`PG Precoder UC`—solid curve, square marker) shows a faster convergence than for small transmit powers. However, it is still outperformed by the tangent cone projection search direction (`PG Precoder Proj`—dashed curve, square marker).

For the same channel realization, Fig. 7.2 shows the absolute and relative sum rate versus the logarithmic transmit power for different algorithms. The combinatorial and the greedy algorithm (`Combinatorial/Greedy`—square marker) described in Sects. 7.2.2 and 7.2.3 achieve exactly the same performance, which means that the greedy variant finds the best stream configuration for the given channel independent of the transmit power. Note that the gradient-projection algorithm was run with 10 iterations for both variants. Moreover, the two search directions (Wirtinger derivative

[4] Since the global maximum of the sum rate is unknown due to the nonconvexity of the problem, we plot the relative distance to the maximally achieved one.

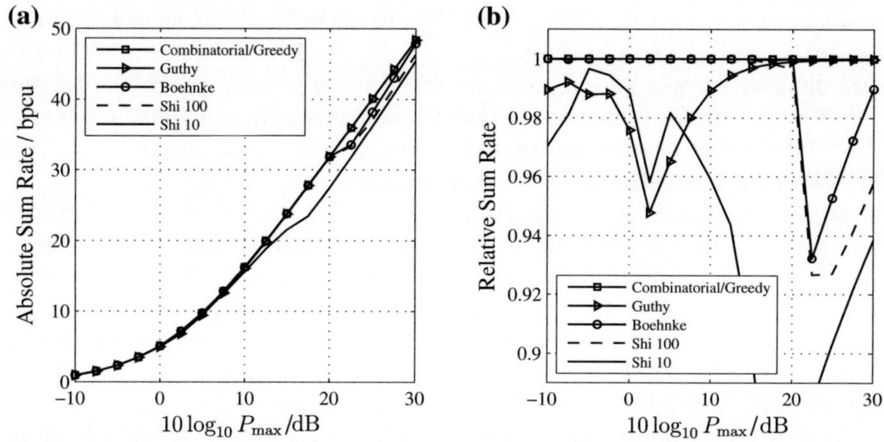

Fig. 7.2 Absolute and relative sum rate for $K = 3$ users and $N = 6$ antennas. **a** Absolute sum rate, **b** relative sum rate

and its tangent cone projection) basically lead to the same result and achieve sum rates which coincide up to the third decimal place. The successive stream allocation method from Guthy et al. in [129] (`Guthy`—triangle-right marker) directly operates in the MIMO BC and allocates the streams one by one. We chose the generalized eigenvalue as a criterion to whom the next stream will be allocated. Moreover, adding streams was stopped as soon as the true sum rate could not be increased. The third algorithm is taken from Böhnke in [62] (`Boehnke`—circle marker). It operates in the dual domain as well and is very similar to our gradient-projection algorithm (with fixed stream configuration) except that conjugate search directions and the on-demand step-size control mechanism are used. For the initialization, every user features as many data streams as his terminal has antennas. In turn, 10 iterations were run. Finally, the geometric-programming-based variant from Shi et al. in [57, 58] is shown which performs an alternating optimization between the precoders, the receive filters, and the power allocation. For the geometric-programming part of the power allocation, we used the *fmincon-geometric* solver of the Yalmip [130] toolbox. The 10 iterations variant is depicted as (`Shi 10`—solid curve, no marker) where the 100 iterations version is depicted as (`Shi 100`—dashed curve, no marker). Figure 7.2a shows the absolute sum rate in bits per channel user (bpcu) versus the logarithmic sum power. The combinatorial and the greedy algorithm (`Combinatorial/Greedy`—square marker) achieve the largest sum rate over the whole power regime. The successive algorithm from Guthy (`Guthy`—triangle-right marker) achieves practically the same performance at large transmit powers above 20 dB. For $10 \log_{10} P_{\max} = 2.5$ dB, about 95 % of the maximum achieved sum rate are obtained, see Fig. 7.2b. Böhnke's algorithm (`Boehnke`—circle marker) works perfect at transmit powers below 20 dB (see Fig. 7.2b) but fails to converge to the right local optimum for transmit powers between 22.5 and 30 dB. There, the optimum stream configuration (also after convergence) is given by $L_1 = L_2 = 2$ and $L_3 = 1$ whereas Böhnke's algorithm converges to a local optimum with $L_1 = L_2 = L_3 = 2$ which features a smaller sum rate. Shi's alternating optimization approach with 10 iterations (`Shi 10`—solid curve, no marker) has a poor performance at large transmit powers. For 100 iterations (`Shi 100`—dashed curve, no marker), this behavior improves and the performance is very good for $10\log_{10} P_{\max} \leq 20$ dB. However, it should be mentioned that the computational complexity of this algorithm is tremendous. The number of monomials in the product-MSE cost function (which is a posynomial) before grouping is given by $(M+1)^M$. For $M = 6$, there are more than 117000 monomials which makes the evaluation of the cost function computationally extremely demanding. Hence, the simulations with 100 iterations took a very long time even for a single channel realization only.

Channel averaged results are shown in Figs. 7.3, 7.4 and 7.5 where the entries of the channel matrix $\boldsymbol{H}$ were chosen as i.i.d. complex Gaussian random variables with unit variance. Averaging has been done over 1000 channel realizations. Due to the enormous complexity of the geometric programming part of the approach in [57, 58], this algorithm is not included in the channel averaged results. All iterative algorithms were again run with 10 iterations per channel realization.

Figure 7.3 shows the results for an $N = 6$ antenna base station that communicates with $K = 3$ two-antenna users. On the left, the relative sum rate (i.e., the achieved sum rate of a particular algorithm divided by the maximally achieved sum rate of all linear schemes) is plotted versus the logarithmic transmit power in Fig. 7.3a for various algorithms. As one can see, the largest relative sum rate of all linear schemes is obtained by the combinatorial approach (`Combinatorial`—solid curve, square marker) over the complete transmit power regime. Since the relative sum rate of this algorithm is equal to one, it achieves the largest absolute sum rate for all transmit powers. Moreover, it practically does not make a difference whether the standard Wirtinger derivative or its projection onto the tangent cone is chosen as search direction since the difference cannot be resolved on this scale. For comparison purposes, the ergodic sum capacity is also depicted (`Dirty Paper Coding`—dashed curve, no marker). At low transmit powers, the absolute sum rate difference between the best linear scheme and dirty paper coding goes to zero since single-stream transmission is optimum in both cases in the low power regime, see Appendix A.15. The reduced complexity successive algorithm from Sect. 7.2.3 (`Greedy`—dashed curve, square marker) reaches the maximum relative sum rate up to one per mill over the whole power regime and the absolute difference to the maximally achieved (channel averaged) sum rate is below 0.04 bits per channel use, see Fig. 7.3b. Böhnke's algorithm [62] (`Boehnke`—circle marker) has two streams per user active at the initialization and seems to converge to the global optimum for small powers below 0 dB. For larger transmit powers, convergence to a local optimum which is not globally optimum is obtained and the relative sum rate reduces to about 0.97 times the maximally achieved one at 30 dB. Above 30 dB, the absolute rate difference to the maximally achieved one decays to zero for increasing transmit powers, see Fig. 7.3b. This is due to the fact that a uniform power allocation is asymptotically optimum with respect to the

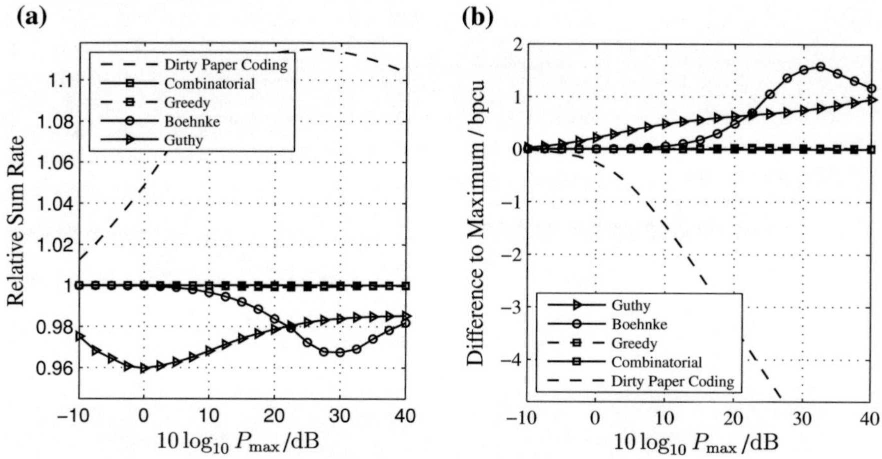

Fig. 7.3 Relative sum rate and absolute difference to (achieved) maximum for $K = 3$ users and $N = 6$. **a** Relative sum rate, **b** absolute difference to maximum sum rate

sum rate when $N \geq M$, see Chap. 8. As the system is fully loaded ($M = N$), the uniform power allocation over all streams becomes optimum at relatively large powers. This changes when only $K = 2$ two-antenna users are served by the $N = 6$ antenna base station, see Fig. 7.5. The successive stream allocation algorithm from Guthy in [129] (`Guthy`—triangle-right marker) is optimal for $10 \log_{10} P_{\max} \to -\infty$dB since single-stream transmission achieves capacity there, see Appendix A.15. At 0 dB, a relative sum rate of 0.96 is achieved. For larger transmit powers, the relative sum rate seems to converge to 0.985 and Fig. 7.3b indicates that the absolute difference to the maximum sum rate increases almost linearly in the logarithmic transmit power.

Removing three antennas from the base station ($N = 3$) leads to the results in Fig. 7.4. In turn, the largest sum rate of all linear schemes is achieved by the combinatorial algorithm from Sect. 7.2.2 (`Combinatorial`—solid curve, square marker) closely followed by the reduced complexity successive algorithm from Sect. 7.2.3 (`Greedy`—dashed curve, square marker). The basic behavior of Guthy's successive stream allocation algorithm (`Guthy`—triangle-right marker) matches that from Fig. 7.3 except that the distance to the maximally achieved sum rate saturates at about 0.4 bits per channel use, see Fig. 7.4b. Böhnke's approach (`Boehnke`—circle marker) still works fine at small transmit powers for $N < M$. However, the initialization with $M = 6$ data streams fails to converge to a local maximum with reasonable sum rate at medium to large transmit powers. Thus, the relative sum rate drops below 0.9 times the maximally achieved one at 40 dB, see Fig. 7.4a, and the difference to the maximum sum rate quickly increases with increasing sum power and reaches 4 bits per channel use at 40 dB, see Fig. 7.4b. For the given setup, the ergodic sum capacity (`Dirty Paper Coding`—dashed curve, no marker) is less than 0.8 bits per

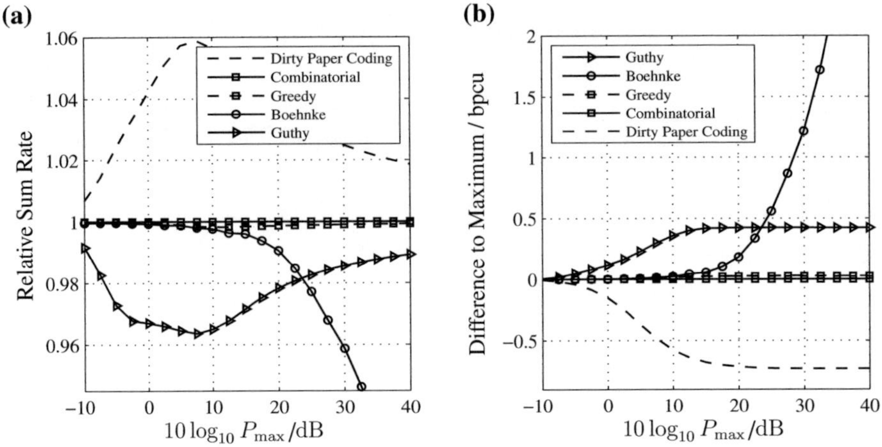

Fig. 7.4 Relative sum rate and absolute difference to (achieved) maximum for $K = 3$ users and $N = 3$. **a** Relative sum rate, **b** absolute difference to maximum sum rate

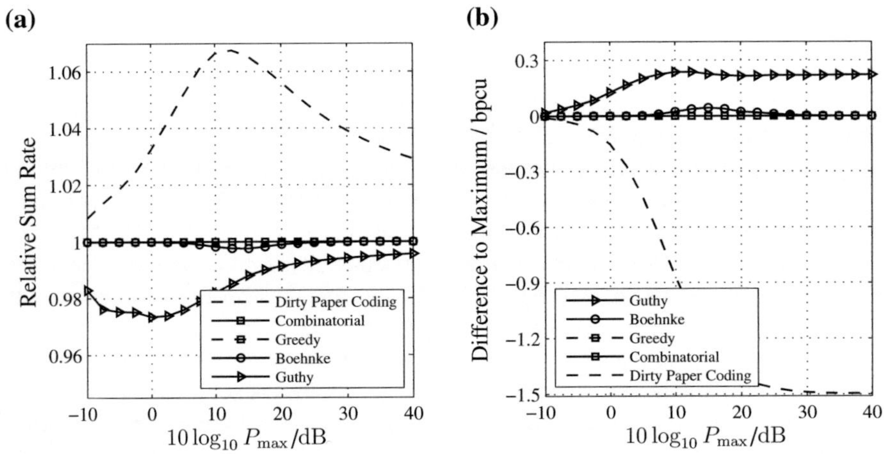

Fig. 7.5 Relative sum rate and absolute difference to (achieved) maximum for $K = 2$ users and $N = 6$. **a** Relative sum rate, **b** absolute difference to maximum sum rate

channel use away from the maximum averaged sum rate of the best linear scheme, see Fig. 7.4b.

Finally, a weaker loaded system with $K = 2$ two-antenna users served by an $N = 6$ antenna base station is investigated in Fig. 7.5. The principle characteristics of the combinatorial (`Combinatorial`—solid curve, square marker) and the successive algorithm (`Greedy`—dashed curve, square marker) remain the same as before. Guthy's algorithm (`Guthy`—triangle-right marker) shows a saturation of the difference to the maximum at around 0.2 bits per channel use for large transmit powers, see Fig. 7.5b. The relative sum rate is always larger than 0.97 times the maximum one for this setup over the whole power regime, see Fig. 7.5a. As mentioned before, scaled identity precoders are asymptotically optimum at large transmit powers for $N \geq M$. Therefore, Böhnke's algorithm (`Boehnke`—circle marker) will obtain the largest possible sum rate at large transmit powers for the given setup. Due to the fact that the system is not fully loaded, the uniform power allocation for the scaled identity precoders yields almost the maximum possible sum rate at moderately large transmit powers and the absolute difference to the maximum sum rate is already below 0.02 bits per channel use for $10 \log_{10} P_{\max} \geq 20$ dB, see Fig. 7.5b. In addition, the ergodic sum capacity is only 1.5bpcu above the maximum sum rate of the best linear scheme.

7.3 Weighted Sum Rate Maximization

The demand for a quickly converging weighted sum rate maximization algorithm discussed in Sect. 6.2.1 seamlessly translates to the case of linear transceivers. Like the standard sum rate, the *weighted* sum rate maximization problem is not convex for more than one user in case of linear transceivers and may feature several stationary points which are not globally optimum. By means of our combinatorial algorithm from Sect. 7.2.2 which probes all reasonable stream configurations for the initialization we try to cover as many different local optima as possible such that the probability of finding the global one is very high. To overcome the potentially large computational complexity of the combinatorial algorithm, the successive variant from Sect. 7.2.3 may be applied as well. In order to let the two algorithms become applicable we just need to compute the Wirtinger derivatives for the modified utility which are required by the precoder-based gradient-projection algorithm from Chap. 5.

7.3.1 Weighted Sum Rate Expression and Wirtinger Derivatives

Given the rate expression $R_k^{\mathrm{MAC}}(\boldsymbol{T})$ of user k in (7.14), the weighted sum rate expression $f(\boldsymbol{T}, \boldsymbol{w})$ with nonnegative user weights $\boldsymbol{w} := [w_1, \ldots, w_K]^{\mathrm{T}} \in \mathbb{R}_{+,0}^K$ reads as

$$
\begin{aligned}
f(\boldsymbol{T}, \boldsymbol{w}) &:= \sum_{k=1}^{K} w_k R_k^{\mathrm{MAC}}(\boldsymbol{T}) \\
&= \Big(\sum_{k=1}^{K} w_k\Big) \log_2 \Big| \underbrace{\mathbf{I} + \sum_{i=1}^{K} \boldsymbol{H}_i \boldsymbol{T}_i \boldsymbol{T}_i^{\mathrm{H}} \boldsymbol{H}_i^{\mathrm{H}}}_{=:\boldsymbol{X}} \Big|
\end{aligned}
$$

$$
-\sum_{k=1}^{K} w_k \log_2 \Big| \underbrace{\mathbf{I} + \sum_{i \neq k} \boldsymbol{H}_i \boldsymbol{T}_i \boldsymbol{T}_i^{\mathrm{H}} \boldsymbol{H}_i^{\mathrm{H}}}_{=:\boldsymbol{X}_k} \Big|
$$

$$
= \sum_{k=1}^{K} w_k \log_2 \Big| \mathbf{I} + \Big(\mathbf{I} + \sum_{i \neq k} \boldsymbol{H}_i \boldsymbol{T}_i \boldsymbol{T}_i^{\mathrm{H}} \boldsymbol{H}_i^{\mathrm{H}}\Big)^{-1} \boldsymbol{H}_k \boldsymbol{T}_k \boldsymbol{T}_k^{\mathrm{H}} \boldsymbol{H}_k^{\mathrm{H}} \Big|. \tag{7.19}
$$

Using the Wirtinger derivatives of $R_j^{\mathrm{MAC}}(\boldsymbol{T})$ and $R_k^{\mathrm{MAC}}(\boldsymbol{T})$ for $k \neq j$ with respect to $\boldsymbol{T}_j^*$ in (7.15) and (7.16), the derivative of the weighted sum rate $f(\boldsymbol{T}, \boldsymbol{w})$ with respect to the conjugate precoder $\boldsymbol{T}_j^*$ can readily be computed to

$$
\frac{\partial_{\mathrm{W}} f(\boldsymbol{T}, \boldsymbol{w})}{\partial_{\mathrm{W}} \boldsymbol{T}_j^*} = \frac{1}{\ln 2} \cdot \boldsymbol{H}_j^{\mathrm{H}} \left[\Big(\sum_{k=1}^{K} w_k\Big) \boldsymbol{X}^{-1} - \sum_{k \neq j} w_k \boldsymbol{X}_k^{-1} \right] \boldsymbol{H}_j \boldsymbol{T}_j. \tag{7.20}
$$

For $\boldsymbol{T} \in \partial\mathcal{C}$, i.e., when $\boldsymbol{T}$ is on the boundary of the constraint set with $\|\boldsymbol{T}\|_{\mathrm{F}}^2 = P_{\max}$, the constraint set is left when moving along the direction of the Wirtinger derivative or its projection onto the tangent cone, see Appendix A.16. Thus, this property is not only valid for the standard sum rate but also for the *weighted* sum rate with arbitrary positive weights $\boldsymbol{w} > 0$. For the Armijo step-size rule, the common scaling factor $s_0^{(n)}$ in iteration n is chosen as in (7.18) except that $R^{\mathrm{MAC}}(\boldsymbol{T})$ must be replaced by the weighted sum rate utility $f(\boldsymbol{T}, \boldsymbol{w})$. By means of (7.20), the two recursion formulas (5.16) and (5.17) of the gradient-projection algorithm can be implemented.

7.3.2 Simulation Results

We start with plotting the rate pairs that are achieved by various weighted sum rate maximization algorithms for a fixed channel realization when sweeping the user weights in a two-user scenario. The composite MAC channel $\boldsymbol{H}$ is taken from (6.20). But since we consider the case of $K = 2$ two-antenna users, only the first 4 columns of $\boldsymbol{H}$ in (6.20) are used. Moreover, only the first N rows will be selected where the number of antennas N at the base station is chosen to be $N = 4$ in Fig. 7.6 and $N = 2$ in Fig. 7.7. For both scenarios, the two transmit powers $10\log_{10} P_{\max} = 0\,\mathrm{dB}$ and $10\log_{10} P_{\max} = +10\,\mathrm{dB}$ are shown. The angle θ of the weight vector $\boldsymbol{w} = [-\sin(\theta), \cos(\theta)]^{\mathrm{T}}$ was swept from $\theta = 0°$ to $\theta = -90°$ with 40 samples and every marker corresponds to (at least) one such angle. All markers are connected by line segments. Since the various weighted sum rate maximization algorithms need not be globally optimal, the resulting region need not be convex. For the given channel realization, the combinatorial stream allocation approach always coincides with the successive one for the four investigated system setups.

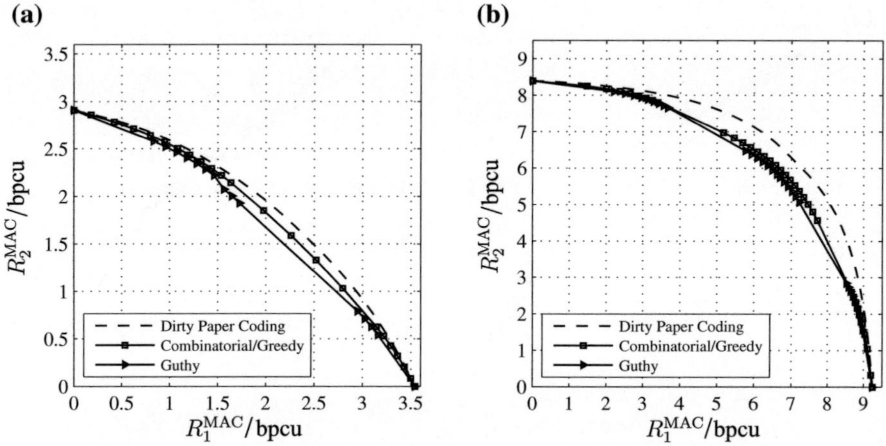

Fig. 7.6 Achievable rate region for $K = 2$ users and $N = 4$ antennas at the base station. **a** $10 \log_{10} P_{\mathrm{max}} = 0\,\mathrm{dB}$, **b** $10 \log_{10} P_{\mathrm{max}} = +10\,\mathrm{dB}$

In Fig. 7.6a, the combinatorial/greedy stream allocation algorithm corresponds to (`Combinatorial/Greedy`—square marker). For every reasonable stream configuration, the gradient-projection algorithm was run with 10 iterations and parameters $\beta = 0.5$ and $\sigma = 0.3$. The tangent cone projection of the Wirtinger derivative was chosen as the search direction but taking the derivative itself practically does not change the result. For the single-user rate pairs where one entry of the weight vector and thus also one entry of the rate pair is zero, the maximum possible single-user rate is achieved by construction.[5] Equivalently, the low-complexity successive algorithm from Guthy et al. in [64] (`Guthy`—triangle-right marker) which directly operates in the broadcast channel always achieves these maximum single-user rates by construction. However, a smaller rate region is obtained which becomes manifest in particular when the two entries of the weight vector are approximately the same. Obviously, the capacity region (`Dirty Paper Coding`—dashed curve) which requires the usage of dirty paper coding is a superset of the rate region that is achievable by means of linear transceivers. Note that Böhnke's algorithm in [62, 124] coincides with our combinatorial stream allocation algorithm for the simple two-user setup with $N = 4$ and is hence not plotted. This means that the initialization with two streams per user converges to the same local optimum as the best local optimum of the combinatorial algorithm for the given setup and the given transmit powers.

Figure 7.6b shows that the difference between the capacity region (`Dirty Paper Coding`—dashed curve) and the rate regions obtained by linear transceivers is more pronounced for a transmit power of 10 dB. For weight vectors whose entries are approximately the same, a slight performance reduction of Guthy's

[5] This holds as long as the canonical unit vectors are not eigenvectors of the channel Gram that do not correspond to the maximum eigenvalue.

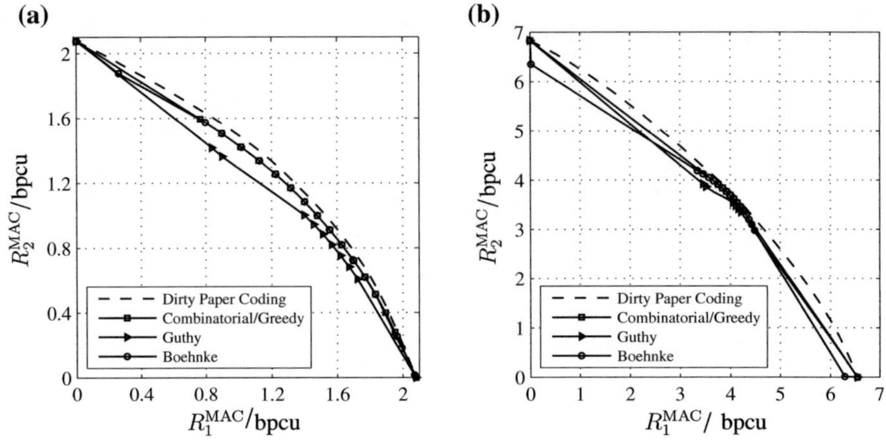

Fig. 7.7 Achievable rate region for $K = 2$ users and $N = 2$ antennas at the base station. **a** $10\log_{10} P_{\max} = 0\,\text{dB}$, **b** $10\log_{10} P_{\max} = +10\,\text{dB}$

algorithm (`Guthy`—triangle-right marker) with respect to the combinatorial/greedy one (`Combinatorial/Greedy`—square marker) can again be seen.

Reducing the number of antennas at the base station to $N = 2$ leads to Fig. 7.7. For this setup, Böhnke's algorithm (`Boehnke`-circle marker) deviates from our combinatorial/greedy one (`Combinatorial/Greedy`—square marker) in particular for larger transmit powers, see Fig. 7.7b. The two rate pairs ($R_i^{\mathrm{MAC}} \approx 6.2$bpcu, $R_{3-i}^{\mathrm{MAC}} \approx 0$bpcu) with $i \in \{1, 2\}$ are obtained due to the fact that 10 iterations did not lead to convergence during Böhnke's algorithm in this case. At low transmit powers (Fig. 7.7a), the best linear scheme leads to rate pairs close to the capacity region (`Dirty Paper Coding`—dashed curve). In Fig. 7.7b, most of the rate pairs returned by the different algorithms group around $R_1^{\mathrm{MAC}} = R_2^{\mathrm{MAC}} = 4$bpcu.

Channel-averaged results of a $K = 3$ users scenario are plotted in Figs. 7.8 and 7.9 for $N = 6$ and $N = 3$ antennas, respectively. Averaging was done over 1000 channel realizations and the weights of the weighted sum rate maximization were set to $\boldsymbol{w} = [0.5, 1.0, 1.5]^{\mathrm{T}}$. The outcomes do not differ much from the results of the sum rate maximization with equal weights in Figs. 7.3 and 7.4. All algorithms achieve the same weighted sum rate for small transmit powers due to the optimality of single-stream transmission in this regime. Böhnke's algorithm (`Boehnke`—circle marker) converges to the right local maximum for small transmit powers despite its initialization with M data streams. For larger powers and only $N = 3$ antennas at the base station, the achieved weighted sum rate strongly deviates from the combinatorial (`Combinatorial`—solid curve, square marker) and the greedy one (`Greedy`—dashed curve, square marker), see Fig. 7.9b. These two essentially lead to the same results over the whole power regime. For the given choice of the weight vector $\boldsymbol{w}$ and the fully loaded system with $M = N$ in Fig. 7.8, only 5 of the 6 possible data

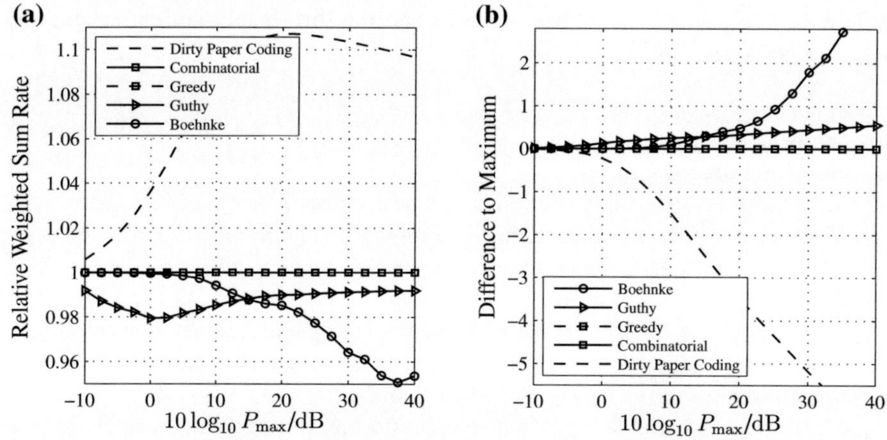

Fig. 7.8 Relative weighted sum rate and absolute difference to maximum for $K = 3$ users and $N = 6$. **a** Relative weighted sum rate, **b** absolute difference to maximum WSR

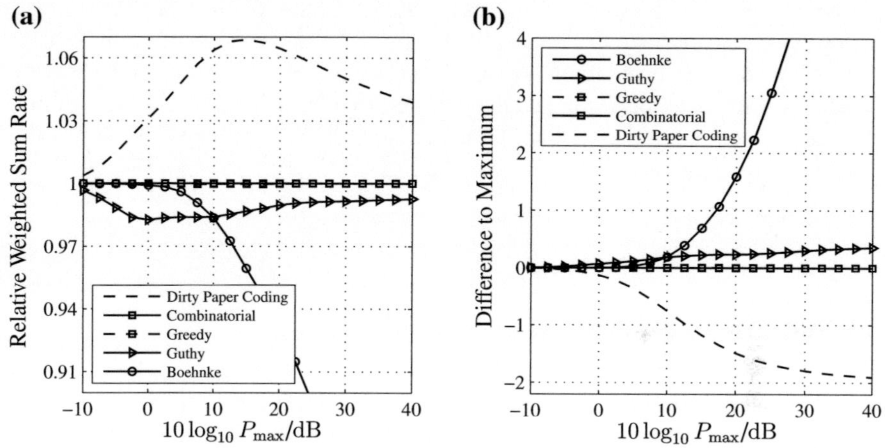

Fig. 7.9 Relative weighted sum rate and absolute difference to maximum for $K = 3$ users and $N = 3$. **a** Relative weighted sum rate, **b** absolute difference to maximum WSR

streams are multiplexed at 40 dB for most channel realizations. Therefore, Böhnke's full rank initialization (`Boehnke`—circle marker) does not converge to the right local optimum such that the difference to the maximally achieved sum rate of the combinatorial algorithm (`Combinatorial`—solid curve, square marker) in Fig. 7.8b will reduce to zero bits per channel use for transmit powers larger than 40 dB. Moreover, the weighted sum rate difference of the dirty paper coding scheme (`Dirty Paper Coding`—dashed curve, no marker) in Fig. 7.8b has not yet converged to a constant for $10 \log_{10} P_{\max} = 40$ dB. In the next chapter, the asymptotic weighted sum rate difference between the best linear scheme and the dirty paper coding scheme for $P_{\max} \to \infty$ is derived.

Synopsis: The proposed combinatorial weighted sum rate maximization algorithm for linear transceivers turns out to feature a superior performance. It outperforms existing algorithms and may be used as a reference to evaluate the efficiency of other strategies with less computational complexity for example. We conjecture that the combinatorial approach leads to the maximum possible (weighted) sum rate under linear filtering for most scenarios although we cannot prove it. A reduction of the complexity with almost unnoticeable performance reduction over the complete power regime is achieved by our successive algorithm. There, data streams are allocated to the different users in a greedy fashion. The gradient-projection based algorithm from Böhnke [124] initializes the precoder of every user with as many data streams as he has antennas. For small transmit powers, convergence to a local optimum with sufficiently large utility is observed. For medium to large transmit powers and systems with less antennas at the base station than at the terminals ($N < M$), however, convergence to a strongly suboptimal stationary point is observed which leads to a considerable loss of the utility. An attractive low-complexity scheme with good performance is presented by Guthy in [64]. His algorithm directly operates in the broadcast channel and allocates streams successively to the users under a zero-forcing constraint.

7.4 Sum Mean Square Error Minimization

For an uncoded broadcast channel system, small raw bit error ratios are desirable. Unfortunately, the functional relationship between the transmit/receive filters and the obtained bit error ratios is highly nonlinear and noncontinuous due to the quantization process which maps the soft values to elements of the modulation alphabet. As a consequence, using any kind of bit error ratio based cost function for the optimization is basically impossible and must be discarded. One way to overcome this difficulty is to let the cost function be based on the soft values *in front of* the quantizer input. Let $\hat{\boldsymbol{s}}_1^{\mathrm{BC}}, \ldots, \hat{\boldsymbol{s}}_K^{\mathrm{BC}}$ be the quantizer inputs of the K users in the broadcast channel as depicted in Fig. 2.1 in Chap. 2. If these estimates are close to the originally sent symbol vectors $\boldsymbol{s}_1, \ldots, \boldsymbol{s}_K$ then the quantized versions of $\hat{\boldsymbol{s}}_1^{\mathrm{BC}}, \ldots, \hat{\boldsymbol{s}}_K^{\mathrm{BC}}$ are likely to coincide with $\boldsymbol{s}_1, \ldots, \boldsymbol{s}_K$. A reasonable cost function may therefore be defined as the expected value of the squared norm between the stacked symbol vector $\boldsymbol{s} := [\boldsymbol{s}_1^{\mathrm{T}}, \ldots, \boldsymbol{s}_K^{\mathrm{T}}]^{\mathrm{T}}$ and its estimate $\hat{\boldsymbol{s}}^{\mathrm{BC}} := [\hat{\boldsymbol{s}}_1^{\mathrm{BC,T}}, \ldots, \hat{\boldsymbol{s}}_K^{\mathrm{BC,T}}]^{\mathrm{T}}$. This sum mean square error metric turns out to be quadratic in both the transmit and receive filters which makes it accessible to various kinds of optimizations like the alternating optimization approach in [131].

There, the quadratic dependency of the cost function on both the transmit and receive filters is used to optimize one end of the link while the other end is kept fixed. For fixed transmitters, the optimum receivers can be computed in closed form and vice versa. A drawback of this approach is its low speed of convergence. Another partitioning of the optimization variables into receive filters, unit norm beamformers, and the power allocation is used by Shi et al. in [15–17] where the MSE duality is used to frequently switch between the BC and the dual MAC. These frequent domain changes have their origin in the circumstance that the receive filters can be computed in closed form for fixed transmit filters and a fixed power allocation. Their normalized versions serve as unit-norm beamformers when changing from the BC to the MAC or vice versa.

7.4.1 Dual MIMO MAC Approach

In this subsection, we propose an algorithm which entirely operates in the dual MIMO MAC. Having solved the sum MSE minimization there, the transmit and receive filters are converted to the original MIMO BC via the level one MAC to BC conversion from Sect. 3.2.4.1. Since only the sum mean square error needs to remain constant, the computational complexity of this conversion is extremely small. The transmit filters in the BC are given by the scaled receive filters in the MAC and vice versa. Moreover, only a single scaling factor has to be computed, see (3.32). The reason for operating in the dual domain is the convexity of the utility. While the complete MSE region without time-sharing is convex only in the two-user case with single-antenna terminals [132], the standard sum mean square error cost function is convex in the dual MAC with respect to the transmit covariance matrices for any number of users and antennas as long as every user multiplexes at least as many streams as he has antennas, i.e., $L_k = M_k \; \forall k$. A proof for this statement will be given later in this section. Note that the sum MSE minimization in the original MIMO BC can also be reformulated as a convex optimization by means of several tricky modifications, see [18, Sect. VI.A]. However, revealing its convexity by means of the dual domain is much more elegant and features only negligible complexity due to the simple MAC to BC conversion.

In the dual MIMO MAC with transmit filters $\boldsymbol{T}_1, \ldots, \boldsymbol{T}_K$ and receive filters $\boldsymbol{G}_1, \ldots, \boldsymbol{G}_K$, the mean square error $\varepsilon_k^{\mathrm{MAC}}$ of user k reads as [see (3.7)]

$$\varepsilon_k^{\mathrm{MAC}}(\boldsymbol{T}, \boldsymbol{G}_k) = L_k + \operatorname{tr}\left(\boldsymbol{G}_k \sum_{i=1}^{K} \boldsymbol{H}_i \boldsymbol{T}_i \boldsymbol{T}_i^{\mathrm{H}} \boldsymbol{H}_i^{\mathrm{H}} \boldsymbol{G}_k + \boldsymbol{G}_k \boldsymbol{G}_k^{\mathrm{H}} - \boldsymbol{G}_k \boldsymbol{H}_k \boldsymbol{T}_k - \boldsymbol{T}_k^{\mathrm{H}} \boldsymbol{H}_k^{\mathrm{H}} \boldsymbol{G}_k^{\mathrm{H}}\right). \tag{7.21}$$

Since the MSEs of the K users are not coupled by the receive filters, $\boldsymbol{G}_k$ can be optimized separately for every user k. Minimizing $\varepsilon_k^{\mathrm{MAC}}$ with respect to $\boldsymbol{G}_k$ leads to the MMSE receive filter

$$G_k^{\text{MMSE}} := T_k^{\text{H}} H_k^{\text{H}} \Big(\mathbf{I} + \sum_{i=1}^{K} H_i T_i T_i^{\text{H}} H_i^{\text{H}} \Big)^{-1} \tag{7.22}$$

and the MMSE

$$\varepsilon_k^{\text{MAC}}(T, G_k^{\text{MMSE}}) = L_k - \operatorname{tr}\Big[T_k^{\text{H}} H_k^{\text{H}} \underbrace{\Big(\mathbf{I} + \sum_{i=1}^{K} H_i T_i T_i^{\text{H}} H_i^{\text{H}} \Big)}_{=:X}{}^{-1} H_k T_k \Big]. \tag{7.23}$$

Note that the inverse X^{-1} does not depend on the user index k due to the alignment of the channel and precoder indices in the dual MAC. This allows us to find a simple expression for the sum MMSE:

$$\begin{aligned}
\sum_{k=1}^{K} \varepsilon_k^{\text{MAC}}(T, G_k^{\text{MMSE}}) &= L - \sum_{k=1}^{K} \operatorname{tr}\Big[\Big(\mathbf{I}_N + \sum_{i=1}^{K} H_i T_i T_i^{\text{H}} H_i^{\text{H}} \Big)^{-1} H_k T_k T_k^{\text{H}} H_k^{\text{H}} \Big] \\
&= L - \operatorname{tr}\Big[X^{-1}(X - \mathbf{I}_N) \Big] \\
&= L - N + \operatorname{tr}\big[X^{-1} \big]
\end{aligned}$$

The sum MMSE can also be expressed in terms of the transmit covariance matrices $Q_1, \ldots, Q_K$, where $Q_k := T_k T_k^{\text{H}}\ \forall k$:

$$\begin{aligned}
\varepsilon^{\text{MAC}}(Q) &:= L - N + \operatorname{tr}\Big[\Big(\mathbf{I} + \sum_{i=1}^{K} H_i Q_i H_i^{\text{H}} \Big)^{-1} \Big] \\
&= L - N + \operatorname{tr}\Big[\big(\mathbf{I} + H Q H^{\text{H}} \big)^{-1} \Big] \qquad (7.24)
\end{aligned}$$

Again, H and Q are defined via $H := [H_1, \ldots, H_K]$ and $Q :=$ blockdiag $\{Q_k\}_{k=1}^{K}$. In the following, we prove that $\varepsilon^{\text{MAC}}(Q)$ is convex in Q by showing that the second derivative of $\varepsilon^{\text{MAC}}(Q + tV)$ with respect to t is nonnegative for any Q, V, and t that satisfy $Q + tV \succeq \mathbf{0}$. Using Appendix A.5, we obtain

$$\frac{\partial \varepsilon^{\text{MAC}}(Q+tV)}{\partial t} = -\operatorname{tr}\Big[\Big(\mathbf{I} + H(Q+tV)H^{\text{H}} \Big)^{-1} H V H^{\text{H}} \Big(\mathbf{I} + H(Q+tV)H^{\text{H}} \Big)^{-1} \Big].$$

Using the substitution $X' := \mathbf{I} + H(Q + tV)H^{\text{H}}$, the second derivative reads as

$$\frac{\partial^2 \varepsilon^{\text{MAC}}(Q + tV)}{\partial t^2} = 2 \cdot \operatorname{tr}\Big[X'^{-1} H V H^{\text{H}} X'^{-1} H V H^{\text{H}} X'^{-1} \Big].$$

For $\boldsymbol{Q} + t\boldsymbol{V} \succcurlyeq \mathbf{0}$, the matrix $\boldsymbol{X}'$ is positive definite such that the positive definite square root $\boldsymbol{X}'^{-\frac{1}{2}}$ exists. Then, the second derivative can be rewritten as

$$\frac{\partial^2 \varepsilon^{\text{MAC}}(\boldsymbol{Q} + t\boldsymbol{V})}{\partial t^2} = 2 \cdot \|\boldsymbol{X}'^{-1}\boldsymbol{H}\boldsymbol{V}\boldsymbol{H}^{\text{H}}\boldsymbol{X}'^{-\frac{1}{2}}\|_{\text{F}}^2$$

from which we conclude that $\varepsilon^{\text{MAC}}(\boldsymbol{Q})$ is convex in $\boldsymbol{Q}$ irrespective of whether $\boldsymbol{Q}$ is block-diagonal or not. Accordingly, the sum power constrained sum MSE minimization

$$\underset{\boldsymbol{Q}_1,\ldots,\boldsymbol{Q}_K}{\text{minimize}} \quad \text{tr}\left[\left(\mathbf{I} + \sum_{i=1}^{K} \boldsymbol{H}_i\boldsymbol{Q}_i\boldsymbol{H}_i^{\text{H}}\right)^{-1}\right] \quad \text{s.t.: } \boldsymbol{Q}_k \succcurlyeq \mathbf{0} \ \ \forall k$$
$$\sum_{k=1}^{K} \text{tr}(\boldsymbol{Q}_k) \leq P_{\max} \qquad (7.25)$$

is a convex optimization that can be solved to global optimality using the covariance-based gradient-projection algorithm from Chap. 5. However, (7.25) is equivalent to the precoder-based sum MSE minimization only for square or wide precoders $\boldsymbol{T}_1, \ldots, \boldsymbol{T}_K$, i.e., only when every user may multiplex at least as many data streams as he has antennas ($L_k \geq M_k \ \forall k$). This is due to the fact that the rank of the outer product $\boldsymbol{T}_k\boldsymbol{T}_k^{\text{H}}$ is upper bounded by $\min\{L_k, M_k\}$. For $L_k < M_k$, the optimum covariance matrix $\check{\boldsymbol{Q}}_k$ that leads to the minimum sum MSE might turn out to have rank M_k which cannot be synthesized by a precoder with only $L_k < M_k$ columns. In such a case, additional rank constraints $\text{rank}(\boldsymbol{Q}_k) \leq L_k$ on the transmit covariance matrices of all users k which intend to transmit less symbols than antennas would have to be included into the optimization (7.25). However, convexity of the constraint set would be lost since the rank constraint does not define a convex set.[6] As soon as there exists a user k with $L_k < M_k$, we propose to switch to the precoder-based gradient-projection algorithm since the problem is no longer convex. Doing so, the rank constraint is automatically fulfilled by the tall structure of the precoder $\boldsymbol{T}_k$ but convergence to the global optimum is no longer ensured.

In order to apply the recursion formulas of the gradient-projection algorithm in (5.23) and (5.24) for the covariance-based scenario and in (5.16) and (5.17) for the precoder-based scenario,[7] we finally compute the Wirtinger derivatives of the sum MMSE cost function with respect to $\boldsymbol{Q}_k^{\text{T}}$ and $\boldsymbol{T}_k^*$. From the element-wise Wirtinger derivative

[6] The sum of two rank-one matrices may have rank two for example.

[7] Note that the negative gradients in (7.26) and (7.27) must be used for the recursion formulas since the sum MMSE must be *minimized* and not maximized.

$$
\begin{aligned}
\frac{\partial_{\mathrm{w}} \varepsilon^{\mathrm{MAC}}(\boldsymbol{Q})}{\partial_{\mathrm{w}}[\boldsymbol{Q}_k^{\mathrm{T}}]_{m,n}} &= -\operatorname{tr}\left[\left(\mathbf{I}+\sum_{i=1}^{K}\boldsymbol{H}_i\boldsymbol{Q}_i\boldsymbol{H}_i^{\mathrm{H}}\right)^{-1}\boldsymbol{H}_k\boldsymbol{e}_n\boldsymbol{e}_m^{\mathrm{T}}\boldsymbol{H}_k^{\mathrm{H}}\left(\mathbf{I}+\sum_{i=1}^{K}\boldsymbol{H}_i\boldsymbol{Q}_i\boldsymbol{H}_i^{\mathrm{H}}\right)^{-1}\right] \\
&= -\boldsymbol{e}_m^{\mathrm{T}}\boldsymbol{H}_k^{\mathrm{H}}\left(\mathbf{I}+\sum_{i=1}^{K}\boldsymbol{H}_i\boldsymbol{Q}_i\boldsymbol{H}_i^{\mathrm{H}}\right)^{-2}\boldsymbol{H}_k\boldsymbol{e}_n
\end{aligned}
$$

we find the derivative with respect to the complete transposed covariance matrix

$$
\frac{\partial_{\mathrm{w}} \varepsilon^{\mathrm{MAC}}(\boldsymbol{Q})}{\partial_{\mathrm{w}} \boldsymbol{Q}_k^{\mathrm{T}}} = -\boldsymbol{H}_k^{\mathrm{H}}\left(\mathbf{I}+\sum_{i=1}^{K}\boldsymbol{H}_i\boldsymbol{Q}_i\boldsymbol{H}_i^{\mathrm{H}}\right)^{-2}\boldsymbol{H}_k. \tag{7.26}
$$

When operating on the precoders, the Wirtinger derivative with respect to the conjugate precoder of user k reads as

$$
\frac{\partial_{\mathrm{w}} \varepsilon^{\mathrm{MAC}}(\boldsymbol{T}\boldsymbol{T}^{\mathrm{H}})}{\partial_{\mathrm{w}} \boldsymbol{T}_k^{*}} = -\boldsymbol{H}_k^{\mathrm{H}}\left(\mathbf{I}+\sum_{i=1}^{K}\boldsymbol{H}_i\boldsymbol{T}_i\boldsymbol{T}_i^{\mathrm{H}}\boldsymbol{H}_i^{\mathrm{H}}\right)^{-2}\boldsymbol{H}_k\boldsymbol{T}_k. \tag{7.27}
$$

Having obtained the optimum precoders that lead to the minimum sum MMSE in the dual MIMO MAC, the respective MMSE receive filters $\boldsymbol{G}_1^{\mathrm{MMSE}}, \ldots, \boldsymbol{G}_K^{\mathrm{MMSE}}$ in (7.22) are computed for all users. Then, the optimum BC precoder $\boldsymbol{P}_k$ of user k is given by (see Sect. 3.2.4.1)

$$
\boldsymbol{P}_k = \varphi \cdot \boldsymbol{G}_k^{\mathrm{MMSE,H}} \quad \forall k
$$

where φ is chosen such that the sum power constraint $\sum_{k=1}^{K}\|\boldsymbol{P}_k\|_{\mathrm{F}}^2 \leq P_{\max}$ holds with equality.

Synopsis: For an uncoded transmission, the sum mean square error between the transmitted symbols and their soft estimates is a reasonable metric for the transceiver design in the MIMO broadcast channel. Tackling this sum MMSE minimization in the dual MIMO MAC immediately reveals the convexity of the problem setting if every user transmits at least as many data symbols per channel use as his terminal has antennas. Such a setting can then be solved to global optimality by means of the covariance-based gradient-projection algorithm from Chap. 5. If there is at least one user transmitting less symbols per channel use than his terminal has antennas, the sum MMSE minimization can only be solved to local optimality due to the lack of convexity. However, the precoder-based gradient-projection algorithm is still applicable. As soon as locally optimal precoders have been found in the dual MAC, the corresponding precoders in the MIMO BC are given by the scaled MMSE receive filters in the dual MAC. The scaling has to be chosen such that the sum power constraint is fulfilled with equality. Since only the sum MSE needs to remain constant during the duality conversion, the complexity of our application-specific level one MSE duality is extremely small—only a single scaling needs to be computed.

Chapter 8
Asymptotic High Power Analysis of the MIMO BC

The detailed analyses of the weighted sum rate maximization problem in the previous two chapters have revealed that the optimum transmit covariance matrices and the accompanied maximum weighted sum rate are usually not available in closed form for a multi-user MIMO broadcast channel system setup. In fact, the global optimum can be found by means of iterative algorithms in case of dirty paper coding (see Chap. 6) but only few qualitative conclusions can be deduced therefrom. For linear transceivers, the situation is even worse. Up to now, a globally optimum algorithm does not seem to exist although we conjecture that the combinatorial stream allocation algorithm presented in Chap. 7 achieves this global optimum in most of the cases.

In the high power regime, however, the weighted sum rate utility can be shown to converge to an asymptote irrespective of whether dirty paper coding or linear filtering is applied. Depending on the antenna configuration of the system, convergence can be observed at rather modest transmit powers which renders an approximate characterization of the performance possible over a vast power regime. In addition, the asymptote can conveniently be described in terms of the antenna configuration and the current channel realization which allows for a deeper understanding of the parameters that actually govern the (asymptotic) weighted sum rate utility and for a deeper understanding of the performance loss of linear transceivers with respect to dirty paper coding. Due to the fact that we can derive the asymptotic behavior for a fixed channel realization, channel-averaged results can quickly be obtained for arbitrary[1] fading models. For the semi-correlated Rayleigh fading, a closed form expression of the asymptotic weighted sum rate will be derived.

In Sect. 8.1, we start by introducing some definitions that allow us to characterize the asymptotic growth rate of a function whose argument tends to infinity. These notations allow for a convenient description of the various types of convergence that will arise during the high power analysis. With these definitions, the basic principle of the affine high power approximation of the weighted sum rate utility is stated in an abstract way followed by a comprehensive literature overview in this field. Section 8.2 deals with the affine approximation of the asymptotic weighted sum rate utility for

[1] The probability of a rank deficient channel matrix must be zero.

R. Hunger, *Analysis and Transceiver Design for the MIMO Broadcast Channel*, Foundations in Signal Processing, Communications and Networking 8,
DOI: 10.1007/978-3-642-31692-0_8, © Springer-Verlag Berlin Heidelberg 2013

a fixed channel realization. More precisely, the asymptotic behavior of the point-to-point MIMO channel, the MIMO BC with dirty paper coding, and the MIMO BC with linear transceivers is investigated. For semi-correlated Rayleigh fading, closed form expressions for the channel-averaged asymptotic utilities are derived, again for the various types of cooperation at the receivers. Note that basically all achievements presented in this chapter result from the analysis in the dual MIMO MAC. Thanks to our novel rate duality for linear transceivers in Sect. 3.3 and our newly formulated duality for systems with interference cancelation in Sect. 4.4, the obtained results immediately translate to the MIMO BC. Moreover, the asymptotically optimum transmission strategy in the dual MIMO MAC can conveniently be converted back to the original MIMO. BC by means of the duality conversion formulas. In particular, the duality for linear transceivers allows us to find a formal proof for the theorem that block-diagonalization is asymptotically optimum in the high power regime if the transmitting base station is equipped with enough antennas. Channel-averaged asymptotic results are finally presented in Sect. 8.3 for semi-correlated Rayleigh fading.

8.1 Overview and Asymptotic Notation

In this section, the affine approximation of the weighted sum rate utility is introduced in an abstract way. The two parameters of the affine approximation are the slope/multiplexing gain and the rate/power offset. Before we present the formal relationship between the weighted sum rate utility and its asymptote, a few definitions on the asymptotic growth rate of a function f with respect to the parameter $P_{\max}$ will be presented in Sect. 8.1.1. Afterwards, the main idea of the affine utility approximation and a comprehensive literature overview are given in Sect. 8.1.2.

8.1.1 Definitions

The following four definitions simplify the characterization of the asymptotic behavior of a function f depending on the parameter $P_{\max}$ when $P_{\max} \to \infty$. Note that the last two definitions are simply derivatives of the first two ones, which are taken from [133, Sect. 3.1] and might already be familiar to the reader. For matrix-valued f and g, all relations are assumed to hold element-wise. Definition 2 is usually known as the big-O-notation:

Definition 8.2 A function f is said not to grow faster (is said not to decay slower) than another function g, if and only if there exist a positive constant $P_0 > 0$ and a finite constant $C < \infty$ for which

$$|f(P_{\max})| \leq C \cdot |g(P_{\max})| \quad \forall P_{\max} > P_0.$$

We shall write $f(P_{\max}) \in O\big(g(P_{\max})\big)$ or $f(P_{\max}) = O\big(g(P_{\max})\big)$ with a slight abuse of notation. Unless $g(P_{\max}) = 0$ for all $P_{\max}$, this is equivalent to

$$\limsup_{P_{\max}\to\infty} \left|\frac{f(P_{\max})}{g(P_{\max})}\right| < \infty.$$

Similarly, the little-o-notation is defined as follows:

Definition 8.3 A function f is said to be asymptotically dominated by g, if and only if

$$\lim_{P_{\max}\to\infty} \frac{f(P_{\max})}{g(P_{\max})} = 0$$

and we shall write $f(P_{\max}) \in o\big(g(P_{\max})\big)$ or $f(P_{\max}) = o\big(g(P_{\max})\big)$.

Definition 4 can be used to introduce the weak and the strong asymptotic equivalence between two functions f and g that both depend on the parameter $P_{\max}$. Both equivalence relationships imply that $f \in O(g)$ and $g \in O(f)$. However, they are more restrictive since they also limit the growth rate of $f - g$. Weak and strong asymptotic equivalence are defined as follows:

Definition 8.4 Two functions f and g are said to be weakly asymptotically equivalent, if and only if

$$f(P_{\max}) = g(P_{\max}) + o\big(g(P_{\max})\big)$$

and we shall use the notation $f(P_{\max}) \doteqdot g(P_{\max})$. This is equivalent to

$$\lim_{P_{\max}\to\infty} \frac{f(P_{\max})}{g(P_{\max})} = 1.$$

Definition 8.5 Two functions f and g are said to be strongly asymptotically equivalent, if and only if

$$f(P_{\max}) = g(P_{\max}) + o(1)$$

and we shall use the notation $f(P_{\max}) \cong g(P_{\max})$. This is equivalent to

$$\lim_{P_{\max}\to\infty} \big[f(P_{\max}) - g(P_{\max})\big] = 0.$$

Note that strong asymptotic equivalence always implies weak asymptotic equivalence. The converse is usually not true.

8.1.2 Basic Principle and Literature Overview

The affine approximation of the spectral efficiency as a function of the logarithmic energy per bit and the noise power density ratio was first proposed in a formal way by Shamai and Verdú in 2001 when they analyzed single-antenna CDMA systems [134]. A few years later, this concept was adopted to the single-user MIMO link under different fading models by Lozano et al. in [135, 136] in order to accurately describe the influence of transmitter- or receivers-sided channel correlations on the ergodic mutual information in the high power regime. Hitherto existing contributions on the asymptotic behavior of the point-to-point MIMO link were solely focusing on the slope of this affine approximation and were thus not able to capture the effects induced by the various correlations and fading types. These effects are reflected in the offset term of the affine approximation.

Let $\check{f}(\boldsymbol{w}, P_{\max})$ denote the maximum weighted sum rate utility in bits per channel use that can be achieved by a given strategy (dirty paper coding or linear filtering for example) for a given transmit power $P_{\max}$ and a given weight vector $\boldsymbol{w}$. Without loss of generality, we assume $\boldsymbol{w} > \mathbf{0}$, i.e., only positive weights are considered. If the scalar weight of a particular user is zero, no power is allocated to him and he can be dropped from the system model. In the logarithmic power domain, the utility $\check{f}(\boldsymbol{w}, P_{\max})$ is observed to converge to an asymptote $\check{f}_{\infty}(\boldsymbol{w}, P_{\max})$ for $P_{\max} \to \infty$ for any reasonable signaling strategy. Therefore, we may write

$$\begin{aligned} \check{f}(\boldsymbol{w}, P_{\max}) &= \check{f}_{\infty}(\boldsymbol{w}, P_{\max}) + o(1) \\ &\cong \check{f}_{\infty}(\boldsymbol{w}, P_{\max}) \end{aligned} \tag{8.1}$$

which means that strong asymptotic equivalence between the utility $\check{f}(\boldsymbol{w}, P_{\max})$ and the affine asymptote $\check{f}_{\infty}(\boldsymbol{w}, P_{\max})$ holds in the sense of Definition 5: the difference between the utility and its asymptote vanishes as $P_{\max} \to \infty$. We prefer to define the affine function $\check{f}_{\infty}(\boldsymbol{w}, P_{\max})$ via

$$\begin{aligned} \check{f}_{\infty}(\boldsymbol{w}, P_{\max}) &:= s_{\infty} \cdot \log_2 P_{\max} + r_{\infty}(\boldsymbol{w}) \\ &= s_{\infty} \cdot \frac{10 \log_{10} P_{\max}}{\underbrace{10 \log_{10} 2}_{\approx 3.01 \text{ dB}}} + r_{\infty}(\boldsymbol{w}) \end{aligned} \tag{8.2}$$

where s_{∞} denotes the slope (multiplexing gain when $\boldsymbol{w} = \mathbf{1}$) of the affine function in the logarithmic domain and $r_{\infty}(\boldsymbol{w})$ denotes the weighted sum rate offset that may depend on the weight vector $\boldsymbol{w}$. Note that s_{∞} represents the increase of the utility per 3 dB power increase in the high power domain. The two parameters s_{∞} and $r_{\infty}(\boldsymbol{w})$ are defined via

$$s_{\infty} := \lim_{P_{\max} \to \infty} \frac{\check{f}(\boldsymbol{w}, P_{\max})}{\log_2 P_{\max}} \tag{8.3}$$

and

$$r_\infty(\boldsymbol{w}) := \lim_{P_{\max}\to\infty} \left[\check{f}(\boldsymbol{w}, P_{\max}) - s_\infty \log_2 P_{\max}\right]. \tag{8.4}$$

A slightly different definition of the high power asymptote that is usually found in the literature makes use of the logarithmic power offset $l_\infty(\boldsymbol{w})$ (in 3dB units) which leads to

$$\check{f}_\infty(\boldsymbol{w}, P_{\max}) = s_\infty \cdot \left[\log_2 P_{\max} - l_\infty(\boldsymbol{w})\right].$$

However, the drawback of this definition results from its inability to handle the case $s_\infty = 0$ in which the utility $\check{f}(\boldsymbol{w}, P_{\max})$ converges to a finite (nonzero) limit as $P_{\max} \to \infty$. Note that the affine approximation concept is applicable for a fixed channel realization as well as for fading channels. In the latter case, a channel-averaged utility is approximated.

Besides the analysis of the multi-user CDMA system with single-antenna terminals, the affine high power approximations were initially applied in a fading single-user point-to-point MIMO system. Early work in this field was done by Gauthier et al. in [137], where he obtained the asymptotic ergodic sum capacity with and without channel state information available at the transmitter for uncorrelated i.i.d. Rayleigh fading. In order to find the asymptotic results, the one in the argument of the logarithm $\log_2(1 + x)$ was neglected for large x.

In a similar fashion, Grant furthered the derivation of the asymptotic ergodic sum capacity of a point-to-point MIMO link with uncorrelated Rayleigh fading when the transmitter has no CSI [138]. To this end, he presented an upper and a lower bound on the ergodic capacity which become asymptotically tight. The lower bound (which we will also make use of) follows from $\left|\mathbf{I} + \boldsymbol{A}\right| > \left|\boldsymbol{A}\right|$ for positive semi-definite $\boldsymbol{A}$ and finally leads to the digamma expressions in the rate offset r_∞ due to the Wishart distribution of $\boldsymbol{A}$, see Sect. 8.3.2.

Using the lower bound on the determinant from Grant, Salo et al. decompose the lower bound (which is asymptotically tight) on the *current* sum capacity for a particular channel realization out of a fading process into three parts: a supremum capacity term for channels with fixed Frobenius norm, a term that mirrors the effect of the deviation of the instantaneous SNR from the average one, and a third term, which is the degradation due to the spread of the eigenvalues of the current channel matrix Gram. In addition, probability density functions for the last two terms are presented.

A more specific fading model with two-sided channel correlations (Kronecker channel model) was investigated by Lozano et al. in [135, 136]. They found a complicated, but closed form expression for the power offset l_∞ of the affine approximation of the channel-averaged sum rate, which still depends on the chosen power allocation at the transmitter. If the receiver is equipped with at least as many antenna elements as the transmitter, the weakly asymptotically optimum power allocation is shown to be uniform. Otherwise, the power allocation is unknown even in the high power limit due to the channel correlations.

A high power analysis of the vector broadcast channel with at most as many single-antenna terminals as antenna elements at the base station was presented by Jindal in [139]. Using results in [140], the sum capacity of the fixed-channel vector BC with dirty paper coding was argued to be strongly asymptotically equivalent to the sum capacity of the equivalent point-to-point link (with uncorrelated noise) when all users are allowed to cooperate. Furthermore, the instantaneous and channel-averaged rate loss of linear zero-forcing with respect to dirty paper coding was derived.

The extension to multi-antenna terminals followed one year later by Lee and Jindal in [141], where the MIMO BC is considered in which the base station is equipped with at least as many antenna elements as the user terminals have in sum. In a first step, the asymptotic sum capacity of dirty paper coding is argued to coincide with the asymptotic sum capacity of the cooperative system (with uncorrelated noise) for the multi-antenna terminals case as well, again using results from [140]. In addition to zero-forcing, block-diagonalization is investigated. However, a closed form expression for the asymptotic sum rate of the block-diagonalization strategy that solely depends on the channel matrix is not given for a fixed channel realization. Instead, the asymptotic rate expression still depends on the choice of the beamforming matrices, which are not exactly specified. In case of uncorrelated Rayleigh fading, the average rate loss of block-diagonalization with respect to dirty paper coding was found by considering the remaining degrees of freedom after the block-diagonalization. Under the limitation of single-antenna terminals and a linear power allocation, the (weakly) asymptotically optimum power allocation *in the dual vector MAC* was derived to be proportional to the user weights both for the case of dirty paper coding and zero-forcing. However, the structure of the beamformers and the power allocation in the original vector BC was not considered.

For multi-antenna terminals, scaled identity matrices whose traces are proportional to the user weights are shown to be weakly asymptotically optimum in the dual MIMO MAC for the weighted sum rate maximization in [142] when restricting the power allocation to be linear. This result is claimed to hold both for dirty paper coding and for block-diagonalization. In turn, the result is derived in the dual MAC and it is unclear how the precoders in the original MIMO BC look like and how the power is allocated to the users. This holds in particular for the linear scheme since at that time, a rate duality for linear transceivers was not available in case of multi-antenna elements.

In 2009, we tackled the asymptotic analysis of the MIMO BC with enough antennas at the base station and linear transceivers in [22,23] by means of the rate duality from Sect. 3.3 both for a fixed channel realization and for semi-correlated Rayleigh fading. Therein, block-diagonalization is proven to asymptotically achieve the maximum possible weighted sum rate under linear filtering. For a particular channel realization, the first precoder-free expression for the asymptotic rate loss of linear transceivers with respect to dirty paper coding was derived that solely depends on the composite channel matrix. In addition, the optimum signaling in the original MIMO BC could be found by means of the novel rate duality which revealed that the weakly asymptotically optimum power allocation in the MIMO BC coincides with the one from the dual MAC. A detailed derivation of those results will be given in the following two sections.

Some other publications in this field that we would like to mention are the results from Shen et al. in [143], where the performance of the TDMA operation mode is used to find a very loose bound on the ratio of the asymptotic DPC sum capacity and the block-diagonalization sum rate. In [144], a single-user MIMO system with isotropic input covariance matrix and MMSE receive filter with separate stream decoding is analyzed in the high power regime. Exact expressions for the channel-averaged sum rate are presented for all transmit powers. Besides correlated and uncorrelated Rayleigh fading, Ricean fading is considered as well.

For realistic scenarios with imperfect channel state information at the transmitter, Jindal showed that the number of feedback bits per user in a vector BC has to grow linearly in the logarithmic power in order to maintain the maximum possible multiplexing gain, see [145].

Synopsis: A closed form expression for the maximum achievable weighted sum rate utility in a multi-user MIMO broadcast channel system with either dirty paper coding or linear filtering is not available. For large transmit powers, however, the utility converges to an affine asymptote in the logarithmic power domain irrespective of whether DPC or linear filtering is applied. For $P_{\max} \to \infty$, the difference between the maximum utility and the affine asymptote converges to zero from above. The slope s_∞ of the affine approximation denotes the increase of the weighted sum rate utility per 3 dB power increase and the rate offset r_∞ denotes the shift of the affine function which is usually a function of the channel properties and the user weights. Depending on the antenna configuration, the affine approximation of the maximum possible weighted sum rate utility may be accurate even at modest transmit powers. Therefore, it can for example be used to obtain an upper bound on the required transmit power to achieve certain rate requirements.

8.2 Fixed Channel Weighted Sum Rate Analysis

In this section, we derive the asymptotes of the weighted sum rate utility for a *fixed* channel realization under different signaling strategies like dirty paper coding and various linear schemes. Besides the maximum possible asymptotic utilities, we find the weakly optimum transmit covariance matrices that lead to the asymptotically optimum values. While the initial analyses are carried out in the dual MIMO MAC, the asymptotically optimum transmission strategies will be converted back to the original MIMO BC by means of the rate dualities from Sects. 3.3 and 4.4. We begin by investigating the point-to-point MIMO case in Sect. 8.2.1, where the individual users are allowed to jointly process their data. Despite its theoretical nature, it serves as a simple introduction to the asymptotic analysis since a closed form expression for the capacity is available for any value of the transmit power. We will furthermore see in Sect. 8.2.2, that cooperation of the users does not lead to a larger asymptotic sum capacity if the noise portions of the individual users are uncorrelated. Finally, Sect. 8.2.3 deals with the asymptotic behavior of the weighted sum rate utility when the transceivers are restricted to be linear.

8.2.1 Cooperative Point-to-Point MIMO System

Under the assumption that the K transmitters in the dual MIMO MAC are able to cooperatively design their transmit signals, the resulting setup can be interpreted as a single-user point-to-point system with M antennas at the transmitter and N antennas at the receiver. Due to the fact that the sum capacity of a point-to-point MIMO system can be expressed in semi-closed form[2] for all transmit powers, the asymptotic analysis of such a system is relatively simple. Nonetheless, it allows us to quantify the loss incurred by the circumstance that the individual users in a MIMO MAC/BC cannot cooperate both for the case of dirty paper coding and linear transceivers.

Let $\boldsymbol{U}\boldsymbol{\Sigma}\boldsymbol{U}^{\mathrm{H}} = \boldsymbol{H}^{\mathrm{H}}\boldsymbol{H}$ denote the sorted eigenvalue decomposition of the composite MAC channel Gram $\boldsymbol{H}^{\mathrm{H}}\boldsymbol{H} \in \mathbb{S}^M$ with diagonal $\boldsymbol{\Sigma} = \mathrm{diag}\{\sigma_i\}_{i=1}^M$ such that $\sigma_1 \geq \sigma_2 \geq \cdots \geq \sigma_M \geq 0$ and with unitary $\boldsymbol{U}$. According to [52], the optimum transmit covariance matrix $\boldsymbol{Q}$ (which need not be block-diagonal due to the cooperation of the transmitters) is given by

$$\boldsymbol{Q} = \boldsymbol{U}\boldsymbol{\Lambda}\boldsymbol{U}^{\mathrm{H}}$$

where $\boldsymbol{\Lambda} = \mathrm{diag}\{\lambda_i\}_{i=1}^M$ contains the powers of the individual modes. These entries of $\boldsymbol{\Lambda}$ follow from the water-filling principle and read as [52]

$$\lambda_i = \left[\nu - \frac{1}{\sigma_i}\right]_+ \quad \forall i \in \{1, \ldots, M\} \tag{8.5}$$

where $[\cdot]_+$ again denotes $\max\{\cdot, 0\}$. The water-level ν is chosen such that the sum power constraint $\sum_{i=1}^M \lambda_i \leq P_{\max}$ holds with equality:

$$\nu = \frac{P_{\max} + \sum_{\ell=1}^{\varrho} \frac{1}{\sigma_\ell}}{\varrho} \tag{8.6}$$

The number of active streams ϱ with positive power allocation in (8.6) is determined such that $\lambda_\varrho > 0$ and $\lambda_i = 0$ for all $i > \varrho$. Inserting $\lambda_1, \ldots, \lambda_M$ from (8.5) into the rate expression of the point-to-point MIMO link, the obtained sum capacity reads as

---

[2] The solution may require up to $\min\{M, N\} - 1$ comparisons in order to determine the optimum water-level.

$$C^{\mathrm{PtP}} = \log_2 \left|\mathbf{I} + \boldsymbol{\Lambda}\boldsymbol{\Sigma}\right| = \sum_{i=1}^{M} \log_2(1+\lambda_i\sigma_i) = \sum_{i=1}^{\varrho} \log_2(\nu\sigma_i). \qquad (8.7)$$

Let $r = \mathrm{rank}(\boldsymbol{H})$ denote the rank of the composite MAC channel matrix. From (8.5) and (8.6), we find that all r modes belonging to the nonzero eigenvalues become active as soon as

$$\lambda_r > 0 \quad \Leftrightarrow \quad P_{\max} > \frac{r}{\sigma_r} - \sum_{\ell=1}^{r} \frac{1}{\sigma_\ell}. \qquad (8.8)$$

In this case, the power allocation of the r active modes is given by

$$\lambda_i = \frac{P_{\max}}{r} + \frac{1}{r}\sum_{\ell=1}^{r} \frac{1}{\sigma_\ell} - \frac{1}{\sigma_i} \quad \forall i \in \{1, \ldots, r\}$$

and is thus weakly (but not strongly, unless $r = 1$ or $\sigma_1, \ldots, \sigma_r$ are identical) asymptotically equivalent in the sense of Definition 4 to

$$\lambda_i \doteq \frac{P_{\max}}{r} \quad \forall i \in \{1, \ldots, r\}$$

and even satisfies

$$\lambda_i = \frac{P_{\max}}{r} + O(1) \quad \forall i \in \{1, \ldots, r\}$$

which is a slightly stronger statement. Obviously, no power is allocated to inactive modes with zero eigenvalue, i.e., $\sigma_i = 0 \Rightarrow \lambda_i = 0$ and thus $\lambda_i = 0 \; \forall i > r$. By construction, the rank r of the composite channel matrix is upper bounded by its dimensions, i.e., $r \leq \min\{M, N\}$. As a consequence, at most N streams will be active independent of the sum of terminal antennas M and thus also independent of the number of users K. As we will see in Sect. 8.2.2, this property does not hold if the individual users may not cooperate and dirty paper coding is applied. Depending on the channel, more than N active modes might be necessary there if $M > N$ in order to obtain the maximum sum rate in the high power regime.

Going back to (8.7), the capacity of the cooperating point-to-point system reads for powers $P_{\max}$ satisfying (8.8) as

$$C^{\mathrm{PtP}}(P_{\max}) = \sum_{i=1}^{r} \log_2(\nu\sigma_i) = r\log_2\left(P_{\max} + \sum_{\ell=1}^{r}\frac{1}{\sigma_\ell}\right) - r\log_2 r + \log_2\left(\prod_{i=1}^{r}\sigma_i\right)$$

and is strongly asymptotically equivalent in the sense of Definition 5 to

$$C^{\mathrm{PtP}}(P_{\max}) \cong r\log_2 P_{\max} - r\log_2 r + \log_2\left(\prod_{i=1}^{r}\sigma_i\right). \qquad (8.9)$$

Note that the capacity converges to this limit from above. As expected, the multiplexing gain is equal to the rank r of the channel matrix $\boldsymbol{H}$ whereas the rate offset corresponds to the logarithm of the product of the nonzero eigenvalues of the channel Gram minus some constant that depends on the rank of $\boldsymbol{H}$. Hence, the asymptote in (8.2) features the two parameters[3]

$$\begin{aligned} s_\infty &= \operatorname{rank}(\boldsymbol{H}) \\ r_\infty(1) &= \log_2\left(\prod_{i=1}^{\operatorname{rank}(\boldsymbol{H})} \sigma_i\right) - \operatorname{rank}(\boldsymbol{H}) \cdot \log_2\big(\operatorname{rank}(\boldsymbol{H})\big). \end{aligned}$$

In addition, any power allocation that is weakly asymptotically equivalent to the uniform power allocation for all nonzero eigenmodes achieves a sum rate that is strongly asymptotically equivalent to the capacity and therefore has the same asymptote. In particular, choosing the transmit covariance matrix $\boldsymbol{Q}$ in the dual MIMO MAC as the scaled projector $\boldsymbol{Q} = \frac{P_{\max}}{r} \cdot \boldsymbol{U}_1\boldsymbol{U}_1^{\mathrm{H}}$ with $\boldsymbol{U}_1$ denoting the first r columns of $\boldsymbol{U}$ belonging to the nonzero eigenvalues $\sigma_1, \ldots, \sigma_r$ asymptotically achieves the sum capacity in the strong sense. Note that the difference between the asymptote and the true capacity reads for all powers $P_{\max}$ that satisfy (8.8) as

$$C^{\mathrm{PtP}}(P_{\max}) - \big[s_\infty \log_2 P_{\max} + r_\infty(1)\big] = r\log_2\left(1 + \frac{\sum_{\ell=1}^{r} \frac{1}{\sigma_\ell}}{P_{\max}}\right)$$

which of course vanishes as $P_{\max} \to \infty$ such that strong asymptotic equivalence holds. Now if the smallest positive eigenvalue σ_r is close to zero, $P_{\max}$ must be large in order to let the asymptote be close to the true capacity.

8.2.1.1 Enough Antennas at the Base Station

In the relevant case with full rank channels and $N \geq M$, the rank of $\boldsymbol{H}$ is $r = M$ and the capacity of the point-to-point system in (8.9) is strongly asymptotically equivalent to

$$C^{\mathrm{PtP}}(P_{\max}) \cong M\log_2 P_{\max} - M\log_2 M + \log_2\left|\boldsymbol{H}^{\mathrm{H}}\boldsymbol{H}\right| \quad \text{for } N \geq M \tag{8.10}$$

and the isotropic input covariance matrix $\boldsymbol{Q} = \frac{P_{\max}}{M} \cdot \mathbf{I}_M$ *of the dual MAC* is weakly asymptotically optimum and thus features the same high power asymptote. Reformulating the MAC channel $\boldsymbol{H}$ of the single super-user in terms of the original BC channel $\boldsymbol{H}^{\mathrm{BC}}$ via [cf. Corollary 1]

[3] The weight vector $\boldsymbol{w}$ reduces to the scalar 1 if the users are allowed to cooperate and thus act as a single super-user.

$$\boldsymbol{H} = \boldsymbol{H}^{\mathrm{BC,H}} \boldsymbol{C}_{\eta}^{-\frac{1}{2},\mathrm{H}} \tag{8.11}$$

where $\boldsymbol{C}_{\eta}$ denotes the *total* covariance matrix of the stacked noise components $\boldsymbol{\eta} = [\boldsymbol{\eta}_1^{\mathrm{T}}, \ldots, \boldsymbol{\eta}_K^{\mathrm{T}}]^{\mathrm{T}}$ in the original BC, the sum capacity is strongly asymptotically equivalent to

$$\begin{aligned} C^{\mathrm{PtP}}(P_{\max}) &\cong M \log_2 P_{\max} - M \log_2 M + \log_2 \left| \boldsymbol{C}_{\eta}^{-\frac{1}{2}} \boldsymbol{H}^{\mathrm{BC}} \boldsymbol{H}^{\mathrm{BC,H}} \boldsymbol{C}_{\eta}^{-\frac{1}{2},\mathrm{H}} \right| \\ &= M \log_2 P_{\max} - M \log_2 M + \log_2 \left| \boldsymbol{H}^{\mathrm{BC}} \boldsymbol{H}^{\mathrm{BC,H}} \right| - \log_2 \left| \boldsymbol{C}_{\eta} \right| \end{aligned} \tag{8.12}$$

where we assumed $\boldsymbol{C}_{\eta}$ to be full rank.[4] From the Fischer inequality [146, Theorem 7.8.3], it follows that

$$\left| \boldsymbol{C}_{\eta} \right| \leq \prod_{k=1}^{K} \left| \boldsymbol{C}_{\eta_k} \right| \tag{8.13}$$

where $\boldsymbol{C}_{\eta_k} = \mathrm{E}[\boldsymbol{\eta}_k \boldsymbol{\eta}_k^{\mathrm{H}}]$ denotes the kth main diagonal block of $\boldsymbol{C}_{\eta}$ and corresponds to the noise covariance matrix at user k in the true MIMO broadcast channel. From (8.13), we can conclude that for fixed noise covariance matrices $\boldsymbol{C}_{\eta_1}, \ldots, \boldsymbol{C}_{\eta_K}$ at the K users in the BC and for $N \geq M$, the worst case total noise covariance matrix $\boldsymbol{C}_{\eta}$ leading to the minimum asymptotic sum capacity, is indeed block-diagonal, i.e., $\boldsymbol{C}_{\eta} = \mathrm{blockdiag}\{\boldsymbol{C}_{\eta_k}\}_{k=1}^{K}$. Thus, the worst case asymptotic behavior of the point-to-point MIMO channel (with cooperating BC users) with at least as many transmit antennas as the user terminals have antennas in sum is given by [cf. (8.12)]

$$C^{\mathrm{PtP}}(P_{\max}) \cong M \log_2 P_{\max} - M \log_2 M + \log_2 \left| \boldsymbol{H}^{\mathrm{BC}} \boldsymbol{H}^{\mathrm{BC,H}} \right| - \sum_{k=1}^{K} \log_2 \left| \boldsymbol{C}_{\eta_k} \right| \tag{8.14}$$

such that the two parameters of the asymptote in (8.2)–(8.4) are

$$\begin{aligned} S_{\infty} &= M \\ r_{\infty}(1) &= -M \log_2 M + \log_2 \left| \boldsymbol{H}^{\mathrm{BC}} \boldsymbol{H}^{\mathrm{BC,H}} \right| - \sum_{k=1}^{K} \log_2 \left| \boldsymbol{C}_{\eta_k} \right|. \end{aligned}$$

For such block-diagonal $\boldsymbol{C}_{\eta}$ with $N \geq M$, cooperation of the K users in the MIMO BC does not increase the asymptotic sum capacity compared to dirty paper coding as we will see in the next subsection. Instead, it features the same asymptote. In the original MIMO BC with cooperating receivers, the weakly optimum transmit covariance matrix which asymptotically achieves the sum capacity can be obtained by means of the duality conversion. Since the K transmitters in the dual MAC are

[4] For rank deficient $\boldsymbol{C}_{\eta}$, the sum capacity is infinite for any $P_{\max} > 0$ if the channel $\boldsymbol{H}^{\mathrm{BC}}$ allows for transmission in the respective subspace.

unified to act as a super-user, the two MAC to BC conversion procedures from Sects. 3.3.2 and 4.4.3 coincide. As mentioned above, the MAC transmit covariance matrix $\boldsymbol{Q} = \frac{P_{\max}}{M} \cdot \mathbf{I}_M$ asymptotically achieves the sum capacity, so we may choose the MAC precoder as

$$\boldsymbol{T} = \sqrt{\frac{P_{\max}}{M}} \cdot \mathbf{I}_M.$$

The matched filter receiver $\boldsymbol{G}^{\mathrm{MF}}$ in the dual MAC reads as [cf. (4.55)]

$$\boldsymbol{G}^{\mathrm{MF}} = \boldsymbol{T}^{\mathrm{H}} \boldsymbol{H}^{\mathrm{H}} \in \mathbb{C}^{M \times N}$$

since no inter-user interference is present when the MAC transmitters are allowed to cooperate. For the precoder parametrization (4.59) in the BC, the eigenvalue decomposition in (4.57) reveals that the unitary eigenbasis $\boldsymbol{V}$ of $\boldsymbol{G}^{\mathrm{MF}} \boldsymbol{H} \boldsymbol{T} = \frac{P_{\max}}{M} \cdot \boldsymbol{H}^{\mathrm{H}} \boldsymbol{H}$ is equal to $\boldsymbol{U}$, i.e., $\boldsymbol{V} = \boldsymbol{U}$. Thus, the BC precoder can be written as [cf. (4.59)]

$$\boldsymbol{P} = \boldsymbol{G}^{\mathrm{MF,H}} \boldsymbol{U} \cdot \operatorname{diag}\{\varphi_j\}_{j=1}^{M}. \tag{8.15}$$

The scaling parameters $\varphi_1^2, \ldots, \varphi_M^2$ are given by [cf. (4.60)]

$$\varphi_j^2 = \frac{\|\boldsymbol{T}\boldsymbol{U}\boldsymbol{e}_j\|_2^2}{\|\boldsymbol{e}_j^{\mathrm{T}} \boldsymbol{U}^{\mathrm{H}} \boldsymbol{G}^{\mathrm{MF}}\|_2^2} = \frac{\frac{P_{\max}}{M}}{\frac{P_{\max}}{M} \cdot \sigma_j} = \frac{1}{\sigma_j}$$

since $\boldsymbol{Z}$ in (4.60) is diagonal due to the single super-user setup. Again, σ_j denotes the j-th eigenvalue of $\boldsymbol{H}^{\mathrm{H}} \boldsymbol{H}$. Then, the transmit covariance matrix $\boldsymbol{S}^{\mathrm{PtP}} := \boldsymbol{P}\boldsymbol{P}^{\mathrm{H}}$ in the BC reads by means of (8.15) as

$$\begin{aligned}
\boldsymbol{S}^{\mathrm{PtP}} &= \boldsymbol{G}^{\mathrm{MF,H}} \boldsymbol{U} \cdot \operatorname{diag}\{\varphi_j^2\}_{j=1}^{M} \cdot \boldsymbol{U}^{\mathrm{H}} \boldsymbol{G}^{\mathrm{MF}} \\
&= \frac{P_{\max}}{M} \cdot \boldsymbol{H} \left(\boldsymbol{H}^{\mathrm{H}} \boldsymbol{H}\right)^{-1} \boldsymbol{H}^{\mathrm{H}} \\
&= \frac{P_{\max}}{M} \cdot \boldsymbol{H}^{\mathrm{BC,H}} \left(\boldsymbol{H}^{\mathrm{BC}} \boldsymbol{H}^{\mathrm{BC,H}}\right)^{-1} \boldsymbol{H}^{\mathrm{BC}}
\end{aligned} \tag{8.16}$$

and is independent of the noise covariance matrix $\boldsymbol{C}_\eta$. We observe that $\boldsymbol{S}^{\mathrm{PtP}}$ is a scaled projector which reduces to $\boldsymbol{S}^{\mathrm{PtP}} = \frac{P_{\max}}{M} \cdot \mathbf{I}_M$ for $M = N$. Equivalently, we could have obtained the same result by directly investigating the point-to-point MIMO system with cooperating BC receivers in the original domain. While this would have been possible for the case of cooperating users, finding the weakly optimum transmit covariance matrices in the BC by means of the duality is essential for noncooperating receivers as in Sects. 8.2.2 and 8.2.3.

8.2.1.2 Base Station with Too Few Antennas

For full rank $\boldsymbol{H}$ with $N < M$, we have $r = N$ and therefore [see (8.9)]

$$C^{\text{PtP}}(P_{\max}) \cong N \log_2 P_{\max} - N \log_2 N + \log_2 \left|\boldsymbol{H}\boldsymbol{H}^{\text{H}}\right| \quad \text{for } N < M. \tag{8.17}$$

Using (8.11), the asymptotic sum capacity can be expressed in terms of the BC channel and the BC noise covariance matrix via

$$C^{\text{PtP}}(P_{\max}) \cong N \log_2 P_{\max} - N \log_2 N + \log_2 \left|\boldsymbol{H}^{\text{BC,H}}\boldsymbol{C}_\eta^{-1}\boldsymbol{H}^{\text{BC}}\right| \quad \text{for } N < M. \tag{8.18}$$

For $N < M$, however, the asymptotically worst case total noise covariance matrix $\boldsymbol{C}_\eta \succ \mathbf{0}$ whose main diagonal blocks are $\boldsymbol{C}_{\eta_1}, \ldots, \boldsymbol{C}_{\eta_K}$ cannot be obtained in closed form and even depends on the composite BC channel matrix $\boldsymbol{H}^{\text{BC}}$. This is in contrast to the case $N \geq M$, where the asymptotically worst case noise covariance matrix $\boldsymbol{C}_\eta$ is always block-diagonal. From the results on the sum capacity of the MIMO BC that have been revisited in Sect. 4.2.2.1, we know that for the worst case noise covariance matrix, the sum capacity of the MIMO BC with dirty paper coding equals the sum capacity of the same system with cooperating receivers. For block-diagonal $\boldsymbol{C}_\eta$, however, the cooperating point-to-point system will in general achieve a larger sum capacity which we expect to hold also for the asymptotes. In the original BC, the transmit covariance matrix

$$\boldsymbol{S}^{\text{PtP}} = \frac{P_{\max}}{N} \cdot \mathbf{I}_N$$

asymptotically achieves the sum capacity in (8.18).

Synopsis: The asymptotic analysis of the point-to-point MIMO system in which the K users are allowed to cooperate and thus act as a single super-user, reveals that the slope s_∞ of the affine sum capacity approximation corresponds to the rank of the composite dual MAC channel matrix $\boldsymbol{H}$ and thus also to the rank of the composite BC channel matrix $\boldsymbol{H}^{\text{BC}}$ since the noise correlation matrix is assumed to have full rank. The rate offset r_∞ equals the logarithm of the nonzero eigenvalues product of the matrix $\boldsymbol{H}^{\text{BC,H}}\boldsymbol{C}_\eta^{-1}\boldsymbol{H}^{\text{BC}}$ minus some constant term that depends on the rank of $\boldsymbol{H}^{\text{BC}}$. If the base station is equipped with at least as many antenna elements as the user terminals have in sum and $\boldsymbol{H}^{\text{BC}}$ is full rank, the weakly optimum transmit covariance matrix $\boldsymbol{S}^{\text{PtP}}$ in the BC which achieves the 'largest' sum capacity asymptote

$$C^{\text{PtP}}(P_{\max}) \cong M \log_2 P_{\max} - M \log_2 M + \log_2 \left|\boldsymbol{H}^{\text{BC}}\boldsymbol{H}^{\text{BC,H}}\right| - \sum_{k=1}^{K} \log_2 \left|\boldsymbol{C}_{\eta_k}\right|$$

corresponds to the scaled projector

$$S^{\mathrm{PtP}} = \frac{P_{\max}}{M} \cdot \boldsymbol{H}^{\mathrm{BC,H}} \left(\boldsymbol{H}^{\mathrm{BC}} \boldsymbol{H}^{\mathrm{BC,H}} \right)^{-1} \boldsymbol{H}^{\mathrm{BC}}$$

independent of the structure of the noise covariance matrix. In addition, the worst case total noise covariance matrix $\boldsymbol{C}_\eta$ which asymptotically achieves the smallest sum capacity is block-diagonal and main diagonal blocks are given by the individual noise covariances $\boldsymbol{C}_{\eta_1}, \ldots, \boldsymbol{C}_{\eta_K}$ of the K receivers in the BC. For systems in which the base station has less transmit antennas than the user terminals have in sum, the transmit covariance matrix

$$S^{\mathrm{PtP}} = \frac{P_{\max}}{N} \cdot \mathbf{I}_N$$

leads to the 'largest' sum capacity asymptote

$$C^{\mathrm{PtP}}(P_{\max}) \cong N \log_2 P_{\max} - N \log_2 N + \log_2 \left| \boldsymbol{H}^{\mathrm{BC,H}} \boldsymbol{C}_\eta^{-1} \boldsymbol{H}^{\mathrm{BC}} \right|.$$

For this antenna configuration, a closed form expression for the worst case noise covariance matrix $\boldsymbol{C}_\eta$ is not available.

8.2.2 MIMO Broadcast Channel with Dirty Paper Coding

In a real MIMO broadcast channel, the K users are obviously not allowed to cooperate. Contrary to the virtual point-to-point MIMO system with cooperating receivers discussed in the preceding subsection, it is basically inevitable to perform the asymptotic analysis of the MIMO BC with dirty paper coding in the dual MIMO MAC. We again distinguish the two cases in which there are either enough degrees of freedom to support full multiplexing for every user ($N \geq M$) or in which the system is overloaded, i.e., in which the user terminals have more antennas in sum than the base station ($N < M$). Interestingly, the latter scenario is much more complicated than the former one and a general closed form expression for the maximum weighted sum rate asymptote has so far never been obtained. We manage to find the weakly optimum power allocation for the vector broadcast channel with single-antenna terminals when the base station is equipped with $N = 2$ antennas. For the sake of readability, we assume that the user weights are sorted in a nondecreasing fashion, i.e., $w_1 \leq w_2 \leq \cdots \leq w_K$. This can always be obtained by simply relabeling the users.

8.2.2.1 Enough Degrees of Freedom

In this subsection, we assume that the base station is equipped with at least as many antennas than the user terminals have in sum ($N \geq M$). Moreover, the MAC channel matrix $\boldsymbol{H} = [\boldsymbol{H}_1, \ldots, \boldsymbol{H}_K]$ is assumed to have full rank. When successive decoding is applied in the dual MIMO MAC, the weighted sum rate utility with user weights $\boldsymbol{w} = [w_1, \ldots, w_K]^{\mathrm{T}}$ and MAC transmit covariance matrices in $\boldsymbol{Q} = \mathrm{blockdiag}\{\boldsymbol{Q}_i\}_{i=1}^{K}$ can be expressed as [see (6.16)]

$$f(\boldsymbol{Q}, \boldsymbol{w}) = \sum_{k=1}^{K} v_k \log_2 \left| \mathbf{I} + \sum_{i \geq k} \boldsymbol{H}_i \boldsymbol{Q}_i \boldsymbol{H}_i^{\mathrm{H}} \right| \tag{8.19}$$

where $v_1 = w_1$ and $v_k = w_k - w_{k-1}$ for $k > 1$. Let user i have L_i active data streams. Then, his transmit covariance matrix can be written in terms of the *reduced* eigenvalue decomposition

$$\boldsymbol{Q}_i = \boldsymbol{U}_i \boldsymbol{\Lambda}_i \boldsymbol{U}_i \quad \forall i \in \{1, \ldots, K\}$$

where $\boldsymbol{\Lambda}_i \in \mathbb{R}_{+,0}^{L_i \times L_i}$ is diagonal with rank $L_i \leq M_i$ and contains the L_i positive eigenvalues as main diagonal elements. The (sub-)unitary eigenbasis $\boldsymbol{U}_i \in \mathbb{C}^{M_i \times L_i}$ satisfies $\boldsymbol{U}_i^{\mathrm{H}} \boldsymbol{U}_i = \mathbf{I}_{L_i}$ and is either square or tall. By means of the (square or tall) selection matrix

$$\boldsymbol{J}_k := \begin{bmatrix} \mathbf{0}_{\sum_{j=1}^{k-1} L_j \times \sum_{j=k}^{K} L_j} \\ \mathbf{I}_{\sum_{j=k}^{K} L_j} \end{bmatrix} \in \{0, 1\}^{L \times \sum_{j=k}^{K} L_j} \tag{8.20}$$

and the two block-diagonal matrices $\boldsymbol{U} := \mathrm{blockdiag}\{\boldsymbol{U}_k\}_{k=1}^{K} \in \mathbb{C}^{M \times L}$ and $\boldsymbol{\Lambda} := \mathrm{blockdiag}\{\boldsymbol{\Lambda}_k\}_{k=1}^{K} \in \mathbb{R}_{+,0}^{L \times L}$, the weighted sum rate utility in (8.19) can be rewritten as

$$\begin{aligned} f(\boldsymbol{U}\boldsymbol{\Lambda}\boldsymbol{U}^{\mathrm{H}}, \boldsymbol{w}) &= \sum_{k=1}^{K} v_k \log_2 \left| \mathbf{I}_N + \boldsymbol{H}\boldsymbol{U}\boldsymbol{J}_k \boldsymbol{J}_k^{\mathrm{T}} \boldsymbol{\Lambda} \boldsymbol{J}_k \boldsymbol{J}_k^{\mathrm{T}} \boldsymbol{U}^{\mathrm{H}} \boldsymbol{H}^{\mathrm{H}} \right| \\ &= \sum_{k=1}^{K} v_k \log_2 \left| \mathbf{I}_{\sum_{j=k}^{K} L_j} + \boldsymbol{J}_k^{\mathrm{T}} \boldsymbol{U}^{\mathrm{H}} \boldsymbol{H}^{\mathrm{H}} \boldsymbol{H} \boldsymbol{U} \boldsymbol{J}_k \cdot \boldsymbol{J}_k^{\mathrm{T}} \boldsymbol{\Lambda} \boldsymbol{J}_k \right|. \end{aligned} \tag{8.21}$$

For the eigenvalue matrix $\boldsymbol{\Lambda}_k$ of user k, we choose the parametrization

$$\boldsymbol{\Lambda}_k = a_k P_{\max}^{\tau_k} \cdot \bar{\boldsymbol{\Lambda}}_k \quad \forall k \tag{8.22}$$

with $a_k > 0$ and $\bar{\boldsymbol{\Lambda}}_k$ has rank L_k with unit trace. Since the power allocated to user k cannot grow faster than the available power $P_{\max}$, the exponent τ_k is upper bound by $\tau_k \leq 1\ \forall k$. However, sublinear power allocation schemes are allowed as well. We first assume that $\tau_k > 0\ \forall k$ and see what happens when $\tau_m = 0$ for some m later on. Given this parametrization, all L_k eigenvalues of $\boldsymbol{\Lambda}_k$ grow beyond all limits when $P_{\max} \to \infty$. Since $\boldsymbol{H}$ is assumed to be full rank and since $\boldsymbol{U}$ can be chosen such that $\boldsymbol{U}^{\mathrm{H}}\boldsymbol{H}^{\mathrm{H}}\boldsymbol{H}\boldsymbol{U}$ is also full rank, the matrix $\boldsymbol{J}_k^{\mathrm{T}}\boldsymbol{U}^{\mathrm{H}}\boldsymbol{H}^{\mathrm{H}}\boldsymbol{H}\boldsymbol{U}\boldsymbol{J}_k$ in (8.21) is regular as well for all k Therefore, we maintain strong asymptotic equivalence when we neglect the identity matrices inside the determinants of (8.21)[5] and obtain

$$
\begin{aligned}
f(\boldsymbol{U}\boldsymbol{\Lambda}\boldsymbol{U}^{\mathrm{H}}, \boldsymbol{w}) &\cong \sum_{k=1}^{K} v_k \log_2 \left|\boldsymbol{J}_k^{\mathrm{T}}\boldsymbol{U}^{\mathrm{H}}\boldsymbol{H}^{\mathrm{H}}\boldsymbol{H}\boldsymbol{U}\boldsymbol{J}_k\right| + \sum_{k=1}^{K} v_k \log_2 \left|\boldsymbol{J}_k^{\mathrm{T}}\boldsymbol{\Lambda}\boldsymbol{J}_k\right| \\
&= \sum_{k=1}^{K} v_k \log_2 \left|\boldsymbol{J}_k^{\mathrm{T}}\boldsymbol{U}^{\mathrm{H}}\boldsymbol{H}^{\mathrm{H}}\boldsymbol{H}\boldsymbol{U}\boldsymbol{J}_k\right| + \sum_{k=1}^{K} v_k \log_2 \textstyle\prod_{i\geq k} \left|\boldsymbol{\Lambda}_i\right|.
\end{aligned}
\tag{8.23}
$$

Note that only the last sum in (8.23) depends on the power allocation policy from (8.22). Due to $v_1 = w_1$ and $v_k = w_k - w_{k-1}$ for $k > 1$, we obtain by means of (8.22)

$$
\begin{aligned}
\sum_{k=1}^{K} v_k \log_2 \prod_{i\geq k} \left|\boldsymbol{\Lambda}_i\right| &= \sum_{k=1}^{K} w_k \log_2 \left|\boldsymbol{\Lambda}_k\right| \\
&= \Big(\sum_{k=1}^{K} w_k L_k \tau_k\Big) \log_2 P_{\max} + \sum_{k=1}^{K} w_k L_k \log_2 a_k \\
&\quad + \sum_{k=1}^{K} w_k \log_2 \left|\bar{\boldsymbol{\Lambda}}_k\right|.
\end{aligned}
\tag{8.24}
$$

In order to maximize the prefactor of $\log_2 P_{\max}$ in above equation, both the number of active data streams L_k and the power exponent τ_k must be chosen as large as possible for every user k. In other words, every user has to multiplex as many data streams as he has antenna elements ($L_k = M_k\ \forall k$) in the dual MAC and the weakly optimum dual MAC power allocation must be *linear* ($\tau_k = 1\ \forall k$). Therefore, the

---

[5] To see this, let $\boldsymbol{A}(x)$ have the eigenvalue decomposition $\boldsymbol{A}(x) = \boldsymbol{V}(x)\boldsymbol{D}(x)\boldsymbol{V}^{\mathrm{H}}(x)$ with diagonal $\boldsymbol{D}(x) = \mathrm{diag}\{d_k(x)\}_{k=1}^{L}$ and unitary $\boldsymbol{V}(x)$. The logarithm of the determinant of $\mathbf{I} + \boldsymbol{A}(x)$ then reads as

$$
\begin{aligned}
\log_2 \left|\mathbf{I} + \boldsymbol{A}(x)\right| &= \sum_{k=1}^{L} \log_2(1 + d_k(x)) = \sum_{k=1}^{L} \log_2 d_k(x) + \sum_{k=1}^{L} \log_2(1 + d_k^{-1}(x)) \cong \log_2 \left|\boldsymbol{D}(x)\right| \\
&= \log_2 \left|\boldsymbol{A}(x)\right|.
\end{aligned}
$$

Strong asymptotic equivalence holds since $d_k^{-1}(x) \to 0\ \forall k$ as $P_{\max} \to \infty$.

limitation to linear power allocation schemes in [141, 142] turns out not to lead to any loss of performance in the asymptotic limit. Now if τ_m was zero for some m, the rate of user m would not grow beyond all limits when $P_{\max} \rightarrow \infty$. At the same time, the individual data rates of all other users $i \neq m$ would not grow to infinity with a higher pre-log factor compared to the case $\tau_m > 0$. Hence, an exponent $\tau_m = 0$ is suboptimal.

Having found out that the weakly optimum power must be allocated linearly to the users, it remains to determine the fractions of $P_{\max}$ that the users have to transmit with in the dual MAC to obtain the 'best' weighted sum rate asymptote. Maximizing the second sum in (8.24) under the constraint $\sum_{k=1}^{K} a_k = 1$ leads with $L_k = M_k\ \forall k$ to

$$a_k = \frac{w_k M_k}{\sum_{j=1}^{K} w_j M_j}. \tag{8.25}$$

Finally, the determinant of the unit-trace matrix $\bar{\boldsymbol{\Lambda}}_k$ in the last sum of (8.24) is maximized by

$$\bar{\boldsymbol{\Lambda}}_k = \frac{1}{M_k} \cdot \mathbf{I}_{M_k}$$

which follows from the arithmetic-geometric inequality. Combining all these results, we conclude that the 'best' weighted sum rate asymptote is achieved if the matrix $\boldsymbol{\Lambda}_k$ in (8.22) and also the transmit covariance matrix $\boldsymbol{Q}_i$ are chosen via

$$\boldsymbol{Q}_k = \boldsymbol{\Lambda}_k = \frac{w_k \cdot P_{\max}}{\sum_{j=1}^{K} w_j M_j} \cdot \mathbf{I}_{M_k} \quad \forall k \in \{1, \ldots, K\}. \tag{8.26}$$

This result is summarized in the following theorem:

Theorem 8.8 *In the MIMO broadcast channel with at least as many antennas at the base station as the K user terminals have in sum, a linear power allocation asymptotically achieves the maximum weighted sum rate when dirty paper coding is applied. Scaled identities as transmit covariance matrices in the dual MAC lead to the 'best' weighted sum rate asymptote.*

Since $L_k = M_k$ holds for all users k, the matrix $\boldsymbol{U}$ is square. Moreover, $\boldsymbol{U}$ is block-diagonal and unitary and thus does not have an impact on the asymptotic utility in (8.23) where it cancels out. Inserting the weakly asymptotically optimum dual MAC covariance matrices from (8.26) into the weighted sum rate expression in (8.23) yields the 'best' asymptote [cf. (8.2)]

$$\begin{aligned}\check{f}_\infty(\boldsymbol{w}, P_{\max}) = &\sum_{k=1}^{K} w_k \log_2 \left|\boldsymbol{Q}_k\right| + \sum_{k=1}^{K-1} w_k \log_2 \frac{\left|\boldsymbol{J}_k^{\mathrm{T}} \boldsymbol{H}^{\mathrm{H}} \boldsymbol{H} \boldsymbol{J}_k\right|}{\left|\boldsymbol{J}_{k+1}^{\mathrm{T}} \boldsymbol{H}^{\mathrm{H}} \boldsymbol{H} \boldsymbol{J}_{k+1}\right|} \\ &+ w_K \log_2 \left|\boldsymbol{H}_K^{\mathrm{H}} \boldsymbol{H}_K\right|\end{aligned}$$

$$
\begin{aligned}
&= \Big(\sum_{k=1}^{K} w_k M_k\Big) \log_2 P_{\max} + \sum_{k=1}^{K} w_k M_k \log_2 \frac{w_k}{\sum_{j=1}^{K} w_j M_j} \\
&\quad + \sum_{k=1}^{K-1} w_k \log_2 \frac{\left|\boldsymbol{J}_k^{\mathrm{T}} \boldsymbol{H}^{\mathrm{H}} \boldsymbol{H} \boldsymbol{J}_k\right|}{\left|\boldsymbol{J}_{k+1}^{\mathrm{T}} \boldsymbol{H}^{\mathrm{H}} \boldsymbol{H} \boldsymbol{J}_{k+1}\right|} + w_K \log_2 \left|\boldsymbol{H}_K^{\mathrm{H}} \boldsymbol{H}_K\right|.
\end{aligned} \tag{8.27}
$$

The quotient of the two determinants in (8.27) can also be expressed in terms of projected channel matrices. Let $\mathcal{I}_k = \{k, k+1, \ldots, K\}$ denote the set of integers ranging from k to K and let $\boldsymbol{H}_{\mathcal{I}_k} = [\boldsymbol{H}_k, \boldsymbol{H}_{k+1}, \ldots, \boldsymbol{H}_K]$ be the matrix containing the respective channel matrices. Then, we have for $k \le K-1$ with the aid of the determinant rule for partitioned matrices [147, (11.9)]

$$
\begin{aligned}
\frac{\left|\boldsymbol{J}_k^{\mathrm{T}} \boldsymbol{H}^{\mathrm{H}} \boldsymbol{H} \boldsymbol{J}_k\right|}{\left|\boldsymbol{J}_{k+1}^{\mathrm{T}} \boldsymbol{H}^{\mathrm{H}} \boldsymbol{H} \boldsymbol{J}_{k+1}\right|} &= \frac{\left|\boldsymbol{H}_{\mathcal{I}_k}^{\mathrm{H}} \boldsymbol{H}_{\mathcal{I}_k}\right|}{\left|\boldsymbol{H}_{\mathcal{I}_{k+1}}^{\mathrm{H}} \boldsymbol{H}_{\mathcal{I}_{k+1}}\right|} \\
&= \frac{\left|\begin{bmatrix} \boldsymbol{H}_k^{\mathrm{H}} \boldsymbol{H}_k & \boldsymbol{H}_k^{\mathrm{H}} \boldsymbol{H}_{\mathcal{I}_{k+1}} \\ \boldsymbol{H}_{\mathcal{I}_{k+1}}^{\mathrm{H}} \boldsymbol{H}_k & \boldsymbol{H}_{\mathcal{I}_{k+1}}^{\mathrm{H}} \boldsymbol{H}_{\mathcal{I}_{k+1}} \end{bmatrix}\right|}{\left|\boldsymbol{H}_{\mathcal{I}_{k+1}}^{\mathrm{H}} \boldsymbol{H}_{\mathcal{I}_{k+1}}\right|} \\
&= \Big|\boldsymbol{H}_k^{\mathrm{H}} \underbrace{\left[\mathbf{I}_N - \boldsymbol{H}_{\mathcal{I}_{k+1}} \left(\boldsymbol{H}_{\mathcal{I}_{k+1}}^{\mathrm{H}} \boldsymbol{H}_{\mathcal{I}_{k+1}}\right)^{-1} \boldsymbol{H}_{\mathcal{I}_{k+1}}^{\mathrm{H}}\right]}_{=: \boldsymbol{\Pi}_{k+1}} \boldsymbol{H}_k\Big| \\
&= \left|\boldsymbol{H}_k^{\mathrm{H}} \boldsymbol{\Pi}_{k+1} \boldsymbol{H}_k\right| \\
&= \left|\bar{\boldsymbol{H}}_k^{\mathrm{H}} \bar{\boldsymbol{H}}_k\right|
\end{aligned} \tag{8.28}
$$

where $\bar{\boldsymbol{H}}_k = \boldsymbol{\Pi}_{k+1} \boldsymbol{H}_k$ denotes the projection of $\boldsymbol{H}_k$ into the null space of $\boldsymbol{H}_{\mathcal{I}_{k+1}}^{\mathrm{H}}$. Replacing the MAC channels $\boldsymbol{H}_1, \ldots, \boldsymbol{H}_K$ by the transformed BC channels via

$$
\boldsymbol{H} = \boldsymbol{H}^{\mathrm{BC,H}} \cdot \operatorname{blockdiag}\{\boldsymbol{C}_{\eta_k}^{-\frac{1}{2},\mathrm{H}}\}_{k=1}^{K} \quad \Leftrightarrow \quad \boldsymbol{H}_k = \boldsymbol{H}_k^{\mathrm{BC,H}} \boldsymbol{C}_{\eta_k}^{-\frac{1}{2},\mathrm{H}} \quad \forall k \tag{8.29}
$$

the asymptote in (8.27) can be reformulated to

$$
\begin{aligned}
\check{f}_\infty(\boldsymbol{w}, P_{\max}) &= \Big(\sum_{k=1}^{K} w_k M_k\Big) \log_2 P_{\max} + \sum_{k=1}^{K} w_k M_k \log_2 \frac{w_k}{\sum_{j=1}^{K} w_j M_j} \\
&\quad + \sum_{k=1}^{K-1} w_k \log_2 \frac{\left|\boldsymbol{J}_k^{\mathrm{T}} \boldsymbol{H}^{\mathrm{BC}} \boldsymbol{H}^{\mathrm{BC,H}} \boldsymbol{J}_k\right|}{\left|\boldsymbol{J}_{k+1}^{\mathrm{T}} \boldsymbol{H}^{\mathrm{BC}} \boldsymbol{H}^{\mathrm{BC,H}} \boldsymbol{J}_{k+1}\right|} + w_K \log_2 \left|\boldsymbol{H}_K^{\mathrm{BC}} \boldsymbol{H}_K^{\mathrm{BC,H}}\right| \\
&\quad - \sum_{k=1}^{K} w_k \log_2 \left|\boldsymbol{C}_{\eta_k}\right|.
\end{aligned} \tag{8.30}
$$

For $\boldsymbol{w} = \mathbf{1}$, i.e., when all weights are equal to unity, the sum capacity $C^{\mathrm{DPC}}(P_{\max})$ of the MIMO BC with dirty paper coding features the asymptote

$$C^{\mathrm{DPC}}(P_{\max}) \cong \check{f}_{\infty}(\mathbf{1}, P_{\max}) = M \log_2 P_{\max} - M \log_2 M + \log_2 \left| \boldsymbol{H}^{\mathrm{BC}} \boldsymbol{H}^{\mathrm{BC,H}} \right| - \sum_{k=1}^{K} \log_2 \left| \boldsymbol{C}_{\eta_k} \right| \tag{8.31}$$

which therefore coincides with the asymptote of the same system with *cooperating* receivers and worst case (block-diagonal) noise covariance matrix from (8.14). This proves the following theorem.

Theorem 8.9 *The high power asymptote of the dirty paper coding sum capacity in a MIMO broadcast channel with at least as many transmit antennas as total receive antennas coincides with the sum capacity asymptote of the same system with cooperating receivers if the total noise covariance matrix is block-diagonal. In other words, the asymptotic worst-case noise is uncorrelated between the individual users.*

For identity noise covariance matrices, (8.31) has also been obtained in [141, 142]. An interesting observation is that the asymptotic sum capacity in (8.31) does not depend on the distribution of the M receive antennas onto the K users if the noise at all receiving antenna elements is uncorrelated and features the same variance.

In order to find the weakly optimum transmit covariance matrices in the BC which asymptotically achieve the weighted sum rate utility from (8.30), we apply the MAC to BC duality conversion from Sect. 4.4.3. Note that the weakly asymptotically optimum transmit strategy in the original MIMO BC has never been addressed in the literature to the best of our knowledge.

To find the BC precoders, we first compute the matched filter receiver $\boldsymbol{G}_k^{\mathrm{MF}}$ of some user k in the dual MAC via (4.55). Using the substitution $\beta := \sum_{j=1}^{K} w_j M_j$, the weakly optimum transmit covariance matrix of user k in (8.26) can be expressed as $\boldsymbol{Q}_k = \beta^{-1} w_k P_{\max} \cdot \mathbf{I}_{M_k}$ such that we apply the duality transformation for the transmit filters

$$\boldsymbol{T}_k = \sqrt{\frac{w_k P_{\max}}{\beta}} \cdot \mathbf{I}_{M_k} \quad \forall k. \tag{8.32}$$

For a convenient derivation of the dual MAC matched filters, we introduce the substitute

$$\boldsymbol{W} := \frac{1}{\beta} \cdot \mathrm{blockdiag}\{w_k \mathbf{I}_{M_k}\}_{k=1}^{K}.$$

Using (4.55) and the precoders in (8.32), we obtain by means of the selection matrix in (8.20)

$$
\begin{aligned}
\boldsymbol{G}_k^{\mathrm{MF}} &= \sqrt{\frac{w_k P_{\max}}{\beta}} \cdot \boldsymbol{H}_k^{\mathrm{H}} \Big[\mathbf{I}_N + \sum_{i>k} \boldsymbol{H}_i \boldsymbol{H}_i^{\mathrm{H}} \cdot \frac{w_i P_{\max}}{\beta}\Big]^{-1} \\
&= \sqrt{\frac{w_k P_{\max}}{\beta}} \cdot \boldsymbol{H}_k^{\mathrm{H}} \Big[\mathbf{I}_N + P_{\max} \boldsymbol{H} \boldsymbol{W}^{\frac{1}{2}} \boldsymbol{J}_{k+1} \boldsymbol{J}_{k+1}^{\mathrm{T}} \boldsymbol{W}^{\frac{1}{2}} \boldsymbol{H}^{\mathrm{H}}\Big]^{-1} \\
&= \sqrt{\frac{w_k P_{\max}}{\beta}} \cdot \boldsymbol{H}_k^{\mathrm{H}} \Big[\mathbf{I}_N - \boldsymbol{H} \boldsymbol{W}^{\frac{1}{2}} \boldsymbol{J}_{k+1} \big(P_{\max}^{-1} \mathbf{I} + \boldsymbol{J}_{k+1}^{\mathrm{T}} \boldsymbol{W}^{\frac{1}{2}} \boldsymbol{H}^{\mathrm{H}} \boldsymbol{H} \boldsymbol{W}^{\frac{1}{2}} \boldsymbol{J}_{k+1}\big)^{-1} \\
&\qquad\qquad \boldsymbol{J}_{k+1}^{\mathrm{T}} \boldsymbol{W}^{\frac{1}{2}} \boldsymbol{H}^{\mathrm{H}}\Big] \\
&\cong \sqrt{\frac{w_k P_{\max}}{\beta}} \cdot \boldsymbol{H}_k^{\mathrm{H}} \Big[\mathbf{I}_N - \boldsymbol{H} \boldsymbol{W}^{\frac{1}{2}} \boldsymbol{J}_{k+1} \big(\boldsymbol{J}_{k+1}^{\mathrm{T}} \boldsymbol{W}^{\frac{1}{2}} \boldsymbol{H}^{\mathrm{H}} \boldsymbol{H} \boldsymbol{W}^{\frac{1}{2}} \boldsymbol{J}_{k+1}\big)^{-1} \\
&\qquad\qquad \boldsymbol{J}_{k+1}^{\mathrm{T}} \boldsymbol{W}^{\frac{1}{2}} \boldsymbol{H}^{\mathrm{H}}\Big] \\
&= \sqrt{\frac{w_k P_{\max}}{\beta}} \cdot \boldsymbol{H}_k^{\mathrm{H}} \Big[\mathbf{I}_N - \boldsymbol{H} \boldsymbol{J}_{k+1} \big(\boldsymbol{J}_{k+1}^{\mathrm{T}} \boldsymbol{H}^{\mathrm{H}} \boldsymbol{H} \boldsymbol{J}_{k+1}\big)^{-1} \boldsymbol{J}_{k+1}^{\mathrm{T}} \boldsymbol{H}^{\mathrm{H}}\Big] \\
&= \sqrt{\frac{w_k P_{\max}}{\beta}} \cdot \boldsymbol{H}_k^{\mathrm{H}} \Big[\mathbf{I}_N - \boldsymbol{H}_{\mathcal{I}_{k+1}} \big(\boldsymbol{H}_{\mathcal{I}_{k+1}}^{\mathrm{H}} \boldsymbol{H}_{\mathcal{I}_{k+1}}\big)^{-1} \boldsymbol{H}_{\mathcal{I}_{k+1}}^{\mathrm{H}}\Big] \\
&= \sqrt{\frac{w_k P_{\max}}{\beta}} \cdot \boldsymbol{H}_k^{\mathrm{H}} \boldsymbol{\Pi}_{k+1}^{\perp} \\
&= \sqrt{\frac{w_k P_{\max}}{\beta}} \cdot \bar{\boldsymbol{H}}_k^{\mathrm{H}} =: \boldsymbol{G}_k^{\prime\mathrm{MF}} \quad \forall k.
\end{aligned}
\tag{8.33}
$$

For the third equality, the matrix inversion lemma from Appendix A2 was used. Strong asymptotic equivalence holds in the fourth line since $\boldsymbol{J}_{k+1}^{\mathrm{T}} \boldsymbol{W}^{\frac{1}{2}} \boldsymbol{H}^{\mathrm{H}} \boldsymbol{H} \boldsymbol{W}^{\frac{1}{2}} \boldsymbol{J}_{k+1}$ is regular due to the full rank assumption of $\boldsymbol{H}$ and since $\boldsymbol{w} > \mathbf{0}$ is assumed.[6] The matrix $\bar{\boldsymbol{H}}_k$, $k \leq K-1$, denotes the projection of $\boldsymbol{H}_k$ onto the null space of $\boldsymbol{H}_{\mathcal{I}_{k+1}}^{\mathrm{H}}$, i.e., onto the null space of all other Hermitian channels with larger indices and $\bar{\boldsymbol{H}}_K = \boldsymbol{H}_K$, i.e., $\boldsymbol{\Pi}_{K+1}^{\perp} = \mathbf{I}$.

---

[6] To see this, we have to show that $\sqrt{P_{\max}}\big[(P_{\max}^{-1}\mathbf{I} + \boldsymbol{A})^{-1} - \boldsymbol{A}^{-1}\big] \to \mathbf{0}$ for regular $\boldsymbol{A}$ as $P_{\max} \to 0$:

$$
\begin{aligned}
\sqrt{P_{\max}}\big[(P_{\max}^{-1}\mathbf{I} + \boldsymbol{A})^{-1} - \boldsymbol{A}^{-1}\big] &= \sqrt{P_{\max}}(P_{\max}^{-1}\mathbf{I} + \boldsymbol{A})^{-1}\big[\mathbf{I} - (P_{\max}^{-1}\mathbf{I} + \boldsymbol{A})\boldsymbol{A}^{-1}\big] \\
&= \sqrt{P_{\max}}(P_{\max}^{-1}\mathbf{I} + \boldsymbol{A})^{-1}\big[-P_{\max}^{-1}\boldsymbol{A}^{-1}\big] \\
&= \frac{-1}{\sqrt{P_{\max}}}\big[\boldsymbol{A}^2 + P_{\max}^{-1}\boldsymbol{A}\big]^{-1} \to \mathbf{0}
\end{aligned}
$$

Convergence to zero follows from the fact that the term inside the last braces goes to $\boldsymbol{A}^2$ whereas its prefactor goes to zero as $P_{\max} \to 0$.

Next, the transmit filters $\boldsymbol{T}_k$ in (8.32) and the receive filters $\boldsymbol{G}_k^{\prime\mathrm{MF}}$ in (8.33) which are strongly asymptotically equivalent to the exact matched filters are converted back to the original broadcast channel by means of (4.59)–(4.60). Note that not exactly the same data rates are achieved in the MIMO BC if the duality conversion is based on $\boldsymbol{G}_k^{\prime\mathrm{MF}}\ \forall k$ instead of the original matched filters. However, the high power asymptotes remain the same as we will see later. The advantage of using $\boldsymbol{G}_k^{\prime\mathrm{MF}}$ instead of $\boldsymbol{G}_k^{\mathrm{MF}}$ for all k results from the fact that the duality conversion drastically simplifies and can even be computed in closed form. We observe that $\boldsymbol{Z}$ in (4.60) is diagonal when $\boldsymbol{G}_k^{\prime\mathrm{MF}}$ are used instead of $\boldsymbol{G}_k^{\mathrm{MF}}\ \forall k$ since the off-diagonal blocks $\boldsymbol{Z}_{k,j}$ with $k > j$ vanish. This results from the fact that $\boldsymbol{G}_j^{\prime\mathrm{MF}}\boldsymbol{H}_k = \boldsymbol{0}$ for $k > j$ due to $\bar{\boldsymbol{H}}_j^{\mathrm{H}}\boldsymbol{H}_k = \boldsymbol{0}$, which itself results from the projector contained in $\bar{\boldsymbol{H}}_j$. Hence, the receive filter $\boldsymbol{G}_j^{\prime\mathrm{MF}}$ completely removes the inter-user interference from users with larger indices $k > j$. For the precoders in the BC, we need the eigenbasis of $\boldsymbol{G}_k^{\prime\mathrm{MF}}\boldsymbol{H}_k\boldsymbol{T}_k$, cf. (4.59) and (4.57). With (8.32) and (8.33), we find

$$\boldsymbol{G}_k^{\prime\mathrm{MF}}\boldsymbol{H}_k\boldsymbol{T}_k = \frac{w_k P_{\max}}{\beta}\bar{\boldsymbol{H}}_k^{\mathrm{H}}\bar{\boldsymbol{H}}_k = \boldsymbol{G}_k^{\prime\mathrm{MF}}\boldsymbol{G}_k^{\prime\mathrm{MF,H}} = \boldsymbol{V}_k\boldsymbol{\Xi}_k\boldsymbol{V}_k^{\mathrm{H}} = \frac{w_k P_{\max}}{\beta}\cdot\boldsymbol{V}_k\boldsymbol{\Sigma}_k\boldsymbol{V}_k^{\mathrm{H}} \tag{8.34}$$

where the diagonal matrix $\boldsymbol{\Sigma}_k = \mathrm{diag}\{\sigma_{k,j}\}_{j=1}^{M_k}$ contains the M_k eigenvalues of $\bar{\boldsymbol{H}}_k^{\mathrm{H}}\bar{\boldsymbol{H}}_k$. The scaling coefficients $\varphi_{k,j}^2$ in (4.59) can be computed separately since $\boldsymbol{Z}$ in (4.60) is diagonal:

$$\varphi_{k,j}^2 = \frac{\|\boldsymbol{T}_k\boldsymbol{V}_k\boldsymbol{e}_j\|_2^2}{\|\boldsymbol{e}_j^{\mathrm{T}}\boldsymbol{V}_k^{\mathrm{H}}\boldsymbol{G}_k^{\prime\mathrm{MF}}\|_2^2} = \frac{\frac{w_k P_{\max}}{\beta}}{\frac{w_k P_{\max}}{\beta}\cdot[\boldsymbol{\Sigma}_k]_{j,j}} = [\boldsymbol{\Sigma}_k^{-1}]_{j,j}$$

Thus, the precoder of user k in the BC reads as [cf. (4.59)]

$$\boldsymbol{P}_k = \boldsymbol{G}_k^{\prime\mathrm{MF,H}}\boldsymbol{V}_k\boldsymbol{\Sigma}_k^{-\frac{1}{2}}$$

and completely suppresses the inter-user interference of other users j whose indices satisfy $j > k$ since

$$\boldsymbol{C}_{\eta_j}^{-\frac{1}{2}}\boldsymbol{H}_j^{\mathrm{BC}}\boldsymbol{G}_k^{\prime\mathrm{MF,H}} = (\boldsymbol{G}_k^{\prime\mathrm{MF}}\boldsymbol{H}_j)^{\mathrm{H}} = \boldsymbol{0}.$$

Hence, the product of the composite BC channel matrix $\boldsymbol{H}^{\mathrm{BC}}$ and the composite precoder matrix $\boldsymbol{P} = [\boldsymbol{P}_1, \ldots, \boldsymbol{P}_K]$ is upper block-triangular.[7] These off-diagonal blocks which denote the inter-user interference are rendered harmless by means of the dirty paper coding scheme. As a consequence, there is no inter-user interference at all in the end-to-end transmission for this choice of the precoders.

[7] The upper block-triangular structure results from the assumption that the weights satisfy $w_1 \leq w_2 \leq \cdots \leq w_K$.

The matrix $\boldsymbol{P}_k$ induces the transmit covariance matrix

$$
\begin{aligned}
\boldsymbol{S}_k^{\mathrm{DPC}}(\boldsymbol{w}) &= \boldsymbol{P}_k \boldsymbol{P}_k^{\mathrm{H}} = \boldsymbol{G}_k^{\prime \mathrm{MF},\mathrm{H}} \boldsymbol{V}_k \boldsymbol{\Sigma}_k^{-1} \boldsymbol{V}_k^{\mathrm{H}} \boldsymbol{G}_k^{\prime \mathrm{MF}} \\
&= \frac{w_k P_{\max}}{\beta} \cdot \bar{\boldsymbol{H}}_k \left(\bar{\boldsymbol{H}}_k^{\mathrm{H}} \bar{\boldsymbol{H}}_k\right)^{-1} \bar{\boldsymbol{H}}_k^{\mathrm{H}} \\
&= \frac{w_k P_{\max}}{\beta} \cdot \bar{\boldsymbol{H}}_k^{\mathrm{BC},\mathrm{H}} \left(\bar{\boldsymbol{H}}_k^{\mathrm{BC}} \bar{\boldsymbol{H}}_k^{\mathrm{BC},\mathrm{H}}\right)^{-1} \bar{\boldsymbol{H}}_k^{\mathrm{BC}} \quad \forall k
\end{aligned}
\tag{8.35}
$$

which is again a scaled projector as in (8.16), but this time for every user. The matrix $\bar{\boldsymbol{H}}_k^{\mathrm{BC}}$ is defined via

$$
\bar{\boldsymbol{H}}_k^{\mathrm{BC}} = \boldsymbol{H}_k^{\mathrm{BC}} \underbrace{\left[\mathbf{I}_N - \boldsymbol{H}_{\mathcal{I}_{k+1}}^{\mathrm{BC},\mathrm{H}} \left(\boldsymbol{H}_{\mathcal{I}_{k+1}}^{\mathrm{BC}} \boldsymbol{H}_{\mathcal{I}_{k+1}}^{\mathrm{BC},\mathrm{H}}\right)^{-1} \boldsymbol{H}_{\mathcal{I}_{k+1}}^{\mathrm{BC}}\right]}_{=:\boldsymbol{\Pi}_{k+1}^{\perp}}. \tag{8.36}
$$

From (8.35), we conclude that the power allocated to user k is given by

$$
\operatorname{tr}(\boldsymbol{S}_k^{\mathrm{DPC}}(\boldsymbol{w})) = \frac{w_k M_k}{\sum_{j=1}^K w_j M_j} \cdot P_{\max}
$$

and coincides with the power allocation of the dual MIMO MAC, see (8.25). Note that $\boldsymbol{S}_1^{\mathrm{DPC}}(\boldsymbol{w}), \ldots, \boldsymbol{S}_K^{\mathrm{DPC}}(\boldsymbol{w})$ do not depend on the noise covariance matrices in the BC. The total transmit covariance matrix $\boldsymbol{S}^{\mathrm{DPC}}(\boldsymbol{w})$ in the BC which asymptotically achieves the maximum weighted sum rate with dirty paper coding reads as

$$
\boldsymbol{S}^{\mathrm{DPC}}(\boldsymbol{w}) = \sum_{k=1}^K \boldsymbol{S}_k^{\mathrm{DPC}}(\boldsymbol{w}) = \frac{P_{\max}}{\beta} \sum_{k=1}^K w_k \bar{\boldsymbol{H}}_k^{\mathrm{BC},\mathrm{H}} \left(\bar{\boldsymbol{H}}_k^{\mathrm{BC}} \bar{\boldsymbol{H}}_k^{\mathrm{BC},\mathrm{H}}\right)^{-1} \bar{\boldsymbol{H}}_k^{\mathrm{BC}}. \tag{8.37}
$$

Note that $\boldsymbol{S}^{\mathrm{DPC}}(\boldsymbol{w})$ from (8.37) coincides with $\boldsymbol{S}^{\mathrm{PtP}}$ from (8.16) for $\boldsymbol{w} = \alpha \mathbf{1}$, i.e., when all weights are equal. This is proven in Appendix A17. In this case, the sum rate achieved by $\boldsymbol{S}^{\mathrm{PtP}}$ in the cooperative system is in general larger than the sum rate achieved by the same transmit covariance matrix $\boldsymbol{S}^{\mathrm{DPC}}(\boldsymbol{w})$ with dirty paper coding *at finite transmit power*. In the limit $P_{\max} \to \infty$, the difference between the two sum rate vanishes and the sum capacity is achieved asymptotically in both cases. To see this, we first insert $\boldsymbol{S}^{\mathrm{PtP}}$ from (8.16) into the mutual information expression of the point-to-point MIMO system (BC with cooperating receivers) and obtain

$$
\begin{aligned}
R^{\mathrm{PtP}}(\boldsymbol{S}^{\mathrm{PtP}}) &= \log_2 \left|\mathbf{I} + \boldsymbol{H}^{\mathrm{BC},\mathrm{H}} \boldsymbol{C}_\eta^{-1} \boldsymbol{H}^{\mathrm{BC}} \boldsymbol{S}^{\mathrm{PtP}}\right| \\
&= \log_2 \left|\mathbf{I} + \boldsymbol{C}_\eta^{-1} \boldsymbol{H}^{\mathrm{BC}} \boldsymbol{S}^{\mathrm{PtP}} \boldsymbol{H}^{\mathrm{BC},\mathrm{H}}\right| \\
&= \log_2 \left|\mathbf{I} + \boldsymbol{C}_\eta^{-1} \boldsymbol{H}^{\mathrm{BC}} \boldsymbol{H}^{\mathrm{BC},\mathrm{H}} \cdot \frac{P_{\max}}{M}\right|
\end{aligned}
$$

$$
\begin{aligned}
&\cong \log_2 \left| \boldsymbol{C}_{\eta}^{-1} \boldsymbol{H}^{\mathrm{BC}} \boldsymbol{H}^{\mathrm{BC,H}} \cdot \frac{P_{\max}}{M} \right| \\
&= M \log_2 P_{\max} - M \log_2 M + \log_2 \left| \boldsymbol{H}^{\mathrm{BC}} \boldsymbol{H}^{\mathrm{BC,H}} \right| - \sum_{k=1}^{K} \log_2 \left| \boldsymbol{C}_{\eta_k} \right|
\end{aligned}
$$

in accordance to (8.14). Note that we assumed $\boldsymbol{C}_{\eta} = \operatorname{blockdiag}\{\boldsymbol{C}_{\eta_k}\}_{k=1}^{K}$ which is the worst case asymptotic total noise covariance matrix. For the true MIMO BC with dirty paper coding where user 1 is encoded last and user K first (which we used during the duality conversion), the rate expression of user k is given by (4.27) in terms of the transmit covariance matrices. Using $\pi[k] = k$ and $\boldsymbol{S}_k^{\mathrm{DPC}}(\mathbf{1})$ from (8.35), the rate of user k reads as

$$
\begin{aligned}
&R_k^{\mathrm{DPC}}(\boldsymbol{S}^{\mathrm{DPC}}(\mathbf{1})) \\
&= \log_2 \frac{\left| \boldsymbol{C}_{\eta_k} + \boldsymbol{H}_k^{\mathrm{BC}} \sum_{i \le k} \boldsymbol{S}_i^{\mathrm{DPC}}(\mathbf{1}) \boldsymbol{H}_k^{\mathrm{BC,H}} \right|}{\left| \boldsymbol{C}_{\eta_k} + \boldsymbol{H}_k^{\mathrm{BC}} \sum_{i < k} \boldsymbol{S}_i^{\mathrm{DPC}}(\mathbf{1}) \boldsymbol{H}_k^{\mathrm{BC,H}} \right|} \\
&= \log_2 \frac{\left| \boldsymbol{C}_{\eta_k} + \boldsymbol{H}_k^{\mathrm{BC}} \boldsymbol{S}_k^{\mathrm{DPC}}(\mathbf{1}) \boldsymbol{H}_k^{\mathrm{BC,H}} \right|}{\left| \boldsymbol{C}_{\eta_k} \right|} \\
&= \log_2 \frac{\left| \boldsymbol{C}_{\eta_k} + \boldsymbol{H}_k^{\mathrm{BC}} \frac{P_{\max}}{M} \boldsymbol{\Pi}_{k+1} \boldsymbol{H}_k^{\mathrm{BC,H}} \left(\boldsymbol{H}_k^{\mathrm{BC}} \boldsymbol{\Pi}_{k+1} \boldsymbol{H}_k^{\mathrm{BC,H}} \right)^{-1} \boldsymbol{H}_k^{\mathrm{BC}} \boldsymbol{\Pi}_{k+1} \boldsymbol{H}_k^{\mathrm{BC,H}} \right|}{\left| \boldsymbol{C}_{\eta_k} \right|} \\
&= \log_2 \frac{\left| \boldsymbol{C}_{\eta_k} + \frac{P_{\max}}{M} \boldsymbol{H}_k^{\mathrm{BC}} \boldsymbol{\Pi}_{k+1} \boldsymbol{H}_k^{\mathrm{BC,H}} \right|}{\left| \boldsymbol{C}_{\eta_k} \right|} \\
&= \log_2 \left| \mathbf{I} + \frac{P_{\max}}{M} \boldsymbol{C}_{\eta_k}^{-1} \boldsymbol{H}_k^{\mathrm{BC}} \boldsymbol{\Pi}_{k+1} \boldsymbol{H}_k^{\mathrm{BC,H}} \right| \\
&\cong M_k \log_2 P_{\max} - M_k \log_2 M + \log_2 \left| \boldsymbol{H}_k^{\mathrm{BC}} \boldsymbol{\Pi}_{k+1} \boldsymbol{H}_k^{\mathrm{BC,H}} \right| - \log_2 \left| \boldsymbol{C}_{\eta_k} \right|
\end{aligned}
$$

since $\boldsymbol{S}_i^{\mathrm{DPC}}(\cdot)\boldsymbol{H}_k = \mathbf{0}$ for $i < k$. In analogy to (8.28), the determinant of the projected channel Gram can be written as

$$
\left| \boldsymbol{H}_k^{\mathrm{BC}} \boldsymbol{\Pi}_{k+1} \boldsymbol{H}_k^{\mathrm{BC,H}} \right| = \begin{cases} \left| \boldsymbol{J}_K^{\mathrm{T}} \boldsymbol{H}^{\mathrm{BC}} \boldsymbol{H}^{\mathrm{BC,H}} \boldsymbol{J}_K \right| & \text{for } k = K \\ \dfrac{\left| \boldsymbol{J}_k^{\mathrm{T}} \boldsymbol{H}^{\mathrm{BC}} \boldsymbol{H}^{\mathrm{BC,H}} \boldsymbol{J}_k \right|}{\left| \boldsymbol{J}_{k+1}^{\mathrm{T}} \boldsymbol{H}^{\mathrm{BC}} \boldsymbol{H}^{\mathrm{BC,H}} \boldsymbol{J}_{k+1} \right|} & \text{for } k < K \end{cases}
$$

which finally leads to the desired result

$$
\sum_{k=1}^{K} R_k^{\mathrm{DPC}}(\boldsymbol{S}_k^{\mathrm{DPC}}(\mathbf{1})) \cong R^{\mathrm{PtP}}(\boldsymbol{S}^{\mathrm{PtP}}).
$$

Therefore, cooperation at the receivers does not increase the asymptotic sum capacity compared to dirty paper coding.

Synopsis: In the high power regime $P_{\max} \rightarrow \infty$ of a MIMO BC with at least as many transmit antennas as the user terminals have in sum, the base station has to assign exactly as many data streams to a particular user as he has antennas in order to maximize the weighted sum rate. In addition, the weakly optimum power allocation per user is linear. Interestingly, the weakly optimum power allocation of the dual MIMO MAC does not differ from the weakly optimum power allocation in the BC. In both domains, the uniform distribution of power over the individual streams of a user is asymptotically optimal. While scaled identity transmit covariance matrices asymptotically achieve the maximum weighted sum rate in the dual MIMO MAC, choosing the transmit covariance matrices of the K users in the original BC as scaled projectors does the same thing in the broadcast channel. Moreover, the BC channel-precoder chain becomes upper block-triangular and the partial interference of users with larger precoder indices is rendered harmless by means of the dirty paper coding scheme. Thus, the complete suppression of the inter-user interference asymptotically achieves the maximum weighted sum rate. For equal weights, the scaled projector $\boldsymbol{S}^{\mathrm{PtP}}$ onto the range of the Hermitian composite BC channel matrix $\boldsymbol{H}^{\mathrm{BC,H}}$ not only achieves the asymptotic sum capacity in case of cooperating receivers, but also in case of the noncooperating receivers in conjunction with dirty paper coding.

8.2.2.2 Overloaded System

While the antenna configuration with enough degrees of freedom discussed the previous subsection allows for a complete characterization of the weighted sum rate maximization problem with dirty paper coding, the overloaded system is much more complicated to handle. For this reason, we will restrict ourselves to the maximization of the asymptotic sum capacity for the case of single-antenna terminals in order to gain some insights to the overloaded scenario. In particular, we analyze if and under which circumstances more than N users must be served in order to obtain the maximum asymptotic sum capacity. The author is not aware of any publication in the literature which theoretically deals with such an overloaded system.

Let $\boldsymbol{h}_1, \ldots, \boldsymbol{h}_K \in \mathbb{C}^N$ denote the dual MAC channel *vectors* of the K users that are transmitting with powers $q_1, \ldots, q_K \in \mathbb{R}_{+,0}$. Then, the sum rate that is obtained by means of successive decoding in the dual vector MAC and that is also achievable by means of dirty paper coding in the original vector BC reads as [cf. (6.4)]

$$R^{\mathrm{MAC}}(\boldsymbol{Q}) = \log_2 \left| \mathbf{I}_N + \sum_{k=1}^{K} q_k \boldsymbol{h}_k \boldsymbol{h}_k^{\mathrm{H}} \right| = \log_2 \left| \mathbf{I}_N + \boldsymbol{H} \boldsymbol{Q} \boldsymbol{H}^{\mathrm{H}} \right| \tag{8.38}$$

where $\boldsymbol{Q} = \mathrm{diag}\{q_k\}_{k=1}^K$ is diagonal and $\boldsymbol{H} = [\boldsymbol{h}_1, \ldots, \boldsymbol{h}_K] \in \mathbb{C}^{N \times K}$ contains the channel vectors of all users. The difficulty of the asymptotic sum capacity analysis results from the fact that the power allocation cannot easily be separated from the channel vectors for $N < M = K$ as this was the case in the previous subsection, where $N \geq M$ held. Clearly, we may rewrite (8.38) via

$$R^{\mathrm{MAC}}(\boldsymbol{Q}) = \log_2 \left| \mathbf{I}_K + \boldsymbol{H}^{\mathrm{H}} \boldsymbol{H} \boldsymbol{Q} \right| \tag{8.39}$$

but the $K \times K$ matrix product $\boldsymbol{H}^{\mathrm{H}} \boldsymbol{H} \boldsymbol{Q}$ has only rank N when $K > N$ since $\boldsymbol{H}$ is an $N \times K$ matrix.[8] Thus, the identity matrix $\mathbf{I}_K$ in (8.39) cannot simply be dropped in order to end up with an expression that is strongly asymptotically equivalent to (8.38) and the impact of the power allocation thus cannot be separated from the channels.

Instead, the $N \times N$ identity matrix $\mathbf{I}_N$ in (8.38) may be dropped if the powers of at least N users goes to infinity as $P_{\max} \to \infty$ and the corresponding N channel vectors are linearly independent. We assume that the channel matrix of any user subset has maximum rank, a property which will later be called regular channel setup in Chap. 9. When the identity in (8.38) is neglected and $K \geq N$ with at least N powers going to infinity as $P_{\max} \to \infty$, we obtain

$$R^{\mathrm{MAC}}(\boldsymbol{Q}) \cong \log_2 \left| \boldsymbol{H} \boldsymbol{Q} \boldsymbol{H}^{\mathrm{H}} \right|. \tag{8.40}$$

In the following, we present a simple proof that $N+1$ active terminals in the dual MAC may achieve a larger sum rate compared to the case when only N terminals actively transmit data to the base station in the dual MAC. In the original vector BC, this means that serving $N+1$ users may (depending on the channels) lead to a larger sum rate than serving only N users. Note that this statement is valid for any $N > 1$ although it is based on a suboptimum power allocation. Afterwards, we investigate the $N = 2$ antenna base station case more explicitly.

For a setup in which as many single-antenna users communicate with the base station as the latter one has antennas, i.e., $K = N$, the asymptotic sum capacity in the dual MAC is achieved by the power allocation [cf. (8.26)]

$$q_k = \frac{P_{\max}}{N} \quad \forall k \in \{1, \ldots, N\} \tag{8.41}$$

which clearly maximizes the right hand side in (8.40). The resulting asymptotic sum rate (which for only N users also denotes the asymptotic sum capacity) reads as

$$\begin{aligned} R^{\mathrm{MAC}}_{\{1,\ldots,N\}}\left(\frac{P_{\max}}{N}\mathbf{I}_N\right) &\cong \log_2 \left| \boldsymbol{H}\boldsymbol{H}^{\mathrm{H}} \cdot \frac{P_{\max}}{N} \right| \\ &= N \log_2 P_{\max} - N \log_2 N + \log_2 \left| \boldsymbol{H}\boldsymbol{H}^{\mathrm{H}} \right| \end{aligned} \tag{8.42}$$

where $\boldsymbol{H} = [\boldsymbol{h}_1, \ldots \boldsymbol{h}_N]$. Now if a further user with channel vector $\boldsymbol{h} = \boldsymbol{h}_{N+1}$ enters the system, a (suboptimum) uniform power allocation over all $N+1$ users

[8] We assume $\boldsymbol{H}$ to have full rank here.

asymptotically achieves the sum rate

$$\begin{aligned}R^{\mathrm{MAC}}_{\{1,\ldots,N+1\}}\Big(\frac{P_{\max}}{N+1}\mathbf{I}_{N+1}\Big) &\cong \log_2\Big|\big(\boldsymbol{H}\boldsymbol{H}^{\mathrm{H}}+\boldsymbol{h}\boldsymbol{h}^{\mathrm{H}}\big)\cdot\frac{P_{\max}}{N+1}\Big|\\ &= N\log_2 P_{\max} - N\log_2(N+1)+\log_2\big|\boldsymbol{H}\boldsymbol{H}^{\mathrm{H}}+\boldsymbol{h}\boldsymbol{h}^{\mathrm{H}}\big|.\end{aligned} \tag{8.43}$$

The difference of the two sum rates in (8.42) and (8.43) therefore asymptotically reads as

$$\begin{aligned}\Delta R &:= R^{\mathrm{MAC}}_{\{1,\ldots,N+1\}}\Big(\frac{P_{\max}}{N+1}\mathbf{I}_{N+1}\Big) - R^{\mathrm{MAC}}_{\{1,\ldots,N\}}\Big(\frac{P_{\max}}{N}\mathbf{I}_{N}\Big)\\ &\cong \log_2\big(1+\boldsymbol{h}^{\mathrm{H}}(\boldsymbol{H}\boldsymbol{H}^{\mathrm{H}})^{-1}\boldsymbol{h}\big) - N\log_2\big(1+N^{-1}\big).\end{aligned} \tag{8.44}$$

Under the premise of letting only N out of the $N+1$ users actively transmit in the dual MAC, some restrictions have to be imposed on the channel vector $\boldsymbol{h}$ of user $N+1$ such that serving *the first* N ones indeed leads to the largest asymptotic sum rate in (8.42). Letting the first N users be the active ones leads to the largest asymptotic sum rate (under the assumption of only N active users) if replacing any of the first N channel vectors $\boldsymbol{h}_1,\ldots,\boldsymbol{h}_N$ by $\boldsymbol{h}$ does not lead to a larger determinant of the resulting channel Gram:

$$\big|\boldsymbol{H}\boldsymbol{H}^{\mathrm{H}}\big| \ge \big|\boldsymbol{H}\boldsymbol{H}^{\mathrm{H}}+\boldsymbol{h}\boldsymbol{h}^{\mathrm{H}}-\boldsymbol{h}_k\boldsymbol{h}_k^{\mathrm{H}}\big| \quad \forall k\in\{1,\ldots,N\} \tag{8.45}$$

Due to $\boldsymbol{h}_k=\boldsymbol{H}\boldsymbol{e}_k$ with $\boldsymbol{e}_k$ denoting the k-th canonical unit vector, above inequality is equivalent to

$$\begin{aligned}1 &\ge \big|\mathbf{I}+\boldsymbol{H}^{-1}\boldsymbol{h}\boldsymbol{h}^{\mathrm{H}}\boldsymbol{H}^{-\mathrm{H}}-\boldsymbol{e}_k\boldsymbol{e}_k^{\mathrm{T}}\big|\\ &= \big|\mathbf{I}+\boldsymbol{H}^{-1}\boldsymbol{h}\boldsymbol{h}^{\mathrm{H}}\boldsymbol{H}^{-\mathrm{H}}\big|\cdot\big|\mathbf{I}-\big(\mathbf{I}+\boldsymbol{H}^{-1}\boldsymbol{h}\boldsymbol{h}^{\mathrm{H}}\boldsymbol{H}^{-\mathrm{H}}\big)^{-1}\boldsymbol{e}_k\boldsymbol{e}_k^{\mathrm{T}}\big|\\ &= \big(1+\boldsymbol{h}^{\mathrm{H}}\boldsymbol{H}^{-\mathrm{H}}\boldsymbol{H}^{-1}\boldsymbol{h}\big)\cdot\big(1-\boldsymbol{e}_k^{\mathrm{T}}(\mathbf{I}+\boldsymbol{H}^{-1}\boldsymbol{h}\boldsymbol{h}^{\mathrm{H}}\boldsymbol{H}^{-\mathrm{H}})^{-1}\boldsymbol{e}_k\big).\end{aligned}$$

By means of the matrix inversion lemma, the second factor can be written as

$$1-\boldsymbol{e}_k^{\mathrm{T}}(\mathbf{I}+\boldsymbol{H}^{-1}\boldsymbol{h}\boldsymbol{h}^{\mathrm{H}}\boldsymbol{H}^{-\mathrm{H}})^{-1}\boldsymbol{e}_k = \frac{|\boldsymbol{e}_k^{\mathrm{T}}\boldsymbol{H}^{-1}\boldsymbol{h}|^2}{1+\boldsymbol{h}^{\mathrm{H}}\boldsymbol{H}^{-\mathrm{H}}\boldsymbol{H}^{-1}\boldsymbol{h}}$$

such that (8.45) is eventually equivalent to

$$1 \ge |\boldsymbol{e}_k^{\mathrm{T}}\boldsymbol{H}^{-1}\boldsymbol{h}|^2 \quad \forall k\in\{1,\ldots,N\}. \tag{8.46}$$

Summing up (8.46) over all k leads to

$$\boldsymbol{h}^{\mathrm{H}}\boldsymbol{H}^{-\mathrm{H}}\boldsymbol{H}^{-1}\boldsymbol{h} \le N \tag{8.47}$$

i.e., we only have an *upper* bound on the Hermitian form inside the first logarithm of (8.44), but no lower bound. Moreover, the upper bound can also be attained by the choice $\boldsymbol{h} = \boldsymbol{H}\boldsymbol{1}$, which also satisfies the stronger conditions in (8.46). Therefore, the first logarithm in (8.44) can grow with $\log_2(1+N)$, whereas the subtrahend is bounded for any N since we have

$$1 \leq N\log_2(1+N^{-1}) = \log_2\left[(1+N^{-1})^N\right] \leq \log_2 \mathrm{e} = (\ln 2)^{-1} \approx 1.44 \quad \forall N \in \mathbb{N}.$$

Moreover, it can be shown that

$$N\log_2(1+N^{-1}) < \log_2(1+N) \quad \forall N > 1.$$

For $N = 1$, the asymptotic difference in (8.44) is either zero (if $|h| = |h_1|$) or negative (if $|h| < |h_1|$) which means that serving two users does not lead to a larger asymptotic sum rate than serving only one user.[9] But for $N \geq 2$, serving $N+1$ users instead of only N may depending on the channel $\boldsymbol{h}$ of the $N+1$-st user lead to a larger asymptotic sum rate even for the suboptimum, uniform power allocation over the $N+1$ users. This brings us to the following corollary:

Corollary 8.3 *In the vector broadcast channel with $N \geq 2$ antennas at the base station, serving more than N single-antenna users may be necessary to achieve the sum capacity.*

For general $K > N$, the determinant $\boldsymbol{H}\boldsymbol{Q}\boldsymbol{H}^{\mathrm{H}}$ in (8.40) with $\boldsymbol{H} = [\boldsymbol{h}_1, \ldots, \boldsymbol{h}_K]$ can be decomposed into the sum of several subdeterminants using a result from Costas-Santos in [148, Theorem 2.3]:

$$(-1)^{K+1} \cdot \left|\sum_{k=1}^{K} q_k \boldsymbol{h}_k \boldsymbol{h}_k^{\mathrm{H}}\right| = \sum_{\substack{i_1, i_2, \ldots, i_K = 0 \\ (i_1, \ldots, i_K) \neq (1, \ldots, 1)}}^{1} (-1)^{i_1 + \ldots + i_K} \left|\sum_{k=1}^{K} i_k q_k \boldsymbol{h}_k \boldsymbol{h}_k^{\mathrm{H}}\right| \quad (8.48)$$

This way, a determinant of K summands can be rewritten as the sum of several determinants with at most $K-1$ summands inside each determinant. Recursively applying (8.48) finally leads to an expression for the determinant $\left|\boldsymbol{H}\boldsymbol{Q}\boldsymbol{H}^{\mathrm{H}}\right|$, in which only determinants of a sum of N rank-one matrices arise. Thus, the power allocation can be separated from the respective channels and the determinant $\left|\boldsymbol{H}\boldsymbol{Q}\boldsymbol{H}^{\mathrm{H}}\right|$ is therefore a polynomial of degree N in K variables. Unfortunately, a closed form expression for the power allocation $q_1, \ldots, q_K$ that maximizes the asymptotic sum rate (where the identity inside the determinant has already been dropped) under a sum power constraint basically remains very difficult for $N > 2$, if not infeasible. Therefore, we analyze the case $N = 2$ explicitly, in which the determinant $\left|\boldsymbol{H}\boldsymbol{Q}\boldsymbol{H}^{\mathrm{H}}\right|$ reduces to a polynomial of degree 2 which can be written as a quadratic form using the power vector $\boldsymbol{q} := [q_1, \ldots, q_K]^{\mathrm{T}} \in \mathbb{R}_{+,0}^{K}$ and the symmetric $K \times K$ matrix $\boldsymbol{C}$:

[9] This statement remains valid even for an optimized (possibly nonuniform) power allocation, see Sect. 4.2.1.

$$\left|\boldsymbol{H}\boldsymbol{Q}\boldsymbol{H}^{\mathrm{H}}\right| = \frac{1}{2}\boldsymbol{q}^{\mathrm{T}}\boldsymbol{C}\boldsymbol{q} \quad \text{for } N = 2. \tag{8.49}$$

The maximization of the asymptotic sum capacity thus corresponds for $N = 2$ to the optimization

$$\underset{\boldsymbol{q}}{\text{maximize}} \frac{1}{2}\boldsymbol{q}^{\mathrm{T}}\boldsymbol{C}\boldsymbol{q} \quad \text{s.t.: } \mathbf{1}^{\mathrm{T}}\boldsymbol{q} \leq P_{\max} \quad \text{and} \quad \boldsymbol{q} \geq \mathbf{0} \tag{8.50}$$

whose associated Lagrangian function can be written as

$$\mathcal{L}(\boldsymbol{q}, \nu, \boldsymbol{\mu}) := \frac{1}{2}\boldsymbol{q}^{\mathrm{T}}\boldsymbol{C}\boldsymbol{q} - \nu\left[\mathbf{1}^{\mathrm{T}}\boldsymbol{q} - P_{\max}\right] + \boldsymbol{\mu}^{\mathrm{T}}\boldsymbol{q}$$

with the Lagrangian multipliers $\nu \geq 0$ and $\boldsymbol{\mu} \geq \mathbf{0}$. From the derivative of $\mathcal{L}(\boldsymbol{q}, \nu, \boldsymbol{\mu})$ with respect to $\boldsymbol{q}$, we see that the optimizer $\check{\boldsymbol{q}}$ of the maximization in (8.50) satisfies[10]

$$\boldsymbol{C}\check{\boldsymbol{q}} = \nu\mathbf{1} - \boldsymbol{\mu}. \tag{8.51}$$

If $\boldsymbol{C}$ is regular and if $\boldsymbol{C}^{-1}\mathbf{1}$ has only positive entries, the optimizer of (8.50) which maximizes the asymptotic sum capacity reads as

$$\check{\boldsymbol{q}} = \frac{P_{\max}}{\mathbf{1}^{\mathrm{T}}\boldsymbol{C}^{-1}\mathbf{1}} \cdot \boldsymbol{C}^{-1}\mathbf{1} \quad \text{for} \quad \boldsymbol{C}^{-1}\mathbf{1} > \mathbf{0}. \tag{8.52}$$

In other words, all K users are active if $\boldsymbol{C}$ is invertible and $\boldsymbol{C}^{-1}\mathbf{1}$ has only positive entries. In this case, we also conclude from (8.52) that the (weakly) optimum power allocation in the dual vector MAC that asymptotically achieves the sum capacity must be linear in $P_{\max}$. With $\check{\boldsymbol{q}}$ from (8.52) being valid, the asymptotic sum capacity can be written as

$$\frac{1}{2}\check{\boldsymbol{q}}^{\mathrm{T}}\boldsymbol{C}\check{\boldsymbol{q}} = \frac{P_{\max}^2}{2 \cdot \mathbf{1}^{\mathrm{T}}\boldsymbol{C}^{-1}\mathbf{1}}. \tag{8.53}$$

Although the powers of all K users linearly grow to infinity in (8.52), it should be noted that only two[11] individual rates grow with $\log_2 P_{\max}$ whereas the rates of the remaining users converge to a finite value as $P_{\max} \to \infty$.

If $\boldsymbol{C}$ is regular but $\boldsymbol{C}^{-1}\mathbf{1}$ has at least one entry that is not positive, at least one entry of $\boldsymbol{\mu}$ is nonzero and less than K (but at least N) users must be served in order to maximize the asymptotic sum capacity. Hence, at least one user must be inactive and we have to check whether $\boldsymbol{C}^{-1}\mathbf{1}$ is positive for the *reduced* system setup with

---

[10] Note that $\boldsymbol{C}$ is assumed to be symmetric.

[11] The rates of those two users grow to infinity that are encoded last. The user that is encoded last does not see any interference at all whereas the user that is encoded next to last sees interference from only one other user. However, it can be rendered harmless by means of filtering since the base station has $N = 2$ antennas.

only $K-1$ active users. Unfortunately, all cases in which one out of the K users is inactive have to be probed. If none of the K vectors $\boldsymbol{C}^{-1}\mathbf{1}$ of the reduced system is strictly positive, at least two users must be active at the optimum. Setting users inactive must be continued until only N users are active.

For $K=3$ users in the system, it is possible to give exact conditions for the channel vectors under which serving all three users achieves the asymptotic sum capacity. The determinant reformulation in (8.48) yields for $K=3$

$$\begin{aligned}\left|\boldsymbol{H}\boldsymbol{Q}\boldsymbol{H}^{\mathrm{H}}\right| &= \left|q_1\boldsymbol{h}_1\boldsymbol{h}_1^{\mathrm{H}}+q_2\boldsymbol{h}_2\boldsymbol{h}_2^{\mathrm{H}}\right|+\left|q_1\boldsymbol{h}_1\boldsymbol{h}_1^{\mathrm{H}}+q_3\boldsymbol{h}_3\boldsymbol{h}_3^{\mathrm{H}}\right|+\left|q_2\boldsymbol{h}_2\boldsymbol{h}_2^{\mathrm{H}}+q_3\boldsymbol{h}_3\boldsymbol{h}_3^{\mathrm{H}}\right|\\ &= q_1q_2c_{1,2}+q_1q_3c_{1,3}+q_2q_3c_{2,3} \qquad (8.54)\\ &= \frac{1}{2}\boldsymbol{q}^{\mathrm{T}}\boldsymbol{C}\boldsymbol{q}\end{aligned}$$

where

$$\boldsymbol{C}:=\begin{bmatrix}0 & c_{1,2} & c_{1,3}\\ c_{1,2} & 0 & c_{2,3}\\ c_{1,3} & c_{2,3} & 0\end{bmatrix}$$

and $c_{i,j}:=\left|[\boldsymbol{h}_i,\boldsymbol{h}_j][\boldsymbol{h}_i,\boldsymbol{h}_j]^{\mathrm{H}}\right|>0$ denotes the determinant of the channel Gram of the two users i and j which is positive since any two channel vectors are assumed to be linearly independent. The inverse of $\boldsymbol{C}$ can be computed as the ratio of the adjoint matrix and the determinant of $\boldsymbol{C}$ which turns out to be positive, i.e., $\boldsymbol{C}$ is regular:

$$\boldsymbol{C}^{-1}\mathbf{1}=\frac{\begin{bmatrix}-c_{2,3}^2 & c_{1,3}c_{2,3} & c_{1,2}c_{2,3}\\ c_{1,3}c_{2,3} & -c_{1,3}^2 & c_{1,2}c_{1,3}\\ c_{1,2}c_{2,3} & c_{1,2}c_{1,3} & -c_{1,2}^2\end{bmatrix}\cdot\mathbf{1}}{2c_{1,2}c_{1,3}c_{2,3}}=\frac{\begin{bmatrix}c_{2,3}(c_{1,3}+c_{1,2}-c_{2,3})\\ c_{1,3}(c_{2,3}+c_{1,2}-c_{1,3})\\ c_{1,2}(c_{2,3}+c_{1,3}-c_{1,2})\end{bmatrix}}{2c_{1,2}c_{1,3}c_{2,3}}.$$

Now if the three inequalities

$$\begin{aligned}c_{1,3}+c_{1,2}&>c_{2,3}\\ c_{2,3}+c_{1,2}&>c_{1,3}\\ c_{2,3}+c_{1,3}&>c_{1,2}\end{aligned}$$

hold simultaneously, all three users must be active in order to achieve the sum capacity asymptotically. Otherwise, only two users may be active with a uniform power allocation of $P_{\max}/2$ per user such that the asymptotic sum capacity reads as

$$\log_2\left|\boldsymbol{H}\boldsymbol{Q}\boldsymbol{H}^{\mathrm{H}}\right|=2\log_2 P_{\max}-2+\log_2\left(\max\{c_{1,2},c_{1,3},c_{2,3}\}\right).$$

Simulation results with i.i.d. zero-mean, unit-variance Gaussian channel entries reveal that all three users must be active in one out of four channel matrices on average.

For $K = 4$ users in the system, (8.48) can also be used to convert the single determinant of a matrix sum with four summands into a sum of determinants. For a shorter notation, we use the substitute $\boldsymbol{A}_k := q_k \boldsymbol{h}_k \boldsymbol{h}_k^{\mathrm{H}}$ and obtain

$$\begin{aligned}\left|\boldsymbol{H}\boldsymbol{Q}\boldsymbol{H}^{\mathrm{H}}\right| = &\left|\boldsymbol{A}_1 + \boldsymbol{A}_2 + \boldsymbol{A}_3\right| + \left|\boldsymbol{A}_1 + \boldsymbol{A}_2 + \boldsymbol{A}_4\right| + \left|\boldsymbol{A}_1 + \boldsymbol{A}_3 + \boldsymbol{A}_4\right| \\ &+ \left|\boldsymbol{A}_2 + \boldsymbol{A}_3 + \boldsymbol{A}_4\right| - \left|\boldsymbol{A}_1 + \boldsymbol{A}_2\right| - \left|\boldsymbol{A}_1 + \boldsymbol{A}_3\right| - \left|\boldsymbol{A}_1 + \boldsymbol{A}_4\right| \\ &- \left|\boldsymbol{A}_2 + \boldsymbol{A}_3\right| - \left|\boldsymbol{A}_2 + \boldsymbol{A}_4\right| - \left|\boldsymbol{A}_3 + \boldsymbol{A}_4\right|.\end{aligned}$$

Using (8.48) a second time in the style of (8.54) to further decompose the determinants containing three summands, we obtain

$$\begin{aligned}\left|\boldsymbol{H}\boldsymbol{Q}\boldsymbol{H}^{\mathrm{H}}\right| &= \left|\boldsymbol{A}_1 + \boldsymbol{A}_2\right| + \left|\boldsymbol{A}_1 + \boldsymbol{A}_3\right| + \left|\boldsymbol{A}_1 + \boldsymbol{A}_4\right| + \left|\boldsymbol{A}_2 + \boldsymbol{A}_3\right| + \left|\boldsymbol{A}_2 + \boldsymbol{A}_4\right| + \left|\boldsymbol{A}_3 + \boldsymbol{A}_4\right| \\ &= q_1 q_2 c_{1,2} + q_1 q_3 c_{1,3} + q_1 q_4 c_{1,4} + q_2 q_3 c_{2,3} + q_2 q_4 c_{2,4} + q_3 q_4 c_{3,4} \\ &= \frac{1}{2} \cdot \boldsymbol{q}^{\mathrm{T}} \boldsymbol{C} \boldsymbol{q}\end{aligned}$$

with

$$\boldsymbol{C} := \begin{bmatrix} 0 & c_{1,2} & c_{1,3} & c_{1,4} \\ c_{1,2} & 0 & c_{2,3} & c_{2,4} \\ c_{1,3} & c_{2,3} & 0 & c_{3,4} \\ c_{1,4} & c_{2,4} & c_{3,4} & 0 \end{bmatrix}.$$

All four users must be active to obtain the asymptotic sum capacity if $\boldsymbol{C}^{-1}\mathbf{1} > \mathbf{0}$. Simulation results with i.i.d. zero mean, unit-variance Gaussian channel entries reveal that all four users must be served in the asymptotic limit $P_{\max} \to \infty$ in about 1.9 % of the cases. Exactly three users must be active in 62.4 % of the cases and only two users must be served in 35.7 % of the cases.

For more than four users, (8.48) could also be used to find the matrix $\boldsymbol{C}$. However, it is more convenient to pursue a different approach which will finally reveal that the rank of $\boldsymbol{C}$ cannot become larger than four even if $K > 4$. For $N = 2$, the determinant in (8.49) can be expressed as

$$\left|\boldsymbol{H}\boldsymbol{Q}\boldsymbol{H}^{\mathrm{H}}\right| = \left(\sum_{k=1}^{K} |h_{k,1}|^2 q_k\right)\left(\sum_{k=1}^{K} |h_{k,2}|^2 q_k\right) - \left|\sum_{k=1}^{K} h_{k,2} h_{k,1}^* q_k\right|^2 \tag{8.55}$$

where $h_{k,i}$ denotes the i-th entry of the channel vector $\boldsymbol{h}_k$ of user k. Let $\boldsymbol{H}^{\mathrm{H}} = [\boldsymbol{a}_1, \boldsymbol{a}_2]$ have the two columns $\boldsymbol{a}_1 \in \mathbb{C}^K$ and $\boldsymbol{a}_2 \in \mathbb{C}^K$. Using the substitutes

$$\begin{aligned}\boldsymbol{c}_1 &:= \boldsymbol{a}_1 \odot \boldsymbol{a}_1^* \in \mathbb{R}^K \\ \boldsymbol{c}_2 &:= \boldsymbol{a}_2 \odot \boldsymbol{a}_2^* \in \mathbb{R}^K \\ \boldsymbol{d} &:= \boldsymbol{a}_1 \odot \boldsymbol{a}_2^* \in \mathbb{C}^K\end{aligned}$$

defined by the element-wise Schur product, the determinant in (8.55) reads as

$$
\begin{aligned}
|\boldsymbol{H}\boldsymbol{Q}\boldsymbol{H}^{\mathrm{H}}| &= \boldsymbol{q}^{\mathrm{T}}\boldsymbol{c}_1\boldsymbol{c}_2^{\mathrm{T}}\boldsymbol{q} - \boldsymbol{q}^{\mathrm{T}}\boldsymbol{d}\boldsymbol{d}^{\mathrm{H}}\boldsymbol{q} \\
&= \frac{1}{2}\boldsymbol{q}^{\mathrm{T}}(\boldsymbol{c}_1\boldsymbol{c}_2^{\mathrm{T}} + \boldsymbol{c}_2\boldsymbol{c}_1^{\mathrm{T}})\boldsymbol{q} - \boldsymbol{q}^{\mathrm{T}}\mathrm{Re}(\boldsymbol{d}\boldsymbol{d}^{\mathrm{H}})\boldsymbol{q} \\
&= \frac{1}{2}\boldsymbol{q}^{\mathrm{T}}\underbrace{\left(\boldsymbol{c}_1\boldsymbol{c}_2^{\mathrm{T}} + \boldsymbol{c}_2\boldsymbol{c}_1^{\mathrm{T}} - \boldsymbol{d}\boldsymbol{d}^{\mathrm{H}} - \boldsymbol{d}^*\boldsymbol{d}^{\mathrm{T}}\right)}_{=\boldsymbol{C}=\boldsymbol{C}^{\mathrm{T}}}\boldsymbol{q}
\end{aligned}
\tag{8.56}
$$

since $\mathrm{Im}(\boldsymbol{d}\boldsymbol{d}^{\mathrm{H}})$ is skew-symmetric and thus vanishes in the quadratic form. As the *symmetric* matrix $\boldsymbol{C}$ corresponds to the sum of four rank-one matrices, its rank is upper bounded by four, i.e., $\mathrm{rank}(\boldsymbol{C}) \leq 4$ irrespective of the number of users K. Hence, the solution in (8.52) does not exist for $K > 4$ since $\boldsymbol{C}$ is not invertible in this case. Moreover, it is unclear, whether (8.51) can hold with $\boldsymbol{\mu} = \boldsymbol{0}$ and $\boldsymbol{q} > \boldsymbol{0}$. We conjecture that this is not the case such that serving more than four users in the vector BC with an $N = 2$ antenna base station is conjectured not to achieve the asymptotic sum capacity. Simulation results support this hypothesis. Finally, the matrix $\boldsymbol{C}$ in (8.56) may also be written in terms of the channel matrix $\boldsymbol{H}$:

$$
\begin{aligned}
\boldsymbol{C} &= \boldsymbol{c}_1\boldsymbol{c}_2^{\mathrm{T}} + \boldsymbol{c}_2\boldsymbol{c}_1^{\mathrm{T}} - \boldsymbol{d}\boldsymbol{d}^{\mathrm{H}} - \boldsymbol{d}^*\boldsymbol{d}^{\mathrm{T}} \\
&= (\boldsymbol{a}_1 \odot \boldsymbol{a}_1^*)(\boldsymbol{a}_2 \odot \boldsymbol{a}_2^*)^{\mathrm{T}} + (\boldsymbol{a}_2 \odot \boldsymbol{a}_2^*)(\boldsymbol{a}_1 \odot \boldsymbol{a}_1^*)^{\mathrm{T}} \\
&\quad - (\boldsymbol{a}_1 \odot \boldsymbol{a}_2^*)(\boldsymbol{a}_1 \odot \boldsymbol{a}_2^*)^{\mathrm{H}} - (\boldsymbol{a}_1^* \odot \boldsymbol{a}_2)(\boldsymbol{a}_1 \odot \boldsymbol{a}_2^*)^{\mathrm{T}} \\
&= (\boldsymbol{a}_1\boldsymbol{a}_2^{\mathrm{T}}) \odot (\boldsymbol{a}_1^*\boldsymbol{a}_2^{\mathrm{H}}) + (\boldsymbol{a}_2\boldsymbol{a}_1^{\mathrm{T}}) \odot (\boldsymbol{a}_2^*\boldsymbol{a}_1^{\mathrm{H}}) \\
&\quad - (\boldsymbol{a}_1\boldsymbol{a}_2^{\mathrm{T}}) \odot (\boldsymbol{a}_2^*\boldsymbol{a}_1^{\mathrm{H}}) - (\boldsymbol{a}_2\boldsymbol{a}_1^{\mathrm{T}}) \odot (\boldsymbol{a}_1^*\boldsymbol{a}_2^{\mathrm{H}}) \\
&= (\boldsymbol{a}_1\boldsymbol{a}_2^{\mathrm{T}}) \odot (\boldsymbol{a}_1^*\boldsymbol{a}_2^{\mathrm{H}} - \boldsymbol{a}_2^*\boldsymbol{a}_1^{\mathrm{H}}) - (\boldsymbol{a}_2\boldsymbol{a}_1^{\mathrm{T}}) \odot (\boldsymbol{a}_1^*\boldsymbol{a}_2^{\mathrm{H}} - \boldsymbol{a}_2^*\boldsymbol{a}_1^{\mathrm{H}}) \\
&= (\boldsymbol{a}_1\boldsymbol{a}_2^{\mathrm{T}} - \boldsymbol{a}_2\boldsymbol{a}_1^{\mathrm{T}}) \odot (\boldsymbol{a}_1^*\boldsymbol{a}_2^{\mathrm{H}} - \boldsymbol{a}_2^*\boldsymbol{a}_1^{\mathrm{H}}) \\
&= (\boldsymbol{H}^{\mathrm{H}}\boldsymbol{e}_1\boldsymbol{e}_2^{\mathrm{T}}\boldsymbol{H}^* - \boldsymbol{H}^{\mathrm{H}}\boldsymbol{e}_2\boldsymbol{e}_1^{\mathrm{T}}\boldsymbol{H}^*) \odot (\boldsymbol{H}^{\mathrm{H}}\boldsymbol{e}_1\boldsymbol{e}_2^{\mathrm{T}}\boldsymbol{H}^* - \boldsymbol{H}^{\mathrm{H}}\boldsymbol{e}_2\boldsymbol{e}_1^{\mathrm{T}}\boldsymbol{H}^*)^* \\
&= \left(\boldsymbol{H}^{\mathrm{H}}\begin{bmatrix} 0 & 1 \\ -1 & 0 \end{bmatrix}\boldsymbol{H}^*\right) \odot \left(\boldsymbol{H}^{\mathrm{H}}\begin{bmatrix} 0 & 1 \\ -1 & 0 \end{bmatrix}\boldsymbol{H}^*\right)^*.
\end{aligned}
$$

Synopsis: In an overloaded vector broadcast channel with more users than transmit antennas at the base station, the analysis of the sum capacity remains a difficult task even in the high power regime. For an $N > 1$ antenna base station, however, it may be necessary to serve $N + 1$ users in order to obtain the asymptotic sum capacity. For $N = 2$ and either $K = 3$ or $K = 4$ users in the system, we can give exact conditions under which exactly three or four users have to be served in order to achieve the asymptotic sum capacity. Besides a closed form expression for the resulting asymptotic sum capacity, the (weakly) optimum power allocation in the dual MAC that leads to this sum rate are presented. Due to the structure of the asymptotic sum capacity computation problem, we conjecture that at most four users must be served by an $N = 2$ antenna base station in order to achieve the asymptotic sum capacity. Simulation results support this hypothesis.

8.2.3 MIMO Broadcast Channel with Linear Transceivers

When linear transceivers are considered, inter-user interference is completely treated as additional noise. As we know from Chap. 7, an algorithm that solves the weighted sum rate maximization problem to global optimality does not seem to exist in the linear case. Fortunately, the asymptotic behavior can be fully characterized under some conditions, for example when the base station is equipped with at least as many antennas as the user terminals have in sum. Contrary to the dirty paper coding setup, there may be at most as many active data streams with powers going to infinity as the base station has transmit antennas when $P_{\max} \rightarrow \infty$ and linear schemes are considered. These at most N active data streams must be distributed to the K users which leads to a user- and stream-selection problem if $M > N$. Otherwise, when $M \leq N$, every user must be served with exactly as many data streams as he has transmit antennas (full rank channels assumed) which will be proven as well. In Sect. 8.2.3.1, we show that at most $\operatorname{rank}(\boldsymbol{H}) \leq N$ streams may have powers that go to infinity to maximize the asymptotic weighted sum rate and that the power allocation must be linear. Given this limitation of the active data streams, the asymptotically optimum power allocation in the dual MAC is discussed in Sect. 8.2.3.2. For a system with enough degrees of freedom ($N \geq M$), the maximum asymptotic weighted sum rate and the corresponding BC transmit covariance matrices are derived in Sect. 8.2.3.3 and an overloaded system with $N < M$ is finally investigated in Sect. 8.2.3.4.

8.2.3.1 Limitation of Active Streams

To see that exactly $\operatorname{rank}(\boldsymbol{H})$ data streams must be active with powers going to infinity with a linear fraction of $P_{\max}$, we focus on the asymptotic pre-log factors of the rate expression of the K users. In the dual MAC, the rate of user k can be written as [see (7.11)]

$$R_k^{\mathrm{MAC}} = \log_2 \left|\mathbf{I} + \sum_{i=1}^{K} \boldsymbol{H}_i \boldsymbol{Q}_i \boldsymbol{H}_i^{\mathrm{H}}\right| - \log_2 \left|\mathbf{I} + \sum_{i \neq k} \boldsymbol{H}_i \boldsymbol{Q}_i \boldsymbol{H}_i^{\mathrm{H}}\right|. \quad (8.57)$$

The transmit covariance matrix of user i may again be represented in terms of the *reduced* eigenvalue decomposition

$$\boldsymbol{Q}_i = \boldsymbol{U}_i \boldsymbol{\Lambda}_i \boldsymbol{U}_i^{\mathrm{H}} \quad \forall i \in \{1, \dots, K\}$$

where $\boldsymbol{\Lambda}_i = \mathrm{diag}\{\lambda_{i,j}\}_{j=1}^{L_i} \in \mathbb{R}_{+,0}^{L_i \times L_i}$ contains the L_i positive eigenvalues as main diagonal elements and $\boldsymbol{U}_i \in \mathbb{C}^{M_i \times L_i}$ is (sub-)unitary with $\boldsymbol{U}_i^{\mathrm{H}} \boldsymbol{U}_i = \mathbf{I}_{L_i}$. Let

$$\boldsymbol{f}_{i,j} := \boldsymbol{H}_i \boldsymbol{U}_i \boldsymbol{e}_j \tag{8.58}$$

denote the product of the channel matrix of user i and his j-th unit-norm beamformer. Then, (8.57) can be rewritten as

$$R_k^{\mathrm{MAC}} = \log_2 \left| \mathbf{I} + \sum_{i=1}^{K} \sum_{j=1}^{L_i} \boldsymbol{f}_{i,j} \boldsymbol{f}_{i,j}^{\mathrm{H}} \lambda_{i,j} \right| - \log_2 \left| \mathbf{I} + \sum_{i \neq k} \sum_{j=1}^{L_i} \boldsymbol{f}_{i,j} \boldsymbol{f}_{i,j}^{\mathrm{H}} \lambda_{i,j} \right|. \tag{8.59}$$

For the eigenvalues of the transmit covariance matrices, we choose the parametrization

$$\lambda_{i,j} = a_{i,j} P_{\max}^{\tau_{i,j}} \quad \forall i \; \forall j \tag{8.60}$$

with $a_{i,j} > 0$ and $0 < \tau_{i,j} \leq 1$. Equivalently, we could make use of the more general power allocation $\lambda_{i,j} \doteq a_{i,j} P_{\max}^{\tau_{i,j}}$ without changing the following result. Clearly, the eigenvalues must satisfy the transmit power constraint

$$\sum_{i=1}^{K} \sum_{j=1}^{L_i} \lambda_{i,j} \leq P_{\max}$$

such that $\tau_{i,j}$ must be upper bound by one since a super-linear growth would consume more power than available. Let $r = \mathrm{rank}(\boldsymbol{H})$ denote the rank of the composite MAC channel matrix $\boldsymbol{H}$. Then it is possible to allocate exactly $L = r$ data streams to the K users and find unit norm beamformers $\boldsymbol{U}_1 \in \mathbb{C}^{M_1 \times L_1}, \dots, \boldsymbol{U}_K \in \mathbb{C}^{M_K \times L_K}$ such that the $L = r$ vectors[12]

$$\boldsymbol{f}_{1,1}, \dots, \boldsymbol{f}_{1,L_1}, \boldsymbol{f}_{2,1}, \dots, \boldsymbol{f}_{2,L_2}, \dots, \boldsymbol{f}_{K,1}, \dots, \boldsymbol{f}_{K,L_K} \tag{8.61}$$

are linearly independent and any subset of them is linearly independent as well. It is shown in Appendix A18 that the two logarithmic determinants in (8.59) have a pre-log factor that corresponds to the sum of the largest exponents $\tau_{\cdot,\cdot}$ whose corresponding vectors $\boldsymbol{f}_{\cdot,\cdot}$ are linearly independent and achieve the maximum rank of all involved vectors inside the determinant. Due to the choice of the vectors $\boldsymbol{f}_{\cdot,\cdot}$ in (8.61), the

[12] Note that $L_i = 0$ might be possible for some users i. In this case, these users are inactive and do not have any active data streams.

first determinant in (8.59) contains all $L = r$ linearly independent vectors such that its logarithmic determinant behaves for $P_{\max} \to \infty$ as [see Appendix A18]

$$\log_2 \left|\mathbf{I} + \sum_{i=1}^{K} \sum_{j=1}^{L_i} \boldsymbol{f}_{i,j} \boldsymbol{f}_{i,j}^{\mathrm{H}} \lambda_{i,j}\right| = \left(\sum_{i=1}^{K} \sum_{j=1}^{L_i} \tau_{i,j}\right) \log_2 P_{\max} + O(1).$$

The second logarithmic determinant that is subtracted in (8.59) contains only $L - L_k$ vectors $\boldsymbol{f}_{\cdot,\cdot}$ which of course are also linearly independent. Thus, the minuend in (8.59) behaves like

$$\log_2 \left|\mathbf{I} + \sum_{i \neq k} \sum_{j=1}^{L_i} \boldsymbol{f}_{i,j} \boldsymbol{f}_{i,j}^{\mathrm{H}} \lambda_{i,j}\right| = \left(\sum_{i \neq k} \sum_{j=1}^{L_i} \tau_{i,j}\right) \log_2 P_{\max} + O(1).$$

As a consequence, the rate of user k under linear filtering reads for a power allocation (8.60) and exactly $L = \operatorname{rank}(\boldsymbol{H})$ active streams that are properly distributed to the K users as

$$R_k^{\mathrm{MAC}} = \left(\sum_{j=1}^{L_k} \tau_{k,j}\right) \log_2 P_{\max} + O(1). \tag{8.62}$$

Since the exponents $\tau_{\cdot,\cdot}$ of the individual streams are not coupled, the maximum pre-log factor of every user is obtained for

$$\tau_{k,j} = 1 \quad \forall k \in \{1, \ldots, K\} \, \forall j \in \{1, \ldots, L_k\}$$

i.e., for a *linear* power allocation. In this case, the rate of user k has the following high-power behavior:

$$R_k^{\mathrm{MAC}} = L_k \log_2 P_{\max} + O(1) \quad \forall k \tag{8.63}$$

In other words, the pre-log factor of a particular user corresponds to his number of active data streams and the pre-log factor of every active data stream is simply one.

In the following, we prove that adding a further data stream ($L' = \operatorname{rank}(\boldsymbol{H}) + 1$) deteriorates the pre-log factors of at least one other data stream and is therefore suboptimal. This finally shows that exactly $L = \operatorname{rank}(\boldsymbol{H})$ data streams must be active with powers that linearly grow in $P_{\max}$.

In the system with $\operatorname{rank}(\boldsymbol{H}) + 1$ active streams, users $1, \ldots, K$ will have $L_1', \ldots, L_K'$ streams. Without loss of generality, we assume that the $r + 1$-st data stream will be added to user K. Obviously, this only makes sense if he had less active streams than antennas so far, i.e., $L_K < M_K$ must have been valid. All user others leave their number of streams unchanged such that

$$L_k' = \begin{cases} L_k & \text{for } k \neq K \\ L_K + 1 & \text{for } k = K. \end{cases}$$

Since the rank of the channel matrix $\boldsymbol{H}$ is r and we yet have $L' = \sum_{k=1}^{K} L'_k = r+1$ active streams, the newly added vector $\boldsymbol{f}_{K,L_{K+1}}$ can be written as a linear combination of all other vectors $\boldsymbol{f}_{1,1}, \ldots, \boldsymbol{f}_{K,L_K}$. From Appendix A18, we know that the $r+1$-st data stream will reduce the pre-log factor of all those streams whose vectors $\boldsymbol{f}_{\cdot,\cdot}$ are necessary to represent $\boldsymbol{f}_{K,L_K+1}$ as a linear combination of them. Therefore, the smallest degradation is obtained if $\boldsymbol{f}_{K,L_K+1}$ is linearly dependent with respect to exactly *one* other vector $\boldsymbol{f}_{\cdot,\cdot}$. Without loss of generality, we assume that $\boldsymbol{f}_{K,L_K+1}$ is a scaled version of $\boldsymbol{f}_{1,1}$. The first determinant in (8.59) consists for L_K replaced by L'_K of $r+1$ vectors $\boldsymbol{f}_{\cdot,\cdot}$ from which we have to pick those r ones that are linearly independent and have the largest sum of the involved exponents $\tau_{\cdot,\cdot}$, see Appendix A18. Hence, we have to select all of them except either $\boldsymbol{f}_{1,1}$ or $\boldsymbol{f}_{K,L_K+1}$ which depends on whether $\tau_{1,1}$ is larger than τ_{K,L_K+1} or not. Thus, the first logarithmic determinant in (8.59) reads as[13]

$$\log_2 \left| \mathbf{I} + \sum_{i=1}^{K} \sum_{j=1}^{L'_i} \boldsymbol{f}_{i,j} \boldsymbol{f}_{i,j}^{\mathrm{H}} \lambda_{i,j} \right| = \left(\max\{\tau_{1,1}, \tau_{K,L_K+1}\} + \sum_{j=2}^{L_1} \tau_{1,j} + \sum_{i=2}^{K} \sum_{j=1}^{L_i} \tau_{i,j} \right) \log_2 P_{\max} + O(1). \quad (8.64)$$

For the first user with $k = 1$, the second logarithmic determinant in (8.59) is given by

$$\log_2 \left| \mathbf{I} + \sum_{i \neq 1} \sum_{j=1}^{L'_i} \boldsymbol{f}_{i,j} \boldsymbol{f}_{i,j}^{\mathrm{H}} \lambda_{i,j} \right| = \left(\tau_{K,L_K+1} + \sum_{i=2}^{K} \sum_{j=1}^{L_i} \tau_{i,j} \right) \log_2 P_{\max} + O(1)$$

since all involved vectors $\boldsymbol{f}_{\cdot,\cdot}$ are linearly independent (user one is missing). As the rate of the first user in the system with $r+1$ active data streams behaves as the difference of above two expressions, we have

$$R_1^{\mathrm{MAC}} = \left([\tau_{1,1} - \tau_{K,L_K+1}]_+ + \sum_{j=2}^{L_1} \tau_{1,j} \right) \log_2 P_{\max} + O(1) \quad (8.65)$$

with $[\cdot]_+ = \max\{0, \cdot\}$. The pre-log factors of the rates of users $2, \ldots, K-1$ are not affected from the $r+1$-st data stream since $\boldsymbol{f}_{K,L_K+1}$ is a linear combination of $\boldsymbol{f}_{1,1}$ alone, i.e.,

$$R_k^{\mathrm{MAC}} = \left(\sum_{j=1}^{L_k} \tau_{k,j} \right) \log_2 P_{\max} + O(1) \quad \forall k \in \{2, \ldots, K-1\}. \quad (8.66)$$

[13] For the system with L' instead of L streams, L_K must be replaced by L'_K in (8.59) for the exact description of the achieved rates since user K now has $L'_K = L_K + 1$ data streams.

However, the asymptotic high power behavior of user K is again influenced by his additional stream. For his rate in (8.59) with L_K replaced by L_K+1, the minuend is given by [cf. Appendix A18]

$$\log_2 \Big|\mathbf{I} + \sum_{i\neq K}\sum_{j=1}^{L_i'} \boldsymbol{f}_{i,j}\boldsymbol{f}_{i,j}^{\mathrm{H}}\lambda_{i,j}\Big| = \Big(\sum_{i=1}^{K-1}\sum_{j=1}^{L_i}\tau_{i,j}\Big)\log_2 P_{\max} + O(1) \qquad (8.67)$$

since the vectors $\boldsymbol{f}_{\cdot,\cdot}$ of the first $K-1$ users are linearly independent. We obtain the asymptotic behavior of the rate of user K as the difference of (8.64) and (8.67):

$$R_K^{\mathrm{MAC}} = \Big([\tau_{K,L_K+1} - \tau_{1,1}]_+ + \sum_{j=1}^{L_K}\tau_{K,j}\Big)\log_2 P_{\max} + O(1) \qquad (8.68)$$

From (8.65), (8.66), and (8.68) we infer that the exponents $\tau_{2,j},\ldots,\tau_{K-1,j}$ must be one for all j in order to maximize the asymptotic weighted sum rate such that the pre-log factor of the rate of every user $k \in \{2,\ldots,K-1\}$ is equal to his number of active streams $L_k' = L_k$. For the first user, the exponents $\tau_{1,2},\ldots,\tau_{1,L_1}$ must also be one in order to maximize (8.65). Equivalently, the first $L_K'-1$ exponents $\tau_{K,1},\ldots,\tau_{K,L_K}$ of user K have to be one to maximize (8.68). It remains to choose the values of $\tau_{1,1}$ and τ_{K,L_K+1}. The part of the asymptotic pre-log factor of the weighted sum rate $\sum_{k=1}^{K} w_k R_k^{\mathrm{MAC}}$ that depends on $\tau_{1,1}$ and τ_{K,L_K+1} reads by means of (8.65) and (8.68) as

$$\begin{aligned} & w_1\big[\tau_{1,1} - \tau_{K,L_K+1}\big]_+ + w_K\big[\tau_{K,L_K+1} - \tau_{1,1}\big]_+ \\ & = \begin{cases} w_1(\tau_{1,1} - \tau_{K,L_K+1}) & \text{for } \tau_{1,1} \geq \tau_{K,L_K+1} \\ w_K(\tau_{K,L_K+1} - \tau_{1,1}) & \text{for } \tau_{1,1} < \tau_{K,L_K+1}. \end{cases} \end{aligned}$$

So for $w_1 > w_K$, the exponent $\tau_{1,1}$ of the first stream must be set to one, whereas the exponent τ_{K,L_K+1} of the last stream should be made as small as possible. Since it has to remain positive, $\tau_{1,1} - \tau_{K,L_K+1}$ is smaller than $\tau_{1,1}$ such that adding the $r+1$-st stream reduces the pre-log factor of the very first stream from one to $1-\tau_{K,L_K+1} < 1$, see (8.65). The pre-log factor of the very last stream is equal to $[\tau_{K,L_K+1}-\tau_{1,1}]_+ = 0$, see (8.68), i.e., this last stream does not contribute to $\log_2 P_{\max}$. In other words, for $w_1 > w_K$, adding the $r+1$-st stream reduces the asymptotic weighted sum rate by reducing its pre-log factor.

If $w_1 < w_K$, the exponent τ_{K,L_K+1} must be set to one and $\tau_{1,1}$ should be made as small as possible. However, it needs to remain positive since we assume $r+1$ active streams with powers going to infinity when $P_{\max} \to \infty$. Since $\tau_{K,L_K+1} > \tau_{1,1}$, the pre-log factor of the first data stream is $[\tau_{1,1}-\tau_{K,L_K+1}]_+ = 0$, see (8.65). The newly added stream has a pre-log factor of $\tau_{K,L_K+1} - \tau_{1,1} < 1$. However, a pre-log factor of one could be achieved for the last stream if the very first stream was completely switched off, i.e., if again only $\operatorname{rank}(\boldsymbol{H}) = r$ streams were active. For $w_1 = w_K$, both argumentations are valid.

Summing up, having more than $r = \operatorname{rank}(\boldsymbol{H})$ active streams with powers going to infinity as $P_{\max} \to \infty$ does not achieve the maximum asymptotic weighted sum rate, which leads to the following theorem:

Theorem 8.10 *Given a MIMO broadcast channel with linear transceivers, the maximum asymptotic weighted sum rate is obtained by exactly* $L = \operatorname{rank}(\boldsymbol{H}^{BC}) = \operatorname{rank}(\boldsymbol{H})$ *active data streams with a linear power allocation. In this case, every of the* L *data streams has a pre-log factor of one. If more than* $\operatorname{rank}(\boldsymbol{H}^{BC})$ *streams are active, neither a linear nor a sub-linear power allocation achieves the maximum asymptotic weighted sum rate.*

Note that this is in sharp contrast to the dirty paper coding scheme where more than $\operatorname{rank}(\boldsymbol{H})$ streams might be necessary to obtain the maximum sum capacity for example, see Sect. 8.2.2.2.

Synopsis: In order to maximize the asymptotic weighted sum rate in a MIMO BC with linear transceivers, exactly $\operatorname{rank}(\boldsymbol{H}^{\mathrm{BC}})$ data streams must be active with a *linear* power allocation. If more than $\operatorname{rank}(\boldsymbol{H}^{\mathrm{BC}})$ streams are dedicated to the users, the performance degrades and the maximum weighted sum rate cannot be achieved in the asymptotic limit. To the best of our knowledge, a proof for the suboptimality of a sub-linear power allocation to possibly more than $\operatorname{rank}(\boldsymbol{H}^{\mathrm{BC}})$ streams has so far not been available in the literature.

8.2.3.2 Asymptotically Optimum Power Allocation

Knowing that the asymptotically optimum power allocation must be linear and that exactly $\operatorname{rank}(\boldsymbol{H})$ streams must be active, we now derive the fractions of the available transmit power $P_{\max}$ that the individual streams must share in order to achieve the maximum asymptotic weighted sum rate. This asymptotically optimum power allocation in the dual MAC is done for a fixed stream allocation $L_1, \dots, L_K$ which satisfies $L = \sum_{k=1}^{K} L_k = \operatorname{rank}(\boldsymbol{H})$. Moreover, the L vectors $\boldsymbol{f}_{\cdot,\cdot}$ that result from the normalized precoders multiplied by the MAC channel matrices must be linearly independent, see the preceding subsection. How these $L = \operatorname{rank}(\boldsymbol{H})$ streams must be distributed to the users will be discussed afterwards in this subsection.

The data rate of user k in the dual MAC can be written in terms of the precoders $\boldsymbol{T}_1, \dots, \boldsymbol{T}_K$ via

$$R_k^{\mathrm{MAC}} = \log_2 \left| \mathbf{I}_{L_k} + \boldsymbol{T}_k^{\mathrm{H}} \boldsymbol{H}_k^{\mathrm{H}} \boldsymbol{X}_k^{-1} \boldsymbol{H}_k \boldsymbol{T}_k \right| \tag{8.69}$$

where $\boldsymbol{T}_k \in \mathbb{C}^{M_k \times L_k}$ $\forall k$ and the noise plus interference covariance matrix $\boldsymbol{X}_k$ seen by user k is defined via

$$\boldsymbol{X}_k := \mathbf{I}_N + \sum_{i \neq k} \boldsymbol{H}_i \boldsymbol{T}_i \boldsymbol{T}_i^{\mathrm{H}} \boldsymbol{H}_i^{\mathrm{H}}. \tag{8.70}$$

Due to above assumption on the distribution of the $L = \text{rank}(\boldsymbol{H})$ streams to the K users and the linear power allocation, the matrix $\boldsymbol{X}_k$ features $L - L_k$ eigenvalues that tend to infinity when $P_{\max} \to \infty$. Having found out that all L_k eigenvalues of $\boldsymbol{T}_k^{\mathrm{H}} \boldsymbol{T}_k$ must scale linearly in $P_{\max}$ for all users k, the precoders can be parametrized via

$$\boldsymbol{T}_k = \sqrt{p_k} \cdot \boldsymbol{T}'_k = \sqrt{a_k P_{\max}} \cdot \boldsymbol{T}'_k \quad \text{with} \quad \|\boldsymbol{T}'_k\|_{\mathrm{F}}^2 = 1 \quad \forall k \tag{8.71}$$

where $p_k = a_k P_{\max}$ denotes the power that is allocated to user k in the dual MAC and the power fractions $a_1, \ldots, a_K$ conform to[14]

$$\sum_{k=1}^{K} a_k = 1 \quad \text{and} \quad 0 \leq a_k \leq 1 \; \forall k.$$

Then, the rate of user k from (8.69) can be written as

$$R_k^{\mathrm{MAC}} = \log_2 \left| \mathbf{I}_{L_k} + P_{\max} \cdot a_k \boldsymbol{T}_k'^{\mathrm{H}} \boldsymbol{H}_k^{\mathrm{H}} \Big(\mathbf{I}_N + P_{\max} \sum_{i \neq k} a_i \boldsymbol{H}_i \boldsymbol{T}'_i \boldsymbol{T}_i'^{\mathrm{H}} \boldsymbol{H}_i^{\mathrm{H}} \Big)^{-1} \boldsymbol{H}_k \boldsymbol{T}'_k \right| \tag{8.72}$$

and it is shown in (A59) in Appendix A19 that its asymptotic behavior reads as

$$R_k^{\mathrm{MAC}} = L_k \log_2 P_{\max} + L_k \log_2 a_k + c_k + o(1) \tag{8.73}$$

and is independent of $\{a_1, \ldots, a_K\} \setminus \{a_k\}$ as long as $a_i > 0$ whenever $L_i > 0$ for all i. The scalar c_k is finite and does not depend on the power allocation. Instead, c_k only depends on the channel matrix $\boldsymbol{H}$ and on the choice of the *normalized* precoders $\boldsymbol{T}'_1, \ldots, \boldsymbol{T}'_K$, see Appendix A19. Hence, (8.73) shows a refined version of the $O(1)$ term in (8.63).

The power allocation that achieves the maximum asymptotic weighted sum rate is now obtained from the following optimization[15]:

$$\underset{a_1, \ldots, a_K}{\text{maximize}} \sum_{k=1}^{K} w_k L_k \log_2 a_k \quad \text{s.t.:} \quad \sum_{k=1}^{K} a_k = 1$$

The solution to above optimization is given by

---

[14] The fraction a_i of some user i may only be zero if $L_i = 0$, i.e., if user i does not have any active stream at all.

[15] The constraints on the positivity of $a_1, \ldots, a_K$ are not required since they are inactive anyway.

$$a_k = \frac{w_k L_k}{\sum_{i=1}^{K} w_i L_i} \quad \forall k. \tag{8.74}$$

Hence, the fraction of power that must be allocated to user k in order to maximize the asymptotic weighted sum rate under a fixed stream allocation solely depends on the number of streams per user and the user weights. This leads to the following theorem:

Theorem 8.11 *Given a stream allocation $L_1, \ldots, L_K$ and normalized precoders $\boldsymbol{T}'_k$ which satisfy*

$$rank\big([\boldsymbol{H}_1\boldsymbol{T}'_1, \ldots, \boldsymbol{H}_K\boldsymbol{T}'_K]\big) = \sum_{k=1}^{K} L_k$$

the dual MAC power allocation to user k that maximizes the asymptotic weighted sum rate reads as

$$p_k = \frac{w_k L_k}{\sum_{i=1}^{K} w_i L_i} P_{max} \quad \forall k.$$

We will see in the next subsection that this type of power allocation achieves the maximum asymptotic weighted sum rate in the original MIMO BC as well.

Up to now, we have proven in the preceding section that exactly $\mathrm{rank}(\boldsymbol{H})$ data streams must be active with a linear power allocation and we derived the optimum fraction of $P_{\max}$ that must be assigned to every user in this subsection. It remains to find the optimum *normalized* precoders and to perform the user selection, i.e., how many streams must be allocated to which user. For the solution of these two problems, we distinguish the two cases of enough degrees of freedom at the base station ($N \geq M$) in Sect. 8.2.3.3 and of an overloaded system ($N < M$) in Sect. 8.2.3.4.

8.2.3.3 Enough Degrees of Freedom

In this subsection, we deal with the system setup in which the base station has at least as many antennas as the user terminals have in sum, i.e., $N \geq M$. Moreover, we assume that the composite channel matrix $\boldsymbol{H} \in \mathbb{C}^{N \times M}$ has full rank M. A rank-deficient scenario is similar to the overloaded case which is discussed in the next subsection.

We know from Sect. 8.2.3.1 that exactly $\mathrm{rank}(\boldsymbol{H}) = M$ streams must be active in order to maximize the asymptotic weighted sum rate. As a consequence, as many data streams must be dedicated to a user as he has antennas, i.e., $L_k = M_k \ \forall k$. In terms of the precoders $\boldsymbol{T}_1, \ldots, \boldsymbol{T}_K$, the rate of user k can be written as

$$R_k^{\mathrm{MAC}} = \log_2 \left| \mathbf{I}_{M_k} + \boldsymbol{T}_k^{\mathrm{H}} \boldsymbol{H}_k^{\mathrm{H}} \Big(\mathbf{I}_N + \sum_{i \neq k} \boldsymbol{H}_i \boldsymbol{T}_i \boldsymbol{T}_i^{\mathrm{H}} \boldsymbol{H}_i^{\mathrm{H}} \Big)^{-1} \boldsymbol{H}_k \boldsymbol{T}_k \right|$$
$$= -\log_2 \left| \mathbf{I}_{M_k} - \boldsymbol{T}_k^{\mathrm{H}} \boldsymbol{H}_k^{\mathrm{H}} \boldsymbol{X}^{-1} \boldsymbol{H}_k \boldsymbol{T}_k \right| \tag{8.75}$$

where the second line follows from the matrix inversion lemma in Appendix A2 and the substitution

$$\boldsymbol{X} := \mathbf{I}_N + \sum_{i=1}^{K} \boldsymbol{H}_i \boldsymbol{T}_i \boldsymbol{T}_i^{\mathrm{H}} \boldsymbol{H}_i^{\mathrm{H}} = \mathbf{I}_N + \boldsymbol{H}\boldsymbol{T}\boldsymbol{T}^{\mathrm{H}}\boldsymbol{H}^{\mathrm{H}}$$

with the block-diagonal matrix $\boldsymbol{T} := \mathrm{blockdiag}\{\boldsymbol{T}_k\}_{k=1}^K \in \mathbb{C}^{M \times M}$. Using the matrix-valued generalization $\boldsymbol{E}_k$ of the canonical unit vector which is defined via

$$\boldsymbol{E}_k^{\mathrm{T}} = \left[\mathbf{0}, \ldots, \mathbf{0}, \mathbf{I}_{M_k}, \mathbf{0}, \ldots, \mathbf{0} \right] \in \{0, 1\}^{M_k \times M} \tag{8.76}$$

with the identity matrix at the k-th block, the rate of user k in (8.75) can alternatively be represented as

$$R_k^{\mathrm{MAC}} = -\log_2 \left| \boldsymbol{E}_k^{\mathrm{T}} (\mathbf{I}_M - \boldsymbol{T}^{\mathrm{H}} \boldsymbol{H}^{\mathrm{H}} \boldsymbol{X}^{-1} \boldsymbol{H} \boldsymbol{T}) \boldsymbol{E}_k \right|$$
$$= -\log_2 \left| \boldsymbol{E}_k^{\mathrm{T}} (\mathbf{I}_M + \boldsymbol{T}^{\mathrm{H}} \boldsymbol{H}^{\mathrm{H}} \boldsymbol{H} \boldsymbol{T})^{-1} \boldsymbol{E}_k \right|. \tag{8.77}$$

Due to the linear power allocation of every of the M active streams the identity matrix inside the inverse of (8.77) may be neglected without losing strong asymptotic equivalence. In combination with the fact that $\boldsymbol{T}_1, \ldots, \boldsymbol{T}_K$ are square matrices, this leads to the asymptotic expression

$$R_k^{\mathrm{MAC}} \cong -\log_2 \left| \boldsymbol{E}_k^{\mathrm{T}} (\boldsymbol{T}^{\mathrm{H}} \boldsymbol{H}^{\mathrm{H}} \boldsymbol{H} \boldsymbol{T})^{-1} \boldsymbol{E}_k \right|$$
$$= -\log_2 \left| \boldsymbol{T}_k^{-1} \boldsymbol{E}_k^{\mathrm{T}} (\boldsymbol{H}^{\mathrm{H}} \boldsymbol{H})^{-1} \boldsymbol{E}_k \boldsymbol{T}_k^{-\mathrm{H}} \right| \tag{8.78}$$
$$= \log_2 \left| \boldsymbol{Q}_k \right| - \log_2 \left| \boldsymbol{E}_k^{\mathrm{T}} (\boldsymbol{H}^{\mathrm{H}} \boldsymbol{H})^{-1} \boldsymbol{E}_k \right|$$

from which we conclude that the rate of user k asymptotically only depends on its own transmit covariance matrix $\boldsymbol{Q}_k$ and the composite channel matrix. From Sect. 8.2.3.2, we know how much power $p_k = \mathrm{tr}(\boldsymbol{Q}_k)$ must be allocated to user k in order to achieve the maximum weighted sum rate. For a fixed trace, the determinant of the matrix $\boldsymbol{Q}_k$ is maximized if all eigenvalues are identical which follows from the arithmetic-geometric mean inequality. This leads to the transmit covariance matrices [see (8.74)]

$$\boldsymbol{Q}_k = \frac{p_k}{M_k} \cdot \mathbf{I}_{M_k} = \frac{w_k \cdot P_{\max}}{\sum_{i=1}^K w_i M_i} \mathbf{I}_{M_k} \quad \forall k \in \{1, \ldots, K\} \tag{8.79}$$

which achieve the maximum weighted sum rate in the dual MAC asymptotically. Interestingly, the MAC covariance matrices in (8.79) coincide with those from (8.26), i.e., the *same* set of MAC transmit covariance matrices asymptotically achieves the maximum weighted sum rate irrespective of whether dirty paper coding or linear filtering is used. Clearly, the two obtained utilities are different in general. Inserting (8.79) into the asymptotic rate expression of user k in (8.78) leads to the asymptotic individual rate

$$R_k^{\mathrm{MAC}} \cong M_k \log_2 P_{\max} + M_k \log_2 \frac{w_k}{\sum_{i=1}^K w_i M_i} - \log_2 \left| \boldsymbol{E}_k^{\mathrm{T}} \left(\boldsymbol{H}^{\mathrm{H}} \boldsymbol{H}\right)^{-1} \boldsymbol{E}_k \right|. \quad (8.80)$$

If all weights are set to one ($\boldsymbol{w} = \mathbf{1}$), the transmit covariance matrix in (8.79) simplifies to $\boldsymbol{Q}_k = \frac{P_{\max}}{M} \mathbf{I}_{M_k}$ and the maximum sum 'capacity' under linear filtering is strongly asymptotically equivalent to

$$C^{\mathrm{Linear}}(P_{\max}) \cong M \log_2 P_{\max} - M \log_2 M - \sum_{k=1}^K \log_2 \left| \boldsymbol{E}_k^{\mathrm{T}} \left(\boldsymbol{H}^{\mathrm{H}} \boldsymbol{H}\right)^{-1} \boldsymbol{E}_k \right|. \quad (8.81)$$

This expression for the asymptotic sum 'capacity' of linear filtering only depends on the composite channel matrix and the available transmit power $P_{\max}$ and has first been published in [22]. When the MAC channels are replaced by the corresponding BC channels and the BC noise covariance matrices via (8.29), the asymptotic sum 'capacity' reads as

$$\begin{aligned} C^{\mathrm{Linear}}(P_{\max}) \cong{}& M \log_2 P_{\max} - M \log_2 M - \sum_{k=1}^K \log_2 \left| \boldsymbol{E}_k^{\mathrm{T}} \left(\boldsymbol{H}^{\mathrm{BC}} \boldsymbol{H}^{\mathrm{BC,H}}\right)^{-1} \boldsymbol{E}_k \right| \\ & - \sum_{k=1}^K \log_2 \left| \boldsymbol{C}_{\eta_k} \right|. \end{aligned} \quad (8.82)$$

Contrary to the asymptotic sum capacity expression for dirty paper coding in (8.31), the asymptotic sum 'capacity' of linear filtering does depend on the distribution of the M antennas to the K users. Thanks to the closed form expression of the sum 'capacity' in (8.82) for a fixed channel realization, channel-averaged results can quickly be simulated for any fading statistic. In Sect. 8.3.4, we will see that analytical expressions for the asymptotic sum 'capacity' are possible under semi-correlated Rayleigh fading.

Compared to the asymptotic sum capacity with dirty paper coding [cf. (8.31)]

$$C^{\mathrm{DPC}}(P_{\max}) \cong M \log_2 P_{\max} - M \log_2 M - \log_2 \left| \left(\boldsymbol{H}^{\mathrm{BC}} \boldsymbol{H}^{\mathrm{BC,H}}\right)^{-1} \right| - \sum_{k=1}^K \log_2 \left| \boldsymbol{C}_{\eta_k} \right|$$

the asymptote of linear transceivers has to face a reduction of the rate offset which reads as

$$C^{\mathrm{DPC}} - C^{\mathrm{Linear}} \cong \sum_{k=1}^{K} \log_2 \left| \boldsymbol{E}_k^{\mathrm{T}} \left(\boldsymbol{H}^{\mathrm{BC}} \boldsymbol{H}^{\mathrm{BC,H}} \right)^{-1} \boldsymbol{E}_k \right| - \log_2 \left| \left(\boldsymbol{H}^{\mathrm{BC}} \boldsymbol{H}^{\mathrm{BC,H}} \right)^{-1} \right|. \tag{8.83}$$

Obviously, this difference is nonnegative due to Fischer's inequality [146, Theorem 7.8.3] and vanishes, if $\boldsymbol{H}^{\mathrm{BC}} \boldsymbol{H}^{\mathrm{BC,H}}$ is block-diagonal, i.e., if the channels are orthogonal and do not introduce inter-user interference.[16]

In the following, the scaled identity transmit covariance matrices of the dual MAC are converted back to the broadcast channel using the rate duality for linear transceivers from Sect. 3.3. According to (8.79), the MAC precoders that achieve the maximum asymptotic weighted sum rate can be chosen as

$$\boldsymbol{T}_k = \sqrt{\frac{p_k}{M_k}} \cdot \mathbf{I}_{M_k} \quad \forall k. \tag{8.84}$$

For the duality conversion in (3.77), the matched filter receivers in the MAC must be computed via (3.73):

$$\boldsymbol{G}_k^{\mathrm{MF}} = \sqrt{\frac{p_k}{M_k}} \boldsymbol{H}_k^{\mathrm{H}} \Big(\mathbf{I}_N + \sum_{i \neq k} \boldsymbol{H}_i \boldsymbol{H}_i^{\mathrm{H}} \frac{p_i}{M_i} \Big)^{-1}$$

In analogy to the dirty paper coding case in (8.33), the matched filter $\boldsymbol{G}_k^{\mathrm{MF}}$ of user k is strongly asymptotically equivalent to

$$\begin{aligned} \boldsymbol{G}_k^{\mathrm{MF}} &\cong \sqrt{\frac{p_k}{M_k}} \cdot \boldsymbol{H}_k^{\mathrm{H}} \boldsymbol{\Pi}_k^{\perp} \\ &= \sqrt{\frac{p_k}{M_k}} \cdot \bar{\boldsymbol{H}}_k^{\mathrm{H}} =: \boldsymbol{G}_k^{\prime\mathrm{MF}} \end{aligned} \tag{8.85}$$

where $\boldsymbol{\Pi}_k^{\perp}$ here denotes the projector onto the null space of the Hermitian channels of *all* users $i \neq k$ and $\bar{\boldsymbol{H}}_k = \boldsymbol{\Pi}_k^{\perp} \boldsymbol{H}_k$. In turn, we use the asymptotic receive filters $\boldsymbol{G}_1^{\prime\mathrm{MF}}, \ldots, \boldsymbol{G}_K^{\prime\mathrm{MF}}$ for the duality conversion instead of the exact matched filters. This way, the duality conversion drastically simplifies and can be computed in closed form without losing the same high-power asymptote in the original BC.

For the MAC to BC conversion for linear transceivers from Sect. 3.3.2, we need the eigenvalue decomposition of the product $\boldsymbol{G}_k^{\prime\mathrm{MF}} \boldsymbol{H}_k \boldsymbol{T}_k$ with $\boldsymbol{T}_k$ from (8.84) and $\boldsymbol{G}_k^{\prime\mathrm{MF}}$ from (8.85):

[16] More precisely, the introduced inter-user interference in the BC can be canceled at the receivers by means of filtering without an increase of the noise power and without a reduction of the desired signal's quality.

$$\begin{aligned}
G_k^{\prime\mathrm{MF}} H_k T_k &= \frac{p_k}{M_k} \cdot H_k^{\mathrm{H}} \Pi_k^{\perp} H_k = G_k^{\prime\mathrm{MF}} G_k^{\prime\mathrm{MF},\mathrm{H}} \\
&= \frac{p_k}{M_k} \cdot \left[E_k^{\mathrm{T}} (H^{\mathrm{H}} H)^{-1} E_k \right]^{-1} \\
&= V_k \Xi_k V_k^{\mathrm{H}} = V_k \Sigma_k V_k^{\mathrm{H}} \cdot \frac{p_k}{M_k}
\end{aligned} \tag{8.86}$$

The second line follows from Appendix A17 for example. Note that V_k is unitary whereas Ξ_k and Σ_k are diagonal. For the complete parametrization of the BC precoders [cf. (3.77)]

$$\begin{aligned}
P_k &= G_k^{\prime\mathrm{MF},\mathrm{H}} V_k \cdot \mathrm{diag}\{\varphi_{k,j}\}_{j=1}^{M_k} \\
&= \sqrt{\frac{p_k}{M_k}} \cdot \Pi_k^{\perp} H_k V_k \cdot \mathrm{diag}\{\varphi_{k,j}\}_{j=1}^{M_k} \quad \forall k
\end{aligned}$$

the scaling coefficients $\varphi_{k,j}$ must be computed for all users k and all streams j via (3.78). Fortunately, the matrix Z in (3.78) is again diagonal since all off-diagonal blocks vanish due to $G_k^{\prime\mathrm{MF}} H_i = \mathbf{0}$ for $i \neq k$, see the projector $\Pi_k^{\perp}$ in (8.85). In other words, the precoder P_k of user k completely removes the inter-user interference of all other users $i \neq k$, from which we can conclude that block-diagonalization indeed achieves the maximum asymptotic weighted sum rate in the broadcast channel, which is subsumed in the following theorem:

Theorem 8.12 *In a MIMO broadcast channel with at least as many transmit antennas at the base station as the user terminals have in sum, block-diagonalization asymptotically achieves the maximum weighted sum rate for full rank channels with linear transceivers.*

Therefore, the squared scaling coefficient $\varphi_{k,j}^2$ reads by means of (8.86) as [cf. (3.78)–(3.80)]

$$\varphi_{k,j}^2 = \frac{\|T_k V_k e_j\|_2^2}{\|e_j^{\mathrm{H}} V_k^{\mathrm{H}} G_k^{\prime\mathrm{MF}}\|_2^2} = \frac{\frac{p_k}{M_k}}{\frac{p_k}{M_k}[\Sigma_k]_{j,j}} = \left[\Sigma_k^{-1}\right]_{j,j}$$

which leads to the BC precoders

$$P_k = \sqrt{\frac{p_k}{M_k}} \cdot \Pi_k^{\perp} H_k V_k \Sigma_k^{-\frac{1}{2}} \quad \forall k. \tag{8.87}$$

The precoder of user k induces the transmit covariance matrix

$$\begin{aligned}
S_k^{\mathrm{Linear}} = P_k P_k^{\mathrm{H}} &= \frac{p_k}{M_k} \cdot \Pi_k^{\perp} H_k V_k \Sigma_k^{-1} V_k^{\mathrm{H}} H_k^{\mathrm{H}} \Pi_k^{\perp} \\
&= \frac{p_k}{M_k} \cdot \Pi_k^{\perp} H_k \left(H_k^{\mathrm{H}} \Pi_k^{\perp} H_k\right)^{-1} H_k^{\mathrm{H}} \Pi_k^{\perp} \\
&= \frac{p_k}{M_k} \cdot \bar{H}_k \left(\bar{H}_k^{\mathrm{H}} \bar{H}_k\right)^{-1} \bar{H}_k^{\mathrm{H}} \quad \forall k
\end{aligned} \tag{8.88}$$

which means that S_k^{Linear} is again a scaled projector just like in the dirty paper coding case, cf. (8.35). However, the matrix $\bar{H}_k$ denotes the projection of H_k onto the null space of *all other* users' Hermitian channel matrices in case of linear filtering. For dirty paper coding, the projection is onto the null space of the Hermitian channels of those users i with $i > k$, and not $i \neq k$. From (8.88), we also observe that $\operatorname{tr}(S_k^{\text{Linear}}) = p_k$ with p_k defined in Theorem 11, i.e., the power allocation for linear transceivers in the dual MAC that achieves the maximum asymptotic weighted sum rate is also valid in the original BC and achieves the maximum asymptotic weighted sum rate there as well. Alternatively, S_k^{Linear} may also be expressed in terms of the composite channel matrix H via (see Appendix A20)

$$S_k^{\text{Linear}} = \frac{p_k}{M_k} \cdot H(H^{\text{H}}H)^{-1}E_k\big[E_k^{\text{T}}(H^{\text{H}}H)^{-1}E_k\big]^{-1}E_k^{\text{T}}(H^{\text{H}}H)^{-1}H^{\text{H}}. \quad (8.89)$$

Replacing the MAC channels by the broadcast channels and the noise covariance matrices in the BC via (8.29), we see that the transmit covariance matrices $S_1^{\text{Linear}}, \ldots, S_K^{\text{Linear}}$ again do not depend on the BC noise covariance matrices:

$$\begin{aligned} S_k^{\text{Linear}} &= \frac{p_k}{M_k} \cdot \bar{H}_k^{\text{BC,H}}\big(\bar{H}_k^{\text{BC}}\bar{H}_k^{\text{BC,H}}\big)^{-1}\bar{H}_k^{\text{BC}} \\ &= \frac{p_k}{M_k} \cdot H^{\text{BC,H}}(H^{\text{BC}}H^{\text{BC,H}})^{-1}E_k\big[E_k^{\text{T}}(H^{\text{BC}}H^{\text{BC,H}})^{-1}E_k\big]^{-1} \\ &\quad E_k^{\text{T}}(H^{\text{BC}}H^{\text{BC,H}})^{-1}H^{\text{BC}} \end{aligned}$$

In turn, $\bar{H}_k^{\text{BC,H}}$ is the projected version of $H_k^{\text{BC,H}}$ and the power p_k allocated to user k is given in Theorem 11. The total transmit covariance matrix S^{Linear} is simply the sum of the individual ones:

$$S^{\text{linear}} = \sum_{k=1}^{K} S_k^{\text{Linear}} \quad (8.90)$$

Even for common weights $w = \mathbf{1}$, however, S^{Linear} cannot be simplified like in the dirty paper coding case.

Synopsis: In a MIMO BC with linear transceivers and at least as many transmit antennas at the base station as the user terminals have in sum, the dual MAC transmit covariance matrices that lead to the best weighted sum rate asymptotes are the same scaled identity matrices as for dirty paper coding. Hence, the best high power asymptotes of the two signaling strategies are obtained by the same set of transmit covariance matrices. Indeed, the two asymptotes have the same slope but usually feature different rate offsets (unless the channels are pairwise orthogonal). For linear transceivers, the sum 'capacity' is strongly asymptotically equivalent to

$$C^{\text{Linear}}(P_{\max}) \cong M \log_2 P_{\max} - M \log_2 M - \sum_{k=1}^{K} \log_2 \left| \boldsymbol{E}_k^{\mathrm{T}} \left(\boldsymbol{H}^{\mathrm{BC}} \boldsymbol{H}^{\mathrm{BC,H}}\right)^{-1} \boldsymbol{E}_k \right|$$
$$- \sum_{k=1}^{K} \log_2 \left| \boldsymbol{C}_{\eta_k} \right|$$

and does (in contrast to dirty paper coding) depend on the distribution of the M antennas to the K users. The duality transformation to the BC reveals, that for the given setup, the best weighted sum rates asymptotes are achieved by block-diagonalization with the same power allocation in the BC as in the dual MAC. Moreover, the transmit covariance matrices of the individual users are again scaled projectors just like in the dirty paper coding case. Even for common weights, however, the total transmit covariance matrix does not simplify as in the DPC case.

8.2.3.4 Overloaded System

From Sect. 8.2.3.1, we know that exactly $\text{rank}(\boldsymbol{H})$ streams must be active in order to maximize the asymptotic weighted sum rate. For full rank channels with $M \leq N$, the distribution of the M active streams to the K users is unique since every user has to multiplex as many data streams as he has antennas, see the preceding subsection. For an overloaded system with $M > N$ and full rank channels, however, distributing the $\text{rank}(\boldsymbol{H}) = N$ streams to the M antennas of the K receivers leads to a user and stream selection problem. From (8.63), we know that a stream belonging to user i contributes to the asymptotic weighted sum rate utility via $w_i \log_2 P_{\max}$ as long as up to $L \leq \text{rank}(\boldsymbol{H})$ linearly independent streams are active with a linear power allocation. For full rank channels, we therefore conclude that the N streams should be distributed to the users with the largest weights to maximize the rate offset of the weighted sum rate asymptote.

Without loss of generality, we may assume that the user weights are sorted in a nonincreasing fashion, i.e., $w_1 \geq w_2 \geq \cdots, \geq w_K$. This can always be achieved by simply relabeling the users. For some antenna and weight configurations, the asymptotic WSR maximization in an overloaded system simplifies to the case of enough degrees of freedom. This happens under the following two conditions:

(1) The N streams can be distributed to the first K' users with the largest weights such that every of the first K' users can apply full multiplexing with $L_k = M_k \; \forall k \in \{1, \ldots, K'\}$. In other words, $\sum_{k=1}^{K'} M_k = N$ must hold.

(2) The weight $w_{K'+1}$ must be strictly smaller than the weight $w_{K'}$ of the active user with the largest index. Then, no streams are allocated to users with indices larger than K', i.e., $L_k = 0\ \forall k > K'$.

When these two conditions are met, choosing the transmit covariance matrices of the first K' users as scaled identity matrices in the dual MAC again achieves the asymptotic WSR and the eigenbasis of the transmit covariance matrix is irrelevant since all eigenvalues are identical. The following example helps to understand the situation.

Example Assume that there are $K = 3$ two-antenna users ($M_1 = M_2 = M_3 = 2$) which are served by an $N = 4$ antenna base station. The user weights shall read as $w_1 = w_2 = 1$ and $w_3 = 0.5$. Then, the $L = N = 4$ streams must be dedicated to the first $K' = 2$ users since user three has the smallest weight ($w_3 < w_1$ and $w_3 < w_2$ with strict inequality) and since $M_1 + M_2 = N = 4$. While the first $K' = 2$ users apply full multiplexing with $L_1 = M_1$ and $L_2 = M_2$, no stream is allocated to user three ($L_3 = 0$).

If above two conditions do not hold, the overloaded system cannot be transformed into a system with enough degrees of freedom. The next three examples show the resulting problems which then arise.

Example

- Again, $K = 3$ two-antenna users with weights $w_1 = w_2 = 1$ and $w_3 = 0.5$ are served by a base station, which now has $N = 5$ antenna elements instead of only four. For the maximum asymptotic WSR, the first two users have to apply full multiplexing with $L_1 = M_1$ and $L_2 = M_2$, whereas user three has only one active data stream. While the power allocation of user three follows from Theorem 11, it is unclear how the power-normalized beamformer (eigenvector of $\boldsymbol{Q}_3$) must be chosen. Unfortunately, a closed form expression for this beamformer is not available such that a simple expression for the WSR asymptote is not possible. However, we presented an iterative algorithm which (locally) maximizes the rate offset of the WSR asymptote with respect to the normalized beamformer in [149].
- When the $K = 3$ two-antenna users are served by an $N = 4$ antenna base station and the user weights are $w_1 = 1$ and $w_2 = w_3 = 0.5$, then user one must be served with $L_1 = M_1 = 2$ streams since he has the largest weight. However, the distribution of the remaining $N - L_1 = 2$ streams to users two and three (with the strictly smaller weights) is not unique. Unfortunately, all three possible permutations have to be checked. The two cases $L_2 = 2, L_3 = 0$ and $L_2 = 0, L_3 = 2$ can be converted to the setup with enough degrees of freedom since either user two or user three applies full multiplexing like user one. The third possibility in which user two and user three each have one active stream ($L_2 = L_3 = 1$) has to be investigated as well. There, the optimum power-normalized beamformers of the last two users which maximize the rate offset of the WSR asymptote again cannot be obtained in closed form.

- When the three weights $w_1 = w_2 = w_3$ are identical, all six different possibilities of distributing $L = N = 4$ streams to the $K = 3$ users (such that at most $M_k = 2\ \forall k$ streams are allocated to a particular user) have to be probed to find the best WSR asymptote.

The analysis of the challenging cases in which there is at least one user that does not apply full multiplexing reveals, that all fully multiplexing users again have to apply scaled identity matrices as transmit covariance matrices in the dual MAC. Unfortunately, the power-normalized precoders belonging to the other users clearly have an impact on the obtained asymptotic WSR. Even worse, neither a closed form solution, nor a globally optimum algorithm that returns the best precoders seems to exist. If only a single user does not apply full multiplexing and has only one active stream instead, the weighted sum rate utility maximization with respect to this unit-norm beamformer boils down to the maximization of a Hermitian form with two different exponents. In [149], we presented an algorithm which finds a local optimum of the asymptotic WSR maximization by means of a sequential principal eigenvector computation. Since the best weighed sum rate asymptote therefore cannot be expressed in closed form if at least one user does not use full multiplexing, we do not investigate this scenario in more details.

8.3 Ergodic Weighted Sum Rate Analysis

The high-power asymptotes of the weighted sum rate utility are analyzed on a larger time scale for the various transmission strategies in this section. For this purpose, the base station is assumed to experience temporally uncorrelated block-fading to which it adapts its transmission due to the availability of full CSI. Moreover, the maximum sum power $P_{\max}$ shall be dissipated for every channel realization and the fading shall be slow enough such that the individual channel realizations may be assumed to be atomic, which means that the rate analysis can be done separately for every channel. Given this type of setup, the channel-averaged asymptotic sum capacity is derived for the virtual point-to-point MIMO system with cooperating receivers in Sect. 8.3.2. The ergodic weighted sum rate asymptotes for the MIMO BC with dirty paper coding are found in Sect. 8.3.3 whereas linear transceivers are handled in Sect. 8.3.4. Since the Wishart and the inverse Wishart distribution play a major role during these ergodic derivations, we first introduce the relevant properties of them in Sect. 8.3.1.

Due to the availability of closed form expressions for the (weighted) sum rate asymptotes given a *fixed* channel realization, channel-averaged expressions can basically be obtained for any kind of fading scenario with arbitrary channel correlations at the antenna elements via simulations. This particularly holds for the linear transceivers case where we have derived asymptotic rate expressions that for the first time solely depend on the composite BC channel matrix and not on any precoders or subspaces that need to be computed separately. Hence, ergodic results can quickly be obtained by means of simulations.

For semi-correlated Rayleigh fading with correlations only at the terminals, closed form expressions for the WSR asymptotes may be derived. In this kind of fading model, the BC channel matrix of some user $k \in \{1, \ldots, K\}$ can be written as (cf. [136, Eq. (23)])

$$\boldsymbol{H}_k^{\mathrm{BC}} = \boldsymbol{\Theta}_k^{\frac{1}{2}} \boldsymbol{\Upsilon}_k \tag{8.91}$$

where $\boldsymbol{\Theta}_k \in \mathbb{S}_+^{M_k}$ describes the channel correlations seen by receiver k in the BC and $\boldsymbol{\Upsilon}_k \in \mathbb{C}^{M_k \times N}$ has zero-mean i.i.d. unit variance complex Gaussian entries such that $\mathrm{vec}(\boldsymbol{\Upsilon}_k) \sim \mathcal{CN}(\mathbf{0}, \mathbf{I}_{NM_k})$, where $\mathrm{vec}(\cdot)$ stacks the columns of its matrix-valued argument. Note that different path loss values which may result from the fact that the K users are not co-located, can easily be incorporated by letting $\boldsymbol{\Theta}_1, \ldots, \boldsymbol{\Theta}_K$ have different traces, which can be seen from

$$\mathrm{E}[\|\boldsymbol{H}_k^{\mathrm{BC}}\|_{\mathrm{F}}^2] = \mathrm{tr}\left(\boldsymbol{\Theta}_k^{\frac{1}{2}}\, \mathrm{E}[\boldsymbol{\Upsilon}_k \boldsymbol{\Upsilon}_k^{\mathrm{H}}] \boldsymbol{\Theta}_k^{\frac{1}{2},\mathrm{H}}\right) = N \cdot \mathrm{tr}(\boldsymbol{\Theta}_k).$$

The composite BC channel matrix can then be written as

$$\begin{aligned} \boldsymbol{H}^{\mathrm{BC}} &= [\boldsymbol{H}_1^{\mathrm{BC,H}}, \ldots, \boldsymbol{H}_K^{\mathrm{BC,H}}]^{\mathrm{H}} \\ &= \mathrm{blockdiag}\{\boldsymbol{\Theta}_k^{\frac{1}{2}}\}_{k=1}^K \cdot [\boldsymbol{\Upsilon}_1^{\mathrm{H}}, \ldots, \boldsymbol{\Upsilon}_K^{\mathrm{H}}]^{\mathrm{H}} \\ &= \boldsymbol{\Theta}^{\frac{1}{2}} \cdot \boldsymbol{\Upsilon} \end{aligned} \tag{8.92}$$

where

$$\boldsymbol{\Upsilon} := [\boldsymbol{\Upsilon}_1^{\mathrm{H}}, \ldots, \boldsymbol{\Upsilon}_K^{\mathrm{H}}]^{\mathrm{H}} \in \mathbb{C}^{M \times N}$$

has again i.i.d. complex Gaussian entries and satisfies $\mathrm{vec}(\boldsymbol{\Upsilon}) \sim \mathcal{CN}(\mathbf{0}, \mathbf{I}_{MN})$. For a shorter notation, we also define the matrix

$$\boldsymbol{\Theta} := \mathrm{blockdiag}\{\boldsymbol{\Theta}_k\}_{k=1}^K \in \mathbb{S}_+^M.$$

Throughout the derivations in Sects. 8.3.2–8.3.4, the Gaussian matrix $\boldsymbol{\Upsilon}$ arises only as an outer product $\boldsymbol{W} := \boldsymbol{\Upsilon}\boldsymbol{\Upsilon}^{\mathrm{H}} \in \mathbb{S}^M$. Thus, it is worthwhile to briefly introduce the distribution and some properties of the matrix $\boldsymbol{W}$ and its inverse $\boldsymbol{W}^{-1}$ for the case that $M \leq N$.

8.3.1 Wishart and Inverse Wishart Distribution

We start with the definition of the Wishart distribution of an $M \times M$ matrix with N degrees of freedom. Note that only the *complex* Wishart distribution is considered here whose properties differ from its real-valued counterpart. Due to the fact that we decouple the correlations $\boldsymbol{\Theta}$ from the Gaussian matrix $\boldsymbol{\Upsilon}$, the Wishart distribution has only two parameters: the dimension of the involved matrix and its degrees of freedom.

Definition 8.6 Let the M-dimensional column vectors $\boldsymbol{y}_1, \dots, \boldsymbol{y}_N \in \mathbb{C}^M$ be independently and identically distributed random variables that are taken from a zero-mean, complex Gaussian distribution with identity covariance matrix via $\boldsymbol{y}_n \sim \mathcal{CN}(\mathbf{0}, \mathbf{I}_M)$ $\forall n \in \{1, \dots, N\}$. Then, the $M \times M$ matrix

$$\boldsymbol{W} := \sum_{n=1}^{N} \boldsymbol{y}_n \boldsymbol{y}_n^{\mathrm{H}} \in \mathbb{S}^M$$

is said to have a (central) complex Wishart distribution with N degrees of freedom and we shall write $\boldsymbol{W} \sim \mathcal{W}_M(N)$. For $N \geq M$ and positive semi-definite $\boldsymbol{W}$, the probability density function of $\boldsymbol{W}$ is given by (e.g., [150, Eq. (2)] or [151, Eq. (2.6)])

$$f_{\mathcal{W}_M(N)}(\boldsymbol{W}) = \frac{|\boldsymbol{W}|^{N-M} \exp[-\operatorname{tr}(\boldsymbol{W})]}{\pi^{\frac{M(M-1)}{2}} \prod_{i=1}^{M} (N-i)!}.$$

For $N < M$, the matrix $\boldsymbol{W}$ is said to have a *singular* Wishart distribution which features a probability density function which differs from the one in Definition 6. However, only the case $N \geq M$ is considered here for which the probability of $\boldsymbol{W}$ being regular is one, see [152, p. 88].

For the ergodic analysis of the high power asymptotes of the point-to-point MIMO system and the MIMO BC with dirty paper coding, the expected value of the logarithmic determinant of $\boldsymbol{W}$ is required. According to [151, Eq. (2.12)], this expectation is given by

$$\boldsymbol{W} \sim \mathcal{W}_M(N) \quad \Rightarrow \quad \mathrm{E}\left[\log_2 |\boldsymbol{W}|\right] = \frac{1}{\ln 2} \cdot \sum_{\ell=0}^{M-1} \psi(N-\ell) \tag{8.93}$$

where $\psi(\cdot)$ denotes the digamma function that for integer arguments is defined as [151, Eq. (2.14)]

$$\psi(n+1) = \psi(n) + \frac{1}{n} = \psi(1) + \sum_{\ell=1}^{n} \frac{1}{\ell} \quad \text{for} \quad n \in \mathbb{N} \tag{8.94}$$

and $\psi(1) = -\gamma \approx -0.577$ is the negative Euler-Mascheroni constant.

In case of linear transceivers, the evaluation of the channel-averaged asymptotes of the user rates requires the inverse of a Wishart distributed matrix $\boldsymbol{W}$. More precisely, the expected logarithmic determinant of a main diagonal block of $\boldsymbol{W}^{-1}$ must be computed, see for example (8.82). Given a Wishart distributed matrix $\boldsymbol{W} \sim \mathcal{W}_M(N)$ with $N \geq M$, the matrix $\boldsymbol{A} = \boldsymbol{W}^{-1}$ is said to have an *inverse* Wishart distribution:

Definition 8.7 Let the $M \times M$ matrix $\boldsymbol{W} \sim \mathcal{W}_M(N)$ be Wishart distributed with $N \geq M$ degrees of freedom and let $\boldsymbol{A} = \boldsymbol{W}^{-1}$ denote the inverse of $\boldsymbol{W}$ which exists with probability one. Then, $\boldsymbol{A}$ is said to be complex inverse Wishart distributed with parameters M and N and its probability density function is given by (e.g., [153, Eq. (2)])

$$f_{\mathcal{W}_M^{-1}(N)}(\boldsymbol{A}) = \frac{|\boldsymbol{A}|^{-N-M} \exp\left[-\operatorname{tr}(\boldsymbol{A}^{-1})\right]}{\pi^{\frac{M(M-1)}{2}} \prod_{i=1}^{M} (N-i)!}$$

and we shall write $\boldsymbol{A} \sim \mathcal{W}_M^{-1}(N)$.

An interesting property of the inverse Wishart distribution is that the marginal distribution of any main diagonal block is again inverse Wishart [153, p. 53]:

Theorem 8.13 *Let the $M \times M$ matrix $\boldsymbol{A} \sim \mathcal{W}_M^{-1}(N)$ be inverse Wishart distributed with parameters M and N. Then, the marginal distribution of any $p \times p$ main diagonal block $\boldsymbol{A}_i$ of the matrix $\boldsymbol{A}$ is again inverse Wishart with $\boldsymbol{A}_i \sim \mathcal{W}_p^{-1}(N-M+p)$. As a consequence, the inverse $\boldsymbol{A}_i^{-1}$ is Wishart distributed with $\boldsymbol{A}_i^{-1} \sim \mathcal{W}_p(N-M+p)$.*

From Theorem 13, we conclude that for $\boldsymbol{W} \sim \mathcal{W}_M(N)$, the inverse of any $M_k \times M_k$ main diagonal block of $\boldsymbol{W}^{-1}$ has the marginal distribution

$$\left(\boldsymbol{E}_k^{\mathrm{T}} \boldsymbol{W}^{-1} \boldsymbol{E}_k\right)^{-1} \sim \mathcal{W}_{M_k}(N-M+M_k).$$

Therefore, it is easy to find a closed form expression for the expected logarithmic determinant of such an $M_k \times M_k$ main diagonal block, cf. (8.93):

$$\begin{aligned} -\mathrm{E}\left[\log_2 \left|\boldsymbol{E}_k^{\mathrm{T}} \boldsymbol{W}^{-1} \boldsymbol{E}_k\right|\right] &= \mathrm{E}\left[\log_2 \left|\left(\boldsymbol{E}_k^{\mathrm{T}} \boldsymbol{W}^{-1} \boldsymbol{E}_k\right)^{-1}\right|\right] \\ &= \frac{1}{\ln 2} \cdot \sum_{\ell=0}^{M_k-1} \psi(N-M+M_k-\ell) \qquad (8.95) \end{aligned}$$

By means of (8.93) and (8.95), the asymptotes of the channel-averaged weighted sum rate utility are computed for the different system setups in the next three subsections.

8.3.2 Point-to-Point MIMO System

The asymptotic sum capacity of the point-to-point MIMO link that results from the MIMO BC when the receivers are allowed to cooperate reads for a fixed channel realizations as [see (8.14)]

$$C^{\mathrm{PtP}}(P_{\max}) \cong M \log_2 P_{\max} - M \log_2 M + \log_2 \left|\boldsymbol{H}^{\mathrm{BC}} \boldsymbol{H}^{\mathrm{BC,H}}\right| - \sum_{k=1}^{K} \log_2 \left|\boldsymbol{C}_{\eta_k}\right|.$$

Given the semi-correlated channel model in (8.92) with $\boldsymbol{H}^{\mathrm{BC}} = \boldsymbol{\Theta}^{\frac{1}{2}}\boldsymbol{\Upsilon}$, the ergodic sum capacity $C_{\mathrm{Erg}}^{\mathrm{PtP}}(P_{\max})$ is strongly asymptotically equivalent to

$$C_{\mathrm{Erg}}^{\mathrm{PtP}}(P_{\max}) \cong M \log_2 P_{\max} - M \log_2 M + \mathrm{E}\left[\log_2 \left|\boldsymbol{\Upsilon}\boldsymbol{\Upsilon}^{\mathrm{H}}\right|\right] + \sum_{k=1}^{K}\left(\log_2 \left|\boldsymbol{\Theta}_k\right| - \log_2 \left|\boldsymbol{C}_{\eta_k}\right|\right)$$

which by means of (8.93) finally leads to

$$C_{\mathrm{Erg}}^{\mathrm{PtP}}(P_{\max}) \cong M \log_2 P_{\max} - M \log_2 M + \frac{1}{\ln 2} \cdot \sum_{\ell=0}^{M-1} \psi(N-\ell) + \sum_{k=1}^{K}\left(\log_2 \left|\boldsymbol{\Theta}_k\right| - \log_2 \left|\boldsymbol{C}_{\eta_k}\right|\right). \tag{8.96}$$

We observe that the channel correlations at the receivers change the rate offset of the sum capacity asymptote via the logarithmic determinant. For a given path loss of a particular user k (i.e., for fixed $\mathrm{tr}(\boldsymbol{\Theta}_k)$), correlated antenna elements reduce the rate offset of the ergodic sum capacity asymptote since they reduce the determinant of $\boldsymbol{\Theta}_k$. The number of antennas N at the base station has an impact on the rate offset only with respect to the digamma function $\psi(\cdot)$.

8.3.3 MIMO Broadcast Channel with Dirty Paper Coding

In principle, asymptotes of the channel-averaged weighted sum rate utility with dirty paper coding applied can be derived in closed form for any positive weight vector $\boldsymbol{w}$ when $N \geq M$. To this end, the result in (8.93) simply has to be applied to (8.30) by using the identity

$$\begin{aligned}\boldsymbol{J}_k^{\mathrm{T}}\boldsymbol{H}^{\mathrm{BC}}\boldsymbol{H}^{\mathrm{BC,H}}\boldsymbol{J}_k &= \boldsymbol{J}_k^{\mathrm{T}}\boldsymbol{\Theta}^{\frac{1}{2}}\boldsymbol{\Upsilon}\boldsymbol{\Upsilon}^{\mathrm{H}}\boldsymbol{\Theta}^{\frac{1}{2},\mathrm{H}}\boldsymbol{J}_k \\ &= \boldsymbol{J}_k^{\mathrm{T}}\boldsymbol{\Theta}^{\frac{1}{2}}\boldsymbol{J}_k \cdot \boldsymbol{J}_k^{\mathrm{T}}\boldsymbol{\Upsilon}\boldsymbol{\Upsilon}^{\mathrm{H}}\boldsymbol{J}_k \cdot \boldsymbol{J}_k^{\mathrm{T}}\boldsymbol{\Theta}^{\frac{1}{2},\mathrm{H}}\boldsymbol{J}_k\end{aligned}$$

and noticing that $\boldsymbol{J}_k^{\mathrm{T}}\boldsymbol{\Upsilon}\boldsymbol{\Upsilon}^{\mathrm{H}}\boldsymbol{J}_k \sim \mathcal{W}_{\sum_{j=k}^{K} L_j}(N)$ holds for all $k \in \{1, \ldots, K\}$. However, since the resulting expressions quickly become cumbersome, we restrict our ergodic analysis to the asymptotic sum capacity. Due to the fact that the asymptotic DPC sum capacity coincides with the asymptotic sum capacity of the same system with cooperating receivers and uncorrelated noise terms, the sum capacity of DPC is asymptotically equivalent to the one in (8.96):

$$C_{\mathrm{Erg}}^{\mathrm{DPC}}(P_{\max}) \cong M \log_2 P_{\max} - M \log_2 M + \frac{1}{\ln 2} \cdot \sum_{\ell=0}^{M-1} \psi(N-\ell) + \sum_{k=1}^{K} \left(\log_2 |\boldsymbol{\Theta}_k| - \log_2 |\boldsymbol{C}_{\eta_k}| \right) \tag{8.97}$$

8.3.4 MIMO Broadcast Channel with Linear Transceivers

In case of linear transceivers with enough degrees of freedom, the transmit covariance matrices that asymptotically achieve the maximum weighted sum rate lead to individual user rates that are given in (8.80). Making use of (8.95), closed form channel-averaged expressions for these individual rates can easily be computed as well. Again, we only present an explicit expression for the sum 'capacity' asymptote of linear transceivers here just like in the DPC case.

When the fixed channel realization $\boldsymbol{H}^{\mathrm{BC}}$ in the asymptotic sum 'capacity' expression in (8.82) is replaced by $\boldsymbol{H}^{\mathrm{BC}} = \boldsymbol{\Theta}^{\frac{1}{2}} \boldsymbol{\Upsilon}$, we obtain for the ergodic sum 'capacity' of linear transceivers

$$C_{\mathrm{Erg}}^{\mathrm{Linear}}(P_{\max}) \cong M \log_2 P_{\max} - M \log_2 M + \sum_{k=1}^{K} \mathrm{E}\left[\log_2 \left| \left(\boldsymbol{E}_k^{\mathrm{T}} \left(\boldsymbol{\Upsilon}\boldsymbol{\Upsilon}^{\mathrm{H}} \right)^{-1} \boldsymbol{E}_k \right)^{-1} \right| \right] + \sum_{k=1}^{K} \left(\log_2 |\boldsymbol{\Theta}_k| - \log_2 |\boldsymbol{C}_{\eta_k}| \right).$$

Using (8.95), this expression simplifies to

$$C_{\mathrm{Erg}}^{\mathrm{Linear}}(P_{\max}) \cong M \log_2 P_{\max} - M \log_2 M + \frac{1}{\ln 2} \cdot \sum_{k=1}^{K} \sum_{\ell=0}^{M_k-1} \psi(N - M + M_k - \ell) + \sum_{k=1}^{K} \left(\log_2 |\boldsymbol{\Theta}_k| - \log_2 |\boldsymbol{C}_{\eta_k}| \right). \tag{8.98}$$

The channel-averaged rate offset reduction ΔR_{Erg} of the two sum capacity asymptotes that has to be faced when switching from dirty paper coding to linear transceivers corresponds to the difference of the right hand sides of (8.97) and (8.98):

$$\Delta R_{\mathrm{Erg}} = \frac{1}{\ln 2} \cdot \left[\sum_{\ell=0}^{M-1} \psi(N-\ell) - \sum_{k=1}^{K} \sum_{\ell=0}^{M_k-1} \psi(N - M + M_k - \ell) \right]$$

$$= \frac{1}{\ln 2} \cdot \left[\sum_{\ell=1}^{M} s(N-\ell) - \sum_{k=1}^{K} \sum_{\ell=1}^{M_k} s(N-M+M_k-\ell) \right] \tag{8.99}$$

From the second line in (8.99) and the definition of the digamma function in (8.94), we see that the Euler–Mascheroni constants cancel out such that ΔR_{Erg} can be expressed in terms of the harmonic number

$$s(n) := \sum_{i=1}^{n} \frac{1}{i}.$$

It can be observed from (8.99) that antenna correlations and different path loss values do not have an impact on the ergodic rate shift of the two sum capacity asymptotes for DPC and linear filtering.

When all terminals are equipped with the same number of antennas $\bar{M} = M_1 = \ldots = M_K$ such that $M = K\bar{M}$, the rate shift ΔR_{Erg} of the two asymptotes in (8.99) can be simplified as follows:

$$\begin{aligned}
\Delta R_{\text{Erg}} &= \frac{1}{\ln 2} \cdot \Bigg[\sum_{\ell=1}^{(K-1)\bar{M}} s(N-\ell) + \sum_{\ell=(K-1)\bar{M}+1}^{K\bar{M}} s(N-\ell) \\
&\quad - K \sum_{\ell=1}^{\bar{M}} s\big(N-K\bar{M}+\bar{M}-\ell\big) \Bigg] \\
&= \frac{1}{\ln 2} \cdot \left[\sum_{\ell=1}^{(K-1)\bar{M}} s(N-\ell) - (K-1) \cdot \sum_{\ell=1}^{\bar{M}} s\big(N-K\bar{M}+\bar{M}-\ell\big) \right]
\end{aligned}$$

The first sum in above difference may be expressed via

$$\sum_{\ell=1}^{(K-1)\bar{M}} s(N-\ell) = \sum_{\ell=1}^{(K-1)\bar{M}} \frac{\ell}{N-\ell} + (K-1)\bar{M} \cdot s\big(N-K\bar{M}+\bar{M}-1\big)$$

such that ΔR_{Erg} may be written as

$$\begin{aligned}
\Delta R_{\text{Erg}} &= \frac{1}{\ln 2} \cdot \Bigg[\sum_{\ell=1}^{(K-1)\bar{M}} \frac{\ell}{N-\ell} + (K-1) \cdot \sum_{\ell=1}^{\bar{M}} \Big[s\big(N-K\bar{M}+\bar{M}-1\big) \\
&\qquad - s\big(N-K\bar{M}+\bar{M}-\ell\big) \Big] \Bigg] \\
&= \frac{1}{\ln 2} \cdot \left[\sum_{\ell=1}^{(K-1)\bar{M}} \frac{\ell}{N-\ell} + (K-1) \cdot \sum_{\ell=1}^{\bar{M}} \sum_{i=\bar{M}-\ell+1}^{\bar{M}-1} \frac{1}{N-K\bar{M}+i} \right].
\end{aligned}$$

By means of the substitution $n := \bar{M} - \ell + 1$, the last double-sum can be rewritten to

$$\sum_{\ell=1}^{\bar{M}} \sum_{i=\bar{M}-\ell+1}^{\bar{M}-1} \frac{1}{N - K\bar{M} + i} = \sum_{n=1}^{\bar{M}} \sum_{i=n}^{\bar{M}-1} \frac{1}{N - K\bar{M} + i}$$
$$= \sum_{n=1}^{\bar{M}-1} \sum_{i=n}^{\bar{M}-1} \frac{1}{N - K\bar{M} + i} = \sum_{\ell=1}^{\bar{M}-1} \frac{\ell}{N - K\bar{M} + \ell}$$

which finally leads to the rate shift

$$\Delta R_{\mathrm{Erg}} = \frac{1}{\ln 2} \cdot \left[\sum_{\ell=1}^{(K-1)\bar{M}} \frac{\ell}{N - \ell} + \sum_{\ell=1}^{\bar{M}-1} \frac{(K-1)\ell}{N - K\bar{M} + \ell} \right] \tag{8.100}$$

which coincides with the results in [141,142], but is stated in a different form. The following observations can be made from (8.100) which hold for the semi-correlated Rayleigh fading:

- For a fixed number of users K and a fixed number $\bar{M}$ of antennas per user, $\Delta R_{\mathrm{Erg}} \to 0$ as $N \to \infty$. In other words, the asymptotic rate loss of linear filtering with respect to dirty paper coding goes to zero as the number of antennas N at the base station goes to infinity while the number of users remains fixed.
- For a fully loaded system with $N = K\bar{M}$, it can be shown that $\Delta R_{\mathrm{Erg}} \geq K - 1$. Hence, when $K \to \infty$ and N scales in the same way, the rate shift ΔR_{Erg} goes to infinity as well.
- For single-antenna terminals with $\bar{M} = 1$, the second sum in (8.100) is zero and we obtain

$$\Delta R_{\mathrm{Erg}} = \frac{1}{\ln 2} \cdot \sum_{\ell=1}^{K-1} \frac{\ell}{N - \ell} \quad \text{for} \quad \bar{M} = 1 \tag{8.101}$$

  in accordance to [139]. In a fully loaded system with $N = K$, (8.101) can be written as

$$\Delta R_{\mathrm{Erg}} = K\big[s(K-1) - 1\big] + 1 \quad \text{for} \quad \bar{M} = 1 \quad \text{and} \quad K = N$$

  which grows[17] with $O(K \ln(K))$ as $K \to \infty$.

In order to acquire a rough intuition for the magnitude of the ergodic rate loss ΔR_{Erg} of linear filtering with respect to dirty paper coding, we evaluate (8.99) for some particular values of the number of users K and the number of antennas $M_1, \ldots, M_K$ at the terminals.

Table 8.1 shows the channel-averaged rate loss ΔR_{Erg} in bpcu and the power loss $\Delta R_{\mathrm{Erg}}/M \cdot (10 \log_{10} 2)$ in dB (inside the parenthesis) that linear filtering has to face

[17] Here, the big-O notation is with respect to K instead of $P_{\max}$.

Table 8.1 Rate loss ΔR_{Erg} in bpcu and power loss in dB (inside the parentheses) of linear filtering with respect to dirty paper coding in a two-user system

M_1, M_2	$N = 2$	$N = 3$	$N = 4$	$N = 5$	$N = 6$
1, 1	1.44 (2.17)	0.72 (1.09)	0.48 (0.72)	0.36 (0.54)	0.29 (0.43)
1, 2	–	2.16 (2.17)	1.20 (1.21)	0.84 (0.84)	0.65 (0.65)
1, 3	–	–	2.64 (1.99)	1.56 (1.18)	1.13 (0.85)
1, 4	–	–	–	3.01 (1.81)	1.85 (1.11)
2, 2	–	–	3.37 (2.53)	2.04 (1.54)	1.49 (1.12)
2, 3	–	–	–	4.21 (2.53)	2.69 (1.62)
2, 4	–	–	–	–	4.86 (2.44)

Table 8.2 Rate loss ΔR_{Erg} in bpcu and power loss in dB (inside the parentheses) of linear filtering with respect to dirty paper coding when all K user terminals have the same number of antennas $\bar{M}$

$K, \bar{M}$	$N = 2$	$N = 3$	$N = 4$	$N = 5$	$N = 6$
2, 1	1.44(2.17)	0.72(1.09)	0.48(0.72)	0.36(0.54)	0.29(0.43)
3, 1	–	3.61(3.62)	1.92(1.93)	1.32(1.33)	1.01(1.01)
4, 1	–	–	6.25(4.70)	3.49(2.62)	2.45(1.85)
5, 1	–	–	–	9.26(5.57)	5.34(3.21)
6, 1	–	–	–	–	12.55(6.30)
2, 2	–	–	3.37(2.53)	2.04(1.54)	1.49(1.12)
2, 3	–	–	–	–	5.34(2.68)
3, 2	–	–	–	–	8.22(4.13)

with respect to dirty paper coding in a two-user system for various values of N, M_1, and M_2. Note that only those entries are filled in which $N \geq M_1 + M_2$ holds since the sum capacity asymptotes can be computed in closed form only for these cases.

Equivalently, the rate and power loss of linear filtering with respect to DPC is shown in Table 8.2 for the case when all K user terminals have the same number of antennas $\bar{M}$. For a constant product $K\bar{M}$, it can be seen that the rate loss is smaller for larger $\bar{M}$, which can also be proven analytically. In addition, we observe the increasing rate loss for $K = N$ when K increases.

Synopsis: Given a semi-correlated Rayleigh fading in a MIMO BC with correlations only at the antennas of the user terminals, the channel-averaged asymptotic weighted sum rate utility can be expressed in closed form both for dirty paper coding and linear filtering when the base station is equipped with enough antenna elements. From the arising expressions, we conclude that antenna correlations at the terminals only lead to a shift of the weighted sum rate asymptotes and affect the two asymptotes of DPC and linear filtering in the same way. Consequently, the channel-averaged asymptotic loss in the WSR utility is invariant with respect to these correlations and only depends on the number of antennas N at the base station and on the distribution of the M receiving antenna elements to the K users. When all users have the same number of antennas and $N \to \infty$, the channel-averaged asymptotic rate loss of linear filtering with respect to DPC goes to zero. For a fully loaded system with the same number of antenna elements on both sides of the link, the ergodic rate loss goes to infinity when the number of users K grows beyond all limits.

8.4 Simulation Results

The asymptotic behavior of the sum capacity is first analyzed for a fixed channel realization in a two-user system with $M_1 = M_2 = \bar{M} = 2$ antennas per user terminal and either $N = 6$ or $N = 4$ antennas at the base station. The MAC channels are taken from (6.20) such that $\boldsymbol{H}_1$ corresponds to the first two columns and $\boldsymbol{H}_2$ to columns three and four of the matrix in (6.20). Moreover, the BC noise covariance matrices are assumed to be identity matrices such that $\boldsymbol{H}_k^{\mathrm{BC}} = \boldsymbol{H}_k^{\mathrm{H}}$ holds for $k \in \{1, 2\}$.

Figure 8.1 shows the case $N = 6$. On the left in Fig. 8.1a, the sum capacity of dirty paper coding (`DPC Sum Capacity`—solid curve,circle marker) is plotted versus the logarithmic sum power in dB. Note that the sum capacity was computed by the gradient-projection algorithm as in Chap. 6. Besides the simulated sum capacity curve, its asymptote from (8.31) is plotted (`DPC Asymptote`—dashed curve, no marker). Since the base station is equipped with more antennas than the user terminals have in sum, convergence to the asymptote already occurs at moderately small transmit powers of about 10 dB. Another consequence of $N > M$ is that the asymptotic sum 'capacity' loss of linear filtering with respect to dirty paper coding is very small. Hence, the results obtained by linear transceivers are not shown in the global view in Fig. 8.1a since they almost overlap with the DPC results on this scale. Instead, a zoomed view is depicted in Fig. 8.1b which also shows the sum 'capacity' of linear transceivers (`Linear Sum 'Capacity'`—solid curve, cross marker) and its asymptote (`Linear Asymptote`—dotted curve, no marker). We can verify the small asymptotic rate loss of about 0.42 bpcu in accordance to (8.83) which translates itself to a power loss of 0.31 dB at a multiplexing gain of $s_\infty = 4$.

Removing the last two antenna elements at the base station leads to the scenario with $N = 4$ in Fig. 8.2. As predicted by (8.31), the DPC asymptote (`DPC Asymptote`—dashed curve, no marker) is shifted down by about 2.24 bpcu which corresponds to a power shift of about 1.7 dB, cf. Figs. 8.1a and 8.2a. As the system is fully loaded ($N = M$), convergence of the DPC sum capacity to its asymptote happens at slightly larger power values of about 12 dB. Moreover, the asymptotic sum capacity loss of linear filtering with respect to DPC is increased to about 0.80 bpcu or 0.6 dB, cf. (8.83) and see Fig. 8.2b.

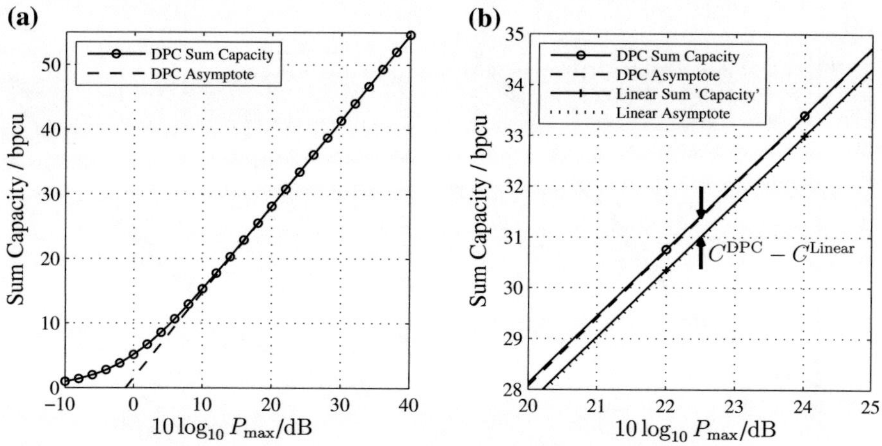

Fig. 8.1 Sum capacity of DPC versus linear filtering and the asymptotes for $N = 6$ antennas. **a** Global view. **b** Zoomed view

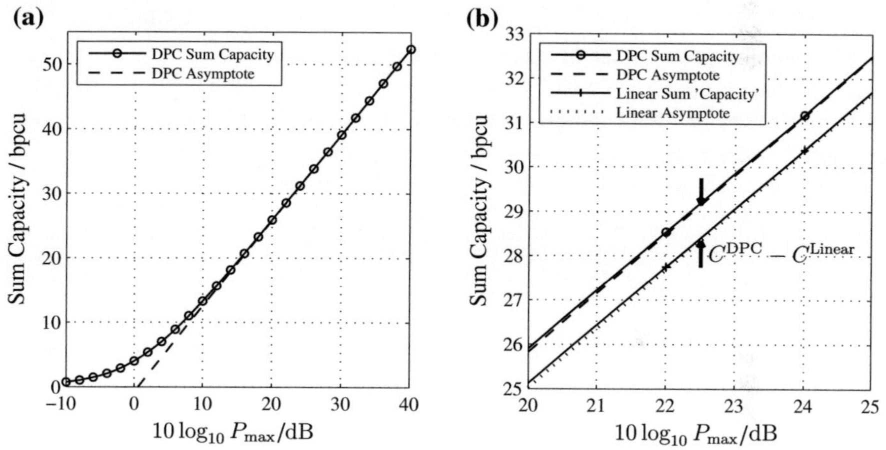

Fig. 8.2 Sum capacity of DPC versus linear filtering and the asymptotes for $N = 4$ antennas. **a** Global view. **b** Zoomed view

Next, we evaluate the channel-averaged sum capacity for Rayleigh fading without receiver-sided correlations. The matrices $\boldsymbol{\Theta}_1$ and $\boldsymbol{\Theta}_2$ in (8.91) are chosen as identity matrices since they only lead to a shift of the ergodic sum capacity curves. Averaging has been done over 1000 channel realizations and the same two-user setup has been simulated in Fig. 8.3. When $N = 6$ antennas are mounted at the base station (Fig. 8.3a), the channel-averaged sum capacity loss of linear filtering (`Linear Sum 'Capacity'`—solid curve, cross marker) with respect to dirty paper coding (`DPC Sum Capacity`—solid curve, circle marker) converges to 1.49 bpcu or 1.12 dB for $P_{\max} \to \infty$, see Table 8.1 and also Fig. 7.5b in Sect. 7.2.4.

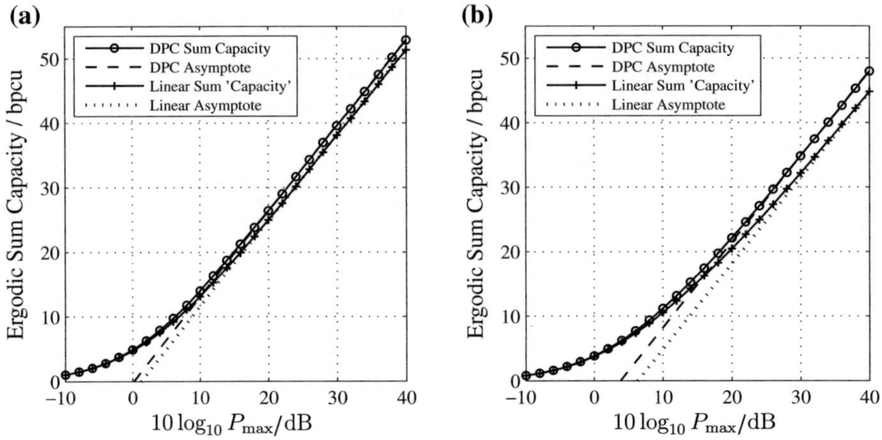

Fig. 8.3 Ergodic sum capacity of DPC versus linear filtering and the corresponding asymptotes. **a** $N = 6$. **b** $N = 4$

By reducing the number of antennas at the base station to $N = 4$ as shown in Fig. 8.3b, the ergodic sum capacity loss is increased to 3.37 bpcu, which corresponds to a power loss of about 2.5 dB, see Table 8.1. It can be observed that the linear scheme converges to its asymptote at much higher transmit powers than its dirty paper coding counterpart when the system is fully loaded.

Chapter 9
Description of the Quality of Service Feasibility Region

The weighted sum rate has so far been the primary utility at which the main focus of our optimizations aimed. Given a maximum available transmit power $P_{\max}$, an algorithm for the maximization of the weighted sum rate finally determined which user subset has to be served in order to achieve the maximum metric. Depending on the choice of the weight vector, the available transmit power, the number of users, and the transmission technique (linear or nonlinear), some users possibly did not receive any active data stream. At a higher level, a master program (scheduler for example) usually updates the user weights such that on a larger time scale, every user may be served at his desired rate.

While this procedure works fine for elastic traffic, it is not applicable for time-critical, inelastic traffic with strict delay constraints. In the latter scenario, Quality-of-Service (QoS) requirements must be guaranteed to every particular user and these requirements must be met instantaneously without the possibility to apply time-sharing and satisfy the needs on a larger time scale. Such QoS constraints are usually formulated in terms of lower bounds on the Shannon rate, the signal-to-interference-and-noise ratios, and bit error ratios, or as upper bounds on the individual mean square errors. Besides assuring the compliance with the QoS constraints for example to enable a voice-communication with tolerable quality, the minimization of the required transmit power is the main objective in order to keep the generated interference to the adjacent cells as small as possible. The resulting optimization is known as *power minimization* and can be written in an abstract way as[1]

$$\text{minimize}\, P_{\max} \quad \text{s.t.: } \mathrm{QoS}_k \geq \mathrm{QoS}_k^{\text{target}} \quad \forall k \in \{1, \ldots, K\} \tag{9.1}$$

where the minimization is with respect to the power allocation and the transmit and receive filters. The central question which naturally arises from above power minimization problem is whether a solution does exist at all, i.e., whether the constraint set is non-empty or not. In other words, it is important to know whether the QoS

[1] If the QoS constraints are formulated as upper bounds like in the MSE case, the larger-than-or-equal-to sign must be reverted.

R. Hunger, *Analysis and Transceiver Design for the MIMO Broadcast Channel*, Foundations in Signal Processing, Communications and Networking 8,
DOI: 10.1007/978-3-642-31692-0_9, © Springer-Verlag Berlin Heidelberg 2013

constraints can be satisfied with a possibly unbounded transmit power or not. As we will see, this need not necessarily be the case. The answer to this question heavily depends on the choice of the transmission strategy. When dirty paper coding is applied, basically any set of QoS requirements is feasibly when the available transmit power is unbounded.[2] This was proven by Schubert et al. in [154]. However, as we have shown in Sect. 7.1.1, the complexity of the dirty paper coding implementation is tremendously high especially when the power of the known side-information becomes large, which happens for large QoS constraints. For the easier to implement linear transceivers, not every set of QoS requirements needs to be feasible even when the available transmit power $P_{\max}$ is allowed to grow beyond all limits. In this chapter, we present a complete description of the QoS feasibility region in the MIMO BC with linear transceivers which also allows for an extraordinary simple feasibility check of given QoS requirements.

Feasibility is first discussed for the vector BC with single-antenna receivers in Sect. 9.2. By means of the MSE duality for linear transceivers in Sect. 3.2, the BC feasibility problem is solved in its dual vector MAC for the regular and singular channels scenario. The former setup is characterized by the property that the channel matrix of any user subset features maximum rank whereas rank-deficient channels are allowed in the latter one. A key component of our analysis is the treatment of the feasibility question in the MMSE domain instead of the SINR domain which most of the publications in the literature focus on.

The extension to the MIMO BC with multi-antenna terminals follows in Sect. 9.3. It turns out that single-stream transmission is sufficient to clarify the feasibility of given QoS requirements. Hence, the treatment of feasibility in the MIMO BC is closely related to the vector BC. Obviously, single-stream transmission is usually suboptimum for the power minimization problem in (9.1). For the feasibility check, however, it will be shown that single-stream transmission is sufficient.

To begin with, we revisit existing results on feasibility and give a brief overview on hitherto feasibility test methods and their drawbacks in Sect. 9.1.

9.1 Literature Overview on Feasibility and Power Minimization

One of the first contributions on power minimization was presented by Yates in 1995 where he introduced a general framework for such a power minimization problem in which the QoS constraints are formulated in terms of the power inequality[3] [155]

$$\boldsymbol{p} \geq \boldsymbol{I}^{\mathrm{FF}}(\boldsymbol{p}, \boldsymbol{\mu}) \tag{9.2}$$

---

[2] All BC channel matrices must be different from the all-zeros matrix.

[3] Lower bounds on the SINRs or upper bounds on the MSEs for example can easily be recast into the power inequality form.

where $\boldsymbol{p} := [p_1, \dots, p_K]^{\mathrm{T}}$ contains the power allocated to the K users and $\boldsymbol{\mu} := [\mu_1, \dots, \mu_K]^{\mathrm{T}}$ the QoS requirements. The kth entry $I_k^{\mathrm{FF}}(\boldsymbol{p}, \mu_k)$ of the vector-valued (possibly nonlinear) interference function $\boldsymbol{I}^{\mathrm{FF}}(\boldsymbol{p}, \boldsymbol{\mu})$ comprises the interference power received by user k when users $1, \dots, K$ transmit with powers $p_1, \dots, p_K$. For particular interference functions $\boldsymbol{I}^{\mathrm{FF}}(\boldsymbol{p}, \boldsymbol{\mu})$ termed *standard*, Yates showed various convergence properties of the standard power control algorithm

$$\boldsymbol{p}^{(n+1)} = \boldsymbol{I}^{\mathrm{FF}}(\boldsymbol{p}^{(n)}, \boldsymbol{\mu}) \tag{9.3}$$

which updates the power allocation in iteration $n+1$ to the value of the interference function in iteration n. A major result is that when (9.3) converges as $n \to \infty$, then this fixed point is unique and the recurrence relation (9.3) converges to the unique optimum for any initialization $\boldsymbol{p}^{(0)} \geq \boldsymbol{0}$. Due to the uniqueness of the fixed point (if it exists!), the optimizer of the minimization

$$\underset{\boldsymbol{p} \geq \boldsymbol{0}}{\text{minimize}} \, \|\boldsymbol{p}\|_1 \quad \text{s.t.:} \; \boldsymbol{p} \geq \boldsymbol{I}^{\mathrm{FF}}(\boldsymbol{p}, \boldsymbol{\mu})$$

returns the only power vector $\check{\boldsymbol{p}} := \boldsymbol{p}^{(\infty)}$ which satisfies the inequality constraint $\check{\boldsymbol{p}} \geq \boldsymbol{I}^{\mathrm{FF}}(\check{\boldsymbol{p}}, \boldsymbol{\mu})$ *with equality*. Unfortunately, conditions for the existence of the fixed point are not given by Yates. Clearly, one could run the standard power control algorithm (9.3) and see whether it converges or not. However, a concrete convergence criterion is not trivial due to the fact that convergence is usually not monotonic for an arbitrary initialization $\boldsymbol{p}^{(0)}$. Monotonic convergence (with nonincreasing entries in $\boldsymbol{p}^{(n)}$) is obtained for a feasible initialization $\boldsymbol{p}^{(0)}$ which satisfies the constraints $\boldsymbol{p}^{(0)} \geq \boldsymbol{I}^{\mathrm{FF}}(\boldsymbol{p}^{(0)}, \boldsymbol{\mu})$, but knowing a feasible initialization would already mean that the constraints are feasible, which is unclear in advance. Another possibility to achieve monotonic convergence was shown to be the all-zeros initialization $\boldsymbol{p}^{(0)} = \boldsymbol{0}$. But this time, the entries of $\boldsymbol{p}^{(n)}$ are increasing in n such that the constraint $\boldsymbol{p}^{(n)} \geq \boldsymbol{I}^{\mathrm{FF}}(\boldsymbol{p}^{(n)}, \boldsymbol{\mu})$ can be satisfied only for $n \to \infty$ (conditioned on the feasibility of the constraints). Moreover, it was shown in [156, Fig. 3], that the number of iterations of the fixed point iteration (9.3) that are required to satisfy the constraints up to a certain tolerance dramatically increases close to the plane that separates feasibility from infeasibility. We can conclude that running Yates' power minimization algorithm and checking whether it 'converges' or not is not suitable to decide whether given QoS constraints are feasible or not.

The concept of interference functions has further been promoted by Boche et al. In [157], special interference functions with the property of being convex or concave are studied. Such an interference model no longer relies on a coupling matrix, but can be seen as an extension that includes adaptive receiver structures. The main result is that every convex interference function can be regarded as an optimum over linear interference functions, which often helps to reveal a hidden convexity of an optimization by changing the optimization variables. For non-convex (non-concave) interference functions, the best possible convex (concave) approximations are derived as well. Going further, Boche et al. derive a relationship between comprehensive feasible sets and an axiomatic framework of interference functions in [158]. For example, the SIR region is a sublevel set of a spectral radius interference function and is hence determined by the properties of the interference function. In [159], an extension to log-convex interference functions is presented, which contain the convex interference functions as a subset. Again, every log-convex interference function can be expressed as a maximum of elementary interference functions. In addition, the reachability of the boundary of a QoS region, that is intimately related to an interference function is discussed. Based on the interference function framework, the largest class of utility and interference functions is derived in [160], for which a transformation of power variables exists that reveals the hidden convexity of the optimization. For any other utility/interference function that does not belong to the derived class, a transformation into a convex/concave formulation is impossible. The results highlight the special property of log-convex interference functions. This powerful contribution is accompanied by the proof that the exponential transformation of the power variables is the unique transformation that reveals a hidden convexity.

In 1999, Bengtsson and Ottersten tackled the power minimization problem with given SINR requirements in the vector BC [161]. For instantaneous channel knowledge, they reformulated the power minimization problem as a second-order cone program which can be solved to global optimality *conditioned on the feasibility* of the constraints. Infeasible constraints were said to be reported by the second order cone program itself. However, it is difficult to adjust the parameters of the solver that control the accuracy of this decision.

Another possibility to detect feasibility results from the distinguished publications from Boche and Schubert in the field of SINR balancing (e.g.,[30, 35, 156]) and SIR balancing (e.g.,[162–165]). Given *relative* SINR targets $\gamma_1, \ldots, \gamma_K > 0$, the SINR balancing optimization reads as

$$\begin{aligned} \text{maximize } \gamma_0 \quad \text{s.t.: } & \text{SINR}_k \geq \gamma_0 \cdot \gamma_k \ \forall k \in \{1, \ldots, K\} \\ & P_{\text{diss}}^{\text{BC}} \leq P_{\max} \end{aligned} \tag{9.4}$$

i.e., the scaling γ_0 of the *absolute* SINR lower bounds is maximized. For $P_{\max} \to \infty$ and a linear power allocation, the SINR balancing is equivalent to the SIR balancing in which the noise component can be neglected and only the ratios of the users' powers are of interest. If the maximum objective $\check{\gamma}_0$ of the SINR balancing in (9.4) is larger than or equal to one, then the SINR targets $\gamma_1, \ldots, \gamma_K$ are feasible. Clearly, $\check{\gamma}_0$ is nondecreasing in the maximum available transmit power $P_{\max}$ such that the largest $\check{\gamma}_0$ is obtained for $P_{\max} \to \infty$, i.e., for the *SIR* balancing. Note that the difference between the optimum objective $\check{\gamma}_0$ of the SINR balancing *with power limitation* and the maximum objective of the SIR balancing can be made as small as desired by increasing the sum power limit $P_{\max}$. For the boundary that separates feasibility from infeasibility, it therefore suffices to analyze the SIR balancing problem and see whether the maximum common SIR scaling is larger than one (then the SINR targets are feasible with finite sum power), equal to one (then the SINR targets are 'feasible' with an infinite amount of power), or smaller than one (then the SINR targets are infeasible even for an infinite amount of power). SIR balancing was first considered amongst others by Montalbano and Slock in [166] for the case $\gamma_1 = \cdots = \gamma_K$ and a heuristic iterative algorithm was presented that is based on the alternating optimization between the power allocation and the receive filter adaptation. The extension towards different relative SINR targets $\gamma_1, \ldots, \gamma_K$ as well as a rigorous convergence analysis of the alternating optimization algorithm was published in 2002 in [162]. An important result from Boche and Schubert is that the maximum scaling $\check{\gamma}_0$ of both the SINR and SIR balancing is closely related to the spectral radius of a scaled coupling matrix (e.g.,[163–166]). Note that this scaling depends on the relative SIR targets $\gamma_1, \ldots, \gamma_K$. For *fixed* receivers, different SIR targets can be tested for feasibility by computing the spectral radius of the scaled coupling matrix but only little information about the structure of the underlying SIR region can be deduced from that. Since the receivers may be adapted to the power allocation, the coupling matrix depends on the choice of the receive filters such that the feasibility check for adaptive receivers can only be performed in an iterative manner. Fortunately, the speed of convergence of the SI(N)R balancing algorithm of Schubert and Boche is much higher than that of Yates' power minimization algorithm.

Nonetheless, the iterative nature of the feasibility check remains problematic since one has to decide for the number of iterations that will be run. If this number is chosen to small, it might happen that a few additional iterations would change an infeasible power allocation to a feasible one. Hence, the decision for feasibility of iterative algorithms is sensitive close to the boundary that separates feasibility from infeasibility.

The (so far existing) lack of a complete description of the QoS feasibility region in the multi-user BC with noncooperative receivers leads to a further problem. Assume that given QoS requirements have turned out to be infeasible. Without any knowledge about the feasibility region, it is unclear how to update the requirements such that they become feasible. Finally, the hitherto existing feasibility checks had to be executed for every channel realization. We will show the somehow surprising result that feasibility of given QoS requirements does *not* depend on the channel realization as long as the channel matrix fulfills some very mild rank conditions.

For system setups different from the multi-antenna broadcast channel, some theoretical results on the feasibility of given QoS requirements have already been published by various authors. Note that some conclusions are only *necessary* conditions that QoS requirements have to satisfy in order to be feasible. Sufficiency is much harder to show. In the following, we give a short overview of those contributions.

In 1999, Viswanath et al. investigated the uplink of a multi-user CDMA system with a single-antenna base station and MMSE receivers [167]. As a first outcome, they proved that the sum of the complementary MMSEs (one minus the MMSE) is upper bounded (with strict inequality) by the processing gain. This limit is valid for *any* sequence and denotes a necessary condition for the feasibility of SINR requirements. Sufficiency for this condition is proven via an explicit signature and power allocation scheme. In other words, there exists a signature for which the SINR requirements can be satisfied if and only if they are chosen such that the complementary MMSEs (also termed effective bandwidths) sum up to less than the processing gain. Translated to the vector BC in which the channel vectors correspond to the CDMA sequences, this means that a channel matrix exists for which the SINR requirements can be fulfilled. However, the channel matrix of the vector BC cannot be altered and is given by the environment. Their observations remain valid in the downlink of the CDMA system as well except that a closed form solution of the power allocation is not possible there.

Palomar et al. analyzed the point-to-point MIMO link with full cooperation at both ends of the link and MMSE receive filtering in [168] for the case that QoS constraints are imposed for every of the L parallel streams. By means of the majorization theory, they showed that given MMSE targets $\mu_1, \ldots, \mu_L$ are feasible if and only if

$$\sum_{j=1}^{L} \mu_j > L - \min\{L, \operatorname{rank}(\boldsymbol{H})\}. \tag{9.5}$$

We will show that basically the same result also holds in the vector BC where cooperation amongst the single-antenna terminals is impossible, which requires a completely different technique for the proof due to the lack of cooperation.

In 2006, Wiesel et al. focused on the MIMO BC with fixed receivers in which every antenna element at the receiving side is interpreted as a virtual user [36]. The resulting setup thus corresponds to a vector BC. Similar to the work of Bengtsson and Ottersten [161], they reformulated the power minimization problem with given SINR requirements as a second order cone program. In addition, they showed that if the minimum SIR over all users is larger than or equal to the common absolute requirement $\bar{\gamma}$, then $\bar{\gamma}$ is smaller than or equal to $(K/\operatorname{rank}(\boldsymbol{H}^{\mathrm{BC}}) - 1)^{-1}$, where K denotes the number of (virtual) users:

$$\min_k \mathrm{SIR}_k \geq \bar{\gamma} \quad \Rightarrow \quad \bar{\gamma} \leq \frac{1}{\frac{K}{\operatorname{rank}(\boldsymbol{H}^{\mathrm{BC}})} - 1} \tag{9.6}$$

Note that (9.6) does by no means say that the converse holds as well, i.e., it does *not* say that if $\bar{\gamma}$ satisfies the rank inequality, then the minimum SIR over all users is at least $\bar{\gamma}$. Thus, (9.6) is only a necessary condition for the feasibility of the common SIR requirement $\bar{\gamma}$, but not a sufficient one, and Wiesel also presents a counterexample in [36].

A multi-user SISO broadcast channel with fading is considered by Jorswieck et al. in [169]. The simple *in*feasibility condition

$$\sum_{k=1}^{K} \mu_k < K - 1 \quad \Rightarrow \quad \boldsymbol{\mu} \text{ is infeasible}$$

of the MSE targets $\mu_1, \ldots, \mu_K$ belonging to the K users is proven. However, if above inequality does not hold, then this does not automatically mean that the MSE targets $\boldsymbol{\mu}$ are feasible.

In the following, we derive a complete description of the QoS feasibility region in the broadcast channel. To this end, we first handle the vector BC with single-antenna terminals and then extend the results to the MIMO BC. The MMSE domain will turn out to serve as the adequate means which allows for a simple analysis. These results have partly been published in [170, 171].

9.2 Feasibility in the Vector Broadcast Channel

As we have seen in Sect. 3.2, the vector BC and the dual vector MAC with a sum power constraint share the same MSE region. Hence, we can tackle the QoS feasibility problem in the dual vector MAC instead where it offers a much simpler structure since the receivers can be optimized independently. As a consequence, the system can completely be characterized in terms of the MAC transmit powers $\boldsymbol{q} := [q_1, \ldots, q_K]^{\mathrm{T}} \geq \mathbf{0}$ of the K users. The MAC channel vectors $\boldsymbol{h}_1, \ldots, \boldsymbol{h}_K$ are stored in the composite channel matrix

$$\boldsymbol{H} := [\boldsymbol{h}_1, \ldots, \boldsymbol{h}_K] \in \mathbb{C}^{N \times K} \tag{9.7}$$

and we assume $\boldsymbol{h}_k \neq \mathbf{0}\ \forall k$, i.e., data transmission from every of the K users to the base station in the dual MAC is possible. When the vector-valued MMSE receive filter[4]

$$\boldsymbol{g}_k^{\mathrm{T}} = \sqrt{q_k}\boldsymbol{h}_k^{\mathrm{H}}\left(\mathbf{I}_N + \boldsymbol{H}\boldsymbol{Q}\boldsymbol{H}^{\mathrm{H}}\right)^{-1} \tag{9.8}$$

of user k is applied with $\boldsymbol{Q} := \mathrm{diag}\{q_k\}_{k=1}^{K} \in \mathbb{R}_{+,0}^{K \times K}$, his MSE in (3.7) boils down to the *M*MSE and reads as[5]

$$\begin{aligned}\varepsilon_k(\boldsymbol{q}) &= 1 - q_k\boldsymbol{h}_k^{\mathrm{H}}\left(\mathbf{I}_N + \boldsymbol{H}\boldsymbol{Q}\boldsymbol{H}^{\mathrm{H}}\right)^{-1}\boldsymbol{h}_k \\ &= \left[\left(\mathbf{I}_K + \boldsymbol{Q}^{\frac{1}{2}}\boldsymbol{H}^{\mathrm{H}}\boldsymbol{H}\boldsymbol{Q}^{\frac{1}{2}}\right)^{-1}\right]_{k,k}\end{aligned} \tag{9.9}$$

where the second line follows from the matrix inversion lemma in Appendix A.2. Note that describing the feasibility in terms of the MMSE is fully equivalent to describing it in any other domain to which a bijective mapping exists. For example, the MMSE is related to the maximum SINR via [cf. (A8) in Appendix A.3]

$$\varepsilon_k(\boldsymbol{q}) = \frac{1}{1 + \mathrm{SINR}_k(\boldsymbol{q})} \tag{9.10}$$

and to the Shannon rate via

[4] The scalar MAC precoders $t_1, \ldots, t_K$ are assumed to be real-valued and nonnegative such that $q_k = t_k^2\ \forall k$.

[5] In this chapter, the *M*MSE is denoted by the variable ε.

$$R_k(\boldsymbol{q}) = -\log_2 \varepsilon_k(\boldsymbol{q}). \tag{9.11}$$

Testing feasibility in the MMSE domain therefore means to find out whether a power allocation $\boldsymbol{q} \geq \boldsymbol{0}$ exists for which the achieved MMSEs $\varepsilon_1(\boldsymbol{q}), \ldots, \varepsilon_K(\boldsymbol{q})$ do not exceed the (maximum) MMSE requirements $\mu_1, \ldots, \mu_K$.

For any power allocation $\boldsymbol{q} \geq \boldsymbol{0}$, the second line in (9.9) reveals that $\varepsilon_k(\boldsymbol{q}) \geq 0 \, \forall k$, whereas $\varepsilon_k(\boldsymbol{q}) \leq 1 \, \forall k$ follows from the first line in (9.9). For $q_k < \infty$, the inequality $\varepsilon_k(\boldsymbol{q}) > 0$ is strict. Conversely, for $q_k \to \infty$, it does *not* automatically follow that $\varepsilon_k(\boldsymbol{q}) \to 0$. Instead, $q_k \to \infty$ is only a *necessary*, but usually not a sufficient condition for $\varepsilon_k(\boldsymbol{q}) \to 0$. We subsume these important results in Lemma 9.1.

Lemma 9.1 *Using the MMSE receive filter in the dual vector MAC with identity noise covariance matrix, the MMSE $\varepsilon_k(\boldsymbol{q})$ of some user k satisfies the box constraint*

$$0 \leq \varepsilon_k(\boldsymbol{q}) \leq 1 \quad \forall k \tag{9.12}$$

for any nonnegative power allocation $\boldsymbol{q} \geq \boldsymbol{0}$. For $q_k < \infty$, the inequality $\varepsilon_k(\boldsymbol{q}) > 0$ is strict. A necessary but usually not sufficient condition for $\varepsilon_k(\boldsymbol{q}) \to 0$ is that $q_k \to \infty$.

Given the individual MMSEs in (9.9), the *sum* MMSE can be expressed as

$$\begin{aligned} \sum_{k\in\mathcal{K}} \varepsilon_k(\boldsymbol{q}) &= K - \operatorname{tr}\left[\boldsymbol{H}\boldsymbol{Q}\boldsymbol{H}^{\mathrm{H}}\left(\mathbf{I}_N + \boldsymbol{H}\boldsymbol{Q}\boldsymbol{H}^{\mathrm{H}}\right)^{-1}\right] \\ &= K - N + \operatorname{tr}\left[\left(\mathbf{I}_N + \boldsymbol{H}\boldsymbol{Q}\boldsymbol{H}^{\mathrm{H}}\right)^{-1}\right] \end{aligned} \tag{9.13}$$

where $\mathcal{K} := \{1, \ldots, K\}$ denotes the set of user indices. Due to the fact that $\mathbf{I}_N + \boldsymbol{H}\boldsymbol{Q}\boldsymbol{H}^{\mathrm{H}} \succ \boldsymbol{0}$, the sum MMSE in lower bounded by $K - N$ for any power allocation $\boldsymbol{q} \geq \boldsymbol{0}$. In addition, since $\operatorname{rank}(\boldsymbol{H}\boldsymbol{Q}\boldsymbol{H}^{\mathrm{H}}) \leq \operatorname{rank}(\boldsymbol{H})$, the trace in the first line of (9.13) is upper bounded by $\operatorname{rank}(\boldsymbol{H})$ such that the sum MMSE is finally lower bounded by $K - \operatorname{rank}(\boldsymbol{H})$. For $\|\boldsymbol{q}\|_1 < \infty$, this inequality is strict due to the noise component. But for any power allocation in which the power of *every* individual user goes to infinity as $\|\boldsymbol{q}\|_1 = P_{\max} \to \infty$, the lower bound on the sum MMSE is obtained:

$$\sum_{k\in\mathcal{K}} \varepsilon_k(\boldsymbol{q}) \to K - \operatorname{rank}(\boldsymbol{H}) \quad \text{for } q_k \to \infty \, \forall k \in \mathcal{K} \tag{9.14}$$

Above limit is due to the fact that for $q_k \to \infty \, \forall k \in \mathcal{K}$, all $\operatorname{rank}(\boldsymbol{H})$ nonzero eigenvalues of the matrix product $\boldsymbol{H}\boldsymbol{Q}\boldsymbol{H}^{\mathrm{H}}$ grow beyond all limits. In particular, any *positive* power allocation of the form

$$\boldsymbol{q} = \boldsymbol{a}P_{\max} \quad \text{with } \boldsymbol{a} > \boldsymbol{0} \text{ and } \|\boldsymbol{a}\|_1 = 1$$

achieves the lower bound in (9.14) when $P_{\max} \to \infty$. Note that (9.14) can be generalized to the case when possibly less than K users have powers that grow beyond all limits. Let $\mathcal{I}$ denote the set of user indices such that the matrix $\boldsymbol{H}_{\mathcal{I}}$ which contains the channel vectors of all users in $\mathcal{I}$ has the same rank as the composite channel matrix $\boldsymbol{H}_{\mathcal{K}} = \boldsymbol{H}$ that contains the channel vectors of all users. This means that the cardinality of $\mathcal{I}$ is at least $|\mathcal{I}| \geq \text{rank}(\boldsymbol{H})$. Then, the sum MMSE lower bound is achieved when all users in $\mathcal{I}$ have powers that tend to infinity. These results on the sum MMSE are combined in the following lemma:

Lemma 9.2 *For any power allocation $\boldsymbol{q} \geq \boldsymbol{0}$ in the dual vector MAC, the sum MMSE is lower bounded by*

$$\sum_{k\in\mathcal{K}} \varepsilon_k(\boldsymbol{q}) \geq K - \mathit{rank}(\boldsymbol{H}). \tag{9.15}$$

Note that above inequality is strict for $\|\boldsymbol{q}\|_1 < \infty$, i.e.,

$$\sum_{k\in\mathcal{K}} \varepsilon_k(\boldsymbol{q}) > K - \mathit{rank}(\boldsymbol{H}) \quad \mathit{for} \quad \|\boldsymbol{q}\|_1 < \infty. \tag{9.16}$$

Let $\mathcal{I}$ be any set of user indices such that the matrix $\boldsymbol{H}_{\mathcal{I}}$ which contains the corresponding channel vectors of all users in $\mathcal{I}$ features the same rank as the matrix $\boldsymbol{H}$, which implies that $|\mathcal{I}| \geq \mathit{rank}(\boldsymbol{H})$. The lower bound in (9.15) is asymptotically achieved for any power allocation satisfying $q_k \to \infty \ \forall k \in \mathcal{I}$, i.e.,

$$\sum_{k\in\mathcal{K}} \varepsilon_k(\boldsymbol{q}) \to K - \mathit{rank}(\boldsymbol{H}) \quad \mathit{as} \ q_k \to \infty \ \forall k \in \mathcal{I}. \tag{9.17}$$

In particular, a power allocation in which all powers $q_1, \ldots, q_K$ tend to infinity achieves the lower bound on the sum MMSE with equality. The linear power allocation $\boldsymbol{q} = \boldsymbol{a} P_{\max}$ with $\boldsymbol{a} > \boldsymbol{0}$, $\|\boldsymbol{a}\|_1 = 1$, and $P_{\max} \to \infty$ is a prominent example for which the lower bound is achieved.

Note that certain target MMSE requirements $\mu_1, \ldots, \mu_K$ are feasible, if and only if there exists a finite power allocation $\boldsymbol{q} \geq \boldsymbol{0}$ with $\|\boldsymbol{q}\|_1 < \infty$ such that

$$\varepsilon_k(\boldsymbol{q}) \leq \mu_k \quad \forall k \in \mathcal{K} \tag{9.18}$$

i.e., for which the MMSE of user k is at most μ_k and this is valid for every user k out of $\mathcal{K}$. By means of (9.9), we can rewrite the QoS constraints in (9.18) to[6]

[6] In the literature, the interference function is usually stated in terms of the SINR requirements $\gamma_1, \ldots, \gamma_K$. As the MMSE receive filter in (9.8) also maximizes the SINR of user k, we find

$$\text{SINR}_k(\boldsymbol{q}) = q_k \boldsymbol{h}_k^{\mathrm{H}} \Big(\mathbf{I} + \sum_{i\neq k} \boldsymbol{h}_i \boldsymbol{h}_i^{\mathrm{H}} q_i\Big)^{-1} \boldsymbol{h}_k.$$

Using the relationship between the MMSE and the maximum SINR in (9.10), the QoS requirement in (9.18) can alternatively to (9.19) be written as

$$q_k \geq \frac{1-\mu_k}{\boldsymbol{h}_k^{\mathrm{H}}\left(\mathbf{I}_N + \boldsymbol{H}\boldsymbol{Q}\boldsymbol{H}^{\mathrm{H}}\right)^{-1}\boldsymbol{h}_k} = I_k^{\mathrm{FF}}(\boldsymbol{q}, \mu_k) \quad \forall k \in \mathcal{K}. \tag{9.19}$$

It is shown in Appendix A.21 that $I_1^{\mathrm{FF}}(\boldsymbol{q}, \mu_1), \ldots, I_K^{\mathrm{FF}}(\boldsymbol{q}, \mu_K)$ fulfill for $0 \leq \mu_k < 1\ \forall k$ the standard interference function conditions that have been defined by Yates in [155]. These properties lead to the following lemma:

Lemma 9.3 *If some MMSE requirements $\boldsymbol{\mu}$ are feasible, i.e., a power allocation $\boldsymbol{q}$ exists for which $\boldsymbol{q} \geq \boldsymbol{I}^{\mathrm{FF}}(\boldsymbol{q}, \boldsymbol{\mu})$, then all other MMSE requirements $\boldsymbol{\mu}' \geq \boldsymbol{\mu}$ for which $\boldsymbol{\mu}' \leq \mathbf{1}$ holds are feasible as well.*

For the proof, we first rule out the case that $\mu_k' = 1$ holds for some k. If so, simply no power is allocated to this user and he can be dropped from the system. Therefore, we may assume $\boldsymbol{\mu}' < \mathbf{1}$ without loss of generality. From the results of Yates in [155], we know that there exists a unique power allocation $\check{\boldsymbol{q}}$ for which $\check{\boldsymbol{q}} = \boldsymbol{I}^{\mathrm{FF}}(\check{\boldsymbol{q}}, \boldsymbol{\mu})$ since $\boldsymbol{\mu}$ is feasible, see the overview in Sect. 9.1. Since $\check{\boldsymbol{\mu}} \geq \boldsymbol{\mu}$, we know from (9.19) that

$$\boldsymbol{I}^{\mathrm{FF}}(\check{\boldsymbol{q}}, \boldsymbol{\mu}') \leq \boldsymbol{I}^{\mathrm{FF}}(\check{\boldsymbol{q}}, \boldsymbol{\mu}) = \check{\boldsymbol{q}}.$$

As a consequence, the power allocation $\check{\boldsymbol{q}}$ achieves MMSEs that are not larger than the requirements $\boldsymbol{\mu}'$ which means that $\boldsymbol{\mu}'$ is also feasible and a unique power allocation $\check{\boldsymbol{q}}' \leq \check{\boldsymbol{q}}$ exists for which $\check{\boldsymbol{q}}' = \boldsymbol{I}^{\mathrm{FF}}(\check{\boldsymbol{q}}', \boldsymbol{\mu}')$.

Lemmas 9.1–9.3 will play a central role for the theorem on the structure of the QoS feasibility region. For its complete description, we need to distinguish the 'regular' channels scenario from the 'singular' one. The former scenario which will be discussed in Sect. 9.2.1 is characterized by the fact that the channel vectors of any nonempty user subset $\mathcal{I} \subseteq \mathcal{K}$ with cardinality $1 \leq |\mathcal{I}| \leq K$ satisfy the rank relation[7]

$$\operatorname{rank}(\boldsymbol{H}_{\mathcal{I}}) = \min\{|\mathcal{I}|, N\} \quad \forall \mathcal{I} \subseteq \mathcal{K}. \tag{9.20}$$

Contrary, if a set $\mathcal{I} \subseteq \mathcal{K}$ exists for which $\operatorname{rank}(\boldsymbol{H}_{\mathcal{I}}) < \min\{|\mathcal{I}|, N\}$, the scenario is termed singular and its QoS feasibility region will be dealt with in Sect. 9.2.2.

(Footnote 6 continued)

$$q_k \geq \frac{\frac{1}{\mu_k} - 1}{\boldsymbol{h}_k^{\mathrm{H}}\left(\mathbf{I} + \sum_{i \neq k} \boldsymbol{h}_i \boldsymbol{h}_i^{\mathrm{H}} q_i\right)^{-1}\boldsymbol{h}_k} \quad \forall k \in \mathcal{K}$$

which one usually finds in the literature. Note that these two (different) representations do not share the same speed of convergence when the standard power control algorithm in (9.3) is applied.

[7] The validity of the regularity condition is independent of whether the MAC or BC channels are tested, see (3.14).

9.2.1 Regular Channels Scenario

It is to be expected that basically any channel matrix that may arise in a practical system satisfies the regular channels condition in (9.20). Under this assumption, $\mathrm{rank}(\boldsymbol{H}) = \min\{N, K\}$ holds and the structure of the QoS feasibility region is given by the following theorem:

Theorem 9.1 *The closure of the feasible MMSE region in the vector broadcast channel whose BC channel vectors fulfill* (9.20) *is the polytope*

$$\mathcal{P} := \left\{\boldsymbol{\mu} \in \mathbb{R}^K \middle| \boldsymbol{0} \leq \boldsymbol{\mu} \leq \boldsymbol{1} \wedge \boldsymbol{1}^{\mathrm{T}}\boldsymbol{\mu} \geq K - N\right\} \tag{9.21}$$

whose bounding half-spaces are the individual box constraints $0 \leq \mu_k \leq 1 \quad \forall k \in \mathcal{K}$ *and the sum MMSE constraint* $\sum_{k\in\mathcal{K}} \mu_k \geq K - N$. *By means of a positive power allocation with finite sum power, any point belonging to the interior* $\mathrm{int}(\mathcal{P})$ *can be achieved. For MMSE requirements equal to one, no power is allocated to the particular user and any point* $\boldsymbol{\mu} \neq \boldsymbol{1}$ *on the boundary* $\partial\mathcal{P}$ *requires an infinite amount of sum power. MMSE requirements outside the polytope* $\mathcal{P}$ *are infeasible even with an infinite amount of sum power.*

Due to the MSE duality, the dual MAC results seamlessly translate to the vector BC, i.e., all results that have so far been proven in the dual vector MAC also hold in the original vector BC. The fact that MMSE requirements outside $\mathcal{P}$ are infeasible follows from (9.12) in Lemmas 9.1 and 9.2. An MMSE requirement μ_k of some user k that is equal to one can be satisfied by simply allocating no power to the respective user k. This follows from (9.9) and also translates to the vector BC.

So far, we have proven the 'necessary' parts of the theorem. It remains to prove that a power allocation with finite sum power exists for any MMSE requirement that belongs to the interior of $\mathcal{P}$ and that requirements on the boundary of the polytope that are different from the all-ones vector require an infinite amount of power. For these proofs, we further distinguish the cases $K \leq N$ (enough degrees of freedom) and $K > N$ (overloaded system).

9.2.1.1 Enough Degrees of Freedom

When the base station is equipped with at least as many antenna elements as there are users in the vector BC, the remaining parts of the proof for Theorem 9.1 are simple. For $K \leq N$, only the box constraints come into play such that the polytope $\mathcal{P} = \{\boldsymbol{\mu}|\boldsymbol{0} \leq \boldsymbol{\mu} \leq \boldsymbol{1}\}$ is eventually the unit hyper-cube in this case. For $\boldsymbol{\mu} \in \mathrm{int}(\mathcal{P})$, $\boldsymbol{\mu} < \boldsymbol{1}$ holds such that some positive power must be allocated to every user, i.e., $\boldsymbol{q} > \boldsymbol{0}$ and thus $\boldsymbol{Q} \succ \boldsymbol{0}$ is necessary to satisfy requirements $\boldsymbol{\mu} < \boldsymbol{1}$, see (9.9). Then, the MMSE of user k in (9.9) obeys the inequality

$$\varepsilon_k(\boldsymbol{q}) \leq q_k^{-1}\left[(\boldsymbol{H}^{\mathrm{H}}\boldsymbol{H})^{-1}\right]_{k,k}$$

such that in order to achieve an MMSE requirement μ_k, the power that has to be allocated to user k is upper bounded by

$$q_k \leq \mu_k^{-1}\left[(\boldsymbol{H}^{\mathrm{H}}\boldsymbol{H})^{-1}\right]_{k,k} \quad \forall k. \tag{9.22}$$

Thus, for any $\boldsymbol{\mu} \in \mathrm{int}(\mathcal{P})$, the requirements $\mu_1, \ldots, \mu_K$ are strictly larger than zero such that q_k in (9.22) is finite for all k and thus, also the sum power $\|\boldsymbol{q}\|_1$ is finite. If $\boldsymbol{q}$ is chosen as the upper bound in (9.22) with $\boldsymbol{\mu} > \boldsymbol{0}$, the power allocation $\boldsymbol{q}$ satisfies $\boldsymbol{q} > \boldsymbol{I}^{\mathrm{FF}}(\boldsymbol{q}, \boldsymbol{\mu})$ and therefore, $\boldsymbol{\mu}$ is feasible. By increasing q_k, the requirement μ_k can be made as close to zero as desired and this is even valid for all users k simultaneously since $K \leq N$. This proves that any requirement $\boldsymbol{\mu} \in \mathrm{int}(\mathcal{P})$ is feasible with finite sum power.

It remains to show that requirements $\boldsymbol{\mu} \in \partial\mathcal{P}$ on the boundary with $\boldsymbol{\mu} \neq \boldsymbol{1}$ can only be achieved asymptotically with an infinite amount of transmit power since at least one entry of $\boldsymbol{\mu}$ is zero for $\boldsymbol{\mu} \in \partial\mathcal{P}$ and $\boldsymbol{\mu} \neq \boldsymbol{1}$. This follows from (9.22) and Lemma 9.1, which means that the bound in (9.22) becomes sharp as $\mu_k \to 0$ and $q_k \to \infty$ is both necessary and sufficient for $\varepsilon_k(\boldsymbol{q}) \to 0$ when $K \leq N$. These steps complete the proof of Theorem 9.1 for the case $K \leq N$.

9.2.1.2 Overloaded System

When more users are present in the system than the base station has antenna elements, the proof that any point in the interior of the polytope $\mathcal{P}$ can be obtained with finite sum power is slightly more challenging as the matrix $\boldsymbol{H}^{\mathrm{H}}\boldsymbol{H}$ is not invertible, i.e., a zero-forcing upper bound as in (9.22) is impossible. For $K > N$, not only the individual box constraints limit the closure of the feasible MMSE region. Instead, the sum MMSE lower bound $\sum_{k\in\mathcal{K}} \varepsilon_k(\boldsymbol{q}) \geq K - N$ from (9.15) in Lemma 9.2 crops the unit hyper-cube.

The basic idea for the proof that there exists a power allocation $\boldsymbol{q} > \boldsymbol{0}$ with $\|\boldsymbol{q}\|_1 < \infty$ for any MMSE requirement $\boldsymbol{\mu} \in \mathrm{int}(\mathcal{P})$ goes as follows: First, we show that any point in the interior of the $K-1$-dimensional bounding polytope

$$\mathcal{P}_\Sigma := \left\{\boldsymbol{\mu} \middle| \boldsymbol{0} \leq \boldsymbol{\mu} \leq \boldsymbol{1} \wedge \boldsymbol{1}^{\mathrm{T}}\boldsymbol{\mu} = K - N\right\}$$

can asymptotically be achieved by a linear power allocation $\boldsymbol{q} = \boldsymbol{a}P_{\max}$ with $\boldsymbol{a} > \boldsymbol{0}$ and $\|\boldsymbol{a}\|_1 = 1$ as $P_{\max} \to \infty$. Furthermore, we show that with finite sum power $\|\boldsymbol{q}\|_1 < \infty$, it is possible to synthesize MMSEs that come as close to any target MMSE $\boldsymbol{\mu} \in \mathrm{int}(\mathcal{P}_\Sigma)$ as desired, but with a strictly larger sum MMSE. The feasibility of all remaining MMSE targets $\boldsymbol{\mu} \in \mathrm{int}(\mathcal{P})$ out of the interior of the whole polytope $\mathcal{P}$ then follows from Lemma 9.3.

For any $\boldsymbol{\mu} \in \text{int}(\mathcal{P}_\Sigma)$, the strict inequalities $0 < \mu_k < 1\ \forall k$ are valid. Using (9.10), the MMSE requirements $\mu_1, \ldots, \mu_K$ can be converted into SINR requirements $\gamma_1, \ldots, \gamma_K$ via

$$\gamma_k = \mu_k^{-1} - 1 \quad \forall k \in \mathcal{K} \tag{9.23}$$

and these SINR targets fulfill $0 < \gamma_k < \infty\ \forall k$ such that the requirements in (9.23) are well-defined for the application of the SINR *balancing* algorithm [see (9.4)] that has extensively been analyzed by Boche and Schubert (e.g., [30, 35, 156]):

$$\begin{aligned} \underset{\boldsymbol{q}\geq \mathbf{0},\gamma_0}{\text{maximize}}\ \gamma_0 \quad \text{s.t.:}\ & \text{SINR}_k(\boldsymbol{q}) \geq \gamma_0 \cdot \gamma_k\ \forall k \in \{1, \ldots, K\} \\ & \|\boldsymbol{q}\|_1 \leq P_{\text{max}} \end{aligned} \tag{9.24}$$

Note that the signal-to-interference-and-noise ratio of user k is defined via (see Footnote 6 on page 249)

$$\text{SINR}_k(\boldsymbol{q}) := q_k \boldsymbol{h}_k^{\text{H}} \Big(\mathbf{I} + \sum_{i\neq k} \boldsymbol{h}_i \boldsymbol{h}_i^{\text{H}} q_i\Big)^{-1} \boldsymbol{h}_k. \tag{9.25}$$

In other words, any SINR requirements that fulfill $0 < \gamma_k < \infty\ \forall k$ (which is a consequence of $\boldsymbol{\mu} \in \text{int}(\mathcal{P}_\Sigma)$) can be balanced. Let $\check{\gamma}_0(P_{\text{max}})$ denote the maximum of the optimization in (9.24). Since $\text{SINR}_k(\alpha\boldsymbol{q}) > \text{SINR}_k(\boldsymbol{q})\ \forall k$ holds with strict inequality[8] for any $\boldsymbol{q} > 0$ when $\boldsymbol{h}_k \neq \mathbf{0}\ \forall k$ and $\alpha > 1$, the maximum $\check{\gamma}_0(P_{\text{max}})$ is strictly increasing in the sum power P_{max}. Consequently, the optimum power allocation $\check{\boldsymbol{q}}$ satisfies $\|\check{\boldsymbol{q}}\|_1 = P_{\text{max}}$ with equality and the largest common scaling $\check{\gamma}_0(P_{\text{max}})$ of the SINR balancing is obtained for $P_{\text{max}} \to \infty$. For $P_{\text{max}} > 0$, the optimum power allocation $\check{\boldsymbol{q}}$ must also be strictly positive (i.e., $\check{\boldsymbol{q}} > \mathbf{0}$) since $\check{\gamma}_0(P_{\text{max}}) > 0$ in this case and since $\gamma_1, \ldots, \gamma_K > 0$ holds for $\boldsymbol{\mu} \in \text{int}(\mathcal{P}_\Sigma)$.

For $P_{\text{max}} \to \infty$, we can even show that the power allocation must be linear in P_{max} for every user when $\boldsymbol{\mu} \in \text{int}(\mathcal{P}_\Sigma)$. Since the constraints in (9.24) hold with equality at the optimum, the maximum SINRs are positive and balanced according to the requirements $\gamma_1, \ldots, \gamma_K$. As a consequence, $P_{\text{max}} \to \infty$ leads to $\check{q}_k \to \infty\ \forall k \in \mathcal{K}$, which we show by contradiction. Let $\mathcal{I}_1 \subset \mathcal{K}$ denote the set of user indices for which $q_k \to \infty\ \forall k \in \mathcal{I}_1$ and let $\mathcal{I}_2 := \mathcal{K} \setminus \mathcal{I}_1$ be the set of user indices whose power allocations remain *finite* when $P_{\text{max}} \to \infty$. For $|\mathcal{I}_1| \geq N$, the SINRs of all users out of $\mathcal{I}_2$ approach zero as $P_{\text{max}} \to \infty$ whereas there is at least one $k \in \mathcal{I}_1$ for which the corresponding SINR remains finite, see (9.25). This contradicts the balancing of all SINRs. For $|\mathcal{I}_1| < N$, the SINRs of all users in $\mathcal{I}_2$ remain finite whereas there is at least one user in $\mathcal{I}_1$, whose SINR grows beyond all limits when $P_{\text{max}} \to \infty$. Again, this contradicts the balancing of all SINRs which proves that $P_{\text{max}} \to \infty$ leads to $\check{q}_k \to \infty\ \forall k \in \mathcal{K}$ in (9.24). Next, we state the power allocation more precisely and show that the growth rates must be the same for all users which means that a *linear* power allocation is required to asymptotically achieve MMSE targets $\boldsymbol{\mu} \in \text{int}(\mathcal{P}_\Sigma)$.

---

[8] The proof for this basically coincides with showing the scalability property in Appendix A.21.

The proof is again by contradiction: we first allow different growth rates $q_k = a_k P_{\max}^{\tau_k}$ with $a_k > 0$ and $\tau_k > 0$ for all users $k \in \mathcal{K}$ and then see that $\tau_1, \ldots, \tau_K$ must be identical from which we conclude that $\tau_k = 1\ \forall k$ must hold in order to make use of the total available transmit power. If not all exponents $\tau_1, \ldots, \tau_K$ are identical, then it is possible to find partitions $\mathcal{I}_1$ and $\mathcal{I}_2$ with $\mathcal{I}_1 \cup \mathcal{I}_2 = \mathcal{K}$ and $\mathcal{I}_1 \cap \mathcal{I}_2 = \emptyset$ such that

$$\tau_{\min,1} := \min_{k\in\mathcal{I}_1} \tau_k > \max_{k\in\mathcal{I}_2} \tau_k.$$

Now if $|\mathcal{I}_1| \geq N$, then the SINR of every user $k \in \mathcal{I}_2$ is given by

$$\begin{aligned}
\mathrm{SINR}_k(\boldsymbol{q}) &= a_k P_{\max}^{\tau_k} \boldsymbol{h}_k^{\mathrm{H}} \Big(\mathbf{I} + \sum_{i\in\mathcal{I}_1} \boldsymbol{h}_i \boldsymbol{h}_i^{\mathrm{H}} a_i P_{\max}^{\tau_i} + \sum_{\substack{i\in\mathcal{I}_2\\ i\neq k}} \boldsymbol{h}_i \boldsymbol{h}_i^{\mathrm{H}} a_i P_{\max}^{\tau_i}\Big)^{-1} \boldsymbol{h}_k \\
&\leq a_k P_{\max}^{\tau_k} \boldsymbol{h}_k^{\mathrm{H}} \Big(\sum_{i\in\mathcal{I}_1} \boldsymbol{h}_i \boldsymbol{h}_i^{\mathrm{H}} a_i P_{\max}^{\tau_{\min,1}}\Big)^{-1} \boldsymbol{h}_k \\
&= a_k P_{\max}^{\tau_k - \tau_{\min,1}} \boldsymbol{h}_k^{\mathrm{H}} \Big(\sum_{i\in\mathcal{I}_1} \boldsymbol{h}_i \boldsymbol{h}_i^{\mathrm{H}} a_i\Big)^{-1} \boldsymbol{h}_k \quad \to 0 \quad \text{as} \quad P_{\max} \to \infty
\end{aligned}$$

due to $\tau_{\min,1} > \tau_k\ \forall k \in \mathcal{I}_2$, which contradicts the balancing since it has to assure strictly positive SINR values even in the limit $P_{\max} \to \infty$. Conversely, if $|\mathcal{I}_1| < N$, it follows from Appendix A.18 that the rate of every user $k \in \mathcal{I}_1$ diverges when $P_{\max} \to \infty$, i.e., the SINRs of all users in $\mathcal{I}_1$ grow beyond all limits. Due to the balancing, the SINRs of *all* users (in $\mathcal{K}$) would have to diverge which implies MMSEs that go to zero. However, this contradicts Lemma 9.2 since $K > N$. Thus, we have shown that all exponents $\tau_1, \ldots, \tau_K$ must be identical when the MMSE targets $\boldsymbol{\mu}$ used throughout the SINR balancing are taken from the interior of $\mathcal{P}_\Sigma$. Furthermore, every exponent must be one in order to fully exploit the sum power budget, i.e., a *linear* power allocation must be used.

Having proven that the optimum power allocation of every user grows beyond all limits when $P_{\max} \to \infty$ with SINR targets that fulfill $0 < \gamma_k < \infty\ \forall k \in \mathcal{K}$, we conclude from Lemma 9.2 that the resulting sum MMSE converges to $K - N$. For the particular choice of the SINR requirements $\gamma_1, \ldots, \gamma_K$ according to the MMSE requirements $\boldsymbol{\mu} \in \mathrm{int}(\mathcal{P}_\Sigma)$ via (9.23), the maximizer $\check{\gamma}_0(P_{\max})$ of the SINR balancing problem (9.24) approaches one as $P_{\max} \to \infty$, i.e.,

$$\lim_{P_{\max}\to\infty} \check{\gamma}_0(P_{\max}) = 1 \quad \text{for} \quad \gamma_k = \mu_k^{-1} - 1 \quad \text{and} \quad \boldsymbol{\mu} \in \mathrm{int}(\mathcal{P}_\Sigma). \tag{9.26}$$

In other words, any MMSE requirement $\boldsymbol{\mu} \in \mathrm{int}(\mathcal{P}_\Sigma)$ can asymptotically be achieved with an infinite amount of sum power. For finite $P_{\max}$, MMSE values that are arbitrarily close to such $\boldsymbol{\mu}$ can be obtained since $\check{\gamma}_0(P_{\max})$ is strictly increasing in $P_{\max}$.

Having shown that any MMSE requirement $\boldsymbol{\mu} \in \mathrm{int}(\mathcal{P}_\Sigma)$ can be achieved asymptotically with an infinite sum power, the feasibility of *all* requirements $\boldsymbol{\mu} \in \mathrm{int}(\mathcal{P})$

in the interior of the whole polytope $\mathcal{P}$ in Theorem 9.1 follows from Lemma 9.3. Points on the boundary that are different from the all-ones vector require an infinite amount of sum power either because one MMSE requirement is zero or the MMSE requirements sum up to the lower bound $K - N$, see Lemmas 9.1 and 9.2. This completes the proof for Theorem 9.1 for the overloaded case.

From Theorem 9.1 and (9.10), the following corollary results that allows to quickly check either SINR or MMSE requirements for their feasibility:

Corollary 9.1 *SINR requirements $\gamma_1, \ldots, \gamma_K$ that fulfill $0 \leq \gamma_k < \infty \; \forall k$ are feasible with finite sum power in a vector broadcast channel with an N-antenna base station and channels that fulfill the regularity condition in (9.20), if and only if*

$$\sum_{k=1}^{K} \frac{1}{1+\gamma_k} > K - N$$

or, equivalently, if and only if

$$\sum_{k=1}^{K} \frac{\gamma_k}{1+\gamma_k} < N$$

i.e., the effective bandwidth must sum up to less than the number N of antennas that are mounted at the base station. Alternatively, MMSE requirements $\mu_1, \ldots, \mu_K$ that fulfill $0 < \mu_k \leq 1 \; \forall k$ are feasible, if and only if

$$\sum_{k=1}^{K} \mu_k > K - N.$$

In other words, the result of Viswanath et al. that has been derived for a multi-user SISO CDMA system in [167] translates to the vector BC if the processing gain is replaced by the number of antennas N at the base station. Interestingly, Corollary 9.1 coincidences with Palomar's result in (9.5) where the receivers are allowed to cooperate (point-to-point link), see Sect. 9.1. Hence, cooperation of the receivers does not enlarge the feasibility region:

Corollary 9.2 *Under the regular channels condition (9.20), cooperation of the receivers in a vector BC does not enlarge the feasible QoS region.*

Note that as a byproduct of the QoS feasibility region, we solved the SIR balancing problem (without considering the noise) when the channel knowledge is instantaneous. SIR balancing can be regarded as a special case of SINR balancing by either setting the noise variance to zero or allowing for an infinite amount of sum power *under the restriction of a linear power allocation*. Since the optimal power allocation was proven to be linear for relative SINR targets $0 < \gamma_k < \infty \; \forall k$, we can easily solve the SIR balancing problem

$$\max_{\boldsymbol{a}>\boldsymbol{0},\gamma_0} \gamma_0 \quad \text{s.t.: } \mathrm{SIR}_k(\boldsymbol{a}) \geq \gamma_0 \cdot \gamma_k \; \forall k \in \mathcal{K} \text{ and } \|\boldsymbol{a}\|_1 = 1 \tag{9.27}$$

with the SIR of user k

$$\mathrm{SIR}_k(\boldsymbol{a}) := a_k \boldsymbol{h}_k^{\mathrm{H}} \Big(\sum_{i \neq k} \boldsymbol{h}_i \boldsymbol{h}_i^{\mathrm{H}} a_i \Big)^{-1} \boldsymbol{h}_k$$

by interpreting it as the SINR balancing problem when $P_{\max} \to \infty$. Despite the many result on the SIR balancing in [162–166],the maximum 'radius' $\check{\gamma}_0$ [best value of (9.27)] was still unknown. From Theorem 9.1, we now know that for $K > N$ and regular channels, $\check{\gamma}_0$ is the unique positive solution of

$$\sum_{k=1}^{K} \frac{1}{1 + \check{\gamma}_0 \gamma_k} = K - N \tag{9.28}$$

which can easily be computed due to the convexity of the involved fractions. In particular, $\check{\gamma}_0 = 1$, if the corresponding MMSE values $\boldsymbol{\mu}$ are taken from $\mathrm{int}(\mathcal{P}_\Sigma)$.

Corollary 9.3 *For channel vectors that fulfill the regularity condition (9.20) the SIR balancing radius $\check{\gamma}_0$ is the unique positive solution of*

$$\sum_{k=1}^{K} \frac{1}{1 + \check{\gamma}_0 \gamma_k} = K - N$$

when $K > N$ and $0 < \gamma_k < \infty \; \forall k$.

It should, however, be noted, that the more general case of channel autocovariance matrices is considered in[162–166] in which the the rank-one outer product $\boldsymbol{h}_i \boldsymbol{h}_i^{\mathrm{H}}$ of every user i can also be replaced by its expectation $\mathrm{E}[\boldsymbol{h}_i \boldsymbol{h}_i^{\mathrm{H}}]$ with possibly larger ranks. Though, our feasibility results are only valid for *instantaneous* channel knowledge (rank-one outer products) which we admittedly consider to be the more relevant case since applying the expectation operator separately in the numerator and denominator in of the SI(N)R quotient (before the optimum receive filter has been applied) makes the relation of the resulting averaged SI(N)R to the Shannon rate or BER problematic.

Synopsis: The closure of the MMSE feasibility region in the vector BC with regular channels can be shown to be a polytope whose bounding half-spaces are the individual box constraints that limit all individual MMSEs to the interval [0, 1] and the lower bound on the sum MMSE, which must be larger than or equal to $K - N$, where K denotes the number of users and N is the number of base station antennas. Any MMSE requirement taken from the interior of this polytope is achievable with finite sum power, requirements outside the polytope are infeasible even with an infinite amount of power. Points lying on the boundary of the polytope which are different from the all-ones vector can only be achieved asymptotically with an infinite amount of sum power. As a consequence, feasibility of given MMSE requirements $\boldsymbol{\mu} > \mathbf{0}$ can quickly be tested by probing whether $\mathbf{1}^{\mathrm{T}}\boldsymbol{\mu}$ is larger than $K - N$ or not. This characterization of the MMSE feasibility region allows for a complete description in other domains like SINR or Shannon rate, which renders a solution of the SIR balancing possible. The maximum 'radius' which scales the relative SIR requirements can easily be computed as the root of a polynomial.

9.2.2 Singular Channels Scenario

If requirement (9.20) for the regular channels condition does not hold, the scenario is termed *singular* due to the existence of a user subset $\mathcal{I}$ with the following properties:

$$\exists \mathcal{I} \subseteq \mathcal{K} : \ \mathrm{rank}(\boldsymbol{H}_{\mathcal{I}}) < \min\{N, |\mathcal{I}|\} \tag{9.29}$$

Although the case of rank-deficient channels usually does not arise in practice, it is interesting to see how it impacts the feasibility region. As we will see, the polytope structure of the closure of the MMSE feasibility region also holds in this case. However, the half-space constraint which results from the sum MMSE lower bound takes a slightly different form and additional half-space constraints may come into play. The channel matrices $\boldsymbol{H}$ that satisfy (9.29) can further be distinguished into two classes:

1. The first class is characterized by

$$\nexists \mathcal{I} \subset \mathcal{K} : \mathrm{rank}(\boldsymbol{H}_{\mathcal{I}}) < \min\{\mathrm{rank}(\boldsymbol{H}), |\mathcal{I}|\} \tag{9.30}$$

   i.e., such a set $\mathcal{I}$ does not exist. An example for this class is the $K = 2$ user setup in which the two channel vectors $\boldsymbol{h}_1$ and $\boldsymbol{h}_2$ are colinear and $N \geq 2$. In this case, (9.29) holds for $\mathcal{I} = \mathcal{K} = \{1, 2\}$, but a subset $\mathcal{I} \subset \mathcal{K}$ which fulfills $\mathrm{rank}(\boldsymbol{H}_{\mathcal{I}}) < \min\{\mathrm{rank}(\boldsymbol{H}), |\mathcal{I}|\}$ does not exist. Using an appropriate dimension-reduction front end at the receiver *in advance* to the MMSE receive filter, this setup can be transformed to a scenario in which the base station is equipped with $\mathrm{rank}(\boldsymbol{H})$ instead of N antenna elements and thus satisfies the regular channels condition. For the first class, Theorem 9.1 is therefore valid if N is replaced by $\mathrm{rank}(\boldsymbol{H})$.

2. The second class can be regarded as the complementary part of the first class, i.e.,
$$\exists \mathcal{I} \subset \mathcal{K} : \text{rank}(\boldsymbol{H}_{\mathcal{I}}) < \min\{\text{rank}(\boldsymbol{H}), |\mathcal{I}|\}. \tag{9.31}$$

   Note that (9.29) follows from (9.31) due to $\text{rank}(\boldsymbol{H}) \leq N$ whereas the converse is not necessarily true. Besides the total sum MMSE constraint, channel representatives of this class feature additional half-space constraints which limit the structure of the MMSE feasibility region. More precisely, an additional constraint on the sum MMSE of the users in $\mathcal{I}$ arises for every subset $\mathcal{I} \subset \mathcal{K}$ that fulfills $\text{rank}(\boldsymbol{H}_{\mathcal{I}}) < \min\{\text{rank}(\boldsymbol{H}), |\mathcal{I}|\}$.

In the following, we first state the theorem on the the MMSE feasibility structure, present two examples, and afterwards prove the theorem.

Theorem 9.2 *The closure of the MMSE feasibility region in the vector broadcast channel is the polytope*

$$\mathcal{P} := \Big\{\boldsymbol{\mu}\Big|\boldsymbol{0} \leq \boldsymbol{\mu} \leq \boldsymbol{1} \wedge \boldsymbol{1}^{\mathrm{T}}\boldsymbol{\mu} \geq K - \mathit{rank}(\boldsymbol{H}). \\ \wedge \sum_{k\in\mathcal{I}} \mu_k \geq |\mathcal{I}| - \mathit{rank}(\boldsymbol{H}_{\mathcal{I}})\ \forall \mathcal{I} \subset \mathcal{K} : \mathit{rank}(\boldsymbol{H}_{\mathcal{I}}) < \min\{\mathit{rank}(\boldsymbol{H}), |\mathcal{I}|\}\Big\} \tag{9.32}$$

whose bounding half-spaces are the individual box constraints $0 \leq \mu_k \leq 1\ \forall k \in \mathcal{K}$, the constraint $\sum_{k\in\mathcal{K}} \mu_k \geq K - \mathit{rank}(\boldsymbol{H})$ on the sum MMSE of all users, and the partial sum MMSE lower bound $\sum_{k\in\mathcal{I}} \mu_k \geq |\mathcal{I}| - \mathit{rank}(\boldsymbol{H}_{\mathcal{I}})$ on the sum MMSE of those users in all subsets $\mathcal{I} \subset \mathcal{K}$, that fulfill $\mathit{rank}(\boldsymbol{H}_{\mathcal{I}}) < \min\{\mathit{rank}(\boldsymbol{H}), |\mathcal{I}|\}$. Requirements outside the polytope $\mathcal{P}$ are infeasible even with an infinite amount of sum power whereas all MMSE requirements belonging to the interior of $\mathcal{P}$ are feasible with finite sum power. Points on the boundary $\partial\mathcal{P}$ that differ from the all-ones vector can asymptotically be achieved by an infinite amount of sum power.

According to Theorem 9.2, the slanted hyper-cube is further cropped by an additional half-space for every subset $\mathcal{I} \subset \mathcal{K}$ whose corresponding channel matrix violates $\text{rank}(\boldsymbol{H}_{\mathcal{I}}) = \min\{\text{rank}(\boldsymbol{H}), |\mathcal{I}|\}$. Note that some of these constraints might be redundant, see the following examples which both belong to the second class.

Examples:

- Let $K = 3$ users be served by an $N = 2$ antenna base station in the vector BC, in which the BC channel vectors of the first two users are colinear such that the dual MAC channels $\boldsymbol{h}_1$ and $\boldsymbol{h}_2$ are colinear as well. Furthermore, the composite MAC channel matrix is assumed to satisfy $\text{rank}(\boldsymbol{H}) = 2$, which implies that $\boldsymbol{h}_3$ is linearly independent of $\boldsymbol{h}_1$ (and thus also of $\boldsymbol{h}_2$). Besides the individual box constraints, the half-space defined by the lower bound on the sum MMSE reads as $\mu_1 + \mu_2 + \mu_3 \geq K - \text{rank}(\boldsymbol{H}) = 1$. In addition, the subset $\mathcal{I} = \{1, 2\}$ satisfies (9.31), so another half-space constraint contributes to the MMSE feasibility region, namely $\mu_1 + \mu_2 \geq |\mathcal{I}| - \text{rank}(\boldsymbol{H}_\mathcal{I}) = 1$. Note that this constraint is stricter than the sum MMSE constraint over all three users since $\mu_1 + \mu_2 \geq 1$ clearly implies $\mu_1 + \mu_2 + \mu_3 \geq 1$ due to $\mu_3 \geq 0$. Summing up, the closure of the feasible MMSE region for this particular channel is given by the individual box constraints plus $\mu_1 + \mu_2 \geq 1$.

- In a second example, $K = 5$ users are served by an $N = 3$ antenna base station. The three channel vectors $\boldsymbol{h}_1$, $\boldsymbol{h}_2$, and $\boldsymbol{h}_3$ shall lie in a two-dimensional plane whereas every channel vector pair out of those three vectors shall be linearly independent. In addition, no further linear dependencies shall exist. Then, $\text{rank}(\boldsymbol{H}) = 3$ implies the sum MMSE constraint of all users $\mu_1 + \mu_2 + \mu_3 + \mu_4 + \mu_5 \geq 2$. The only subset $\mathcal{I}$ satisfying (9.31) is $\mathcal{I} = \{1, 2, 3\}$, and this leads to the additional condition $\mu_1 + \mu_2 + \mu_3 \geq 1$.

The proof for Theorem 9.2 is even more technical than the one for the feasibility region of the regular channels scenario and will take the rest of this subsection. In particular, *nonlinear* power allocations may become necessary. We start by showing that the additional half-space constraints in (9.32) which arise from subsets $\mathcal{I} \subset \mathcal{K}$ that satisfy $\text{rank}(\boldsymbol{H}_\mathcal{I}) < \min\{\text{rank}(\boldsymbol{H}), |\mathcal{I}|\}$, are necessary for the feasibility. Afterwards, sufficiency is proven as well.

Necessity of the additional constraints in (9.32) immediately follows from (9.15) in Lemma 9.2 if the total set $\mathcal{K}$ is replaced by the set $\mathcal{I}$, and if K is replaced by $|\mathcal{I}|$:

$$\sum_{k \in \mathcal{I}} \varepsilon_k(\boldsymbol{q}) \geq |\mathcal{I}| - \text{rank}(\boldsymbol{H}_\mathcal{I}) \quad \forall \mathcal{I} \subseteq \mathcal{K} \tag{9.33}$$

This corresponds to the setup in which users that do not belong to the set $\mathcal{I}$ are passive and thus can be dropped. Moreover, the lower bound cannot be reduced if those formerly passive users actively transmit and thus generate interference. Note that (9.33) is necessary for *every* (nonempty) subset $\mathcal{I} \subseteq \mathcal{K}$. However, some of these $2^{|\mathcal{K}|} - 1$ lower bounds are redundant and can thus be dropped. As we will see, besides $\mathcal{I} = \mathcal{K}$, only those sets $\mathcal{I}$ remain that satisfy $\text{rank}(\boldsymbol{H}_\mathcal{I}) < \min\{\text{rank}(\boldsymbol{H}), |\mathcal{I}|\}$ in accordance to (9.32).

It is obvious that $\text{rank}(\boldsymbol{H}_\mathcal{I})$ cannot be larger than $\text{rank}(\boldsymbol{H})$ since $\boldsymbol{H}$ contains all columns of $\boldsymbol{H}_\mathcal{I}$ for every $\mathcal{I}$. Hence, we have to show that the constraints which result from the two cases $\text{rank}(\boldsymbol{H}) = \text{rank}(\boldsymbol{H}_\mathcal{I})$ and $\text{rank}(\boldsymbol{H}_\mathcal{I}) = |\mathcal{I}|$ are already included in the constraints in (9.32). In the latter case ($\text{rank}(\boldsymbol{H}_\mathcal{I}) = |\mathcal{I}|$), the resulting constraint $\sum_{k \in \mathcal{I}} \varepsilon_k(\boldsymbol{q}) \geq 0$ in (9.33) is a consequence of the individual box constraints and is thus superfluous due to $\varepsilon_k(\boldsymbol{q}) \geq 0 \ \forall k \in \mathcal{K}$. The former case $\text{rank}(\boldsymbol{H}_\mathcal{I}) = \text{rank}(\boldsymbol{H})$ is again only interesting for a positive right hand side of (9.33), i.e., when $|\mathcal{I}| > \text{rank}(\boldsymbol{H}_\mathcal{I}) = \text{rank}(\boldsymbol{H})$. In this case, the constraints

$$\begin{gathered}\sum_{k \in \mathcal{I}} \varepsilon_k(\boldsymbol{q}) \geq |\mathcal{I}| - \text{rank}(\boldsymbol{H}_\mathcal{I}) = |\mathcal{I}| - \text{rank}(\boldsymbol{H}) \\ \forall \mathcal{I} \subset \mathcal{K} : \text{rank}(\boldsymbol{H}_\mathcal{I}) = \text{rank}(\boldsymbol{H}) \text{ and } |\mathcal{I}| > \text{rank}(\boldsymbol{H})\end{gathered} \tag{9.34}$$

can be shown to be already included in the lower bound constraint on the *total* sum MMSE of *all* users

$$\sum_{k\in\mathcal{K}} \varepsilon_k(\boldsymbol{q}) \geq K - \text{rank}(\boldsymbol{H}) =: d. \tag{9.35}$$

The proof for this statement goes as follows. Since $\varepsilon_k(\boldsymbol{q}) \leq 1 \ \forall k \in \mathcal{K}$, (9.35) implicates for every $\mathcal{I} \subset \mathcal{K} : |\mathcal{I}| \geq K - d + 1 = \text{rank}(\boldsymbol{H}) + 1$ that

$$\sum_{k\in\mathcal{I}} \varepsilon_k(\boldsymbol{q}) \geq K - \text{rank}(\boldsymbol{H}) - (K - |\mathcal{I}|) = |\mathcal{I}| - \text{rank}(\boldsymbol{H}). \tag{9.36}$$

But for those $\mathcal{I}$ for which $\text{rank}(\boldsymbol{H}_{\mathcal{I}}) = \text{rank}(\boldsymbol{H})$, (9.36) exactly matches (9.34), so (9.34) is a direct consequence of (9.35). In other words, the $2^{|\mathcal{K}|} - 1$ constraints in (9.33) define the same constraint set as in (9.32).

So far we have shown the necessity of the bounding half-spaces in (9.32). Before passing over to the proof for their sufficiency, we analyze how the sum MMSE of users in a subset $\mathcal{I} \subset \mathcal{K}$ depends on the power allocation. This will later be useful for the sufficiency proof. To this end, let $\boldsymbol{H}$ be such that (9.29) holds and let $r := \text{rank}(\boldsymbol{H})$. Let $\boldsymbol{V} \in \mathbb{C}^{N\times r}$ be a basis for the range of $\boldsymbol{H}$ such that $\boldsymbol{V}^{\text{H}}\boldsymbol{V} = \mathbf{I}_r$. Using $\boldsymbol{V}^{\text{H}}$ as the first stage of the receiver in the dual MAC, the N-dimensional observation $\boldsymbol{y} = \boldsymbol{H}\boldsymbol{Q}^{\frac{1}{2}}\boldsymbol{s} + \boldsymbol{n}$ is transformed to

$$\boldsymbol{y}' := \boldsymbol{V}^{\text{H}}\boldsymbol{H}\boldsymbol{Q}^{\frac{1}{2}}\boldsymbol{s} + \boldsymbol{V}^{\text{H}}\boldsymbol{n} = \boldsymbol{\Phi}\boldsymbol{Q}^{\frac{1}{2}}\boldsymbol{s} + \boldsymbol{n}' \tag{9.37}$$

where $\boldsymbol{\Phi} := \boldsymbol{V}^{\text{H}}\boldsymbol{H} \in \mathbb{C}^{r\times K}$. Since $\boldsymbol{n}$ has an identity covariance matrix, $\boldsymbol{y}'$ is a sufficient statistic and the covariance matrix of the transformed noise $\boldsymbol{n}' := \boldsymbol{V}^{\text{H}}\boldsymbol{n} \in \mathbb{C}^r$ is again the identity matrix. Thus, we have reduced[9] the dimension of the observation from N to r and we can interpret $\boldsymbol{\Phi}$ as the transformed composite channel. If $\boldsymbol{\Phi}$ satisfies the regularity condition, then (9.30) is valid and the original channel $\boldsymbol{H}$ belongs to class one. As a consequence, the feasibility region follows from Theorem 9.1 with $\boldsymbol{H}$ replaced by $\boldsymbol{\Phi}$, or, equivalently, with N replaced by r. The more interesting case (class two) arises for $\text{rank}(\boldsymbol{\Phi}) < r$, in which a user subset $\mathcal{I}_1 \subset \mathcal{K}$ exists for which $\text{rank}(\boldsymbol{H}_{\mathcal{I}_1}) < \min\{\text{rank}(\boldsymbol{H}), |\mathcal{I}_1|\}$. Without loss of generality, we may assume that $\mathcal{I}_1 = \{1, \ldots, |\mathcal{I}_1|\}$, which can always be achieved by relabeling the users. In the following, we investigate how the sum MMSE of those users in $\mathcal{I}_1$ depends on the power allocation.

Let the transformed channel matrix $\boldsymbol{\Phi} = [\boldsymbol{\Phi}_{\mathcal{I}_1}, \boldsymbol{\Phi}_{\mathcal{I}_2}]$ be partitioned into $\boldsymbol{\Phi}_{\mathcal{I}_1} \in \mathbb{C}^{r\times|\mathcal{I}_1|}$ and $\boldsymbol{\Phi}_{\mathcal{I}_2} \in \mathbb{C}^{r\times|\mathcal{I}_2|}$, where $\mathcal{I}_2 := \mathcal{K} \setminus \mathcal{I}_1$. Note that $\boldsymbol{\Phi}_{\mathcal{I}_1}$ is rank deficient due to the choice of $\mathcal{I}_1$, which satisfies $\text{rank}(\boldsymbol{H}_{\mathcal{I}_1}) < \min\{r, |\mathcal{I}_1|\}$. The transmit covariance matrices of the users in $\mathcal{I}_1$ and $\mathcal{I}_2$ shall read as

$$\boldsymbol{Q}_{\mathcal{I}_i} = \boldsymbol{A}_{\mathcal{I}_i}\boldsymbol{P}_{\max}^{\mathcal{T}_i} \quad \forall i \in \{1, 2\} \tag{9.38}$$

[9] Clearly, the dimension of the observation only reduces, if $r = \text{rank}(\boldsymbol{H}) < N$.

with diagonal $\boldsymbol{A}_{\mathcal{I}_i} \succ \mathbf{0}\ \forall i$ and $0 < \tau_i \leq 1\ \forall i$. Nonlinear power allocations are allowed by choosing τ_1 different from τ_2. Under these assumptions, the MMSE matrix $\boldsymbol{M}$ whose main diagonal elements correspond to the MMSEs of all individual users reads as [cf. (9.9)]

$$
\begin{aligned}
\boldsymbol{M} &:= \left[\mathbf{I}_K + \boldsymbol{Q}^{\frac{1}{2}}\boldsymbol{\Phi}^{\mathrm{H}}\boldsymbol{\Phi}\boldsymbol{Q}^{\frac{1}{2}}\right]^{-1} \\
&= \begin{bmatrix} \mathbf{I}_{|\mathcal{I}_1|} + \boldsymbol{Q}_{\mathcal{I}_1}^{\frac{1}{2}}\boldsymbol{\Phi}_{\mathcal{I}_1}^{\mathrm{H}}\boldsymbol{\Phi}_{\mathcal{I}_1}\boldsymbol{Q}_{\mathcal{I}_1}^{\frac{1}{2}} & \boldsymbol{Q}_{\mathcal{I}_1}^{\frac{1}{2}}\boldsymbol{\Phi}_{\mathcal{I}_1}^{\mathrm{H}}\boldsymbol{\Phi}_{\mathcal{I}_2}\boldsymbol{Q}_{\mathcal{I}_2}^{\frac{1}{2}} \\ \boldsymbol{Q}_{\mathcal{I}_2}^{\frac{1}{2}}\boldsymbol{\Phi}_{\mathcal{I}_2}^{\mathrm{H}}\boldsymbol{\Phi}_{\mathcal{I}_1}\boldsymbol{Q}_{\mathcal{I}_1}^{\frac{1}{2}} & \mathbf{I}_{|\mathcal{I}_2|} + \boldsymbol{Q}_{\mathcal{I}_2}^{\frac{1}{2}}\boldsymbol{\Phi}_{\mathcal{I}_2}^{\mathrm{H}}\boldsymbol{\Phi}_{\mathcal{I}_2}\boldsymbol{Q}_{\mathcal{I}_2}^{\frac{1}{2}} \end{bmatrix}^{-1} \\
&= \begin{bmatrix} \mathbf{I}_{|\mathcal{I}_1|} + P_{\max}^{\tau_1}\boldsymbol{Y}_{\mathcal{I}_1}^{\mathrm{H}}\boldsymbol{Y}_{\mathcal{I}_1} & P_{\max}^{\frac{\tau_1+\tau_2}{2}}\boldsymbol{Y}_{\mathcal{I}_1}^{\mathrm{H}}\boldsymbol{Y}_{\mathcal{I}_2} \\ P_{\max}^{\frac{\tau_1+\tau_2}{2}}\boldsymbol{Y}_{\mathcal{I}_2}^{\mathrm{H}}\boldsymbol{Y}_{\mathcal{I}_1} & \mathbf{I}_{|\mathcal{I}_2|} + P_{\max}^{\tau_2}\boldsymbol{Y}_{\mathcal{I}_2}^{\mathrm{H}}\boldsymbol{Y}_{\mathcal{I}_2} \end{bmatrix}^{-1}
\end{aligned}
$$

where $\boldsymbol{Y}_{\mathcal{I}_i} := \boldsymbol{\Phi}_{\mathcal{I}_i}\boldsymbol{A}_{\mathcal{I}_i}^{\frac{1}{2}} \in \mathbb{C}^{r\times|\mathcal{I}_i|}\ \forall i$. Applying the matrix inversion lemma from Appendix A.2 two times, we compute the MMSE matrix $\boldsymbol{M}_{\mathcal{I}_1}$, which is the upper left $|\mathcal{I}_1| \times |\mathcal{I}_1|$ block of $\boldsymbol{M}$, to

$$
\begin{aligned}
\boldsymbol{M}_{\mathcal{I}_1} &:= \left[\mathbf{I}_{|\mathcal{I}_1|} + \boldsymbol{Y}_{\mathcal{I}_1}^{\mathrm{H}}\left[\mathbf{I}_r - \boldsymbol{Y}_{\mathcal{I}_2}\left(\mathbf{I}_{|\mathcal{I}_2|} + \boldsymbol{Y}_{\mathcal{I}_2}^{\mathrm{H}}\boldsymbol{Y}_{\mathcal{I}_2}P_{\max}^{\tau_2}\right)^{-1}\boldsymbol{Y}_{\mathcal{I}_2}^{\mathrm{H}}P_{\max}^{\tau_2}\right]\boldsymbol{Y}_{\mathcal{I}_1}P_{\max}^{\tau_1}\right]^{-1} \\
&= \left[\mathbf{I}_{|\mathcal{I}_1|} + \boldsymbol{Y}_{\mathcal{I}_1}^{\mathrm{H}}\left(\mathbf{I}_r + \boldsymbol{Y}_{\mathcal{I}_2}\boldsymbol{Y}_{\mathcal{I}_2}^{\mathrm{H}}P_{\max}^{\tau_2}\right)^{-1}\boldsymbol{Y}_{\mathcal{I}_1}P_{\max}^{\tau_1}\right]^{-1} \\
&= \mathbf{I}_{|\mathcal{I}_1|} - \boldsymbol{Y}_{\mathcal{I}_1}^{\mathrm{H}}\left[\mathbf{I}_r \cdot P_{\max}^{-\tau_1} + \boldsymbol{Y}_{\mathcal{I}_2}\boldsymbol{Y}_{\mathcal{I}_2}^{\mathrm{H}}P_{\max}^{\tau_2-\tau_1} + \boldsymbol{Y}_{\mathcal{I}_1}\boldsymbol{Y}_{\mathcal{I}_1}^{\mathrm{H}}\right]^{-1}\boldsymbol{Y}_{\mathcal{I}_1}.
\end{aligned} \tag{9.39}
$$

To see how $\boldsymbol{M}_{\mathcal{I}_1}$ behaves for $P_{\max} \to \infty$, we distinguish three cases depending on the choice of τ_1 and τ_2:

1. $\tau_1 > \tau_2$: For this type of power allocation, $\boldsymbol{M}_{\mathcal{I}_1}$ converges for $P_{\max} \to \infty$ to[10]

$$
\begin{aligned}
\boldsymbol{M}_{\mathcal{I}_1} &\to \mathbf{I}_{|\mathcal{I}_1|} - \boldsymbol{Y}_{\mathcal{I}_1}^{\mathrm{H}}\left(\boldsymbol{Y}_{\mathcal{I}_1}\boldsymbol{Y}_{\mathcal{I}_1}^{\mathrm{H}}\right)^{+}\boldsymbol{Y}_{\mathcal{I}_1} \\
&= \mathbf{I}_{|\mathcal{I}_1|} - \boldsymbol{A}_{\mathcal{I}_1}^{\frac{1}{2}}\boldsymbol{\Phi}_{\mathcal{I}_1}^{\mathrm{H}}\left(\boldsymbol{\Phi}_{\mathcal{I}_1}\boldsymbol{A}_{\mathcal{I}_1}\boldsymbol{\Phi}_{\mathcal{I}_1}^{\mathrm{H}}\right)^{+}\boldsymbol{\Phi}_{\mathcal{I}_1}\boldsymbol{A}_{\mathcal{I}_1}^{\frac{1}{2}} \\
&= \mathbf{I}_{|\mathcal{I}_1|} - \boldsymbol{\Pi}_{\boldsymbol{Y}_{\mathcal{I}_1}^{\mathrm{H}}} = \boldsymbol{\Pi}_{\boldsymbol{Y}_{\mathcal{I}_1}^{\mathrm{H}}}^{\perp}
\end{aligned} \tag{9.40}
$$

   where $\boldsymbol{\Pi}_{\boldsymbol{Y}_{\mathcal{I}_1}^{\mathrm{H}}}$ denotes the projector onto the range of $\boldsymbol{Y}_{\mathcal{I}_1}^{\mathrm{H}}$. As we can conclude from (9.40), the MMSE matrix of the users in the set $\mathcal{I}_1$ with the largest exponent asymptotically does not depend on users with smaller power exponents. Hence, a system in which the users in $\mathcal{I}_1$ are isolated and do not see interference from any other users features the same asymptotic MMSE matrix due to

$$
\mathbf{I}_{|\mathcal{I}_1|} - \boldsymbol{Q}_{\mathcal{I}_1}^{\frac{1}{2}}\boldsymbol{\Phi}_{\mathcal{I}_1}^{\mathrm{H}}\left(\mathbf{I} + \boldsymbol{\Phi}_{\mathcal{I}_1}\boldsymbol{Q}_{\mathcal{I}_1}\boldsymbol{\Phi}_{\mathcal{I}_1}^{\mathrm{H}}\right)^{-1}\boldsymbol{\Phi}_{\mathcal{I}_1}\boldsymbol{Q}_{\mathcal{I}_1}^{\frac{1}{2}} \to \mathbf{I}_{|\mathcal{I}_1|} - \boldsymbol{A}_{\mathcal{I}_1}^{\frac{1}{2}}\boldsymbol{\Phi}_{\mathcal{I}_1}^{\mathrm{H}}\left(\boldsymbol{\Phi}_{\mathcal{I}_1}\boldsymbol{A}_{\mathcal{I}_1}\boldsymbol{\Phi}_{\mathcal{I}_1}^{\mathrm{H}}\right)^{+}\boldsymbol{\Phi}_{\mathcal{I}_1}\boldsymbol{A}_{\mathcal{I}_1}^{\frac{1}{2}}.
$$

[10] Remember that the $r \times r$ matrix $\boldsymbol{Y}_{\mathcal{I}_1}\boldsymbol{Y}_{\mathcal{I}_1}^{\mathrm{H}}$ is rank deficient, which leads to the pseudo-inverse!

The sum MMSE of all users in $\mathcal{I}_1$ asymptotically reads as

$$\lim_{P_{\max}\to\infty} \mathrm{tr}(\boldsymbol{M}_{\mathcal{I}_1}) = |\mathcal{I}_1| - \mathrm{rank}(\boldsymbol{Y}_{\mathcal{I}_1}) = |\mathcal{I}_1| - \mathrm{rank}(\boldsymbol{\Phi}_{\mathcal{I}_1}) = |\mathcal{I}_1| - \mathrm{rank}(\boldsymbol{H}_{\mathcal{I}_1}) \tag{9.41}$$

i.e., the lower bound in (9.33) is achieved by this kind of nonlinear power allocation for *any* $\boldsymbol{A}_{\mathcal{I}_1} \succ \boldsymbol{0}$. Note that the lower bound in (9.41) is also achieved if the diagonal elements of $\boldsymbol{Q}_{\mathcal{I}_1}$ have different exponents in $P_{\max}$ as long as they remain larger than τ_2.
The sum MMSE of the users in the complementary group $\mathcal{I}_2$ with the smaller power exponent is given by

$$\mathrm{tr}(\boldsymbol{M}_{\mathcal{I}_2}) = \mathrm{tr}(\boldsymbol{M}) - \mathrm{tr}(\boldsymbol{M}_{\mathcal{I}_1}).$$

Due to (9.38), the powers of *all* users tend to infinity as $P_{\max} \to \infty$. Consequently, the total sum MMSE of all users (in $\mathcal{K}$) achieves its lower bound $|\mathcal{K}| - r$, see Lemma 9.2. In combination with (9.41), this yields

$$\begin{aligned}
\lim_{P_{\max}\to\infty} \mathrm{tr}(\boldsymbol{M}_2) &= \lim_{P_{\max}\to\infty} \left[\mathrm{tr}(\boldsymbol{M}) - \mathrm{tr}(\boldsymbol{M}_1)\right] \\
&= |\mathcal{I}_1| + |\mathcal{I}_2| - r - \left[|\mathcal{I}_1| - \mathrm{rank}(\boldsymbol{\Phi}_{\mathcal{I}_1})\right] \\
&= |\mathcal{I}_2| - r + \mathrm{rank}(\boldsymbol{\Phi}_{\mathcal{I}_1}).
\end{aligned} \tag{9.42}$$

Because of the rank inequality

$$r = \mathrm{rank}([\boldsymbol{\Phi}_{\mathcal{I}_1}, \boldsymbol{\Phi}_{\mathcal{I}_2}]) \leq \mathrm{rank}(\boldsymbol{\Phi}_{\mathcal{I}_1}) + \mathrm{rank}(\boldsymbol{\Phi}_{\mathcal{I}_2})$$

the asymptotic sum MMSE of the users in $\mathcal{I}_2$ is lower bounded by

$$\lim_{P_{\max}\to\infty} \mathrm{tr}(\boldsymbol{M}_{\mathcal{I}_2}) \geq |\mathcal{I}_2| - \mathrm{rank}(\boldsymbol{\Phi}_{\mathcal{I}_2}) \tag{9.43}$$

where equality only holds for $r{:=}\mathrm{rank}([\boldsymbol{\Phi}_{\mathcal{I}_1}, \boldsymbol{\Phi}_{\mathcal{I}_2}]) = \mathrm{rank}(\boldsymbol{\Phi}_{\mathcal{I}_1}) + \mathrm{rank}(\boldsymbol{\Phi}_{\mathcal{I}_2})$:

$$\lim_{P_{\max}\to\infty} \mathrm{tr}(\boldsymbol{M}_{\mathcal{I}_2}) \begin{cases} = |\mathcal{I}_2| - \mathrm{rank}(\boldsymbol{\Phi}_{\mathcal{I}_2}) \text{ if } r = \mathrm{rank}(\boldsymbol{\Phi}_{\mathcal{I}_1}) + \mathrm{rank}(\boldsymbol{\Phi}_{\mathcal{I}_2}) \\ > |\mathcal{I}_2| - \mathrm{rank}(\boldsymbol{\Phi}_{\mathcal{I}_2}) \text{ if } r < \mathrm{rank}(\boldsymbol{\Phi}_{\mathcal{I}_1}) + \mathrm{rank}(\boldsymbol{\Phi}_{\mathcal{I}_2}) \end{cases} \tag{9.44}$$

2. $\tau_1 = \tau_2$: When the two exponents are equivalent (and thus usually equal to one in order to fully exploit the available power budget), the matrix $\boldsymbol{M}_{\mathcal{I}_1}$ in (9.2.2) converges for $P_{\max} \to \infty$ to

$$\begin{aligned}
\boldsymbol{M}_{\mathcal{I}_1} &\to \mathbf{I}_{|\mathcal{I}_1|} - \boldsymbol{Y}_{\mathcal{I}_1}^{\mathrm{H}}\left[\boldsymbol{Y}_{\mathcal{I}_1}\boldsymbol{Y}_{\mathcal{I}_1}^{\mathrm{H}} + \boldsymbol{Y}_{\mathcal{I}_2}\boldsymbol{Y}_{\mathcal{I}_2}^{\mathrm{H}}\right]^{-1}\boldsymbol{Y}_{\mathcal{I}_1} \\
&= \mathbf{I}_{|\mathcal{I}_1|} - \boldsymbol{A}_{\mathcal{I}_1}^{\frac{1}{2}}\boldsymbol{\Phi}_{\mathcal{I}_1}^{\mathrm{H}}\left[\boldsymbol{\Phi}\boldsymbol{A}\boldsymbol{\Phi}^{\mathrm{H}}\right]^{-1}\boldsymbol{\Phi}_{\mathcal{I}_1}\boldsymbol{A}_{\mathcal{I}_1}^{\frac{1}{2}}
\end{aligned}$$

where $\boldsymbol{A} := \mathrm{blockdiag}\{\boldsymbol{A}_1, \boldsymbol{A}_2\} \succ \boldsymbol{0}$. This time, the sum MMSE of all users in $\mathcal{I}_1$ asymptotically reads as

$$\lim_{P_{\max}\to\infty} \mathrm{tr}(\boldsymbol{M}_{\mathcal{I}_1}) = |\mathcal{I}_1| - \mathrm{tr}\big(\boldsymbol{\Phi}_{\mathcal{I}_1}\boldsymbol{A}_{\mathcal{I}_1}\boldsymbol{\Phi}_{\mathcal{I}_1}^{\mathrm{H}}\big[\boldsymbol{\Phi}\boldsymbol{A}\boldsymbol{\Phi}^{\mathrm{H}}\big]^{-1}\big). \tag{9.45}$$

Due to $\mathrm{rank}(\boldsymbol{\Phi}_{\mathcal{I}_1}) < r$, $\mathrm{rank}(\boldsymbol{\Phi}) = r$, and $\boldsymbol{\Phi}\boldsymbol{A}\boldsymbol{\Phi}^{\mathrm{H}} = \boldsymbol{\Phi}_{\mathcal{I}_1}\boldsymbol{A}_{\mathcal{I}_1}\boldsymbol{\Phi}_{\mathcal{I}_1}^{\mathrm{H}} + \boldsymbol{\Phi}_{\mathcal{I}_2}\boldsymbol{A}_{\mathcal{I}_2}\boldsymbol{\Phi}_{\mathcal{I}_2}^{\mathrm{H}}$, the inequality $\lim_{P_{\max}\to\infty} \mathrm{tr}(\boldsymbol{M}_{\mathcal{I}_1}) \geq |\mathcal{I}_1| - \mathrm{rank}(\boldsymbol{\Phi}_{\mathcal{I}_1})$ is strict, unless $\mathrm{rank}(\boldsymbol{\Phi}_{\mathcal{I}_1}) = r - \mathrm{rank}(\boldsymbol{\Phi}_{\mathcal{I}_2})$.

3. $\tau_1 < \tau_2$: For this configuration, the asymptotic behavior of $\boldsymbol{M}_{\mathcal{I}_1}$ can be obtained by means of the substitution

$$\boldsymbol{\Omega} := \mathbf{I}\cdot P_{\max}^{-\tau_1} + \boldsymbol{Y}_{\mathcal{I}_2}\boldsymbol{Y}_{\mathcal{I}_2}^{\mathrm{H}} P_{\max}^{\tau_2-\tau_1}.$$

Using the matrix inversion Lemma from Appendix A.2, (9.2.2) can be reformulated to

$$\boldsymbol{M}_{\mathcal{I}_1} = \mathbf{I} - \boldsymbol{Y}_{\mathcal{I}_1}^{\mathrm{H}}\big[\boldsymbol{\Omega} + \boldsymbol{Y}_{\mathcal{I}_1}\boldsymbol{Y}_{\mathcal{I}_1}^{\mathrm{H}}\big]^{-1}\boldsymbol{Y}_{\mathcal{I}_1} = \big[\mathbf{I} + \boldsymbol{Y}_{\mathcal{I}_1}^{\mathrm{H}}\boldsymbol{\Omega}^{-1}\boldsymbol{Y}_{\mathcal{I}_1}\big]^{-1}$$

For $P_{\max} \to \infty$, the inverse $\boldsymbol{\Omega}^{-1}$ behaves due to $\tau_1 < \tau_2$ as

$$\boldsymbol{\Omega}^{-1} = P_{\max}^{\tau_1}\big[\mathbf{I} + \boldsymbol{Y}_{\mathcal{I}_2}\boldsymbol{Y}_{\mathcal{I}_2}^{\mathrm{H}} P_{\max}^{\tau_2}\big]^{-1} \cong P_{\max}^{\tau_1}\boldsymbol{\Pi}_{\boldsymbol{Y}_{\mathcal{I}_2}^{\mathrm{H}}}^{\perp}$$

such that $\boldsymbol{M}_{\mathcal{I}_1}$ eventually converges to

$$\boldsymbol{M}_{\mathcal{I}_1} \cong \big[\mathbf{I} + P_{\max}^{\tau_1}\boldsymbol{Y}_{\mathcal{I}_1}^{\mathrm{H}}\boldsymbol{\Pi}_{\boldsymbol{Y}_{\mathcal{I}_2}^{\mathrm{H}}}^{\perp}\boldsymbol{Y}_{\mathcal{I}_1}\big]^{-1} \to \boldsymbol{\Pi}_{\boldsymbol{\Pi}_{\boldsymbol{Y}_{\mathcal{I}_2}^{\mathrm{H}}}^{\perp}\boldsymbol{Y}_{\mathcal{I}_1}}^{\perp}$$

i.e., to the projector onto the null-space of $\boldsymbol{\Pi}_{\boldsymbol{Y}_{\mathcal{I}_2}^{\mathrm{H}}}^{\perp}\boldsymbol{Y}_{\mathcal{I}_1}$. Interestingly, the asymptotic MMSE matrix of the users in $\mathcal{I}_1$ that now share the smallest power exponent can again (just like in the case where τ_1 is the largest exponent) be interpreted as the asymptotic MMSE matrix of a system where *only* the $|\mathcal{I}_1|$ users in $\mathcal{I}_1$ are present. However, the channel matrix $\boldsymbol{\Phi}_{\mathcal{I}_1}$ of these users is first projected onto the null-space of $\boldsymbol{Y}_{\mathcal{I}_2}^{\mathrm{H}} = \boldsymbol{A}_{\mathcal{I}_2}^{\frac{1}{2}}\boldsymbol{\Phi}_{\mathcal{I}_2}^{\mathrm{H}}$. Thus, in the limit $P_{\max} \to \infty$, the MMSE matrix $\boldsymbol{M}_{\mathcal{I}_1}$ coincides with the asymptotic MMSE matrix of an isolated system in which solely the users in $\mathcal{I}_1$ are present with the channel matrix $\bar{\boldsymbol{\Phi}}_{\mathcal{I}_1} := \boldsymbol{\Pi}_{\boldsymbol{Y}_{\mathcal{I}_2}^{\mathrm{H}}}^{\perp}\boldsymbol{\Phi}_{\mathcal{I}_1}$:

$$\big[\mathbf{I} + P_{\max}^{\tau_1}\boldsymbol{A}_{\mathcal{I}_1}^{\frac{1}{2}}\bar{\boldsymbol{\Phi}}_{\mathcal{I}_1}^{\mathrm{H}}\bar{\boldsymbol{\Phi}}_{\mathcal{I}_1}\boldsymbol{A}_{\mathcal{I}_1}^{\frac{1}{2}}\big] \cong \boldsymbol{M}_{\mathcal{I}_1}$$

This leads to an asymptotic sum MMSE

$$\mathrm{tr}(\boldsymbol{M}_{\mathcal{I}_1}) \to |\mathcal{I}| - \mathrm{rank}(\boldsymbol{\Pi}_{\boldsymbol{Y}_{\mathcal{I}_2}^{\mathrm{H}}}^{\perp}\boldsymbol{Y}_{\mathcal{I}_1}) = |\mathcal{I}| - \mathrm{rank}(\boldsymbol{\Pi}_{\boldsymbol{Y}_{\mathcal{I}_2}^{\mathrm{H}}}^{\perp}\boldsymbol{\Phi}_{\mathcal{I}_1})$$

which obviously is lower bounded by $|\mathcal{I}| - \text{rank}(\boldsymbol{\Phi}_{\mathcal{I}_1})$. However, this lower bound is only attained for $\text{rank}([\boldsymbol{\Phi}_{\mathcal{I}_1}, \boldsymbol{\Phi}_{\mathcal{I}_2}]) = \text{rank}(\boldsymbol{\Phi}_{\mathcal{I}_1}) + \text{rank}(\boldsymbol{\Phi}_{\mathcal{I}_2})$, i.e., when the projector does not lead to a rank reduction of $\boldsymbol{\Phi}_{\mathcal{I}_1}$, see the first case with $\tau_1 > \tau_2$.

We conclude that the sum MMSE of those users in $\mathcal{I}_1$ can achieve its lower bound $|\mathcal{I}| - \text{rank}(\boldsymbol{\Phi}_{\mathcal{I}_1})$ in the asymptotic limit $P_{\max} \to \infty$ only in two cases: Either, if the two user sets $\mathcal{I}_1$ and $\mathcal{I}_2$ are decoupled by chance, which means that $\text{rank}([\boldsymbol{\Phi}_{\mathcal{I}_1}, \boldsymbol{\Phi}_{\mathcal{I}_2}]) = \text{rank}(\boldsymbol{\Phi}_{\mathcal{I}_1}) + \text{rank}(\boldsymbol{\Phi}_{\mathcal{I}_2})$. This means that in the asymptotic limit, the power allocation $\boldsymbol{A}_{\mathcal{I}_1}$ of the users in $\mathcal{I}_1$ does not have an impact on the obtained MMSEs of the users in $\mathcal{I}_2$.[11] In this case, it does not matter whether τ_1 is larger than, smaller than, or equal to τ_2.

Or, if the power exponent of all users in $\mathcal{I}_1$ is strictly larger than the largest exponent of the users in the complementary group $\mathcal{I}_2$ (which means $\tau_1 > \tau_2$ in the discussed case). For this choice of the power exponents, the users in $\mathcal{I}_1$ *always* achieve their lower sum MMSE bound $|\mathcal{I}_1| - \text{rank}(\boldsymbol{\Phi}_{\mathcal{I}_1})$ for $P_{\max} \to \infty$ irrespective of the rank of $\boldsymbol{\Phi}_{\mathcal{I}_2}$.

Above result can also be generalized to more than two disjoint sets. Let $\mathcal{I}_1, \ldots, \mathcal{I}_F$ be F pairwise disjoint sets such that $\cup_{i=1}^{F} \mathcal{I}_i = \mathcal{K}$ and let the distinct power exponents associated to the F sets be $\tau_1, \ldots, \tau_F$. Without loss of generality, we may assume $\tau_1 > \cdots > \tau_F > 0$. The MMSE matrix $\boldsymbol{M}_{\mathcal{I}_1}$ of the users in $\mathcal{I}_1$ asymptotically only depends on $\boldsymbol{\Phi}_{\mathcal{I}_1}$ and $\boldsymbol{A}_{\mathcal{I}_1}$ since τ_1 is the largest exponent, see (9.40). In general, the MMSE matrix $\boldsymbol{M}_{\mathcal{I}_i}$ of users in some set $\mathcal{I}_i$ only depends on those matrices $\boldsymbol{\Phi}_{\mathcal{I}_f}$ and $\boldsymbol{A}_{\mathcal{I}_f}$, where $f \leq i$, i.e., only those groups $\mathcal{I}_f$ have an impact, whose exponents τ_f are larger than or equal to τ_i. More precisely, the asymptotic MMSE matrix $\boldsymbol{M}_{\mathcal{I}_i}$ of the users in $\mathcal{I}_i$ is equivalent to the one of a system in which the users $\mathcal{I}_i$ are isolated, but their channel matrix $\boldsymbol{\Phi}_{\mathcal{I}_i}$ has first been projected into the null-space of $[\boldsymbol{Y}_{\mathcal{I}_1}, \ldots, \boldsymbol{Y}_{\mathcal{I}_{i-1}}]^{\mathrm{H}}$.

Using these preliminary observations, we start the proof of Theorem 9.2 by checking the feasibility of given MMSE requirements $\boldsymbol{\mu}'$. The proof consists of several steps:

1. First, we rule out the case $\mu_i' = 1$ for some i since this case is trivial. If $\mu_i' = 1$ for some i, then no power is allocated to those users (i.e., $q_i = 0$) and they are dropped from the system. Hence, we can assume $\boldsymbol{\mu} < \mathbf{1}$ without loss of generality in the following. Moreover, $\boldsymbol{\mu}' > \mathbf{0}$ is assumed to be valid, since we are interested in the feasibility of MMSE tuples in the interior of the polytope $\mathcal{P}$.
2. If the requirements $\boldsymbol{\mu}'$ are not element of the polytope $\mathcal{P}$ in (9.32), they violate at least one necessary condition on the lower bound of the sum MMSE of a particular user subset. Hence, $\boldsymbol{\mu}'$ is infeasible even in the asymptotic limit with an infinite amount of power, if $\boldsymbol{\mu}' \notin \mathcal{P}$.

[11] Think of the regular channels scenario with $K \leq N$ users. There, any two disjoint user subsets $\mathcal{I}_1$ and $\mathcal{I}_2$ are decoupled and the MMSE receivers converge to the zero-forcing receivers in the limit, which completely decouple the two user sets.

3. The dimension-reduction filter at the receiver can be assumed to be already applied such that the resulting channel matrix $\boldsymbol{\Phi} \in \mathbb{C}^{r \times K}$ has $r = \mathrm{rank}(\boldsymbol{H})$ rows. Moreover, $\boldsymbol{\Phi}$ does not satisfy the regularity condition in (9.20), otherwise Theorem 9.2 would coincide with Theorem 9.1 which was already proven in Sect. 9.2.1.
4. According to Lemma 9.3, feasibility of $\boldsymbol{\mu}'$ immediately results from the feasibility of some stricter requirements $\boldsymbol{\mu} \leq \boldsymbol{\mu}'$. Instead of proving that $\boldsymbol{\mu}' \in \mathcal{P}$ is feasible, we may alternatively show the (asymptotic) feasibility of more demanding requirements $\boldsymbol{\mu} \leq \boldsymbol{\mu}'$. This will simplify the proof since we can thus change the requirements such that the power of every user will eventually grow beyond all limits as $P_{\max} \to \infty$.
5. The stricter MMSE requirements $\boldsymbol{\mu} \leq \boldsymbol{\mu}'$ are obtained by successively scaling individual entries of $\boldsymbol{\mu}'$ by some factor smaller[12] than one until no entry of $\boldsymbol{\mu}$ can further be reduced without leaving the polytope $\mathcal{P}$. Let $\mathcal{I}_0 := \mathcal{K}$ and let $\mathcal{I}_1, \ldots, \mathcal{I}_F \subset \mathcal{K}$ denote user subsets for which $\mathrm{rank}(\boldsymbol{\Phi}_{\mathcal{I}}) < \min\{r, |\mathcal{I}|\}$. These F subsets induce the inequalities

$$\sum_{k \in \mathcal{I}_i} \mu_k \geq |\mathcal{I}_i| - \mathrm{rank}(\boldsymbol{\Phi}_{\mathcal{I}_i}) \quad \forall i \in \{1, \ldots, F\} \tag{9.46}$$

whereas $\mathcal{I}_0$ induces the lower bound

$$\sum_{k \in \mathcal{I}_0} \mu_k \geq |\mathcal{I}_0| - r. \tag{9.47}$$

In the first step, we find the largest $\alpha \leq 1$ such that $\alpha\boldsymbol{\mu}'$ satisfies all $F+1$ inequalities in (9.46) and (9.47), but at least one of them with equality. Then, $\boldsymbol{\mu}$ is set to $\alpha\boldsymbol{\mu}'$. Assume that the constraint belonging to $\mathcal{I}_1$ holds with equality. Then, the requirements of the users in $\mathcal{I}_1$ may not further be reduced in later steps since their requirements are tight. Hence, further steps may only reduce those entries of $\boldsymbol{\mu}$ that belong to the remaining users $\mathcal{K} \setminus \mathcal{I}_1$. In the second round, we find the largest $\alpha \leq 1$ which scales the requirements of users in $\mathcal{K} \setminus \mathcal{I}_1$ such that again all lower bounds are not violated, but at least one additional lower bound is achieved with equality. This procedure is continued until no user remains whose requirement may be scaled. A pseudo-code implementation is shown in Algorithm 9.1.

After the termination of the algorithm, the resulting $\boldsymbol{\mu}$ may contain entries which are zero. They belong to *decoupled* users, which are defined as follows:

Definition 9.1 A user k in the vector broadcast channel is called decoupled, if the rank of the channel matrix reduces by one if he is dropped from the system, i.e.,

$$\mathrm{rank}(\boldsymbol{\Phi}) = \mathrm{rank}(\boldsymbol{\Phi}_{\mathcal{K} \setminus \{k\}}) + 1. \tag{9.48}$$

[12] In some steps, the scaling may be equal to one.

Note that the MMSE of a decoupled user k goes to zero when his transmit power grows beyond all limits ($q_k \to \infty$) irrespective of the power allocation of all other users, see the discussion of the three power allocation cases before. Additionally, a decoupled user does not impact the asymptotic MMSE values of all other users if their powers grow beyond all limits. For example, in a regular channel scenario with at most $K \leq N$ users, every user is decoupled. Hence, MMSEs that are zero are possible for every user in the asymptotic limit. As a consequence, one could assign the smallest power exponent to all decoupled users such that they do not have an impact on the remaining ones but their MMSEs can still be driven to zero. Thus, we may assume without loss of generality that no decoupled users are present in the following such that we can assume $\mathbf{0} < \boldsymbol{\mu} < \mathbf{1}$.

Another property of Algorithm 9.1 is, that the more demanding constraints $\boldsymbol{\mu} \leq \boldsymbol{\mu}'$ attain the lower bound on the total sum MMSE of all users, i.e., $\mathbf{1}^{\mathrm{T}}\boldsymbol{\mu} = K - r$, which we show by contradiction. Assume that the total sum MMSE was above its lower bound, i.e., $\mathbf{1}^{\mathrm{T}}\boldsymbol{\mu} > K - r$, and none of its entries $\mu_1, \ldots, \mu_K$ could be reduced due to the fact that some other constraints on the sum MMSE of user subsets would be violated. Then, all requirements $\mu_1, \ldots, \mu_K$ would have to be involved in the binding constraints since otherwise, the total sum MMSE could be lowered by reducing those elements that are not involved in the binding constraints. However, if the users in the sets $\mathcal{I}_1$ and $\mathcal{I}_2$ achieve their lower sum MMSE bound individually (per set), then the sum MMSE of the union $\mathcal{I}_1 \cup \mathcal{I}_2$ also achieves its lower bound with equality:

Lemma 9.4 *If the users of the two sets $\mathcal{I}_1$ and $\mathcal{I}_2$ attain their lower sum MMSE bounds, i.e.,*

$$\sum_{k \in \mathcal{I}_i} \varepsilon_k(\boldsymbol{q}) = |\mathcal{I}_i| - \mathrm{rank}(\boldsymbol{\Phi}_{\mathcal{I}_i}) \quad \forall i \in \{1, 2\}$$

then so does the union $\mathcal{I}_1 \cup \mathcal{I}_2$ and the intersection $\mathcal{I}_1 \cap \mathcal{I}_2$ (if it is nonempty):

$$\sum_{k \in \mathcal{I}_1 \cup \mathcal{I}_2} \varepsilon_k(\boldsymbol{q}) = |\mathcal{I}_1 \cup \mathcal{I}_2| - \mathrm{rank}(\boldsymbol{\Phi}_{\mathcal{I}_1 \cup \mathcal{I}_2})$$

$$\sum_{k \in \mathcal{I}_1 \cap \mathcal{I}_2} \varepsilon_k(\boldsymbol{q}) = |\mathcal{I}_1 \cap \mathcal{I}_2| - \mathrm{rank}(\boldsymbol{\Phi}_{\mathcal{I}_1 \cap \mathcal{I}_2})$$

According to Lemma 9.4 (which is proven in Appendix A.22), the total sum MMSE constraint must also be binding.

6. Knowing that the total sum MMSE inequality is binding and that $\boldsymbol{\mu} < \mathbf{1}$, it follows that the powers of all users need to grow beyond all limits as $P_{\max} \to \infty$. The fact that at least r users (whose channel vectors must have rank r) must have diverging powers results from Lemma 9.2 and the fact that *all* users need to have diverging powers is a consequence of $\boldsymbol{\mu} < \mathbf{1}$ and the definition of the sum MMSE in (9.9). As before, we exploit the fact that *any* SINR requirements that fulfill $0 < \gamma_1, \ldots, \gamma_K < \infty$ can be balanced under a sum power limit $P_{\max}$, see (9.24). Due to $\boldsymbol{\mu} < \mathbf{1}$ and $\boldsymbol{\mu} > \mathbf{0}$ (all decoupled users have been removed),

Algorithm 9.1 MMSE Requirement Update Algorithm

Require: MMSE requirements $\boldsymbol{\mu}' \in \mathcal{P}$, user sets $\mathcal{I}_0, \ldots, \mathcal{I}_F$

$\boldsymbol{\mu} \leftarrow \boldsymbol{\mu}'$ — start with initial requirements
$\mathcal{Z} \leftarrow \mathcal{K}$ — in the first round, all requirements will be rescaled
while $\mathcal{Z} \neq \emptyset$ **do**
  for $i = 0$ to F **do**
    $\mathcal{X}_i \leftarrow \mathcal{I}_i \cap \mathcal{Z}$ — users in $\mathcal{I}_i$ whose requirements may be altered
    $\mathcal{Y}_i \leftarrow \mathcal{I}_i \setminus \mathcal{X}_i$ — users in $\mathcal{I}_i$ whose requirements may *not* be altered
    if $\mathcal{X}_i \neq \emptyset$ **then**
      $\beta_i \leftarrow \frac{|\mathcal{I}_i| - \mathrm{rank}(\boldsymbol{\Phi}_{\mathcal{I}_i}) - \sum_{k \in \mathcal{Y}_i} \mu_k}{\sum_{k \in \mathcal{X}_i} \mu_k}$ — factor by which users in $\mathcal{X}_i$ would have to be scaled such that the constraint resulting from $\mathcal{I}_i$ becomes tight
    end if
  end for
  $\alpha \leftarrow \max_{i \in \{0,\ldots,F\}} \beta_i$ s.t.: $\mathcal{X}_i \neq \emptyset$ — pick largest factor such that no constraint is violated
  $k \leftarrow \mathrm{argmax}_{i \in \{0,\ldots,F\}} \beta_i$ s.t.: $\mathcal{X}_i \neq \emptyset$ — find index of the corresponding user set
  $\boldsymbol{\mu}_{\mathcal{Z}} \leftarrow \alpha \boldsymbol{\mu}_{\mathcal{Z}}$ — rescale the requirements of the users in $\mathcal{Z}$
  $\mathcal{Z} \leftarrow \mathcal{Z} \setminus \mathcal{I}_k$ — requirements of users in $\mathcal{I}_k$ may not be reduced further as their constraint is tight
end while

the MMSE targets $\boldsymbol{\mu}$ translate to SINR targets that fulfill $0 < \gamma_1, \ldots, \gamma_K < \infty$, see (9.10). Furthermore, we know that the optimum power allocation is unique and that the maximum common scalar $\check{\gamma}_0(P_{\max})$ is increasing in $P_{\max}$. Hence, the maximum $\check{\gamma}_0(P_{\max})$ is obtained for $P_{\max} \to \infty$ when the sum MMSE lower bound is attained. In this case, $\check{\gamma}_0(P_{\max}) \to 1$ holds due to the choice of $\boldsymbol{\mu}$, which means that the requirements $\boldsymbol{\mu}$ are asymptotically achievable as $P_{\max} \to \infty$. For finite $P_{\max}$, the maximum common scaling $\check{\gamma}_0(P_{\max})$ can be made as close to one as desired by increasing $P_{\max}$. Thus, the requirement $\boldsymbol{\mu} \in \partial\mathcal{P}$ on the boundary of the polytope can be approached from the interior as close as desired with finite sum power. In conjunction with Lemma 9.3, we conclude that *any* point in the interior of the polytope $\mathcal{P}$ can be achieved with finite $P_{\max}$. As a consequence, any point on the boundary that is different from the all-ones vector can asymptotically be achieved with an infinite amount of power. This concludes the proof for Theorem 9.2.

Besides the feasibility of points in the interior of the polytope $\mathcal{P}$, our analysis also delivers further interesting result on the power allocation. In particular, it is shown that the power allocation must be nonlinear in some cases. Conditions under which this may occur will be accompanied by an example for a better understanding. Let $K = 5$ user communicate with an $N = 3$ antenna base station and let the MAC channel matrix be given by

$$\boldsymbol{H} = \begin{bmatrix} 1 & 1 & 2 & 3 & 4 \\ 2 & 3 & 5 & 5 & 6 \\ 1 & 2 & 3 & 7 & 8 \end{bmatrix}.$$

As one can see, $\boldsymbol{h}_3 = \boldsymbol{h}_1 + \boldsymbol{h}_2$, but the rank r of $\boldsymbol{H}$ is equal to the number of antennas N, such that we do not need the first stage $\boldsymbol{V}$ of the receive filter in (9.37), i.e., $\boldsymbol{\Phi} = \boldsymbol{H}$. In addition to the lower bound on the sum MMSE of all users in $\mathcal{I}_0 := \mathcal{K}$

$$\varepsilon_1(\boldsymbol{q}) + \varepsilon_2(\boldsymbol{q}) + \varepsilon_3(\boldsymbol{q}) + \varepsilon_4(\boldsymbol{q}) + \varepsilon_5(\boldsymbol{q}) + \geq 2 \tag{9.49}$$

the only set $\mathcal{I}$ that fulfills $\mathrm{rank}(\boldsymbol{H}_{\mathcal{I}}) < \min\{\mathrm{rank}(\boldsymbol{H}), |\mathcal{I}|\}$ is the set $\mathcal{I}_1 := \{1, 2, 3\}$, which induces the lower bound

$$\varepsilon_1(\boldsymbol{q}) + \varepsilon_2(\boldsymbol{q}) + \varepsilon_3(\boldsymbol{q}) \geq 1. \tag{9.50}$$

Let the MMSE requirements read as

$$\boldsymbol{\mu}' = [0.5, 0.5, 0.5, 0.8, 0.8]^{\mathrm{T}}.$$

Algorithm 9.1 will transform these requirements to

$$\boldsymbol{\mu} = \left[\frac{1}{3}, \frac{1}{3}, \frac{1}{3}, \frac{1}{2}, \frac{1}{2}\right]^{\mathrm{T}} \tag{9.51}$$

such that both (9.49) and (9.50) are binding. Assume that both the users of the set $\mathcal{I}_1 = \{1, 2, 3\}$ and the users of the set $\mathcal{I}_2 := \mathcal{K} \setminus \mathcal{I}_1 = \{4, 5\}$ share a linear power allocation with the same exponents $\tau_1 = \tau_2 = 1$. Then, from (9.45) and the discussion below, it follows that the sum MMSE of the users in $\mathcal{I}_1$ will be *larger* than its lower bound $|\mathcal{I}_1| - \mathrm{rank}(\boldsymbol{\Phi}_1) = 1$ even in the limit $P_{\mathrm{max}} \to \infty$ due to the fact that $2 = \mathrm{rank}(\boldsymbol{\Phi}_{\mathcal{I}_1}) \neq r - \mathrm{rank}(\boldsymbol{\Phi}_{\mathcal{I}_2}) = 1$. In other words, the requirements cannot be obtained by means of a linear power allocation. The only possibility to let the sum MMSE of the first three users in $\mathcal{I}_1$ attain its lower bound of one is that the power exponent $\tau_1 \leq 1$ of all users in $\mathcal{I}_1$ must be larger than the exponent τ_2 of the users in $\mathcal{I}_2$, see (9.41). And that is exactly what the SINR balancing algorithm in [35] does when the MMSE requirements $\boldsymbol{\mu}$ from (9.51) are converted to SINR requirements via (9.23) and P_{max} is increased beyond all limits.[13] Note that the converted SINR requirements that belong to (9.51) can also be used as SIR requirements (without noise). For SIR balancing, however, only the ratios of the individual powers play a role (the 1-norm of the power vector is set to a fixed value), which means that essentially only linear power allocations can be handled. Since the power allocation to achieve the MMSE requirements in (9.51) is nonlinear with a power exponent of the users in $\mathcal{I}_2 = \{4, 5\}$ being smaller than one, the last two entries of the SIR balancing 'power' vector will asymptotically go to zero during the iterations of the SIR balancing algorithm although their SIR requirement is larger than zero. Hence,

[13] The power exponent τ_2 of the users in $\mathcal{I}_2$ that results from the SINR balancing algorithm is approximately $\tau_2 \approx 0.5$ whereas the power exponent τ_1 of the users in $\mathcal{I}_1$ is around $\tau_1 \approx 1$. This apparently leads to the largest common scaling γ_0.

the requirements in (9.51) are not directly achievable as SIRs. The fact that some SIR requirements are not effectively achievable has also been observed in [165].

Above strategy can easily be generalized to more than two binding constraint sets. The users that arise most often in the sets $\mathcal{I}_0, \ldots, \mathcal{I}_F$ share the highest power exponent one, whereas the power exponent of the other users is increasing in the number of sets they belong to. If only the total sum MMSE constraint in (9.49) is binding, which happens for example when $\boldsymbol{\mu} = [0.4, 0.4, 0.4, 0.4, 0.4]^{\mathrm{T}}$, then all users share the same power exponent one.

Theorem 9.2 directly leads to the *supremum* of the SIR balancing algorithm, which not necessarily need to be attainable:

Corollary 9.4 *Let $\mathcal{I}_0 = \{1, \ldots, K\}$ be the total user set and let $\mathcal{I}_1, \ldots, \mathcal{I}_F$ be those F different subsets of $\mathcal{I}_0$ for which rank$(\boldsymbol{H}_{\mathcal{I}}) < \min\{\text{rank}(\boldsymbol{H}), |\mathcal{I}|\}$. Given relative SIR requirements that satisfy $0 < \gamma_1, \ldots, \gamma_K < \infty$, define g_i as the unique positive solution of*

$$\sum_{k \in \mathcal{I}_i} \frac{1}{1 + g_i \gamma_k} = |\mathcal{I}_i| - \mathit{rank}(\boldsymbol{H}_{\mathcal{I}_i}) \quad \forall i \in \{0, \ldots, F\}.$$

Then, the supremum $\check{\gamma}_0^{\sup}$ of the SIR balancing algorithm is given by

$$\check{\gamma}_0^{\sup} = \min_{i \in \{0, \ldots, F\}} g_i .$$

Synopsis: The closure of the MMSE feasibility region in the vector BC with a fixed channel that need not necessarily satisfy the regular channels condition is again a polytope. Besides the individual box constraints and the lower bound on the total sum MMSE of all users in $\mathcal{K}$, additional half-space constraints arise for every user subset $\mathcal{I} \subset \mathcal{K}$ whose corresponding channel matrix $\boldsymbol{H}_{\mathcal{I}}$ features a rank that is smaller than $\min\{|\mathcal{I}|, \text{rank}(\boldsymbol{H})\}$. Every MMSE tuple taken from the interior of the resulting polytope can be achieved with finite sum power whereas MMSE points outside the polytope are infeasible even with an infinite amount of sum power. For MMSE requirements where more than one half-space constraint is binding, the power allocation must be nonlinear with different growth rates. A consequence of this surprising result is that some targets of the SIR balancing (where the noise is not considered) are not effectively achievable.

9.3 Feasibility in the MIMO Broadcast Channel

Having solved the feasibility problem of the vector BC, we now switch to the MIMO BC with multi-antenna terminals. For the QoS metric per user, we choose the Shannon rate that can be achieved with linear transceivers. While the power minimization problem in (9.1) becomes much more difficult when switching from single-antenna to multi-antenna terminals (since one has to decide for the number of active streams per user and the corresponding beamformers), the feasibility problem interestingly remains very simple. This is due to the surprising fact that single-stream transmission suffices to test whether given rate requirements are feasible or not. In Sect. 9.3.1, we discuss the feasibility problem conditioned on single-stream transmission, which will turn out to be closely related to the feasibility in the *vector* BC. Optimality of single-stream transmission (with respect to feasibility) is afterwards proven in Sect. 9.3.2.

9.3.1 Feasibility Under Single-Stream Transmission

As mentioned above, the QoS requirements are formulated in terms of lower bounds $\varrho_1, \dots, \varrho_K$ on the data rates $R_1, \dots, R_K$ when multi-antenna terminals are considered. Feasibility of given requirements $\varrho_1, \dots, \varrho_K$ is then equivalent to the existence of a rate tuple which fulfills

$$R_k \geq \varrho_k \quad \forall k \in \mathcal{K}.$$

Without loss of generality, the rate requirements are assumed to be positive, i.e, $\varrho_k > 0\,\forall k$, since users with vanishing requirements can be dropped from the system. Thanks to the rate duality for linear transceivers in Sect. 3.3, the feasibility problem can equivalently be tackled in the dual MIMO MAC where it is easier to solve. In addition to the power allocation, the design of the dual MAC transmitters also comprises the choice of the power-normalized precoding matrices. Due to single-stream transmission, these precoding matrices reduce to precoding vectors, and the power normalized precoder of user k in the dual MAC will be denoted by $\boldsymbol{u}_k$, where $\|\boldsymbol{u}_k\|_2 = 1$ is valid for all $k \in \mathcal{K}$. Using the channel *matrices* $\boldsymbol{H}_1, \dots, \boldsymbol{H}_K$, the received signal in the dual MAC can then be written as

$$\boldsymbol{y} = \sum_{k=1}^{K} \boldsymbol{H}_k \boldsymbol{u}_k \sqrt{q}_k s_k + \boldsymbol{n} = \boldsymbol{H} \boldsymbol{U} \boldsymbol{Q}^{\frac{1}{2}} \boldsymbol{s} + \boldsymbol{n}$$

where $\boldsymbol{Q} = \mathrm{diag}\{q_k\}_{k=1}^K$ denotes the matrix containing the power allocations of the K users. The composite channel matrix $\boldsymbol{H}$ is defined via

$$\boldsymbol{H} := [\boldsymbol{H}_1, \dots, \boldsymbol{H}_K] \in \mathbb{C}^{N \times M} \tag{9.52}$$

and the tall block-diagonal, power-normalized composite precoding matrix reads as

$$\boldsymbol{U} := \mathrm{blockdiag}\{\boldsymbol{u}_k\}_{k=1}^K \in \mathbb{C}^{M \times K}. \tag{9.53}$$

Since every user has only one active data stream, this MIMO MAC setup can be interpreted as a *vector* MAC with single-antenna transmitters. To this end, we define the equivalent vector MAC composite channel matrix

$$\boldsymbol{H}' := [\boldsymbol{h}_1', \ldots, \boldsymbol{h}_K'] = \boldsymbol{H}\boldsymbol{U} = [\boldsymbol{H}_1\boldsymbol{u}_1, \ldots, \boldsymbol{H}_K\boldsymbol{u}_K] \in \mathbb{C}^{N\times K} \tag{9.54}$$

as the product of the MIMO composite channel matrix $\boldsymbol{H}$ and the composite precoding matrix $\boldsymbol{U}$. Another consequence of the single-stream transmission is that the Shannon rate is directly related to the MMSE via (9.11). Therefore, feasibility of rate requirements $\varrho_1, \ldots, \varrho_K$ under single-stream transmission in a multi-user MIMO system with given composite channel matrix $\boldsymbol{H}$ is equivalent to feasibility of the transformed MMSE requirements $\mu_1 = 2^{-\varrho_1}, \ldots, \mu_K = 2^{-\varrho_K}$ in the vector channel with channel matrix $\boldsymbol{H}'$. The only difference is that we may optimize over the normalized precoding vectors $\boldsymbol{u}_1, \ldots, \boldsymbol{u}_K$. For the derivation of the feasible rate tuples, we again distinguish the regular channels scenario and the singular channels scenario. In contrast to the vector MAC, the regular channels scenario is defined slightly different due to the fact that $\boldsymbol{H}'$ results from the bigger matrix $\boldsymbol{H}$.

9.3.1.1 Regular Channels Scenario

The regular channels scenario definition of the *transformed* vector MAC channel matrix $\boldsymbol{H}'$ is defined via [cf. (9.20)]

$$\operatorname{rank}(\boldsymbol{H}'_{\mathcal{I}}) = \min\{|\mathcal{I}|, N\} \quad \forall \mathcal{I} \subseteq \mathcal{K}. \tag{9.55}$$

This translates to the regularity condition of the MIMO composite channel matrix $\boldsymbol{H}$ in the following way: The $N \times M$ composite MIMO channel matrix fulfills the regular channels condition, if and only if

$$\operatorname{rank}(\boldsymbol{H}_{\mathcal{I}}) \geq \min\{|\mathcal{I}|, N\} \quad \forall \mathcal{I} \subseteq \mathcal{K}. \tag{9.56}$$

Note that $\boldsymbol{H}_{\mathcal{I}}$ is now defined to comprise the channel *matrices* of all users in $\mathcal{I}$. For example, if $\mathcal{I} = \{1, 3\}$, then $\boldsymbol{H}_{\mathcal{I}} = [\boldsymbol{H}_1, \boldsymbol{H}_3]$, whereas for the transformed vector channel, only the respective *columns* are selected, i.e., $\boldsymbol{H}'_{\mathcal{I}} = [\boldsymbol{h}'_1, \boldsymbol{h}'_3]$. The relationship between (9.55) and (9.56) is as follows: The transformed vector MAC channel matrix $\boldsymbol{H}'$ satisfies the regularity condition, if and only if (9.56) is valid. Because only then, it is possible to find unit-norm precoders $\boldsymbol{u}_1, \ldots, \boldsymbol{u}_K$ such that $\boldsymbol{H}' = \boldsymbol{H}\boldsymbol{U}$ satisfies (9.55). Interestingly, the rank of the MIMO matrix $\boldsymbol{H}_{\mathcal{I}}$ can be *larger* than the minimum of $|\mathcal{I}|$ and the number of antennas N at the base station. This happens for example, if $N = 3$, $K = 2$, and a composite MIMO channel matrix $\boldsymbol{H}$ of rank three (this requires at least $M_k = 3$ antennas per terminal). Another aspect is that (9.56) may be valid even if the users share exactly the same channel matrix. For example, let $K = 3$, $N = 3$, and let the identical channel matrix of the three users $\boldsymbol{H}_1 = \boldsymbol{H}_2 = \boldsymbol{H}_3$ have rank three. Then, (9.56) is valid and choosing the

kth unit-norm precoder $\boldsymbol{u}_k$ as the kth canonical unit vector for example leads to the transformed vector MAC channel matrix $\boldsymbol{H}' = \boldsymbol{H}_1$ of rank three which fulfills the regular channels scenario condition.

Assuming MAC channel matrices $\boldsymbol{H}$ that fulfill (9.56), it is possible to find unit-norm precoders $\boldsymbol{u}_1, \ldots, \boldsymbol{u}_K$ such that the resulting vector MAC features a channel matrix $\boldsymbol{H}' = \boldsymbol{H}\boldsymbol{U}$ which fulfills (9.55). For the transformed vector MAC, the MMSE feasibility region is given in Theorem 9.1 in Sect. 9.2.1. Using (9.11), this leads to the following set of feasible rate tuples:

Theorem 9.3 *Given MIMO MAC channel matrices $\boldsymbol{H}_1, \ldots, \boldsymbol{H}_K$ that satisfy (9.56), any rate requirements $0 \leq \varrho_1, \ldots, \varrho_K < \infty$ that obey*

$$\sum_{k=1}^{K} 2^{-\varrho_k} > K - N \tag{9.57}$$

can be achieved with finite sum power under single-stream transmission. Rate requirements that obtain the right hand side of (9.57) with equality can only be achieved asymptotically with an infinite amount of power. Requirements for which $\sum_{k=1}^{K} 2^{-\varrho_k} < K - N$ cannot be met even with an infinite amount of power.

In turn, the feasibility test of given rate requirements $\varrho_1, \ldots, \varrho_K < \infty$ is extremely simple conditioned on single-stream transmission.

9.3.1.2 Singular Channels Scenario

As soon as there exists a subset $\mathcal{I} \subseteq \mathcal{K}$ for which $\text{rank}(\boldsymbol{H}_\mathcal{I}) < \min\{|\mathcal{I}|, N\}$, the scenario is singular and thus limits the set of feasible rate requirements. If (9.56) is violated, it is impossible to find unit-norm beamformers $\boldsymbol{u}_1, \ldots, \boldsymbol{u}_K$ such that the transformed composite channel matrix $\boldsymbol{H}'$ of the *vector* MAC fulfills the regular channels scenario in (9.20). For a user subset $\mathcal{I}$ for which $\text{rank}(\boldsymbol{H}_\mathcal{I}) < \min\{|\mathcal{I}|, N\}$, the maximum rank of the *transformed* vector channel $\boldsymbol{H}'_\mathcal{I}$ is equivalent to $\text{rank}(\boldsymbol{H}_\mathcal{I})$. As a consequence, $\boldsymbol{H}'$ belongs to the singular channels scenario, for which an additional lower bound on the sum MMSE of the users in $\mathcal{I}$ arises:

$$\sum_{k\in\mathcal{I}} \mu_i \geq |\mathcal{I}| - \text{rank}(\boldsymbol{H}'_\mathcal{I}) = |\mathcal{I}| - \text{rank}(\boldsymbol{H}_\mathcal{I}) \quad \forall \mathcal{I} \subset \mathcal{K} : \text{rank}(\boldsymbol{H}_\mathcal{I}) < \min\{N, |\mathcal{I}|\}$$

In analogy to Theorem 9.2, the feasibility of rate requirements given an arbitrary MIMO MAC channel matrix $\boldsymbol{H}$ results from the next theorem:

Theorem 9.4 *Given rate requirements $0 \leq \varrho_1, \ldots, \varrho_K < \infty$ are feasible with finite sum power and single-stream transmission in a system with the dual MIMO MAC composite channel matrix $\boldsymbol{H}$, if and only if*

$$\sum_{k \in \mathcal{K}} 2^{-\varrho_k} > K - rank(\boldsymbol{H}) \tag{9.58}$$

and

$$\sum_{k \in \mathcal{I}} 2^{-\varrho_k} > |\mathcal{I}| - rank(\boldsymbol{H}_{\mathcal{I}}) \quad \forall \mathcal{I} \subset \mathcal{K} : rank(\boldsymbol{H}_{\mathcal{I}}) < \min\{|\mathcal{I}|, rank(\boldsymbol{H})\}. \tag{9.59}$$

9.3.2 Optimality of Single-Stream Transmission

The validity of Theorems 9.3 and 9.4 is so far conditioned on single-stream transmission per user. In this subsection, it is proven that allowing for several active streams per user does not enlarge the set of feasible rate tuples. Clearly, it may lower the required transmit power to achieve the rate targets, but surprisingly, it does not have an impact on the feasibility of them.

For the proof, which is basically taken from [171], we assume that given rate requirements $0 \leq \varrho_1, \ldots, \varrho_K < \infty$ cannot be attained under single-stream transmission with finite sum power.[14] If $\boldsymbol{H}$ fulfills the regular channels condition, this is equivalent to

$$\sum_{k \in \mathcal{K}} \mu_k = K - N - \alpha \tag{9.60}$$

where $\alpha \geq 0$ and $\mu_k = 2^{-\varrho_k}$. Next, we add a second data stream to some user j. As we know from the rate duality of linear transceivers in Sect. 3.3, the precoder of user j can be rotated by means of unitary matrices such that the two streams of user j do not interfere. Hence, treating the outcomes of the two receive filters, that belong to the two data streams of user j, separately does not lead to a rate reduction for user j. As a consequence, the rate requirement ϱ_j can be expressed in terms of the MMSE requirements $\mu_j^{(1)}$ and $\mu_j^{(2)}$ of the two streams:

$$\varrho_j = -\log_2 \mu_j^{(1)} - \log_2 \mu_j^{(2)}$$

Remember that conditioned on single-stream transmission, the relationship between the MMSE requirement μ_j of user j and his rate requirement ϱ_j was given by

$$\mu_j = 2^{-\varrho_j}.$$

Combining these two results, it follows that the product $\mu_j^{(1)} \mu_j^{(2)}$ of the MMSE requirements that belong to the two streams must be identical to the MMSE requirement under single-stream transmission:

---

[14] A rate requirement $\varrho_k = \infty$ will always require an infinite amount of sum power.

$$\mu_j^{(1)}\mu_j^{(2)} = \mu_j \tag{9.61}$$

Due to the fact that it is always possible to decorrelate the two data streams such that they do not see any interference, we may treat the two data streams as two different (virtual) users in a vector MAC setup. According to Theorem 9.1, the lower bound on the sum MMSE of the $K+1$ virtual users is given by $K+1-N$ and it is not achieved with finite sum power. If the requirements $\mu_k \forall k \neq j$, $\mu_j^{(1)}$, and $\mu_j^{(2)}$ of the $K+1$ virtual users were feasible, they would satisfy

$$\sum_{k\neq j}\mu_k + \mu_j^{(1)} + \mu_j^{(2)} > K+1-N. \tag{9.62}$$

In conjunction with (9.60), feasibility would mean

$$\mu_j^{(1)} + \mu_j^{(2)} - \mu_j > 1+\alpha$$

which can be written due to (9.61) as

$$\big(1-\mu_j^{(1)}\big)\big(1-\mu_j^{(2)}\big) + \alpha < 0.$$

Obviously, the MMSE requirements $\mu_j^{(1)}$ and $\mu_j^{(2)}$ are upper bounded by one, such that $\alpha < 0$ would be a necessary condition, if the requirements $\mu_1, \ldots, \mu_K$ were possible. However, $\alpha \geq 0$ must be valid in order to let the requirements be infeasible by single-stream transmission, see (9.60), which is a contradiction. In other words, any rate requirements $\varrho_1, \ldots, \varrho_K$ that are infeasible under single-stream transmission per user remain infeasible, if we allow for a second stream at some user j. This result can be generalized to more than one additional data stream since every further stream can be interpreted as an additional virtual user with a single data stream.

In (9.62), we assumed that the composite transformed vector MAC channel matrix of the system with $K+1$ virtual users still satisfies the regular channels condition. While this need not necessarily be true, we have observed that even under this favorable assumption, the formerly infeasible requirements cannot become feasible by allowing for more than one data stream per user. If they remain infeasible under the (possibly erroneous) assumption of a channel matrix that satisfies the regular channels condition, they will even more be infeasible for a channel matrix belonging to the singular channels scenario.

Synopsis: In order to find out whether given rate requirements $0 \leq \varrho_1, \ldots, \varrho_K < \infty$ are feasible in a MIMO BC using linear transceivers, it surprisingly suffices to check the feasibility under the assumption of single-stream transmission per user. While allowing several streams per user may obviously be beneficial to minimize the required amount of transmit power in order to attain the rate targets, it does not lead to a larger set of feasible rate requirements. With this knowledge, the feasibility problem in the MIMO BC can be converted to an equivalent feasibility problem in the vector BC with single-antenna terminals, where it has been solved completely in the previous section. Due to the simple relationship between the Shannon rate and the MMSE, finite rate requirements $\varrho_1, \ldots, \varrho_K$ are feasible if and only if

$$\sum_{k \in \mathcal{K}} 2^{-\varrho_k} > K - N$$

where N denotes the number of base station antennas and the MIMO channel matrix has to fulfill some very mild rank conditions. If these conditions are violated, the feasibility test looks slightly different and may feature additional lower bounds on the sum MMSE of individual user subsets, see Theorem 9.4.

Appendix

A.1 Inverse Nonnegativity of Nonsingular M-Matrices

Let $\boldsymbol{M} \in \mathbb{R}^{L\times L}$ be a real-valued $L \times L$ matrix with positive main diagonal elements and nonpositive off-diagonal elements. Furthermore, let $\boldsymbol{M}$ be strictly column diagonally dominant (see for example [43, Definition 4.5] for the definition) and let $\alpha > 0$ denote the largest main diagonal element. Then, we can express $\boldsymbol{M}$ as the difference of two matrices:

$$\boldsymbol{M} = \alpha \mathbf{I}_L - \boldsymbol{S} = \alpha(\mathbf{I}_L - \alpha^{-1}\boldsymbol{S}) \tag{A.1}$$

By construction, the subtrahend $\boldsymbol{S} \geq \mathbf{0}$ has only nonnegative entries and the largest column sum is strictly smaller than α due to the column diagonally dominance of $\boldsymbol{M}$. Since the spectral radius of a matrix is the infimum of all induced matrix norms [172, p. 9], the spectral radius $\rho(\boldsymbol{S})$ of the matrix $\boldsymbol{S}$ is smaller than or equal to its 1-norm (largest column sum of $\boldsymbol{S}$ in this case):

$$\rho(\boldsymbol{S}) \leq \|\boldsymbol{S}\|_1 < \alpha$$

This clearly implies

$$\rho(\alpha^{-1}\boldsymbol{S}) < 1. \tag{A.2}$$

The inverse of $\boldsymbol{M}$ can be written as

$$\boldsymbol{M}^{-1} = \alpha^{-1}\sum_{k=0}^{\infty}(\alpha^{-1}\boldsymbol{S})^k \geq 0 \tag{A.3}$$

which follows from the identity

$$(\mathbf{I} - \boldsymbol{X})\big(\mathbf{I} + \boldsymbol{X} + \boldsymbol{X}^2 + \cdots + \boldsymbol{X}^k\big) = \mathbf{I} - \boldsymbol{X}^{k+1} \quad k \in \mathbb{N}_0$$

R. Hunger, *Analysis and Transceiver Design for the MIMO Broadcast Channel*,
Foundations in Signal Processing, Communications and Networking 8,
DOI: 10.1007/978-3-642-31692-0, © Springer-Verlag Berlin Heidelberg 2013

when $k \to \infty$. The series in (A.3) is convergent due to (A.2) such that $\boldsymbol{X}^{k+1} \to \boldsymbol{0}$. Moreover, all entries of $\boldsymbol{M}^{-1}$ are nonnegative because $\boldsymbol{S} \geq \boldsymbol{0}$. The strict diagonal dominance implies that $\boldsymbol{M}$ has full rank. In combination with (A.3), this leads to the property that there is at least one nonzero and hence strictly positive element per row of $\boldsymbol{M}^{-1}$, otherwise, $\boldsymbol{M}^{-1}$ would be rank deficient. Summing up, the inverse of a strictly column diagonally dominant matrix $\boldsymbol{M}$ where all main diagonal elements are positive and all off-diagonal elements are nonpositive, has only nonnegative entries, where at least one positive entry per row must exist.

A.2 Matrix Inversion Lemma

Given the four matrices $\boldsymbol{A}$, $\boldsymbol{B}$, $\boldsymbol{C}$, and $\boldsymbol{D}$, where $\boldsymbol{A}$ and $\boldsymbol{C}$ are both square and invertible, the following identity is called the matrix inversion lemma [173, p. 571]:

$$\left(\boldsymbol{A} + \boldsymbol{B}\boldsymbol{C}\boldsymbol{D}\right)^{-1} = \boldsymbol{A}^{-1} - \boldsymbol{A}^{-1}\boldsymbol{B}\left(\boldsymbol{C}^{-1} + \boldsymbol{D}\boldsymbol{A}^{-1}\boldsymbol{B}\right)^{-1}\boldsymbol{D}\boldsymbol{A}^{-1} \tag{A.4}$$

A.3 Minimum MSE ↔ Maximum SINR Relationship

In Fig. A.1, the data symbol $s \in \mathbb{C}$ with unit variance propagates over the channel $\boldsymbol{h} \in \mathbb{C}^N$ and is perturbed by the interference $\boldsymbol{i} \in \mathbb{C}^N$ and the noise $\boldsymbol{n} \in \mathbb{C}^N$. The three quantities s, $\boldsymbol{i}$, and $\boldsymbol{n}$ are assumed to be zero-mean and pairwise independent. At the receiver side, the filter $\boldsymbol{g}^{\mathrm{T}} \in \mathbb{C}^{1\times N}$ generates the symbol estimate $\hat{s}$. Given this system model, the SINR depending on $\boldsymbol{g}$ reads as

$$\mathrm{SINR}(\boldsymbol{g}) := \frac{\boldsymbol{g}^{\mathrm{T}}\boldsymbol{h}\boldsymbol{h}^{\mathrm{H}}\boldsymbol{g}^*}{\boldsymbol{g}^{\mathrm{T}}(\boldsymbol{C}_{\boldsymbol{i}} + \boldsymbol{C}_{\boldsymbol{n}})\boldsymbol{g}^*}$$

where $\boldsymbol{C}_{\boldsymbol{i}} := \mathrm{E}[\boldsymbol{i}\boldsymbol{i}^{\mathrm{H}}]$ and $\boldsymbol{C}_{\boldsymbol{n}} := \mathrm{E}[\boldsymbol{n}\boldsymbol{n}^{\mathrm{H}}]$ are the interference and the noise covariance matrix, respectively. The receive filter that maximizes $\mathrm{SINR}(\boldsymbol{g})$ can be parametrized as

$$\boldsymbol{g}_{\mathrm{maxSINR}}^{\mathrm{T}} := \alpha \boldsymbol{h}^{\mathrm{H}}\left(\boldsymbol{C}_{\boldsymbol{i}} + \boldsymbol{C}_{\boldsymbol{n}}\right)^{-1} \tag{A.5}$$

with $\alpha \in \mathbb{C} \setminus \{0\}$, and achieves the maximum SINR

$$\gamma := \mathrm{SINR}(\boldsymbol{g}_{\mathrm{maxSINR}}) = \boldsymbol{h}^{\mathrm{H}}\left(\boldsymbol{C}_{\boldsymbol{i}} + \boldsymbol{C}_{\boldsymbol{n}}\right)^{-1}\boldsymbol{h}. \tag{A.6}$$

Equivalently, we can define the MSE between s and $\hat{s}$ via

$$\varepsilon(\boldsymbol{g}) := 1 - \boldsymbol{g}^{\mathrm{T}}\boldsymbol{h} - \boldsymbol{h}^{\mathrm{H}}\boldsymbol{g}^* + \boldsymbol{g}^{\mathrm{T}}\left(\boldsymbol{h}\boldsymbol{h}^{\mathrm{H}} + \boldsymbol{C}_{\boldsymbol{i}} + \boldsymbol{C}_{\boldsymbol{n}}\right)\boldsymbol{g}^*.$$

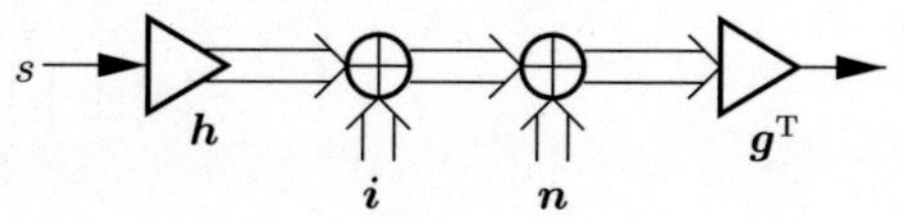

Fig. A.1 Single stream transmission with interference $\boldsymbol{i}$ and noise $\boldsymbol{n}$

This metric is minimized by the receive filter

$$\boldsymbol{g}_{\text{minMSE}}^{\mathrm{T}} := \boldsymbol{h}^{\mathrm{H}}\left(\boldsymbol{h}\boldsymbol{h}^{\mathrm{H}} + \boldsymbol{C}_{\boldsymbol{i}} + \boldsymbol{C}_{\boldsymbol{n}}\right)^{-1} = \frac{1}{1 + \boldsymbol{h}^{\mathrm{H}}\left(\boldsymbol{C}_{\boldsymbol{i}} + \boldsymbol{C}_{\boldsymbol{n}}\right)^{-1}\boldsymbol{h}}\boldsymbol{h}^{\mathrm{H}}\left(\boldsymbol{C}_{\boldsymbol{i}} + \boldsymbol{C}_{\boldsymbol{n}}\right)^{-1}. \tag{A.7}$$

The second equality in (A.7) is achieved by invoking the matrix inversion lemma from Appendix A.2 and shows that $\boldsymbol{g}_{\text{minMSE}}$ not only minimizes the MSE, but also maximizes the SINR, cf. (A.5). Inserting $\boldsymbol{g}_{\text{minMSE}}$ into the MSE expression yields

$$\mu := \varepsilon(\boldsymbol{g}_{\text{minMSE}}) = 1 - \boldsymbol{h}^{\mathrm{H}}\left(\boldsymbol{h}\boldsymbol{h}^{\mathrm{H}} + \boldsymbol{C}_{\boldsymbol{i}} + \boldsymbol{C}_{\boldsymbol{n}}\right)^{-1}\boldsymbol{h} = \frac{1}{1+\gamma} \tag{A.8}$$

stating that there is a one-to-one mapping between the *minimum* MSE and the *maximum* SINR.

A.4 Hadamard's Inequality for Positive Semi-Definite Matrices

Any $N \times N$ positive semi-definite matrix $\boldsymbol{A} \in \mathbb{S}^N$ can be decomposed via

$$\boldsymbol{A} = \boldsymbol{L}\boldsymbol{D}\boldsymbol{L}^{\mathrm{H}}$$

where $\boldsymbol{L}$ is unit lower triangular with ones on the main diagonal and the entry in the ith row and jth column is $\ell_{i,j}$ for $i > j$. The matrix $\boldsymbol{D}$ is diagonal with nonnegative main diagonal entries $d_1, \ldots, d_N$ and d_i denotes the quotient of the ith and the $i-1$th leading minor (the zeroth minor is defined to be 1). Due to the unit lower triangular structure of $\boldsymbol{L}$, the determinant of $\boldsymbol{A}$ then reads as

$$|\boldsymbol{A}| = |\boldsymbol{D}|.$$

The ith main diagonal element of $\boldsymbol{A}$ can be expressed as

$$[\boldsymbol{A}]_{i,i} = d_i + \sum_{j<i} |\ell_{i,j}|^2 d_j \geq d_i$$

where equality only holds if $d_j \ell_{i,j} = 0\, \forall j < i$. Therefore, the following inequality termed *Hadamard's* inequality between the determinant and the product of the main diagonal elements holds:

$$|\boldsymbol{A}| \leq \prod_{i=1}^{N} [\boldsymbol{A}]_{i,i} \tag{A.9}$$

For full rank $\boldsymbol{A}$, the elements $d_1, \ldots, d_N$ are positive, and equality only holds when $\boldsymbol{L}$ is the identity matrix, i.e., when $\boldsymbol{A}$ is diagonal.

A.5 Derivative of a Matrix Inverse

The derivative of the matrix inverse $\boldsymbol{X}^{-1}(\alpha)$ with respect to its scalar parameter α reads as

$$\frac{\mathrm{d}}{\mathrm{d}\alpha}\boldsymbol{X}^{-1}(\alpha) = -\boldsymbol{X}^{-1}(\alpha) \cdot \left(\frac{\mathrm{d}\boldsymbol{X}(\alpha)}{\mathrm{d}\alpha}\right) \cdot \boldsymbol{X}^{-1}(\alpha).$$

For the proof, we apply the product rule to the derivative of the identity matrix:

$$\mathbf{0} = \frac{\mathrm{d}}{\mathrm{d}\alpha}\mathbf{I} = \frac{\mathrm{d}}{\mathrm{d}\alpha}\left[\boldsymbol{X}(\alpha)\boldsymbol{X}^{-1}(\alpha)\right] = \left(\frac{\mathrm{d}\boldsymbol{X}(\alpha)}{\mathrm{d}\alpha}\right) \cdot \boldsymbol{X}^{-1}(\alpha) + \boldsymbol{X}(\alpha) \cdot \frac{\mathrm{d}}{\mathrm{d}\alpha}\boldsymbol{X}^{-1}(\alpha)$$

Left-hand side multiplication by $\boldsymbol{X}^{-1}(\alpha)$ yields the desired result.

A.6 Derivative of a Matrix Determinant

The derivative of the determinant of an $N \times N$ matrix $\boldsymbol{X}(\alpha)$ with respect to the scalar parameter α evaluated at $\alpha = \alpha_0$ reads for full rank $\boldsymbol{X}(\alpha_0)$ as

$$\frac{\mathrm{d}}{\mathrm{d}\alpha}\left|\boldsymbol{X}(\alpha)\right|\Big|_{\alpha=\alpha_0} = \left|\boldsymbol{X}(\alpha_0)\right| \cdot \mathrm{tr}\left[\boldsymbol{X}^{-1}(\alpha_0) \cdot \frac{\mathrm{d}}{\mathrm{d}\alpha}\boldsymbol{X}(\alpha)\Big|_{\alpha=\alpha_0}\right].$$

For the proof, the derivative is rewritten as a limit process:

$$\begin{aligned}
\frac{\mathrm{d}}{\mathrm{d}\alpha}\left|\boldsymbol{X}(\alpha)\right|\Big|_{\alpha=\alpha_0} &= \lim_{\Delta\alpha\to 0} \frac{1}{\Delta\alpha}\Big[\left|\boldsymbol{X}(\alpha_0+\Delta\alpha)\right| - \left|\boldsymbol{X}(\alpha_0)\right|\Big] \\
&= \lim_{\Delta\alpha\to 0} \frac{1}{\Delta\alpha}\Big[\left|\boldsymbol{X}(\alpha_0) + \Delta\alpha\boldsymbol{Y}(\alpha_0)\right| - \left|\boldsymbol{X}(\alpha_0)\right|\Big] \\
&= \left|\boldsymbol{X}(\alpha_0)\right| \cdot \lim_{\Delta\alpha\to 0} \frac{1}{\Delta\alpha}\Big[\left|\mathbf{I} + \Delta\alpha\boldsymbol{X}^{-1}(\alpha_0)\boldsymbol{Y}(\alpha_0)\right| - 1\Big]
\end{aligned}$$

The matrix $\boldsymbol{Y}(\alpha_0) := \frac{\mathrm{d}}{\mathrm{d}\alpha}\boldsymbol{X}(\alpha)|_{\alpha=\alpha_0}$ is defined as the derivative of $\boldsymbol{X}(\alpha)$ with respect to α evaluated at $\alpha = \alpha_0$. Making use of the Schur-decomposition [45, Theorem 7.1.3, p. 335]

$$\boldsymbol{Q}(\alpha_0)\boldsymbol{\Lambda}(\alpha_0)\boldsymbol{Q}^{\mathrm{H}}(\alpha_0) = \boldsymbol{X}^{-1}(\alpha_0)\boldsymbol{Y}(\alpha_0)$$

with unitary $\boldsymbol{Q}(\alpha_0)$ and upper triangular $\boldsymbol{\Lambda}(\alpha_0)$ whose diagonal elements represent the eigenvalues $\lambda_1(\alpha_0), \ldots, \lambda_N(\alpha_0)$, we obtain

$$\begin{aligned}
\frac{\mathrm{d}}{\mathrm{d}\alpha}\left|\boldsymbol{X}(\alpha)\right|\Big|_{\alpha=\alpha_0} &= \left|\boldsymbol{X}(\alpha_0)\right| \cdot \lim_{\Delta\alpha\to 0} \frac{1}{\Delta\alpha}\left[\prod_{i=1}^{N}\left(1+\Delta\alpha\lambda_i(\alpha_0)\right) - 1\right] \\
&= \left|\boldsymbol{X}(\alpha_0)\right| \cdot \sum_{i=1}^{N}\lambda_i(\alpha_0) = \left|\boldsymbol{X}(\alpha_0)\right| \cdot \mathrm{tr}\big(\boldsymbol{\Lambda}(\alpha_0)\big) \\
&= \left|\boldsymbol{X}(\alpha_0)\right| \cdot \mathrm{tr}\left[\boldsymbol{X}^{-1}(\alpha_0) \cdot \frac{\mathrm{d}}{\mathrm{d}\alpha}\boldsymbol{X}(\alpha)\Big|_{\alpha=\alpha_0}\right]
\end{aligned}$$

which completes the proof. For rank-deficient $\boldsymbol{X}(\alpha_0)$, the derivative of the determinant reads as

$$\frac{\mathrm{d}}{\mathrm{d}\alpha}\left|\boldsymbol{X}(\alpha)\right|\Big|_{\alpha=\alpha_0} = \lim_{\Delta\alpha\to 0}\frac{\left|\boldsymbol{X}(\alpha_0+\Delta\alpha)\right|}{\Delta\alpha}$$

if the limit is independent of whether $\Delta\alpha > 0$ or $\Delta\alpha < 0$. Otherwise, the determinant is not differentiable at $\alpha = \alpha_0$.

A.7 Optimum Decoding Order in the MIMO MAC

In the MIMO MAC with successive decoding, the optimum decoding order of the weighted sum rate maximization directly follows from the user weights. The main idea for the following proof can also be found in [67]. However, only the single antenna base station case is treated there and for the proof, the authors refer to an unpublished reference that never appeared.

Let $w_1, \ldots, w_K$ denote these nonnegative weights associated to the K users in the MIMO MAC. Assuming the natural decoding order $1, 2, \ldots, K$, user 1 has to cope with interference from all other users $2, \ldots, K$ whereas user K does not see any interference at all. Then, the rate R_{j-1} of some user $j-1$ with $j \in \{2, \ldots, K\}$ can be expressed as

$$R_{j-1} := \log_2\left|\boldsymbol{C}_{\boldsymbol{n}} + \sum_{i\geq j-1}\boldsymbol{A}_i\right| - \log_2\left|\boldsymbol{C}_{\boldsymbol{n}} + \sum_{i\geq j}\boldsymbol{A}_i\right| \tag{A.10}$$

where $\boldsymbol{C}_{\boldsymbol{n}} \succ \mathbf{0}$ denotes the full rank noise covariance matrix. Using the transmit covariance matrix $\boldsymbol{Q}_i$ of user i, the substitution $\boldsymbol{A}_i$ is defined via $\boldsymbol{A}_i := \boldsymbol{H}_i\boldsymbol{Q}_i\boldsymbol{H}_i^{\mathrm{H}} \succeq \mathbf{0}$. Equivalently, the rate of user j who is decoded *right after* user $j-1$ reads as

$$R_j := \log_2\left|\boldsymbol{C_n} + \sum_{i\geq j} \boldsymbol{A}_i\right| - \log_2\left|\boldsymbol{C_n} + \sum_{i>j} \boldsymbol{A}_i\right| \tag{A.11}$$

i.e., user j sees interference only from users $i > j$. We now change the decoding order of user $j-1$ and user j such that user $j-1$ is decoded *right after* user j. The remaining decoding order is left unchanged such that the total order is $1, 2, \ldots, j-2, j, j-1, j+1, \ldots, K$. Thus, only the two rate expressions of user j and user $j-1$ might vary whereas interchanging the decoding order for those two users has no impact on the rate expressions of all other users $i \notin \{j-1, j\}$. For this setup, user j has to face interference from user $j-1$ and users $i > j$ and his data rate for the modified decoding order reads as

$$R'_j := \log_2\left|\boldsymbol{C_n} + \sum_{i\geq j-1} \boldsymbol{A}_i\right| - \log_2\left|\boldsymbol{C_n} + \sum_{i>j} \boldsymbol{A}_i + \boldsymbol{A}_{j-1}\right|. \tag{A.12}$$

For user $j-1$, the interference from user j has already been subtracted leaving only users $i > j$ as interferers. His data rate can be expressed as

$$R'_{j-1} := \log_2\left|\boldsymbol{C_n} + \sum_{i>j} \boldsymbol{A}_i + \boldsymbol{A}_{j-1}\right| - \log_2\left|\boldsymbol{C_n} + \sum_{i>j} \boldsymbol{A}_i\right|. \tag{A.13}$$

Next, we compare the performance of the two decoding orders by subtracting their weighted sum rates. Since the two decoding orders only differ for user j and $j-1$, the difference of the weighted sum rates solely depends on the rates of those two users. Combining (A.10)–(A.13), the subtraction of the weighted sum rates reads as

$$\begin{aligned}\sum_{k=1}^{K} w_k\left(R_k - R'_k\right) &= \left(w_j - w_{j-1}\right)\log_2\left(\frac{\left|\boldsymbol{C_n} + \sum_{i\geq j} \boldsymbol{A}_i\right| \cdot \left|\boldsymbol{C_n} + \sum_{i>j} \boldsymbol{A}_i + \boldsymbol{A}_{j-1}\right|}{\left|\boldsymbol{C_n} + \sum_{i>j} \boldsymbol{A}_i\right| \cdot \left|\boldsymbol{C_n} + \sum_{i\geq j-1} \boldsymbol{A}_i\right|}\right) \\ &= \left(w_j - w_{j-1}\right)\log_2\left(\frac{\left|\boldsymbol{\Omega}_j + \boldsymbol{A}_j\right| \cdot \left|\boldsymbol{\Omega}_j + \boldsymbol{A}_{j-1}\right|}{\left|\boldsymbol{\Omega}_j\right| \cdot \left|\boldsymbol{\Omega}_j + \boldsymbol{A}_{j-1} + \boldsymbol{A}_j\right|}\right)\end{aligned} \tag{A.14}$$

where $\boldsymbol{\Omega}_j := \boldsymbol{C_n} + \sum_{i>j} \boldsymbol{A}_i \succ \boldsymbol{0}$. Note that the minuend on the left-hand side in (A.14) corresponds to the weighted sum rate of the natural decoding order $1, 2, \ldots, K$, and the decoding order of users j and $j-1$ is reversed in the subtrahend. With a further substitution[1] $\boldsymbol{\Phi}_{j,i} := \boldsymbol{\Omega}_j^{-\frac{1}{2}} \boldsymbol{A}_i \boldsymbol{\Omega}_j^{-\frac{1}{2}} \succeq \boldsymbol{0}$, we can reformulate (A.14):

[1] The Hermitian matrix square root is used throughout Appendix A.7.

$$\sum_{k=1}^{K} w_k\big(R_k - R_k'\big) = \big(w_j - w_{j-1}\big)\log_2\left(\frac{\big|\mathbf{I}+\boldsymbol{\Phi}_{j,j}\big|\cdot\big|\mathbf{I}+\boldsymbol{\Phi}_{j,j-1}\big|}{\big|\mathbf{I}+\boldsymbol{\Phi}_{j,j}+\boldsymbol{\Phi}_{j,j-1}\big|}\right) \tag{A.15}$$

$$= \big(w_j - w_{j-1}\big)\log_2\left(\frac{\big|\mathbf{I}+\boldsymbol{\Phi}_{j,j}\big|}{\big|\mathbf{I}+\boldsymbol{\Phi}_{j,j}^{\frac{1}{2}}\big(\mathbf{I}+\boldsymbol{\Phi}_{j,j-1}\big)^{-1}\boldsymbol{\Phi}_{j,j}^{\frac{1}{2}}\big|}\right)$$

By means of Appendix A.5 and A.6 we compute for $\alpha \geq 0$ the derivative

$$\begin{aligned}
&\frac{\mathrm{d}}{\mathrm{d}\alpha}\log_2\Big|\mathbf{I}+\boldsymbol{\Phi}_{j,j}^{\frac{1}{2}}\big(\mathbf{I}+\alpha\boldsymbol{\Phi}_{j,j-1}\big)^{-1}\boldsymbol{\Phi}_{j,j}^{\frac{1}{2}}\Big| \\
&\quad = -\frac{1}{\ln 2}\Big\|\Big[\mathbf{I}+\boldsymbol{\Phi}_{j,j}^{\frac{1}{2}}\big(\mathbf{I}+\alpha\boldsymbol{\Phi}_{j,j-1}\big)^{-1}\boldsymbol{\Phi}_{j,j}^{\frac{1}{2}}\Big]^{-\frac{1}{2}} \\
&\qquad \times \boldsymbol{\Phi}_{j,j}^{\frac{1}{2}}\big(\mathbf{I}+\alpha\boldsymbol{\Phi}_{j,j-1}\big)^{-1}\boldsymbol{\Phi}_{j,j-1}^{\frac{1}{2}}\Big\|_{\mathrm{F}}^{2} \leq 0 \quad \forall \alpha \geq 0
\end{aligned}$$

to see that the denominator in (A.15) is nonincreasing in $\boldsymbol{\Phi}_{j,j-1} \succcurlyeq \mathbf{0}$. The condition for being constant and not decreasing is that the Frobenius norm and thus its argument must be zero, i.e.,

$$\mathbf{0} = \boldsymbol{\Phi}_{j,j}^{\frac{1}{2}}\big(\mathbf{I}+\alpha\boldsymbol{\Phi}_{j,j-1}\big)^{-1}\boldsymbol{\Phi}_{j,j-1}^{\frac{1}{2}} = \boldsymbol{\Phi}_{j,j}^{\frac{1}{2}}\boldsymbol{\Phi}_{j,j-1}^{\frac{1}{2}}\big(\mathbf{I}+\alpha\boldsymbol{\Phi}_{j,j-1}\big)^{-1} \;\Leftrightarrow\; \boldsymbol{\Phi}_{j,j}\boldsymbol{\Phi}_{j,j-1} = \mathbf{0}.$$

The second equality holds since $\boldsymbol{\Phi}_{j,j-1}^{\frac{1}{2}}$ and the inverse have the same eigenbasis and thus commute. From this, we can conclude that the denominator in the logarithm of (A.15) is either smaller than the numerator (when $\boldsymbol{\Phi}_{j,j}\boldsymbol{\Phi}_{j,j-1} \neq \mathbf{0}$) or equal to the numerator (when $\boldsymbol{\Phi}_{j,j}\boldsymbol{\Phi}_{j,j-1} = \mathbf{0}$). Thus, the weighted sum rate difference of the two different decoding orders in (A.15) obeys the following relationship:

$$\sum_{k=1}^{K} w_k\big(R_k - R_k'\big)\begin{cases} = 0 & \text{for } \boldsymbol{\Phi}_{j,j}\boldsymbol{\Phi}_{j,j-1} = \mathbf{0} \quad \text{or} \quad w_j = w_{j-1} \\ > 0 & \text{for } \boldsymbol{\Phi}_{j,j}\boldsymbol{\Phi}_{j,j-1} \neq \mathbf{0} \quad \text{and } w_j > w_{j-1} \\ < 0 & \text{for } \boldsymbol{\Phi}_{j,j}\boldsymbol{\Phi}_{j,j-1} \neq \mathbf{0} \quad \text{and } w_j < w_{j-1} \end{cases}$$

It can be observed that the decoding order of user j and user $j-1$ does not matter in two cases. First, when both weights are equal, i.e., when $w_j = w_{j-1}$, which also explains that for the standard sum rate with equal weights, the decoding order is irrelevant. Second, the decoding order does not matter if by chance

$$\boldsymbol{\Phi}_{j,j}\boldsymbol{\Phi}_{j,j-1} = \mathbf{0} \;\Leftrightarrow\; \boldsymbol{H}_j\boldsymbol{Q}_j\boldsymbol{H}_j^{\mathrm{H}}\boldsymbol{\Omega}_{j,j}^{-\frac{1}{2}}\boldsymbol{\Omega}_{j,j-1}^{-\frac{1}{2}}\boldsymbol{H}_{j-1}\boldsymbol{Q}_{j-1}\boldsymbol{H}_{j-1}^{\mathrm{H}} = \mathbf{0}.$$

In this case, the two decoding orders do not only achieve the same weighted sum rate, but also the same rate pair, see (A.10)–(A.13).

For the most likely case $\boldsymbol{\Phi}_{j,j}\boldsymbol{\Phi}_{j,j-1} \neq \mathbf{0}$, the difference of the two weighted sum rates is positive for $w_j > w_{j-1}$, whereas it is negative for $w_j < w_{j-1}$. Hence, we come to the conclusion, that user j should be decoded *after* user $j-1$ if he has the larger weight ($w_j > w_{j-1}$) and *before* user $j-1$, if he has the smaller weight ($w_j < w_{j-1}$). In a nutshell, the user with the smaller weight has to be decoded first to achieve a larger weighted sum rate. Since this result is applicable for all consecutively decoded users j and $j-1$, the overall decoding order should be in nondecreasing priorities, i.e., the user with the smallest weight is decoded first and the user with the largest weight is decoded last.

A.8 Proof that the Weighted Sum Rate is a Nondecreasing Function

From Appendix A.7, we know that the decoding order in the MIMO MAC must be in nondecreasing priorities. Without loss of generality, we may assume that the weights satisfy $w_1 \le w_2 \le \cdots \le w_K$ which can always be achieved by relabeling the users. Then, the decoding order $1, 2, \ldots, K$ is optimum. Given the rate expression in (A.11), the weighted sum rate in the MIMO MAC can for a general (full rank) noise covariance matrix $\boldsymbol{C_n}$ be expressed as

$$\sum_{k=1}^{K} w_k R_k = \sum_{k=1}^{K} v_k \log_2 \left| \boldsymbol{C_n} + \sum_{i \ge k} \boldsymbol{H}_i \boldsymbol{Q}_i \boldsymbol{H}_i^{\mathrm{H}} \right| - w_K \log_2 \left| \boldsymbol{C_n} \right| \qquad \text{(A.16)}$$

where $v_k := w_k - w_{k-1} \ge 0 \; \forall k \in \{2, \ldots, K\}$ and $v_1 = w_1 \ge 0$. Above equation can be reformulated by isolating the transmit covariance matrix $\boldsymbol{Q}_j$ of some user $j \in \{1, \ldots, K\}$ such that the weighted sum rate is explicitly written as a function f_j depending on $\boldsymbol{Q}_j$:

$$\begin{aligned} f_j(\boldsymbol{Q}_j, \boldsymbol{w}) := \sum_{k=1}^{K} w_k R_k = \sum_{k=1}^{j} v_k \log_2 \left| \boldsymbol{Z}_{k,j} + \boldsymbol{H}_j \boldsymbol{Q}_j \boldsymbol{H}_j^{\mathrm{H}} \right| + \sum_{k=j+1}^{K} v_k \log_2 \left| \boldsymbol{Z}_{k,j} \right| \\ - w_K \log_2 \left| \boldsymbol{C_n} \right| \end{aligned}$$

The substitution $\boldsymbol{Z}_{k,j}$ is defined via

$$\boldsymbol{Z}_{k,j} := \boldsymbol{C_n} + \sum_{\substack{i \ge k \\ i \ne j}} \boldsymbol{H}_i \boldsymbol{Q}_i \boldsymbol{H}_i^{\mathrm{H}}.$$

To see that the weighted sum rate is a nondecreasing function in the transmit covariance matrix $\boldsymbol{Q}_j$ of the jth user, we replace $\boldsymbol{Q}_j$ by $\boldsymbol{Q}'_j := \boldsymbol{Q}_j + \alpha \boldsymbol{D}_j$, where $\alpha \ge 0$ and $\boldsymbol{D}_j \succeq \boldsymbol{0}$, i.e., starting from $\boldsymbol{Q}_j$ we move along the direction $\boldsymbol{D}_j$. The derivative of $f_j(\boldsymbol{Q}_j, \boldsymbol{w})$ along $\boldsymbol{D}_j$ reads for $\alpha \ge 0$ as

$$\frac{\partial f_j(\boldsymbol{Q}_j + \alpha \boldsymbol{D}_j, \boldsymbol{w})}{\partial \alpha} = \sum_{k=1}^{j} \frac{v_k}{\ln 2} \left\| \left(\boldsymbol{Z}_{k,j} + \boldsymbol{H}_j \boldsymbol{Q}_j' \boldsymbol{H}_j\right)^{-\frac{1}{2}} \boldsymbol{H}_j \boldsymbol{D}_j^{\frac{1}{2}} \right\|_{\mathrm{F}}^2 \geq 0$$

where we used Appendix A.6. As the derivative is nonnegative, the weighted sum rate is nondecreasing in $\boldsymbol{Q}_j$ for every $j \in \{1, \ldots, K\}$. For $w_j > w_1$ or $w_1 > 0$, at least one element out of the set $\{v_1, \ldots, v_j\}$ is positive. Then, $f_j(\boldsymbol{Q}_j)$ is strictly increasing, if and only if $\boldsymbol{H}_j \boldsymbol{D}_j \neq \mathbf{0}$.

A.9 Proof that the Weighted Sum Rate is a Concave Function

For the optimum decoding order, we show that the weighted sum rate is concave with respect to the transmit covariance matrices. Without loss of generality, we again assume that the weights satisfy $w_1 \leq w_2 \leq \cdots \leq w_K$. For an identity noise covariance matrix[2] the weighted sum rate can be expressed as [see (6.16) or (A.16)]

$$f(\boldsymbol{Q}, \boldsymbol{w}) := \sum_{k=1}^{K} v_k \log_2 \left| \mathbf{I} + \sum_{i \geq k} \boldsymbol{H}_i \boldsymbol{Q}_i \boldsymbol{H}_i^{\mathrm{H}} \right|$$

with $v_k := w_k - w_{k-1} \geq 0 \, \forall k \in \{2, \ldots, K\}$ and $v_1 = w_1 \geq 0$. To show that $f(\boldsymbol{Q}, \boldsymbol{w})$ is concave in $\boldsymbol{Q}$, it suffices to show that every single summand is concave since they are summed up with nonnegative weights $v_1, \ldots, v_K$. Let the jth summand be given by

$$f_j(\boldsymbol{Q}, \boldsymbol{w}) = v_j \log_2 \left| \mathbf{I} + \sum_{i \geq j} \boldsymbol{H}_i \boldsymbol{Q}_i \boldsymbol{H}_i^{\mathrm{H}} \right| = v_j \log_2 \left| \mathbf{I} + \boldsymbol{H} \boldsymbol{S}_j \boldsymbol{Q} \boldsymbol{S}_j \boldsymbol{H}^{\mathrm{H}} \right|$$

with $\boldsymbol{H} := [\boldsymbol{H}_1, \ldots, \boldsymbol{H}_K] \in \mathbb{C}^{N \times M}$, $\boldsymbol{Q} := \mathrm{blockdiag}\{\boldsymbol{Q}_k\}_{k=1}^{K} \in \mathbb{S}^M$, and the selection matrix

$$\boldsymbol{S}_j := \begin{bmatrix} \mathbf{0}_{\sum_{\ell=1}^{j-1} M_\ell \times \sum_{\ell=1}^{j-1} M_\ell} & \mathbf{0} \\ \mathbf{0} & \mathbf{I}_{\sum_{\ell=j}^{K} M_\ell} \end{bmatrix} \in \{0, 1\}^{M \times M}.$$

Using this representation, the proof that $f_j(\boldsymbol{Q}, \boldsymbol{w})$ is concave in $\boldsymbol{Q}$ basically coincides with the one for the concavity of the sum rate in Sect. 6.1.2. Thus, concavity is proven by showing that $f_j(\boldsymbol{Q} + t\boldsymbol{V}, \boldsymbol{w})$ is concave when restricted to any line segment $\boldsymbol{Q} + t\boldsymbol{V} \succcurlyeq \mathbf{0}$ that intersects its domain [54, Definition 3.1.1, p. 67]. Using Appendix A.6, we compute the partial derivative with respect to t:

$$\frac{\partial f_j(\boldsymbol{Q} + t\boldsymbol{V}, \boldsymbol{w})}{\partial t} = \frac{v_j}{\ln 2} \cdot \mathrm{tr}\left[\left(\mathbf{I} + \boldsymbol{H} \boldsymbol{S}_j (\boldsymbol{Q} + t\boldsymbol{V}) \boldsymbol{S}_j \boldsymbol{H}^{\mathrm{H}}\right)^{-1} \boldsymbol{H} \boldsymbol{S}_j \boldsymbol{V} \boldsymbol{S}_j \boldsymbol{H}^{\mathrm{H}} \right]$$

---

[2] The concavity statement of course remains valid for correlated noise.

The second derivative then follows from Appendix A.5:

$$\frac{\partial^2 f_j(\boldsymbol{Q}+t\boldsymbol{V},\boldsymbol{w})}{\partial t^2} = -\frac{v_j}{\ln 2}\cdot \operatorname{tr}\left[\left(\left(\mathbf{I}+\boldsymbol{H}\boldsymbol{S}_j(\boldsymbol{Q}+t\boldsymbol{V})\boldsymbol{S}_j\boldsymbol{H}^{\mathrm{H}}\right)^{-1}\boldsymbol{H}\boldsymbol{S}_j\boldsymbol{V}\boldsymbol{S}_j\boldsymbol{H}^{\mathrm{H}}\right)^2\right]$$

For $\boldsymbol{Q}+t\boldsymbol{V}\succcurlyeq \mathbf{0}$, the positive semi-definite square root $(\mathbf{I}+\boldsymbol{H}\boldsymbol{S}_j(\boldsymbol{Q}+t\boldsymbol{V})\boldsymbol{S}_j\boldsymbol{H}^{\mathrm{H}})^{-\frac{1}{2}}$ exists and the second derivative can be rewritten to

$$\begin{aligned}\frac{\partial^2 f_j(\boldsymbol{Q}+t\boldsymbol{V},\boldsymbol{w})}{\partial t^2} = -\frac{v_j}{\ln 2}\cdot \Big\| &\left(\mathbf{I}+\boldsymbol{H}\boldsymbol{S}_j(\boldsymbol{Q}+t\boldsymbol{V})\boldsymbol{S}_j\boldsymbol{H}^{\mathrm{H}}\right)^{-\frac{1}{2}} \\ &\times \boldsymbol{H}\boldsymbol{S}_j\boldsymbol{V}\boldsymbol{S}_j\boldsymbol{H}^{\mathrm{H}}\left(\mathbf{I}+\boldsymbol{H}\boldsymbol{S}_j(\boldsymbol{Q}+t\boldsymbol{V})\boldsymbol{S}_j\boldsymbol{H}^{\mathrm{H}}\right)^{-\frac{1}{2}}\Big\|_{\mathrm{F}}^2 \le 0.\end{aligned}$$

From this, we conclude that $f_j(\boldsymbol{Q}+t\boldsymbol{V},\boldsymbol{w})$ is concave in t for all t satisfying $\boldsymbol{Q}+t\boldsymbol{V}\succcurlyeq \mathbf{0}$ and for all $j\in\{1,\ldots,K\}$. Thus, $f_j(\boldsymbol{Q},\boldsymbol{w})$ is concave in $\boldsymbol{Q}$ for all j from which the concavity of the whole weighted sum rate expression follows.

A.10 Wirtinger Derivative

This section gives a brief introduction to complex differentials and the Wirtinger derivative. Here, we refer only to those results of [174] which are relevant in the context of this. For a complete survey, see [174] and references therein.

The real-valued differentiability concept translates to the complex-valued case only for holomorphic functions satisfying the Cauchy–Riemann equations. Since all nontrivial (nonconstant) maps from the complex domain to the real domain are not holomorphic, complex differentiability cannot be applied for those maps. Unfortunately, many optimizations in the context of signal processing and communications involve mappings from the complex to the real domain excluding the complex differentiability concept as a tool to solve them. A remedy to this circumstance is the Wirtinger derivative which is also applicable to functions that are not holomorphic.

Let $f(z)$ be a univariate function which maps the complex-valued scalar $z\in\mathbb{C}$ to the complex domain $\mathbb{C}$. Then, we can associate a bivariate function $F(x,y)$ to the function $f(z)$ with $x\in\mathbb{R}$ and $y\in\mathbb{R}$ such that

$$F(x,y) = f(z)|_{z=x+\mathrm{j}y} = U(x,y)+\mathrm{j}V(x,y)$$

with $U(\cdot,\cdot)\in\mathbb{R}$ and $V(\cdot,\cdot)\in\mathbb{R}$. Under the assumption that the real-valued function $U(x,y)$ and $V(x,y)$ are differentiable with respect to x and y, the total differential $\mathrm{d}F(x,y)$ reads as [175]

$$\begin{aligned} \mathrm{d}F(x,y) &= \frac{\partial F(x,y)}{\partial x}\mathrm{d}x + \frac{\partial F(x,y)}{\partial y}\mathrm{d}y \\ &= \frac{\partial U(x,y)}{\partial x}\mathrm{d}x + \mathrm{j}\frac{\partial V(x,y)}{\partial x}\mathrm{d}x + \frac{\partial U(x,y)}{\partial y}\mathrm{d}y + \mathrm{j}\frac{\partial V(x,y)}{\partial y}\mathrm{d}y. \end{aligned} \tag{A.17}$$

Making use of

$$\begin{aligned} \mathrm{d}z &= \mathrm{d}x + \mathrm{j}\mathrm{d}y \\ \mathrm{d}z^* &= \mathrm{d}x - \mathrm{j}\mathrm{d}y \end{aligned}$$

the two differentials $\mathrm{d}x$ and $\mathrm{d}y$ can be expressed as

$$\begin{aligned} \mathrm{d}x &= \frac{1}{2}(\mathrm{d}z + \mathrm{d}z^*) \\ \mathrm{d}y &= \frac{1}{2\mathrm{j}}(\mathrm{d}z - \mathrm{d}z^*). \end{aligned} \tag{A.18}$$

Inserting (A.18) into (A.17) and rearranging the terms, we obtain

$$\begin{aligned} \mathrm{d}F(x,y) &= \frac{1}{2}\left[\frac{\partial U(x,y)}{\partial x} + \frac{\partial V(x,y)}{\partial y} + \mathrm{j}\left(\frac{\partial V(x,y)}{\partial x} - \frac{\partial U(x,y)}{\partial y}\right)\right]\mathrm{d}z \\ &+ \frac{1}{2}\left[\frac{\partial U(x,y)}{\partial x} - \frac{\partial V(x,y)}{\partial y} + \mathrm{j}\left(\frac{\partial V(x,y)}{\partial x} + \frac{\partial U(x,y)}{\partial y}\right)\right]\mathrm{d}z^* \\ &= \frac{1}{2}\left[\frac{\partial}{\partial x}F(x,y) - \mathrm{j}\frac{\partial}{\partial y}F(x,y)\right]\mathrm{d}z + \frac{1}{2}\left[\frac{\partial}{\partial x}F(x,y) + \mathrm{j}\frac{\partial}{\partial y}F(x,y)\right]\mathrm{d}z^* \\ &= \frac{1}{2}\left[\frac{\partial}{\partial x} - \mathrm{j}\frac{\partial}{\partial y}\right]F(x,y)\mathrm{d}z + \frac{1}{2}\left[\frac{\partial}{\partial x} + \mathrm{j}\frac{\partial}{\partial y}\right]F(x,y)\mathrm{d}z^*. \end{aligned} \tag{A.19}$$

From the first equality in (A.19), we conclude that the differential $\mathrm{d}F(x,y)$ of the bivariate function F associated to a *holomorphic* function f does not depend on $\mathrm{d}z^*$ since its prefactor is zero when the Cauchy–Riemann equations are satisfied.

The last equality in (A.19) motivates the introduction of the two operators $\frac{\partial_\mathrm{w}}{\partial_\mathrm{w} z}$ and $\frac{\partial_\mathrm{w}}{\partial_\mathrm{w} z^*}$ as they lead to a very nice description of the differential $\mathrm{d}f$ where the real-valued partial derivatives are hidden [176]. Using the so defined Wirtinger derivatives [177]

$$\begin{aligned} \frac{\partial_\mathrm{w}}{\partial_\mathrm{w}(\cdot)} &:= \frac{1}{2}\left[\frac{\partial}{\partial \mathrm{Re}(\cdot)} - \mathrm{j}\frac{\partial}{\partial \mathrm{Im}(\cdot)}\right] \\ \frac{\partial_\mathrm{w}}{\partial_\mathrm{w}(\cdot)^*} &:= \frac{1}{2}\left[\frac{\partial}{\partial \mathrm{Re}(\cdot)} + \mathrm{j}\frac{\partial}{\partial \mathrm{Im}(\cdot)}\right] \end{aligned} \tag{A.20}$$

the differential $\mathrm{d}f(z)$ can be expressed as

$$\mathrm{d}f(z) = \frac{\partial_\mathrm{w} f(z)}{\partial_\mathrm{w} z}\mathrm{d}z + \frac{\partial_\mathrm{w} f(z)}{\partial_\mathrm{w} z^*}\mathrm{d}z^* \tag{A.21}$$

and satisfies $\mathrm{d}F(x, y) = \mathrm{d}f(z = x + \mathrm{j}y)$. Note that the definition in (A.20) is only valid when ∂_W is applied with respect to a complex-valued argument. For an extension to the case of a real-valued argument, see Appendix A.12. Moreover, the Wirtinger derivative coincides with the complex differentiation in case of holomorphic functions.

For the Wirtinger derivatives, the common rules for differentiation known from real-valued analysis concerning the sum, product, quotient, and composition of two functions hold as well. In particular,

$$\frac{\partial_\mathrm{W}}{\partial_\mathrm{W} z} z^* = \frac{\partial_\mathrm{W}}{\partial_\mathrm{W} z^*} z = 0 \tag{A.22}$$

which means that z^* can be regarded constant when differentiating with respect to z, as well as z can be regarded constant when differentiating with respect to z^*. Two examples are

$$\frac{\partial_\mathrm{W}}{\partial_\mathrm{W} z} |z|^2 = \frac{\partial_\mathrm{W}}{\partial_\mathrm{W} z} (z z^*) = z^*$$

$$\frac{\partial_\mathrm{W}}{\partial_\mathrm{W} z} \exp(-|z|^2) = \frac{\partial_\mathrm{W}}{\partial_\mathrm{W} z} \exp(-z z^*) = -z^* \exp(|z|^2).$$

A.11 Differential of a Real-Valued Function with Matrix-Valued Unstructured Complex Argument

If the function $f : z \mapsto f(z)$ maps from $z \in \mathbb{C}$ to $f(z) \in \mathbb{R}$, i.e., from the complex to the real space, the differential $\mathrm{d}f(z, \mathrm{d}z)$ from (A.21) reduces to

$$\begin{aligned} \mathrm{d}f(z, \mathrm{d}z) &= \frac{\partial_\mathrm{W} f(z)}{\partial_\mathrm{W} z} \mathrm{d}z + \frac{\partial_\mathrm{W} f^*(z)}{\partial_\mathrm{W} z^*} \mathrm{d}z^* \\ &= \frac{\partial_\mathrm{W} f(z)}{\partial_\mathrm{W} z} \mathrm{d}z + \left(\frac{\partial_\mathrm{W} f(z)}{\partial_\mathrm{W} z} \mathrm{d}z \right)^* \\ &= 2\mathrm{Re} \left(\frac{\partial_\mathrm{W} f(z)}{\partial_\mathrm{W} z} \mathrm{d}z \right) \end{aligned} \tag{A.23}$$

since $f(z) = f^*(z)$. We can easily generalize (A.23) to the case of a matrix-valued, unstructured[3] argument $\mathbf{Z} \in \mathbb{C}^{M \times N}$. If $f : \mathbf{Z} \mapsto f(\mathbf{Z}) \in \mathbb{R}$ is a mapping to $\mathbb{R}$, the total differential $\mathrm{d}f(\mathbf{Z}, \mathrm{d}\mathbf{Z})$ reads as

[3] The individual entries of $\mathbf{Z}$ are assumed to be independent. $\mathbf{Z}$ does not have a special structure like being Hermitian or anything else.

$$
\begin{aligned}
\mathrm{d}f(\boldsymbol{Z}, \mathrm{d}\boldsymbol{Z}) &= \sum_{m=1}^{M}\sum_{n=1}^{N} 2\mathrm{Re}\left(\frac{\partial_{\mathrm{W}} f(\boldsymbol{Z})}{\partial_{\mathrm{W}}[\boldsymbol{Z}]_{m,n}} \cdot \mathrm{d}[\boldsymbol{Z}]_{m,n}\right) \\
&= \sum_{m=1}^{M}\sum_{n=1}^{N} 2\mathrm{Re}\left(\left[\frac{\partial_{\mathrm{W}} f(\boldsymbol{Z})}{\partial_{\mathrm{W}}\boldsymbol{Z}}\right]_{m,n} \cdot \left[\mathrm{d}\boldsymbol{Z}\right]_{m,n}\right) \\
&= \sum_{m=1}^{M}\sum_{n=1}^{N} 2\mathrm{Re}\left(\left[\frac{\partial_{\mathrm{W}} f(\boldsymbol{Z})}{\partial_{\mathrm{W}}\boldsymbol{Z}^{\mathrm{T}}}\right]_{n,m} \cdot \left[\mathrm{d}\boldsymbol{Z}\right]_{m,n}\right) \\
&= 2\mathrm{Re}\left(\sum_{m=1}^{M}\sum_{n=1}^{N} \left[\frac{\partial_{\mathrm{W}} f(\boldsymbol{Z})}{\partial_{\mathrm{W}}\boldsymbol{Z}^{\mathrm{T}}}\right]_{n,m} \cdot \left[\mathrm{d}\boldsymbol{Z}\right]_{m,n}\right) \\
&= 2\mathrm{Re}\left[\mathrm{tr}\left(\frac{\partial_{\mathrm{W}} f(\boldsymbol{Z})}{\partial_{\mathrm{W}}\boldsymbol{Z}^{\mathrm{T}}} \cdot \mathrm{d}\boldsymbol{Z}\right)\right]
\end{aligned}
\tag{A.24}
$$

where we used the definition for the derivative of a scalar function f with respect to a matrix-valued argument $\boldsymbol{Z} \in \mathbb{C}^{M\times N}$

$$
\frac{\partial_{\mathrm{W}} f(\boldsymbol{Z})}{\partial_{\mathrm{W}}\boldsymbol{Z}} = \begin{bmatrix} \frac{\partial_{\mathrm{W}} f(\boldsymbol{Z})}{\partial_{\mathrm{W}}[\boldsymbol{Z}]_{1,1}} & \cdots & \frac{\partial_{\mathrm{W}} f(\boldsymbol{Z})}{\partial_{\mathrm{W}}[\boldsymbol{Z}]_{1,N}} \\ \vdots & \ddots & \vdots \\ \frac{\partial_{\mathrm{W}} f(\boldsymbol{Z})}{\partial_{\mathrm{W}}[\boldsymbol{Z}]_{M,1}} & \cdots & \frac{\partial_{\mathrm{W}} f(\boldsymbol{Z})}{\partial_{\mathrm{W}}[\boldsymbol{Z}]_{M,N}} \end{bmatrix} \in \mathbb{C}^{M\times N} \tag{A.25}
$$

that holds for the Wirtinger derivative as well as for the standard partial derivative.

A.12 Differential of a Real-Valued Function with Matrix-Valued Hermitian Argument

For a real-valued function $f : \boldsymbol{Z} \mapsto f(\boldsymbol{Z}) \in \mathbb{R}$ with complex-valued, but *Hermitian* argument $\boldsymbol{Z} = \boldsymbol{Z}^{\mathrm{H}} \in \mathbb{C}^{M\times M}$, the differential $\mathrm{d}f(\boldsymbol{Z}, \mathrm{d}\boldsymbol{Z})$ looks slightly different to (A.24) in Appendix A.11. This results from the fact that $\boldsymbol{Z} \in \mathbb{C}^{M\times M}$ does actually not have M^2 independent parameters due to its Hermitian structure. Instead, only the M main diagonal elements and the $\frac{M^2-M}{2}$ strictly upper (lower) entries can be varied freely as the strictly lower (upper) part of $\boldsymbol{Z}$ is the Hermitian version of the strictly upper (lower) part. Hence, only $\frac{M^2+M}{2}$ instead of M^2 degrees of freedom are available and the total differential is given by the sum of these $\frac{M^2+M}{2}$ individual differentials. Using (A.17) and (A.21), we obtain with $\mathrm{d}\boldsymbol{Z} = (\mathrm{d}\boldsymbol{Z})^{\mathrm{H}}$

$$
\mathrm{d}f(\boldsymbol{Z}, \mathrm{d}\boldsymbol{Z}) = \sum_{m=1}^{M} \frac{\partial f(\boldsymbol{Z})}{\partial[\boldsymbol{Z}]_{m,m}} \cdot \mathrm{d}[\boldsymbol{Z}]_{m,m}
$$

$$+\sum_{m=1}^{M}\sum_{n>m}\left(\frac{\partial f(\mathbf{Z})}{\partial \mathrm{Re}\big([\mathbf{Z}]_{m,n}\big)}\cdot \mathrm{dRe}\big([\mathbf{Z}]_{m,n}\big)+\frac{\partial f(\mathbf{Z})}{\partial \mathrm{Im}\big([\mathbf{Z}]_{m,n}\big)}\cdot \mathrm{dIm}\big([\mathbf{Z}]_{m,n}\big)\right) \quad \text{(A.26)}$$

$$=\sum_{m=1}^{M}\frac{\partial f(\mathbf{Z})}{\partial [\mathbf{Z}]_{m,m}}\cdot \mathrm{d}[\mathbf{Z}]_{m,m}$$

$$+\sum_{m=1}^{M}\sum_{n>m}\left(\frac{\partial_{\mathrm{W}} f(\mathbf{Z})}{\partial_{\mathrm{W}}[\mathbf{Z}]_{m,n}}\cdot \mathrm{d}[\mathbf{Z}]_{m,n}+\frac{\partial_{\mathrm{W}} f(\mathbf{Z})}{\partial_{\mathrm{W}}[\mathbf{Z}]^{*}_{m,n}}\cdot \mathrm{d}[\mathbf{Z}]^{*}_{m,n}\right).$$

Since $[\mathbf{Z}]^{*}_{m,n}=[\mathbf{Z}]_{n,m}$ holds due to the Hermitian structure of $\mathbf{Z}$, we further obtain

$$\mathrm{d}f(\mathbf{Z},\mathrm{d}\mathbf{Z})=\sum_{m=1}^{M}\frac{\partial f(\mathbf{Z})}{\partial [\mathbf{Z}]_{m,m}}\cdot \mathrm{d}[\mathbf{Z}]_{m,m}$$

$$+\sum_{m=1}^{M}\sum_{n>m}\left(\frac{\partial_{\mathrm{W}} f(\mathbf{Z})}{\partial_{\mathrm{W}}[\mathbf{Z}]_{m,n}}\cdot \mathrm{d}[\mathbf{Z}]_{m,n}+\frac{\partial_{\mathrm{W}} f(\mathbf{Z})}{\partial_{\mathrm{W}}[\mathbf{Z}]_{n,m}}\cdot \mathrm{d}[\mathbf{Z}]_{n,m}\right)$$

$$=\sum_{m=1}^{M}\frac{\partial f(\mathbf{Z})}{\partial [\mathbf{Z}]_{m,m}}\cdot \mathrm{d}[\mathbf{Z}]_{m,m}$$

$$+\sum_{m=1}^{M}\sum_{n>m}\frac{\partial_{\mathrm{W}} f(\mathbf{Z})}{\partial_{\mathrm{W}}[\mathbf{Z}]_{m,n}}\cdot \mathrm{d}[\mathbf{Z}]_{m,n}+\sum_{m=1}^{M}\sum_{n<m}+\frac{\partial_{\mathrm{W}} f(\mathbf{Z})}{\partial_{\mathrm{W}}[\mathbf{Z}]_{m,n}}\cdot \mathrm{d}[\mathbf{Z}]_{m,n} \quad \text{(A.27)}$$

$$=\sum_{m=1}^{M}\frac{\partial f(\mathbf{Z})}{\partial [\mathbf{Z}]_{m,m}}\cdot \mathrm{d}[\mathbf{Z}]_{m,m}+\sum_{m=1}^{M}\sum_{n\neq m}\frac{\partial_{\mathrm{W}} f(\mathbf{Z})}{\partial_{\mathrm{W}}[\mathbf{Z}]_{m,n}}\cdot \mathrm{d}[\mathbf{Z}]_{m,n}$$

$$=\sum_{m=1}^{M}\sum_{n=1}^{M}\frac{\partial_{\mathrm{W}} f(\mathbf{Z})}{\partial_{\mathrm{W}}[\mathbf{Z}]_{m,n}}\cdot \mathrm{d}[\mathbf{Z}]_{m,n}$$

$$=\mathrm{tr}\left(\frac{\partial_{\mathrm{W}} f(\mathbf{Z})}{\partial_{\mathrm{W}}\mathbf{Z}^{\mathrm{T}}}\cdot \mathrm{d}\mathbf{Z}\right)$$

where we defined the Wirtinger derivative with respect to a *real-valued* argument to be equivalent to the standard partial derivative in the next to last equality. Note that the Wirtinger derivative has to be computed carefully due to the Hermitian structure of $\mathbf{Z}$. During the computation of the Wirtinger derivative of $f(\mathbf{Z})$ with respect to $[\mathbf{Z}]_{m,n}$, the entry $[\mathbf{Z}]_{n,m}=[\mathbf{Z}]^{*}_{m,n}$ can clearly be regarded to be constant, see (A.22). However, $\mathbf{Z}^{*}$ also depends on $[\mathbf{Z}]_{m,n}$. For example, the Wirtinger derivative of $\|\mathbf{Z}\|_{\mathrm{F}}^{2}=\mathrm{tr}(\mathbf{Z}^{\mathrm{H}}\mathbf{Z})$ with respect to $\mathbf{Z}^{\mathrm{T}}$ reads for Hermitian $\mathbf{Z}=\mathbf{Z}^{\mathrm{H}}$ as

$$\frac{\partial_{\mathrm{W}}\|\mathbf{Z}\|_{\mathrm{F}}^{2}}{\partial_{\mathrm{W}}\mathbf{Z}^{\mathrm{T}}}=\mathbf{Z}+\mathbf{Z}=2\mathbf{Z}. \quad \text{(A.28)}$$

A.13 Lipschitz Constant Based Step-Size Rule Derivation

The Lipschitz constant based step-size rule is derived in this section for the covariance matrix based case. The basic idea is similar to the one in [93]. Let $f : \boldsymbol{Q} \mapsto f(\boldsymbol{Q})$ be a real-valued utility for positive semi-definite arguments $\boldsymbol{Q} \succcurlyeq \boldsymbol{0}$. We are interested in a sufficient condition on the step-size s to let the utility still be increasing when moving along the unconstrained gradient of f with step-size s starting from $\boldsymbol{Q} \in \mathcal{C}$ out of the constraint set. Obviously, before passed to the utility f, the gradient step is first projected onto the constraint set. In the following, we use

$$\boldsymbol{V} := \frac{\partial_{\mathrm{W}} f(\boldsymbol{Q})}{\partial_{\mathrm{W}} \boldsymbol{Q}^{\mathrm{T}}}$$

for a shorter notation. Let

$$h(t) := f\big(\boldsymbol{Q} + t\left[\mathcal{P}_{\mathcal{C}}\left(\boldsymbol{Q} + s \cdot \boldsymbol{V}\right) - \boldsymbol{Q}\right]\big).$$

We want to have an upper bound on s such that $h(1) - h(0) > 0$, i.e., that the update yields an increase of the utility. By means of the second part of the fundamental theorem of calculus and the chain rule we obtain

$$\begin{aligned}
h(1) - h(0) &= \int_0^1 \frac{\partial h(t)}{\partial t} \mathrm{d}t \\
&= \int_0^1 \operatorname{tr}\left[\frac{\partial_{\mathrm{W}} f(\boldsymbol{Q}')}{\partial_{\mathrm{W}} \boldsymbol{Q}'^{\mathrm{T}}}\bigg|_{\boldsymbol{Q}'=\boldsymbol{Q}+t[\mathcal{P}_{\mathcal{C}}(\boldsymbol{Q}+s\boldsymbol{V})-\boldsymbol{Q}]} \cdot \left[\mathcal{P}_{\mathcal{C}}(\boldsymbol{Q}+s\boldsymbol{V}) - \boldsymbol{Q}\right]\right] \mathrm{d}t \\
&= \int_0^1 \operatorname{tr}\left[\left(\frac{\partial_{\mathrm{W}} f(\boldsymbol{Q}')}{\partial_{\mathrm{W}} \boldsymbol{Q}'^{\mathrm{T}}}\bigg|_{\boldsymbol{Q}'=\boldsymbol{Q}+t[\mathcal{P}_{\mathcal{C}}(\boldsymbol{Q}+s\boldsymbol{V})-\boldsymbol{Q}]} - \boldsymbol{V} + \boldsymbol{V}\right) \cdot \left[\mathcal{P}_{\mathcal{C}}(\boldsymbol{Q}+s\boldsymbol{V}) - \boldsymbol{Q}\right]\right] \mathrm{d}t \\
&= \operatorname{tr}\Big(\boldsymbol{V}\left[\mathcal{P}_{\mathcal{C}}(\boldsymbol{Q}+s\boldsymbol{V}) - \boldsymbol{Q}\right]\Big) \\
&\quad + \int_0^1 \operatorname{tr}\left[\left(\frac{\partial_{\mathrm{W}} f(\boldsymbol{Q}')}{\partial_{\mathrm{W}} \boldsymbol{Q}'^{\mathrm{T}}}\bigg|_{\boldsymbol{Q}'=\boldsymbol{Q}+t[\mathcal{P}_{\mathcal{C}}(\boldsymbol{Q}+s\boldsymbol{V})-\boldsymbol{Q}]} - \boldsymbol{V}\right) \cdot \left[\mathcal{P}_{\mathcal{C}}(\boldsymbol{Q}+s\boldsymbol{V}) - \boldsymbol{Q}\right]\right] \mathrm{d}t.
\end{aligned} \tag{A.29}$$

A lower bound on the first summand of (A.29) follows from the properties of the orthogonal projection operator. At $\mathcal{P}_{\mathcal{C}}(\boldsymbol{Q} + s\boldsymbol{V})$, we may define a separating hyperplane[4] which separates the set $\mathcal{C}$ from $\boldsymbol{Q} + s\boldsymbol{V}$. Thus, we obtain

$$\operatorname{tr}\left[\big(\boldsymbol{Y} - \mathcal{P}_{\mathcal{C}}(\boldsymbol{Q}+s\boldsymbol{V})\big)\big(\boldsymbol{Q} + s\boldsymbol{V} - \mathcal{P}_{\mathcal{C}}(\boldsymbol{Q}+s\boldsymbol{V})\big)\right] \leq 0 \quad \forall \boldsymbol{Y} \in \mathcal{C} \tag{A.30}$$

[4] Clearly, the gradient step is assumed to leave the constraint set $\mathcal{C}$ for the separating hyperplane interpretation. In any case, (A.30) remains valid.

which can be interpreted in such a way that $\boldsymbol{Y} - \mathcal{P}_{\mathcal{C}}(\boldsymbol{Q} + s\boldsymbol{V})$ makes for any $\boldsymbol{Y} \in \mathcal{C}$ an obtuse angle to $\boldsymbol{Q} + s\boldsymbol{V} - \mathcal{P}_{\mathcal{C}}(\boldsymbol{Q} + s\boldsymbol{V})$. Choosing $\boldsymbol{Y} = \boldsymbol{Q}$, we obtain[5]

$$s \cdot \mathrm{tr}\big[\boldsymbol{V}\big(\mathcal{P}_{\mathcal{C}}(\boldsymbol{Q} + s\boldsymbol{V}) - \boldsymbol{Q}\big)\big] \geq \|\mathcal{P}_{\mathcal{C}}(\boldsymbol{Q} + s\boldsymbol{V}) - \boldsymbol{Q}\|_{\mathrm{F}}^2. \tag{A.31}$$

A lower bound on the second summand of (A.29) results from a lower bound on the integrand and the Lipschitz assumption on the Wirtinger derivative of f:

$$\begin{aligned}
&\int_0^1 \mathrm{tr}\left[\left(\left.\frac{\partial_{\mathrm{W}} f(\boldsymbol{Q}')}{\partial_{\mathrm{W}} \boldsymbol{Q}'^{\mathrm{T}}}\right|_{\boldsymbol{Q}'=\boldsymbol{Q}+t[\mathcal{P}_{\mathcal{C}}(\boldsymbol{Q}+s\boldsymbol{V})-\boldsymbol{Q}]} - \boldsymbol{V}\right) \cdot \big[\mathcal{P}_{\mathcal{C}}(\boldsymbol{Q} + s\boldsymbol{V}) - \boldsymbol{Q}\big]\right] \mathrm{d}t \\
&\geq -\int_0^1 \left\| \left.\frac{\partial_{\mathrm{W}} f(\boldsymbol{Q}')}{\partial_{\mathrm{W}} \boldsymbol{Q}'^{\mathrm{T}}}\right|_{\boldsymbol{Q}'=\boldsymbol{Q}+t[\mathcal{P}_{\mathcal{C}}(\boldsymbol{Q}+s\boldsymbol{V})-\boldsymbol{Q}]} - \frac{\partial_{\mathrm{W}} f(\boldsymbol{Q})}{\partial_{\mathrm{W}} \boldsymbol{Q}^{\mathrm{T}}} \right\|_{\mathrm{F}} \cdot \|\mathcal{P}_{\mathcal{C}}(\boldsymbol{Q} + s\boldsymbol{V}) - \boldsymbol{Q}\|_{\mathrm{F}} \mathrm{d}t \\
&\geq -\int_0^1 L_0 \|t[\mathcal{P}_{\mathcal{C}}(\boldsymbol{Q} + s\boldsymbol{V}) - \boldsymbol{Q}]\|_{\mathrm{F}} \cdot \|\mathcal{P}_{\mathcal{C}}(\boldsymbol{Q} + s\boldsymbol{V}) - \boldsymbol{Q}\|_{\mathrm{F}} \mathrm{d}t \\
&= -\frac{L_0}{2} \|\mathcal{P}_{\mathcal{C}}(\boldsymbol{Q} + s\boldsymbol{V}) - \boldsymbol{Q}\|_{\mathrm{F}}^2
\end{aligned} \tag{A.32}$$

For the second inequality, we used the Lipschitz assumption on the gradient of f

$$\left\| \left.\frac{\partial_{\mathrm{W}} f(\boldsymbol{Q}')}{\partial_{\mathrm{W}} \boldsymbol{Q}'^{\mathrm{T}}}\right|_{\boldsymbol{Q}'=\boldsymbol{X}} - \left.\frac{\partial_{\mathrm{W}} f(\boldsymbol{Q}')}{\partial_{\mathrm{W}} \boldsymbol{Q}'^{\mathrm{T}}}\right|_{\boldsymbol{Q}'=\boldsymbol{Y}} \right\|_{\mathrm{F}} \leq L_0 \cdot \|\boldsymbol{X} - \boldsymbol{Y}\|_{\mathrm{F}}.$$

For twice continuously differentiable $f(\boldsymbol{Q})$, the constant L_0 can be chosen as the maximum magnitude of the second derivative over the set $\mathcal{C}$. Multiplying (A.29) by s and using (A.31) and (A.32), we obtain

$$s \cdot \big(h(1) - h(0)\big) \geq \left[1 - \frac{sL_0}{2}\right] \cdot \|\mathcal{P}_{\mathcal{C}}(\boldsymbol{Q} + s\boldsymbol{V}) - \boldsymbol{Q}\|_{\mathrm{F}}^2. \tag{A.33}$$

As long as $\boldsymbol{Q}$ is not a stationary point and $s > 0$, $\|\mathcal{P}_{\mathcal{C}}(\boldsymbol{Q} + s\boldsymbol{V}) - \boldsymbol{Q}\|_{\mathrm{F}}^2 > 0$ holds. If (A.33) shall be positive, $0 < s < 2/L_0$ must hold. The slightly stronger condition $\varepsilon \leq s \leq 2/L_0 - \varepsilon$ with $\varepsilon > 0$ ensures that the steps are sufficiently large and sufficiently small over the iterations and coincides with (5.26).

---

[5] Note that a slightly less tight (and thus more conservative) bound with an additional factor of 0.5 on the right hand side of (A.31) could also be derived by exploiting that $\mathcal{P}_{\mathcal{C}}(\boldsymbol{Q} + s\boldsymbol{V})$ minimizes the distance of $\boldsymbol{Q} + s\boldsymbol{V}$ to the set $\mathcal{C}$. However, we make use of the tighter bound (A.31) in the following.

A.14 Eigenvalue Perturbation

Let $\lambda_1, \ldots, \lambda_M \in \mathbb{R}_{+,0}$ denote the eigenvalues of the positive semi-definite $M \times M$ matrix $\boldsymbol{Q} \in \mathbb{S}^M$ and let $\boldsymbol{u}_1, \ldots, \boldsymbol{u}_M$ be the respective unit-norm eigenvectors. If all eigenvalues are distinct (which we assume in the beginning) and thus have algebraic multiplicity one, the unit-norm eigenvectors are uniquely determined up to a unimodular complex scalar. The mth eigenvalue and the mth eigenvector satisfy

$$\boldsymbol{Q}\boldsymbol{u}_m = \lambda_m \boldsymbol{u}_m.$$

For a small perturbation $\alpha \boldsymbol{V}$ with Hermitian $\boldsymbol{V} = \boldsymbol{V}^{\mathrm{H}}$, we want to know how the eigenvalue $\lambda'(\alpha)$ of $\boldsymbol{Q}'(\alpha) := \boldsymbol{Q} + \alpha \boldsymbol{V}$ varies around $\alpha = 0$ in its first-order Taylor approximation. For general α, let $\lambda'_m(\alpha)$ and $\boldsymbol{u}'_m(\alpha)$ denote the mth eigenvalue and eigenvector of $\boldsymbol{Q}'(\alpha)$ such that

$$\boldsymbol{Q}'(\alpha)\boldsymbol{u}'_m(\alpha) = \lambda'_m(\alpha)\boldsymbol{u}'_m(\alpha) \tag{A.34}$$

holds. If the algebraic multiplicity of all eigenvalues is one, both the eigenvalues and the eigenvectors vary continuously in α [178, Theorem 1]. Hence, we may differentiate (A.34) with respect to α and evaluate the result at $\alpha = 0$:

$$\boldsymbol{V}\boldsymbol{u}_m + \boldsymbol{Q} \cdot \frac{\partial}{\partial \alpha}\boldsymbol{u}'_m(\alpha)\big|_{\alpha=0} = \frac{\partial}{\partial \alpha}\lambda'_m(\alpha)\big|_{\alpha=0} \cdot \boldsymbol{u}_m + \lambda_m \cdot \frac{\partial}{\partial \alpha}\boldsymbol{u}'_m(\alpha)\big|_{\alpha=0}$$

Left-hand side multiplication by $\boldsymbol{u}_m^{\mathrm{H}}$ and rearranging the terms yields

$$\frac{\partial}{\partial \alpha}\lambda'_m(\alpha)\big|_{\alpha=0} = \boldsymbol{u}_m^{\mathrm{H}}\boldsymbol{V}\boldsymbol{u}_m + \left[\boldsymbol{u}_m^{\mathrm{H}}\boldsymbol{Q} - \lambda_m \boldsymbol{u}_m^{\mathrm{H}}\right] \cdot \frac{\partial}{\partial \alpha}\boldsymbol{u}'_m(\alpha)\big|_{\alpha=0} = \boldsymbol{u}_m^{\mathrm{H}}\boldsymbol{V}\boldsymbol{u}_m \tag{A.35}$$

which means that the Taylor expansion of $\lambda'_m(\alpha)$ around $\alpha = 0$ reads as

$$\begin{aligned} \lambda'_m(\alpha) &= \lambda_m + \frac{\partial}{\partial \alpha}\lambda'_m(\alpha)\big|_{\alpha=0} \cdot \alpha + e(\alpha) \\ &= \lambda_m + \boldsymbol{u}_m^{\mathrm{H}}\boldsymbol{V}\boldsymbol{u}_m \cdot \alpha + e(\alpha) \end{aligned} \tag{A.36}$$

with $e(\alpha)/\alpha \to 0$ as $\alpha \to 0$. In other words, a differential perturbation $\mathrm{d}\boldsymbol{V}$ of the matrix $\boldsymbol{Q}$ leads to the differential perturbation $\boldsymbol{u}_m^{\mathrm{H}}\mathrm{d}\boldsymbol{V}\boldsymbol{u}_m$ of the mth eigenvalue if its algebraic multiplicity is one.

For eigenvalues whose algebraic multiplicity is larger than one, the problem is slightly more challenging. This results from the fact that in this case, the corresponding eigenvectors are not uniquely determined. An illustrative example is the identity matrix for which the eigenbasis is any unitary matrix. In this case, the respective eigenvectors must be chosen appropriately to let them vary continuously in α.

Let one possible eigenvalue decomposition of $\boldsymbol{Q}$ (that is for example returned by MATLAB) read as

$$\boldsymbol{Q} = \boldsymbol{U}\boldsymbol{\Lambda}\boldsymbol{U}^{\mathrm{H}} = [\boldsymbol{U}_1, \boldsymbol{U}_0] \cdot \begin{bmatrix} \boldsymbol{\Lambda}_1 & \mathbf{0} \\ \mathbf{0} & \boldsymbol{\Lambda}_0 \end{bmatrix} \cdot \begin{bmatrix} \boldsymbol{U}_1^{\mathrm{H}} \\ \boldsymbol{U}_0^{\mathrm{H}} \end{bmatrix} \tag{A.37}$$

where $\boldsymbol{\Lambda}_1 \in \mathbb{R}_{+,0}^{M-A\times M-A}$ contains the $M-A$ eigenvalues of $\boldsymbol{Q}$ that are different from λ and the matrix $\boldsymbol{\Lambda}_0 = \lambda \mathbf{I}_A$ contains the eigenvalue λ with algebraic and geometric multiplicity A. The eigenvectors belonging to $\boldsymbol{\Lambda}_1$ are stored in in the subunitary matrix $\boldsymbol{U}_1 \in \mathbb{C}^{M\times M-A}$.

The (fixed) matrix $\boldsymbol{U}_0 \in \mathbb{C}^{M\times A}$ is a possible subunitary basis for the eigen-space that is spanned by the eigenvalue λ whose geometric and algebraic multiplicity is A. Hence, $\boldsymbol{U}_0^{\mathrm{H}}\boldsymbol{U}_0 = \mathbf{I}_A$ and $\boldsymbol{U}_0^{\mathrm{H}}\boldsymbol{U}_1 = \mathbf{0}$ hold. Note that $\boldsymbol{U}_0 \cdot \boldsymbol{B}$ is also a subunitary basis for the eigen-space spanned by the eigenvalue λ for any unitary $\boldsymbol{B} \in \mathbb{C}^{A\times A}$ that satisfies $\boldsymbol{B}\boldsymbol{B}^{\mathrm{H}} = \mathbf{I}_A$. Note that from all possible subunitary matrices $\boldsymbol{U}_0\boldsymbol{B}$, only a specific one with particular (fixed) $\boldsymbol{B}$ varies continuously when $\boldsymbol{Q}$ is exposed to the perturbation $\alpha\boldsymbol{V}$ [179]. If the matrix $\boldsymbol{U}_0$ is right-hand multiplied by the unitary matrix $\boldsymbol{B}$, the identity $\boldsymbol{Q}\boldsymbol{U} = \boldsymbol{U}\boldsymbol{\Lambda}$ in (A.37) yields

$$\boldsymbol{Q}\boldsymbol{U}_0\boldsymbol{B} = \boldsymbol{U}_0\boldsymbol{B}\boldsymbol{\Lambda}_0. \tag{A.38}$$

Let $\boldsymbol{Q}'(\alpha) := \boldsymbol{Q} + \alpha\boldsymbol{V}$ again be the perturbed matrix $\boldsymbol{Q}$ that satisfies

$$\boldsymbol{Q}'(\alpha)\boldsymbol{U}_0'(\alpha)\boldsymbol{B} = \boldsymbol{U}_0'(\alpha)\boldsymbol{B}\boldsymbol{\Lambda}_0'(\alpha)$$

with subunitary $\boldsymbol{U}_0'(\alpha)$ and diagonal $\boldsymbol{\Lambda}_0'(\alpha)$. Differentiating above equation with respect to α and evaluating the result at $\alpha = 0$ yields

$$\boldsymbol{V}\boldsymbol{U}_0\boldsymbol{B} + \boldsymbol{Q} \cdot \left.\frac{\partial \boldsymbol{U}_0'(\alpha)}{\partial \alpha}\right|_{\alpha=0} \cdot \boldsymbol{B} = \left.\frac{\partial \boldsymbol{U}_0'(\alpha)}{\partial \alpha}\right|_{\alpha=0} \cdot \boldsymbol{B}\boldsymbol{\Lambda}_0 + \boldsymbol{U}_0\boldsymbol{B} \cdot \left.\frac{\partial \boldsymbol{\Lambda}_0'(\alpha)}{\partial \alpha}\right|_{\alpha=0}$$

where $\frac{\partial \boldsymbol{\Lambda}_0'(\alpha)}{\partial \alpha}|_{\alpha=0}$ must be diagonal. Left-hand multiplication by $\boldsymbol{B}^{\mathrm{H}}\boldsymbol{U}_0^{\mathrm{H}}$ and rearranging the terms leads to

$$\boldsymbol{B}^{\mathrm{H}}\boldsymbol{U}_0^{\mathrm{H}}\boldsymbol{V}\boldsymbol{U}_0\boldsymbol{B} + \boldsymbol{B}^{\mathrm{H}}\boldsymbol{U}_0^{\mathrm{H}}\boldsymbol{Q} \cdot \left.\frac{\partial \boldsymbol{U}_0'(\alpha)}{\partial \alpha}\right|_{\alpha=0} \cdot \boldsymbol{B} - \boldsymbol{B}^{\mathrm{H}}\boldsymbol{U}_0^{\mathrm{H}} \cdot \left.\frac{\partial \boldsymbol{U}_0'(\alpha)}{\partial \alpha}\right|_{\alpha=0} \cdot \boldsymbol{B}\boldsymbol{\Lambda}_0 = \left.\frac{\partial \boldsymbol{\Lambda}_0'(\alpha)}{\partial \alpha}\right|_{\alpha=0}.$$

Since $\boldsymbol{\Lambda}_0$ is a scaled identity matrix, it commutes with any other matrix and we may shift it to obtain

$$\boldsymbol{B}^{\mathrm{H}}\boldsymbol{U}_0^{\mathrm{H}}\boldsymbol{V}\boldsymbol{U}_0\boldsymbol{B} + \left[\boldsymbol{B}^{\mathrm{H}}\boldsymbol{U}_0^{\mathrm{H}}\boldsymbol{Q} - \boldsymbol{\Lambda}_0\boldsymbol{B}^{\mathrm{H}}\boldsymbol{U}_0^{\mathrm{H}}\right] \cdot \left.\frac{\partial \boldsymbol{U}_0'(\alpha)}{\partial \alpha}\right|_{\alpha=0} \cdot \boldsymbol{B} = \left.\frac{\partial \boldsymbol{\Lambda}_0'(\alpha)}{\partial \alpha}\right|_{\alpha=0}.$$

By definition, the term inside the brackets is zero [see (A.38)] and we end up with

$$\left.\frac{\partial \boldsymbol{\Lambda}_0'(\alpha)}{\partial \alpha}\right|_{\alpha=0} = \boldsymbol{B}^{\mathrm{H}} \boldsymbol{U}_0^{\mathrm{H}} \boldsymbol{V} \boldsymbol{U}_0 \boldsymbol{B} \tag{A.39}$$

which must be diagonal. Hence, $\boldsymbol{B}$ must be chosen as the unitary eigenbasis of $\boldsymbol{U}_0^{\mathrm{H}} \boldsymbol{V} \boldsymbol{U}_0$, and the partial derivatives of all A eigenvalues that are equal to λ correspond to the eigenvalues of $\boldsymbol{U}_0^{\mathrm{H}} \boldsymbol{V} \boldsymbol{U}_0$. Given the perturbation $\alpha \boldsymbol{V}$, the first-order Taylor approximation of the eigenvalue λ with multiplicity A around $\alpha = 0$ reads as

$$\lambda'(\alpha) = \lambda + \mathrm{eig}_i(\boldsymbol{U}_0^{\mathrm{H}} \boldsymbol{V} \boldsymbol{U}_0) \cdot \alpha + e_i(\alpha) \quad \forall i \in \{1, \ldots, A\} \tag{A.40}$$

where the error term $e_i(\alpha)$ satisfies $e_i(\alpha)/\alpha \to 0$ when $\alpha \to 0$.

For the determination of the tangent cone, we are particularly interested in the sensitivity of those eigenvalues that are zero ($\lambda = 0$) when exposed to a perturbation $\alpha \boldsymbol{V}$. They need to remain nonnegative in the first-order Taylor approximation. This can be achieved by forcing the direction $\boldsymbol{V}$ to satisfy

$$\boldsymbol{U}_0^{\mathrm{H}} \boldsymbol{V} \boldsymbol{U}_0 \succeq \boldsymbol{0}. \tag{A.41}$$

A.15 Optimality of Single-Stream Transmission in the Low Power Regime

In this section, the optimality of single-stream transmission in the lower power regime is derived for the weighted sum rate maximization. The result is independent of whether successive decoding is applied or not. The maximum possible data rate at which user k can convey data in the dual MAC with arbitrarily small error probability can be written as

$$R_k^{\mathrm{MAC}}(\boldsymbol{Q}) := \log_2 \left| \boldsymbol{X}_k + \boldsymbol{H}_k \boldsymbol{Q}_k \boldsymbol{H}_k^{\mathrm{H}} \right| - \log_2 \left| \boldsymbol{X}_k \right|.$$

The substitution $\boldsymbol{X}_k$ depends on whether successive decoding is applied or not. In case of interference cancelation with the natural decoding order $1, \ldots, K$, the substitution $\boldsymbol{X}_k$ is given by [cf. (6.15)]

$$\boldsymbol{X}_k = \mathbf{I} + \sum_{i>k} \boldsymbol{H}_i \boldsymbol{Q}_i \boldsymbol{H}_i^{\mathrm{H}}$$

whereas for linear filtering, we have [see (7.11)]

$$\boldsymbol{X}_k = \mathbf{I} + \sum_{i \neq k} \boldsymbol{H}_i \boldsymbol{Q}_i \boldsymbol{H}_i^{\mathrm{H}}.$$

To see how $R_k^{\mathrm{MAC}}(\boldsymbol{Q})$ behaves for small transmit powers $P_{\max} = \sum_{k=1}^{K} \mathrm{tr}(\boldsymbol{Q}_k)$, we compute the derivative of $R_k^{\mathrm{MAC}}(\alpha \boldsymbol{Q})$ with respect to α by means of Appendix A.6 and evaluate the result at $\alpha = 0$:

$$
\begin{aligned}
R_k^{\prime\mathrm{MAC}}(\boldsymbol{Q}) &:= \left.\frac{\partial R_k^{\mathrm{MAC}}(\alpha \boldsymbol{Q}))}{\partial \alpha}\right|_{\alpha=0} \\
&= \frac{1}{\ln 2} \cdot \mathrm{tr}\big(\boldsymbol{X}_k - \mathbf{I} + \boldsymbol{H}_k \boldsymbol{Q}_k \boldsymbol{H}_k^{\mathrm{H}}\big) - \frac{1}{\ln 2} \cdot \mathrm{tr}\big(\boldsymbol{X}_k - \mathbf{I}\big) \\
&= \frac{1}{\ln 2} \cdot \mathrm{tr}\big(\boldsymbol{H}_k \boldsymbol{Q}_k \boldsymbol{H}_k^{\mathrm{H}}\big) \\
&= \frac{1}{\ln 2} \cdot \mathrm{tr}\big(\boldsymbol{H}_k^{\mathrm{H}} \boldsymbol{H}_k \boldsymbol{Q}_k\big)
\end{aligned}
$$

As one can see, interference is negligible compared to the noise in the lower power regime and the rate expressions tend to

$$
R_k^{\mathrm{MAC}}(\boldsymbol{Q}) \to R_k^{\prime\mathrm{MAC}}(\boldsymbol{Q}) = \frac{1}{\ln 2} \cdot \mathrm{tr}\big(\boldsymbol{H}_k^{\mathrm{H}} \boldsymbol{H}_k \boldsymbol{Q}_k\big) \quad \text{as} \quad P_{\max} \to 0 \tag{A.42}
$$

independent of the structure of $\boldsymbol{X}_k$. Hence, the weighted sum rate maximization in the dual MAC will for $P_{\max} \to 0$ feature the same solution as the optimization

$$
\begin{aligned}
\underset{\boldsymbol{Q}_1,\ldots,\boldsymbol{Q}_K}{\text{maximize}} \; \sum_{k=1}^{K} w_k R_k^{\prime\mathrm{MAC}}(\boldsymbol{Q}) \quad \text{s.t.:} \;& \boldsymbol{Q}_k \succeq \mathbf{0} \;\; \forall k \\
& \sum_{k=1}^{K} \mathrm{tr}(\boldsymbol{Q}_k) \le P_{\max}.
\end{aligned} \tag{A.43}
$$

Using the complementary slackness condition for the positive semi-definiteness constraint reveals that the optimum covariance matrix $\check{\boldsymbol{Q}}_k$ commutes with $\boldsymbol{H}_k^{\mathrm{H}}\boldsymbol{H}_k$ for every k such that they share the same eigenbasis. Let $\lambda_{k,i}$ be the ith eigenvalue of $\boldsymbol{Q}_k$ and let $\theta_{k,i}$ denote the ith eigenvalue of $\boldsymbol{H}_k^{\mathrm{H}}\boldsymbol{H}_k$. Then, (A.43) can be rewritten to

$$
\begin{aligned}
\underset{\lambda_{1,1},\ldots,\lambda_{K,M_K}}{\text{maximize}} \; \sum_{k=1}^{K}\sum_{i=1}^{M_K} w_k \lambda_{k,i}\theta_{k,i} \quad \text{s.t.:} \;& \lambda_{k,i} \ge 0 \;\; \forall k \forall i \\
& \sum_{k=1}^{K}\sum_{i=1}^{M_k} \lambda_{k,i} \le P_{\max}.
\end{aligned}
$$

Assume that the maximum product $w_{\check{k}}\theta_{\check{k},\check{i}}$ is attained by a single user $\check{k}$ and a single eigenvalue index $\check{i}$. In this case, the optimizer to above maximization is unique and is given by

$$\lambda_{k,i} = \begin{cases} P_{\max} & \text{for } (k,i) = (\check{k}, \check{i}) \\ 0 & \text{for } (k,i) \neq (\check{k}, \check{i}). \end{cases} \tag{A.44}$$

In the (unlikely) case when the maximum product $w_{\check{k}}\theta_{\check{k},\check{i}}$ is attained by more than one pair (k,i), the optimizer of the maximization is not unique and the available power $P_{\max}$ may be distributed freely among those pairs (k,i) that achieve the maximum product $w_{\check{k}}\theta_{\check{k},\check{i}}$. However, (A.44) remains a possible optimizer. This proves that single-stream transmission is asymptotically optimum in the low power regime $P_{\max} \to 0$ irrespective of whether interference cancelation is applied or not. As a consequence, the convex hull of the underlying rate region must be a polyhedron with $K+1$ vertices since the single-user rate points optimize the weighted sum rate for any nonnegative weight vector.

A.16 Proof for the Necessity of the Constraint Set Projection for Linear Transceivers

In case of linear transceivers, the Wirtinger derivative of the weighted sum rate utility $f(\boldsymbol{T}, \boldsymbol{w})$ with respect to the conjugate precoder of user j is given by [see (7.20)]

$$\boldsymbol{V}_j := \frac{\partial_{\mathrm{W}} f(\boldsymbol{T}, \boldsymbol{w})}{\partial_{\mathrm{W}} \boldsymbol{T}_j^*} = \frac{1}{\ln 2} \cdot \boldsymbol{H}_j^{\mathrm{H}} \left[\left(\sum_{k=1}^{K} w_k \right) \boldsymbol{X}^{-1} - \sum_{k \neq j} w_k \boldsymbol{X}_k^{-1} \right] \boldsymbol{H}_j \boldsymbol{T}_j. \tag{A.45}$$

We show that moving in the direction of this Wirtinger derivative with given stepsize $s > 0$ and positive weights $\boldsymbol{w} > \boldsymbol{0}$ leaves the constraint set for any $\boldsymbol{H} = [\boldsymbol{H}_1, \ldots, \boldsymbol{H}_K]$ and any block-diagonal $\boldsymbol{T} = \mathrm{blockdiag}\{\boldsymbol{T}_k\}_{k=1}^K$ on the boundary of the constraint set that satisfy $\boldsymbol{H}\boldsymbol{T} \neq \boldsymbol{0}$. If $\boldsymbol{T}$ is not a stationary point of the weighted sum rate maximization, this statement is also valid if we move along the direction of the projection of the Wirtinger derivative onto the tangent cone.

Assuming $\|\boldsymbol{T}\|_{\mathrm{F}}^2 = P_{\max}$, the update $\boldsymbol{T} + s \cdot \boldsymbol{V}$ with block-diagonal $\boldsymbol{V} := \mathrm{blockdiag}\{\boldsymbol{V}_k\}_{k=1}^K$ consumes the power

$$\begin{aligned} \|\boldsymbol{T} + s \cdot \boldsymbol{V}\|_{\mathrm{F}}^2 &= \sum_{j=1}^{K} \|\boldsymbol{T}_j + s \cdot \boldsymbol{V}_j\|_{\mathrm{F}}^2 \\ &= P_{\max} + s^2 \cdot \|\boldsymbol{V}\|_{\mathrm{F}}^2 \\ &\quad + \frac{2s}{\ln 2} \cdot \underbrace{\sum_{j=1}^{K} \mathrm{tr}\left[\boldsymbol{T}_j^{\mathrm{H}} \boldsymbol{H}_j^{\mathrm{H}} \left[\left(\sum_{k=1}^{K} w_k \right) \boldsymbol{X}^{-1} - \sum_{k \neq j} w_k \boldsymbol{X}_k^{-1} \right] \boldsymbol{H}_j \boldsymbol{T}_j \right]}_{=:t}. \end{aligned} \tag{A.46}$$

For small s, this squared norm is governed by the term that is linear in s as long as its prefactor is nonzero. Thus, we first evaluate the linear term:

$$\begin{aligned}
t &= \left(\sum_{k=1}^{K} w_k\right)\sum_{j=1}^{K}\mathrm{tr}\big[\boldsymbol{X}^{-1}\boldsymbol{H}_j\boldsymbol{T}_j\boldsymbol{T}_j^{\mathrm{H}}\boldsymbol{H}_j^{\mathrm{H}}\big] - \sum_{j=1}^{K}\sum_{k\neq j} w_k\mathrm{tr}\big[\boldsymbol{X}_k^{-1}\boldsymbol{H}_j\boldsymbol{T}_j\boldsymbol{T}_j^{\mathrm{H}}\boldsymbol{H}_j^{\mathrm{H}}\big] \\
&= \sum_{k=1}^{K} w_k\cdot\mathrm{tr}\big[\boldsymbol{X}^{-1}\boldsymbol{H}\boldsymbol{T}\boldsymbol{T}^{\mathrm{H}}\boldsymbol{H}^{\mathrm{H}}\big] - \sum_{k=1}^{K} w_k\mathrm{tr}\big[\boldsymbol{X}_k^{-1}\boldsymbol{H}\boldsymbol{T}\boldsymbol{T}^{\mathrm{H}}\boldsymbol{H}^{\mathrm{H}}\big] \qquad \text{(A.47)} \\
&\quad + \sum_{k=1}^{K} w_k\mathrm{tr}\big[\boldsymbol{X}_k^{-1}\boldsymbol{H}_k\boldsymbol{T}_k\boldsymbol{T}_k^{\mathrm{H}}\boldsymbol{H}_k^{\mathrm{H}}\big]
\end{aligned}$$

Since $\boldsymbol{H}\boldsymbol{T}\boldsymbol{T}^{\mathrm{H}}\boldsymbol{H}^{\mathrm{H}} = \boldsymbol{X} - \mathbf{I} = \boldsymbol{X}_k + \boldsymbol{H}_k\boldsymbol{T}_k\boldsymbol{T}_k^{\mathrm{H}}\boldsymbol{H}_k^{\mathrm{H}} - \mathbf{I}$, we may reformulate (A.47) to

$$t = \sum_{k=1}^{K} w_k\mathrm{tr}\big[\boldsymbol{X}_k^{-1} - \boldsymbol{X}^{-1}\big] \geq 0$$

such that t is nonnegative. In fact, $t = 0$ is only possible if $\boldsymbol{H}_k\boldsymbol{T}_k = \mathbf{0}\ \forall k$ which follows from $\boldsymbol{X}_k^{-1} \succcurlyeq \boldsymbol{X}^{-1}$ and $\boldsymbol{X} \neq \boldsymbol{X}_k$ for $\boldsymbol{H}_k\boldsymbol{T}_k \neq \mathbf{0}$ due to $\boldsymbol{X} = \boldsymbol{X}_k + \boldsymbol{H}_k\boldsymbol{T}_k\boldsymbol{T}_k^{\mathrm{H}}\boldsymbol{H}_k^{\mathrm{H}} \succcurlyeq \boldsymbol{X}_k$. Hence, we have proven that $t > 0$ for $\boldsymbol{H}\boldsymbol{T} \neq \mathbf{0}$ which means that the term in (A.46) that is linear in s is positive in this case. By definition, the prefactor $\|\boldsymbol{V}\|_{\mathrm{F}}^2$ of the term that is quadratic in s is nonnegative. For $\boldsymbol{H}\boldsymbol{T} = \mathbf{0}$, the Wirtinger derivative $\boldsymbol{V} = \mathbf{0}$ vanishes as well. As a consequence,

$$\|\boldsymbol{T} + s\cdot\boldsymbol{V}\|_{\mathrm{F}}^2 > \|\boldsymbol{T}\|_{\mathrm{F}}^2 \quad \text{for} \quad s > 0 \text{ and } \boldsymbol{H}\boldsymbol{T} \neq \mathbf{0}.$$

In different words, moving along the direction of the Wirtinger derivative with a positive step-size increases the power consumption unless $\boldsymbol{H}\boldsymbol{T} = \mathbf{0}$. For $\|\boldsymbol{T}\|_{\mathrm{F}}^2 = P_{\max}$, i.e., when $\boldsymbol{T}$ is on the boundary of the constraint set, this means that the constraint set is left by such a step. Hence, the projection back onto the constraint set is mandatory.

If the projection of the Wirtinger derivative onto the tangent cone is used as the search direction then the constraint set is left as well unless this projection returns the zero matrix, i.e., unless a stationary point is already achieved. This results from the fact that the tangent cone projection corresponds to the projection onto the tangent plane if a step along the original Wirtinger derivative would leave the constraint set, see Sect. 5.3.1.2. And the tangent plane lies outside the constraint set except from the point where it touches the set. Thus, more power is consumed for the projected Wirtinger derivative search direction as well.

A.17 Projector Decomposition

In this section, we show that the transmit covariance matrix $\boldsymbol{S}^{\text{PtP}}$ in (8.16) which asymptotically achieves the sum capacity in a point-to-point MIMO system (MIMO BC with cooperating receivers) can be decomposed into a sum of projectors. This way, it can be shown to coincide with the transmit covariance matrix $\boldsymbol{S}^{\text{DPC}}(\alpha\mathbf{1})$ in (8.37) that asymptotically achieves the sum capacity in the dirty paper coding system with noncooperating receivers.

From (8.16), we know that the transmit covariance matrix $\boldsymbol{S}^{\text{PtP}}$ can be expressed in terms of the dual MAC channel matrix $\boldsymbol{H}$ via

$$\boldsymbol{S}^{\text{PtP}} = \frac{P_{\max}}{M} \cdot \boldsymbol{H}\left(\boldsymbol{H}^{\text{H}}\boldsymbol{H}\right)^{-1}\boldsymbol{H}^{\text{H}}. \tag{A.48}$$

Let $\boldsymbol{H}$ be partitioned such that $\boldsymbol{H} = [\boldsymbol{A}, \boldsymbol{B}]$, where $\boldsymbol{B} = \boldsymbol{H}_K$ denotes the dual MAC channel matrix of user K and $\boldsymbol{A} = [\boldsymbol{H}_1, \ldots, \boldsymbol{H}_{K-1}]$ contains the MAC channels of the first $K-1$ users. By means of the projector substitutes

$$\begin{aligned}
\boldsymbol{\Pi}_{\boldsymbol{B}}^{\perp} &= \mathbf{I} - \boldsymbol{B}\left(\boldsymbol{B}^{\text{H}}\boldsymbol{B}\right)^{-1}\boldsymbol{B}^{\text{H}} \\
\boldsymbol{\Pi}_{\boldsymbol{B}} &= \boldsymbol{B}\left(\boldsymbol{B}^{\text{H}}\boldsymbol{B}\right)^{-1}\boldsymbol{B}^{\text{H}} = \mathbf{I} - \boldsymbol{\Pi}_{\boldsymbol{B}}^{\perp}
\end{aligned}$$

and the inversion rules for block-partitioned matrices [147, Eq. (11.8)], the projector in (A.48) can be written as

$$\begin{aligned}
\boldsymbol{H}\left(\boldsymbol{H}^{\text{H}}\boldsymbol{H}\right)^{-1}\boldsymbol{H}^{\text{H}} &= \left[\boldsymbol{A}\ \boldsymbol{B}\right]\begin{bmatrix} \boldsymbol{A}^{\text{H}}\boldsymbol{A} & \boldsymbol{A}^{\text{H}}\boldsymbol{B} \\ \boldsymbol{B}^{\text{H}}\boldsymbol{A} & \boldsymbol{B}^{\text{H}}\boldsymbol{B} \end{bmatrix}^{-1}\begin{bmatrix} \boldsymbol{A}^{\text{H}} \\ \boldsymbol{B}^{\text{H}} \end{bmatrix} \\
&= \left[\boldsymbol{A}\ \boldsymbol{B}\right]\begin{bmatrix} \boldsymbol{X}_{1,1} & \boldsymbol{X}_{1,2} \\ \boldsymbol{X}_{2,1} & \boldsymbol{X}_{2,2} \end{bmatrix}\begin{bmatrix} \boldsymbol{A}^{\text{H}} \\ \boldsymbol{B}^{\text{H}} \end{bmatrix}
\end{aligned} \tag{A.49}$$

with

$$\begin{aligned}
\boldsymbol{X}_{1,1} &= (\boldsymbol{A}^{\text{H}}\boldsymbol{\Pi}_{\boldsymbol{B}}^{\perp}\boldsymbol{A})^{-1} \\
\boldsymbol{X}_{1,2} &= -(\boldsymbol{A}^{\text{H}}\boldsymbol{\Pi}_{\boldsymbol{B}}^{\perp}\boldsymbol{A})^{-1}\boldsymbol{A}^{\text{H}}\boldsymbol{B}(\boldsymbol{B}^{\text{H}}\boldsymbol{B})^{-1} \\
\boldsymbol{X}_{2,1} &= -(\boldsymbol{B}^{\text{H}}\boldsymbol{B})^{-1}\boldsymbol{B}^{\text{H}}\boldsymbol{A}(\boldsymbol{A}^{\text{H}}\boldsymbol{\Pi}_{\boldsymbol{B}}^{\perp}\boldsymbol{A})^{-1} \\
\boldsymbol{X}_{2,2} &= (\boldsymbol{B}^{\text{H}}\boldsymbol{B})^{-1}\left(\mathbf{I} + \boldsymbol{B}^{\text{H}}\boldsymbol{A}(\boldsymbol{A}^{\text{H}}\boldsymbol{\Pi}_{\boldsymbol{B}}^{\perp}\boldsymbol{A})^{-1}\boldsymbol{A}^{\text{H}}\boldsymbol{B}(\boldsymbol{B}^{\text{H}}\boldsymbol{B})^{-1}\right).
\end{aligned}$$

By multiplying out the matrix-valued Hermitian form in (A.49), we obtain

$$\begin{aligned}\boldsymbol{H}\big(\boldsymbol{H}^{\mathrm{H}}\boldsymbol{H}\big)^{-1}\boldsymbol{H}^{\mathrm{H}} &= \boldsymbol{A}(\boldsymbol{A}^{\mathrm{H}}\boldsymbol{\Pi}_{\boldsymbol{B}}^{\perp}\boldsymbol{A})^{-1}\boldsymbol{A}^{\mathrm{H}} + \boldsymbol{\Pi}_{\boldsymbol{B}} + \boldsymbol{\Pi}_{\boldsymbol{B}}\boldsymbol{A}(\boldsymbol{A}^{\mathrm{H}}\boldsymbol{\Pi}_{\boldsymbol{B}}^{\perp}\boldsymbol{A})^{-1}\boldsymbol{A}^{\mathrm{H}}\boldsymbol{\Pi}_{\boldsymbol{B}} \\ &\quad - \boldsymbol{A}(\boldsymbol{A}^{\mathrm{H}}\boldsymbol{\Pi}_{\boldsymbol{B}}^{\perp}\boldsymbol{A})^{-1}\boldsymbol{A}^{\mathrm{H}}\boldsymbol{\Pi}_{\boldsymbol{B}} - \boldsymbol{\Pi}_{\boldsymbol{B}}\boldsymbol{A}(\boldsymbol{A}^{\mathrm{H}}\boldsymbol{\Pi}_{\boldsymbol{B}}^{\perp}\boldsymbol{A})^{-1}\boldsymbol{A}^{\mathrm{H}} \\ &= \boldsymbol{A}(\boldsymbol{A}^{\mathrm{H}}\boldsymbol{\Pi}_{\boldsymbol{B}}^{\perp}\boldsymbol{A})^{-1}\boldsymbol{A}^{\mathrm{H}}\big[\mathbf{I} - \boldsymbol{\Pi}_{\boldsymbol{B}}\big] + \boldsymbol{\Pi}_{\boldsymbol{B}} \\ &\quad + \boldsymbol{\Pi}_{\boldsymbol{B}}\boldsymbol{A}(\boldsymbol{A}^{\mathrm{H}}\boldsymbol{\Pi}_{\boldsymbol{B}}^{\perp}\boldsymbol{A})^{-1}\boldsymbol{A}^{\mathrm{H}}\big[\boldsymbol{\Pi}_{\boldsymbol{B}} - \mathbf{I}\big] \\ &= \big[\mathbf{I} - \boldsymbol{\Pi}_{\boldsymbol{B}}\big]\boldsymbol{A}(\boldsymbol{A}^{\mathrm{H}}\boldsymbol{\Pi}_{\boldsymbol{B}}^{\perp}\boldsymbol{A})^{-1}\boldsymbol{A}^{\mathrm{H}}\big[\mathbf{I} - \boldsymbol{\Pi}_{\boldsymbol{B}}\big] + \boldsymbol{\Pi}_{\boldsymbol{B}} \\ &= \boldsymbol{\Pi}_{\boldsymbol{B}} + \boldsymbol{\Pi}_{\boldsymbol{B}}^{\perp}\boldsymbol{A}(\boldsymbol{A}^{\mathrm{H}}\boldsymbol{\Pi}_{\boldsymbol{B}}^{\perp}\boldsymbol{A})^{-1}\boldsymbol{A}^{\mathrm{H}}\boldsymbol{\Pi}_{\boldsymbol{B}}^{\perp}.\end{aligned} \tag{A.50}$$

For $K = 2$, $\boldsymbol{A} = \boldsymbol{H}_1$ and $\boldsymbol{B} = \boldsymbol{H}_2$ such that the scaled projector in (A.50) reads as

$$\begin{aligned}\boldsymbol{S}^{\mathrm{PtP}} &= \frac{P_{\max}}{M}\cdot\boldsymbol{H}\big(\boldsymbol{H}^{\mathrm{H}}\boldsymbol{H}\big)^{-1}\boldsymbol{H}^{\mathrm{H}} \\ &= \frac{P_{\max}}{M}\cdot\big[\boldsymbol{H}_2\big(\boldsymbol{H}_2^{\mathrm{H}}\boldsymbol{H}_2\big)^{-1}\boldsymbol{H}_2^{\mathrm{H}} + \bar{\boldsymbol{H}}_1\big(\bar{\boldsymbol{H}}_1^{\mathrm{H}}\bar{\boldsymbol{H}}_1\big)^{-1}\bar{\boldsymbol{H}}_1^{\mathrm{H}}\big] = \boldsymbol{S}^{\mathrm{DPC}}(\alpha\mathbf{1})\end{aligned}$$

with the projected channel matrix

$$\bar{\boldsymbol{H}}_1 = \boldsymbol{\Pi}_{\boldsymbol{B}}^{\perp}\boldsymbol{H}_1 = \big[\mathbf{I} - \boldsymbol{H}_2\big(\boldsymbol{H}_2^{\mathrm{H}}\boldsymbol{H}_2\big)^{-1}\boldsymbol{H}_2^{\mathrm{H}}\big]\boldsymbol{H}_1.$$

For $K > 2$, the matrix $\boldsymbol{A}$ must further be partitioned and decomposed recursively, but the equality $\boldsymbol{S}^{\mathrm{PtP}} = \boldsymbol{S}^{\mathrm{DPC}}(\alpha\mathbf{1})$ remains valid for all $\alpha > 0$, i.e., as long as all weights are positive and identical.

A.18 Determinant Growth Rate

The growth rates of the logarithmic determinant of a sum of rank-one matrices is analyzed in this section. In particular, we are interested in how the function

$$u(P_{\max}) := \log_2\Big|\mathbf{I}_N + \sum_{\ell=1}^{L}\boldsymbol{f}_\ell\boldsymbol{f}_\ell^{\mathrm{H}}\lambda_\ell\Big| \tag{A.51}$$

grows with $P_{\max}$ when $\lambda_\ell \to \infty$ as $P_{\max} \to \infty$. Let the scalars $\lambda_1, \ldots, \lambda_L$ be parametrized via

$$\lambda_\ell = a_\ell P_{\max}^{\tau_\ell} \quad \forall\ell \in \{1, \ldots, L\} \tag{A.52}$$

with $a_\ell > 0$ and $0 < \tau_\ell \leq 1$ $\forall\ell$. Note that above power allocation may also be replaced by $\lambda_\ell \doteq a_\ell P_{\max}^{\tau_\ell}$ without changing the following results. Furthermore, let $r = \mathrm{rank}(\boldsymbol{F})$ denote the rank of the matrix $\boldsymbol{F} := [\boldsymbol{f}_1, \ldots, \boldsymbol{f}_L] \in \mathbb{C}^{N\times L}$, which obviously satisfies $r \leq \min\{N, L\}$. Then, we can pick r linearly independent columns from $\boldsymbol{F}$ with indices $\pi[1], \ldots, \pi[r]$ and store them in

$$\boldsymbol{F}_1 = [\boldsymbol{f}_{\pi[1]}, \ldots, \boldsymbol{f}_{\pi[r]}] \in \mathbb{C}^{N\times r}$$

such that $\operatorname{rank}(\boldsymbol{F}_1) = r$. For $L > r$, this choice is not unique and we select $\pi[1], \ldots, \pi[r]$ such that $\sum_{i=1}^{r} \tau_{\pi[i]}$ is maximized subject to $\operatorname{rank}(\boldsymbol{F}_1) = r$:

$$\{\pi[1], \ldots, \pi[r]\} = \underset{\pi'[1],\ldots,\pi'[r]\in\{1,\ldots,L\}}{\operatorname{argmax}} \sum_{i=1}^{r} \tau_{\pi'[i]} \quad \text{s.t.: } \operatorname{rank}([\boldsymbol{f}_{\pi'[1]}, \ldots, \boldsymbol{f}_{\pi'[r]}]) = r \tag{A.53}$$

The remaining $L - r$ columns of $\boldsymbol{F}$ that are not contained in $\boldsymbol{F}_1$ are grouped in the matrix

$$\boldsymbol{F}_2 = [\boldsymbol{f}_{\pi[r+1]}, \ldots, \boldsymbol{f}_{\pi[L]}] \in \mathbb{C}^{N\times L-r}$$

such that $\pi[1], \ldots \pi[L]$ is eventually a permutation of $1, \ldots, L$. Using $\boldsymbol{\Lambda}_1 := \operatorname{diag}\{\lambda_{\pi[i]}\}_{i=1}^{r}$ and $\boldsymbol{\Lambda}_2 := \operatorname{diag}\{\lambda_{\pi[i]}\}_{i=r+1}^{L}$, the function $u(P_{\max})$ in (A.51) can be reformulated as

$$\begin{aligned} u(P_{\max}) &= \log_2 \left|\mathbf{I}_N + \boldsymbol{F}_1\boldsymbol{\Lambda}_1\boldsymbol{F}_1^{\mathrm{H}} + \boldsymbol{F}_2\boldsymbol{\Lambda}_2\boldsymbol{F}_2^{\mathrm{H}}\right| \\ &= \log_2 \left|\mathbf{I}_N + \boldsymbol{F}_1\boldsymbol{\Lambda}_1\boldsymbol{F}_1^{\mathrm{H}}\right| + \log_2 \left|\mathbf{I}_N + \left(\mathbf{I}_N + \boldsymbol{F}_1\boldsymbol{\Lambda}_1\boldsymbol{F}_1^{\mathrm{H}}\right)^{-1}\boldsymbol{F}_2\boldsymbol{\Lambda}_2\boldsymbol{F}_2^{\mathrm{H}}\right|. \end{aligned} \tag{A.54}$$

Since $\boldsymbol{F}_1$ has full rank and all diagonal entries of $\boldsymbol{\Lambda}_1$ go to infinity as $P_{\max} \to \infty$ due to $\tau_\ell > 0\ \forall \ell$, the first summand in (A.54) can be written as

$$\begin{aligned} \log_2 \left|\mathbf{I}_N + \boldsymbol{F}_1\boldsymbol{\Lambda}_1\boldsymbol{F}_1^{\mathrm{H}}\right| &= \log_2 \left|\mathbf{I}_r + \boldsymbol{F}_1^{\mathrm{H}}\boldsymbol{F}_1\boldsymbol{\Lambda}_1\right| \\ &= \log_2 \left|\boldsymbol{\Lambda}_1\right| + \log_2 \left|\boldsymbol{F}_1^{\mathrm{H}}\boldsymbol{F}_1\right| + o(1) \\ &= \left(\sum_{i=1}^{r} \tau_{\pi[i]}\right) \log_2 P_{\max} + \sum_{i=1}^{r} \log_2 a_{\pi[i]} + \log_2 \left|\boldsymbol{F}_1^{\mathrm{H}}\boldsymbol{F}_1\right| + o(1) \\ &= \left(\sum_{i=1}^{r} \tau_{\pi[i]}\right) \log_2 P_{\max} + O(1). \end{aligned} \tag{A.55}$$

For the evaluation of the second summand in (A.54), we exploit the fact that $\boldsymbol{F}_2$ can be written as $\boldsymbol{F}_2 = \boldsymbol{F}_1\boldsymbol{K}$ with $\boldsymbol{K} \in \mathbb{C}^{r\times L-r}$ since $\operatorname{rank}(\boldsymbol{F}) = \operatorname{rank}(\boldsymbol{F}_1)$. By means of the matrix inversion lemma from Appendix A.2, we obtain

$$\begin{aligned} &\boldsymbol{\Lambda}_2^{\frac{1}{2}}\boldsymbol{F}_2^{\mathrm{H}}\left(\mathbf{I}_N + \boldsymbol{F}_1\boldsymbol{\Lambda}_1\boldsymbol{F}_1^{\mathrm{H}}\right)^{-1}\boldsymbol{F}_2\boldsymbol{\Lambda}_2^{\frac{1}{2}} \\ &= \boldsymbol{\Lambda}_2^{\frac{1}{2}}\boldsymbol{K}^{\mathrm{H}}\boldsymbol{F}_1^{\mathrm{H}}\left[\mathbf{I}_N - \boldsymbol{F}_1\left(\boldsymbol{\Lambda}_1^{-1} + \boldsymbol{F}_1^{\mathrm{H}}\boldsymbol{F}_1\right)^{-1}\boldsymbol{F}_1^{\mathrm{H}}\right]\boldsymbol{F}_1\boldsymbol{K}\boldsymbol{\Lambda}_2^{\frac{1}{2}} \\ &= \boldsymbol{\Lambda}_2^{\frac{1}{2}}\boldsymbol{K}^{\mathrm{H}}\boldsymbol{F}_1^{\mathrm{H}}\boldsymbol{F}_1\left[\mathbf{I}_r - \left(\boldsymbol{\Lambda}_1^{-1} + \boldsymbol{F}_1^{\mathrm{H}}\boldsymbol{F}_1\right)^{-1}\boldsymbol{F}_1^{\mathrm{H}}\boldsymbol{F}_1\right]\boldsymbol{K}\boldsymbol{\Lambda}_2^{\frac{1}{2}} \\ &= \boldsymbol{\Lambda}_2^{\frac{1}{2}}\boldsymbol{K}^{\mathrm{H}}\boldsymbol{F}_1^{\mathrm{H}}\boldsymbol{F}_1\left[\mathbf{I}_r + \boldsymbol{\Lambda}_1\boldsymbol{F}_1^{\mathrm{H}}\boldsymbol{F}_1\right]^{-1}\boldsymbol{K}\boldsymbol{\Lambda}_2^{\frac{1}{2}} \end{aligned} \tag{A.56}$$

$$= \boldsymbol{\Lambda}_2^{\frac{1}{2}} \boldsymbol{K}^{\mathrm{H}} \left[(\boldsymbol{F}_1^{\mathrm{H}} \boldsymbol{F}_1)^{-1} + \boldsymbol{\Lambda}_1 \right]^{-1} \boldsymbol{K} \boldsymbol{\Lambda}_2^{\frac{1}{2}}$$

which leads to

$$\log_2 \left| \mathbf{I} + \boldsymbol{\Lambda}_2^{\frac{1}{2}} \boldsymbol{K}^{\mathrm{H}} \left[(\boldsymbol{F}_1^{\mathrm{H}} \boldsymbol{F}_1)^{-1} + \boldsymbol{\Lambda}_1 \right]^{-1} \boldsymbol{K} \boldsymbol{\Lambda}_2^{\frac{1}{2}} \right| = O(1)$$

due to the following reasoning. Either the smallest exponent $\min_{i \in \{1,\ldots,r\}} \tau_{\pi[i]}$ of the elements in $\boldsymbol{\Lambda}_1$ is at least as big as the largest exponent $\max_{i \in \{r+1,\ldots,L\}} \tau_{\pi[i]}$ of the diagonal elements of $\boldsymbol{\Lambda}_2$ such that (A.56) is finite. Or, if some $\tau_{\pi[m]}$ with $m \in \{r+1,\ldots,L\}$ is larger than some $\tau_{\pi[n]}$ with $n \in \{1,\ldots,r\}$, the respective entry $[\boldsymbol{K}]_{n,m-r}$ is zero due to the choice of $\pi[1],\ldots,\pi[r]$ in (A.53). Hence, a *positive* exponent $\frac{1}{2}(\tau_{\pi[m]} - \tau_{\pi[n]})$ of $P_{\max}$ never arises in any entry of the matrix product $\boldsymbol{\Lambda}_1^{-0.5} \boldsymbol{K} \boldsymbol{\Lambda}_2^{0.5}$ which governs (A.56) for large $P_{\max}$.

As a consequence, the function $u(P_{\max})$ in (A.51) behaves for $P_{\max} \to \infty$ and a power allocation in (A.52) as

$$u(P_{\max}) = \left(\sum_{i=1}^{r} \tau_{\pi[i]} \right) \log_2 P_{\max} + O(1) \tag{A.57}$$

where $\pi[1],\ldots,\pi[r]$ follow from (A.53). In other words, the pre-log factor is given by the sum of those largest exponents whose corresponding vectors $\boldsymbol{f}_{\pi[1]},\ldots,\boldsymbol{f}_{\pi[r]}$ are linearly independent. The following examples lead to a better understanding. *Real-valued examples*:

- $L = 4$ streams and $N = 2$ antennas at the base station with

$$\boldsymbol{F} = \begin{bmatrix} 2 & 7 & 3 & -1 \\ 1 & 4 & 2 & -1 \end{bmatrix}$$

  and $\tau_1 = 0.9$, $\tau_2 = 0.5$, $\tau_3 = 0.6$, and $\tau_4 = 0.4$. Then, $\boldsymbol{f}_1$ and $\boldsymbol{f}_3$ span the complete $\mathbb{R}^2$ and have the largest sum of the two involved exponents τ_1 and τ_3, i.e.,

$$\boldsymbol{F}_1 = [\boldsymbol{f}_1, \boldsymbol{f}_3] = \begin{bmatrix} 2 & 3 \\ 1 & 2 \end{bmatrix} \quad \text{and} \quad \boldsymbol{F}_2 = [\boldsymbol{f}_2, \boldsymbol{f}_4] = \begin{bmatrix} 7 & -1 \\ 4 & -1 \end{bmatrix} = \boldsymbol{F}_1 \cdot \underbrace{\begin{bmatrix} 2 & 1 \\ 1 & -1 \end{bmatrix}}_{=\boldsymbol{K}}.$$

  Since $\min\{\tau_1, \tau_3\} = \tau_3 = 0.6$ is larger than $\max\{\tau_2, \tau_4\} = \tau_2 = 0.5$ and $\boldsymbol{\Lambda}_1^{-1}$ thus decays faster to zero than the diagonal elements of $\boldsymbol{\Lambda}_2$ tend to infinity, the matrix in (A.56) is strongly asymptotically equivalent to the zero matrix.
- $L = 4$ streams and $N = 2$ antennas at the base station with

$$\boldsymbol{F} = \begin{bmatrix} 2 & 4 & 3 & -1 \\ 1 & 2 & 2 & -1 \end{bmatrix}$$

and $\tau_1 = 0.9$, $\tau_2 = 0.8$, $\tau_3 = 0.6$, and $\tau_4 = 0.4$. Again, $\boldsymbol{f}_1$ and $\boldsymbol{f}_3$ span $\mathbb{R}^2$ and have the largest sum of the two involved exponents τ_1 and τ_3. Clearly, $\tau_2 > \tau_3$, but $\boldsymbol{f}_1$ and $\boldsymbol{f}_2$ are linearly dependent and thus do not span $\mathbb{R}^2$. We find with $\pi[1] = 1$, $\pi[2] = 3$, $\pi[3] = 2$, and $\pi[4] = 4$ that

$$\boldsymbol{F}_1 = [\boldsymbol{f}_1, \boldsymbol{f}_3] = \begin{bmatrix} 2 & 3 \\ 1 & 2 \end{bmatrix} \quad \text{and} \quad \boldsymbol{F}_2 = [\boldsymbol{f}_2, \boldsymbol{f}_4] = \begin{bmatrix} 4 & -1 \\ 2 & -1 \end{bmatrix} = \boldsymbol{F}_1 \cdot \underbrace{\begin{bmatrix} 2 & 1 \\ 0 & -1 \end{bmatrix}}_{=\boldsymbol{K}}.$$

This time, however, $\min\{\tau_{\pi[1]}, \tau_{\pi[2]}\} = \tau_{\pi[2]} = \tau_3 = 0.6$ is not larger than or equal to $\max\{\tau_{\pi[3]}, \tau_{\pi[4]}\} = \tau_{\pi[3]} = \tau_2 = 0.8$. Although $\tau_{\pi[2]} = \tau_3 < \tau_{\pi[3]} = \tau_2$, the difference $\tau_2 - \tau_3 > 0$ never occurs as an exponent in any entry of the matrix product in (A.56) because $[\boldsymbol{K}]_{2,1} = 0$ such that all entries of the matrix in (A.56) remain finite when $P_{\max} \to \infty$. This can be understood better by noticing that (A.56) is mainly governed by the matrix product $\boldsymbol{\Lambda}_1^{-0.5} \boldsymbol{K} \boldsymbol{\Lambda}_2^{0.5}$ for large $P_{\max}$. The lower left entry of this product reads as

$$\begin{aligned}
[\boldsymbol{\Lambda}_1^{-\frac{1}{2}} \boldsymbol{K} \boldsymbol{\Lambda}_2^{\frac{1}{2}}]_{2,1} &= a_3^{-\frac{1}{2}} P_{\max}^{-\frac{\tau_3}{2}} \cdot [\boldsymbol{K}]_{2,1} \cdot a_2^{\frac{1}{2}} P_{\max}^{\frac{\tau_2}{2}} \\
&= \sqrt{\frac{a_2}{a_3}} \cdot P_{\max}^{\frac{\tau_2 - \tau_3}{2}} \cdot [\boldsymbol{K}]_{2,1} = 0
\end{aligned}$$

and is equal to zero due to $[\boldsymbol{K}]_{2,1} = 0$. Thus, the matrix in (A.56) is strongly asymptotically equivalent to the zero matrix for this choice of $\boldsymbol{F}$ and $\tau_1, \ldots, \tau_4$ as well.

A.19 Asymptotic Rate Expression

For the precoder parametrization in (8.71), the rate expression in (8.72) can be written as

$$R_k^{\text{MAC}} = \log_2 \left| \mathbf{I}_{L_k} + P_{\max} a_k \boldsymbol{B}_k^{\text{H}} \left(\mathbf{I}_N + P_{\max} \boldsymbol{C}_k \boldsymbol{A}_k \boldsymbol{C}_k^{\text{H}} \right)^{-1} \boldsymbol{B}_k \right| \tag{A.58}$$

where $\boldsymbol{A}_k = \text{blockdiag}\{a_i \mathbf{I}_{L_i}\}_{i \neq k} \in \mathbb{C}^{L-L_k \times L-L_k}$ is diagonal and contains the positive values $\{a_1, \ldots, a_K\} \setminus \{a_k\}$ as main diagonal elements. The matrix $\boldsymbol{B}_k \in \mathbb{C}^{N \times L_k}$ has rank L_k, the matrix $\boldsymbol{C}_k \in \mathbb{C}^{N \times L-L_k}$ has rank $L - L_k$, and $[\boldsymbol{C}_k, \boldsymbol{B}_k] \in \mathbb{C}^{N \times L}$ has rank L. To see how (A.58) behaves when $P_{\max} \to \infty$, we make use of the singular value decomposition

$$\boldsymbol{C}_k = \boldsymbol{U} \boldsymbol{\Sigma} \boldsymbol{V}^{\text{H}} = \boldsymbol{U}_1 \boldsymbol{\Sigma}_1 \boldsymbol{V}_1^{\text{H}}$$

with unitary $\boldsymbol{U} = [\boldsymbol{U}_1, \boldsymbol{U}_0] \in \mathbb{C}^{N \times N}$ and unitary $\boldsymbol{V} = [\boldsymbol{V}_1, \boldsymbol{V}_0] \in \mathbb{C}^{L-L_k \times L-L_k}$. Note that $\boldsymbol{U}_1 \in \mathbb{C}^{N \times L-L_k}$ is a basis for the range of $\boldsymbol{C}_k$ and that $\boldsymbol{\Sigma}_1 \in \mathbb{R}_{+,0}^{L-L_k}$ is diagonal. Using this decomposition, we obtain

$$\begin{aligned}
&P_{\max} a_k \boldsymbol{B}_k^{\mathrm{H}} \big(\mathbf{I}_N + P_{\max} \boldsymbol{C}_k \boldsymbol{A}_k \boldsymbol{C}_k^{\mathrm{H}}\big)^{-1} \boldsymbol{B}_k \\
&\quad = P_{\max} a_k \boldsymbol{B}_k^{\mathrm{H}} \boldsymbol{U}_0 \boldsymbol{U}_0^{\mathrm{H}} \boldsymbol{B}_k \\
&\qquad + a_k \boldsymbol{B}_k^{\mathrm{H}} \boldsymbol{U}_1 \big(\mathbf{I}_{L-L_k} P_{\max}^{-1} + \boldsymbol{\Sigma}_1 \boldsymbol{V}_1^{\mathrm{H}} \boldsymbol{A}_k \boldsymbol{V}_1 \boldsymbol{\Sigma}_1\big)^{-1} \boldsymbol{U}_1^{\mathrm{H}} \boldsymbol{B}_k \\
&\quad \cong P_{\max} a_k \boldsymbol{B}_k^{\mathrm{H}} \boldsymbol{U}_0 \boldsymbol{U}_0^{\mathrm{H}} \boldsymbol{B}_k \\
&\qquad + a_k \boldsymbol{B}_k^{\mathrm{H}} \boldsymbol{U}_1 \big(\boldsymbol{\Sigma}_1 \boldsymbol{V}_1^{\mathrm{H}} \boldsymbol{A}_k \boldsymbol{V}_1 \boldsymbol{\Sigma}_1\big)^{-1} \boldsymbol{U}_1^{\mathrm{H}} \boldsymbol{B}_k \\
&\quad = P_{\max} a_k \boldsymbol{B}_k^{\mathrm{H}} \boldsymbol{\Pi}_{\boldsymbol{C}_k}^{\perp} \boldsymbol{B}_k + a_k \boldsymbol{B}_k^{\mathrm{H}} \big(\boldsymbol{C}_k \boldsymbol{A}_k \boldsymbol{C}_k^{\mathrm{H}}\big)^{+} \boldsymbol{B}_k
\end{aligned}$$

where $\boldsymbol{\Pi}_{\boldsymbol{C}_k}^{\perp} = \boldsymbol{U}_0 \boldsymbol{U}_0^{\mathrm{H}}$ denotes the projector onto the null space of $\boldsymbol{C}_k^{\mathrm{H}}$ and $(\cdot)^{+}$ is the pseudo-inverse. Since $\operatorname{rank}([\boldsymbol{C}_k, \boldsymbol{B}_k]) = L = \operatorname{rank}(\boldsymbol{H})$, the rank of $\boldsymbol{B}_k^{\mathrm{H}} \boldsymbol{\Pi}_{\boldsymbol{C}_k}^{\perp} \boldsymbol{B}_k$ remains L_k despite the projection. Since only the first summand in above strong asymptotic equivalence scales with $P_{\max}$, the second, constant term may be dropped without losing strong asymptotic equivalence when the logarithm is applied to the determinant. Hence, we have

$$\begin{aligned}
R_k^{\mathrm{MAC}} &= \log_2 \left|\mathbf{I}_{L_k} + P_{\max} a_k \boldsymbol{B}_k^{\mathrm{H}} \big(\mathbf{I}_N + P_{\max} \boldsymbol{C}_k \boldsymbol{A}_k \boldsymbol{C}_k^{\mathrm{H}}\big)^{-1} \boldsymbol{B}_k\right| \\
&\cong \log_2 \left|P_{\max} a_k \boldsymbol{B}_k^{\mathrm{H}} \boldsymbol{\Pi}_{\boldsymbol{C}}^{\perp} \boldsymbol{B}_k\right| \\
&= L_k \log_2 P_{\max} + L_k \log_2 a_k + \log_2 \left|\boldsymbol{B}_k^{\mathrm{H}} \boldsymbol{\Pi}_{\boldsymbol{C}_k}^{\perp} \boldsymbol{B}_k\right|
\end{aligned} \tag{A.59}$$

which only depends on a_k but not on any a_i with $i \neq k$. In addition, the third summand in (A.59) only depends on the channel matrices and the *normalized* precoders, but not on the power allocation.

A.20 Transmit Covariance Matrix Reformulation

In this section, an alternative representation of the transmit covariance matrix $\boldsymbol{S}_k^{\mathrm{Linear}}$ in (8.88) is derived. To this end, we have to show that

$$\boldsymbol{\Pi}_k^{\perp} \boldsymbol{H}_k = \boldsymbol{H}(\boldsymbol{H}^{\mathrm{H}} \boldsymbol{H})^{-1} \boldsymbol{E}_k \big[\boldsymbol{E}_k^{\mathrm{T}} (\boldsymbol{H}^{\mathrm{H}} \boldsymbol{H})^{-1} \boldsymbol{E}_k\big]^{-1} \tag{A.60}$$

is valid. Let $\boldsymbol{J}_k \in \{0,1\}^{M \times M - M_k}$ denote the cropped $M \times M$ identity matrix from which columns $1 + \sum_{i=1}^{k-1} M_i, \ldots, \sum_{i=1}^{k} M_i$ have been removed. Then, the projector onto the null space of the Hermitian channel matrices of all users $i \neq k$ can be written as

$$\boldsymbol{\Pi}_k^{\perp} = \mathbf{I}_N - \boldsymbol{H} \boldsymbol{J}_k \big(\boldsymbol{J}_k^{\mathrm{T}} \boldsymbol{H}^{\mathrm{H}} \boldsymbol{H} \boldsymbol{J}_k\big)^{-1} \boldsymbol{J}_k^{\mathrm{T}} \boldsymbol{H}^{\mathrm{H}}$$

which leads to

$$\begin{aligned}
\boldsymbol{\Pi}_k^{\perp}\boldsymbol{H}\boldsymbol{E}_k &= \left[\mathbf{I}_N - \boldsymbol{H}\boldsymbol{J}_k\left(\boldsymbol{J}_k^{\mathrm{T}}\boldsymbol{H}^{\mathrm{H}}\boldsymbol{H}\boldsymbol{J}_k\right)^{-1}\boldsymbol{J}_k^{\mathrm{T}}\boldsymbol{H}^{\mathrm{H}}\right]\boldsymbol{H}\boldsymbol{E}_k \\
&= \boldsymbol{H}\left[\mathbf{I}_N - \boldsymbol{J}_k\left(\boldsymbol{J}_k^{\mathrm{T}}\boldsymbol{H}^{\mathrm{H}}\boldsymbol{H}\boldsymbol{J}_k\right)^{-1}\boldsymbol{J}_k^{\mathrm{T}}\boldsymbol{H}^{\mathrm{H}}\boldsymbol{H}\right]\boldsymbol{E}_k \\
&= \boldsymbol{H}(\boldsymbol{H}^{\mathrm{H}}\boldsymbol{H})^{-1}\left[\boldsymbol{H}^{\mathrm{H}}\boldsymbol{H} - \boldsymbol{H}^{\mathrm{H}}\boldsymbol{H}\boldsymbol{J}_k\left(\boldsymbol{J}_k^{\mathrm{T}}\boldsymbol{H}^{\mathrm{H}}\boldsymbol{H}\boldsymbol{J}_k\right)^{-1}\boldsymbol{J}_k^{\mathrm{T}}\boldsymbol{H}^{\mathrm{H}}\boldsymbol{H}\right]\boldsymbol{E}_k \\
&= \boldsymbol{H}(\boldsymbol{H}^{\mathrm{H}}\boldsymbol{H})^{-1}\boldsymbol{H}^{\mathrm{H}}\underbrace{\left[\mathbf{I}_N - \boldsymbol{H}\boldsymbol{J}_k\left(\boldsymbol{J}_k^{\mathrm{T}}\boldsymbol{H}^{\mathrm{H}}\boldsymbol{H}\boldsymbol{J}_k\right)^{-1}\boldsymbol{J}_k^{\mathrm{T}}\boldsymbol{H}^{\mathrm{H}}\right]}_{\boldsymbol{\Pi}_k^{\perp}}\boldsymbol{H}\boldsymbol{E}_k \\
&= \boldsymbol{H}(\boldsymbol{H}^{\mathrm{H}}\boldsymbol{H})^{-1}\boldsymbol{E}_k\boldsymbol{H}_k^{\mathrm{H}}\boldsymbol{\Pi}_k^{\perp}\boldsymbol{H}_k \\
&= \boldsymbol{H}(\boldsymbol{H}^{\mathrm{H}}\boldsymbol{H})^{-1}\boldsymbol{E}_k\left[\boldsymbol{E}_k^{\mathrm{T}}(\boldsymbol{H}^{\mathrm{H}}\boldsymbol{H})^{-1}\boldsymbol{E}_k\right]^{-1}
\end{aligned}$$

which proves (A.60). Note that the next to last equality in above set of equations follows from $\boldsymbol{H}_i^{\mathrm{H}}\boldsymbol{\Pi}_k^{\perp} = \mathbf{0}$ for $i \neq k$ such that $\boldsymbol{H}^{\mathrm{H}}\boldsymbol{\Pi}_k^{\perp} = \boldsymbol{E}_k\boldsymbol{H}_k^{\mathrm{H}}\boldsymbol{\Pi}_k^{\perp}$. The very last equality is due to

$$\boldsymbol{H}_k^{\mathrm{H}}\boldsymbol{\Pi}_k^{\perp}\boldsymbol{H}_k = \left[\boldsymbol{E}_k^{\mathrm{T}}(\boldsymbol{H}^{\mathrm{H}}\boldsymbol{H})^{-1}\boldsymbol{E}_k\right]^{-1}$$

from (8.86). The identity in (A.60) can then be used to convert $\boldsymbol{S}_k^{\mathrm{Linear}}$ from (8.88) to the expression in (8.89).

A.21 Standard Interference Function Framework

In [155], Yates defines an interference function $\boldsymbol{I}^{\mathrm{FF}}(\boldsymbol{q}, \boldsymbol{\mu})$ to be standard, in the following way:

Definition 9 An interference function $\boldsymbol{I}^{\mathrm{FF}}(\boldsymbol{q}, \boldsymbol{\mu})$ is termed standard, if the following three properties are valid for any power allocation $\boldsymbol{q} \geq \mathbf{0}$ and any $\boldsymbol{\mu} \in \mathcal{M}$:

1. Nonnegativity: $\boldsymbol{I}^{\mathrm{FF}}(\boldsymbol{q}, \boldsymbol{\mu}) \geq \mathbf{0}$
2. Monotonicity: If $\boldsymbol{q} \geq \boldsymbol{q}'$ then $\boldsymbol{I}^{\mathrm{FF}}(\boldsymbol{q}, \boldsymbol{\mu}) \geq \boldsymbol{I}^{\mathrm{FF}}(\boldsymbol{q}', \boldsymbol{\mu})$
3. Scalability: For all $\alpha > 1$, we have $\alpha\boldsymbol{I}^{\mathrm{FF}}(\boldsymbol{q}, \boldsymbol{\mu}) > \boldsymbol{I}^{\mathrm{FF}}(\alpha\boldsymbol{q}, \boldsymbol{\mu})$

In the following, we show that the vector-valued interference function $\boldsymbol{I}^{\mathrm{FF}}(\boldsymbol{q}, \boldsymbol{\mu})$ whose individual entries are given by [see (9.19)]

$$I_k^{\mathrm{FF}}(\boldsymbol{q}, \mu_k) = \frac{1 - \mu_k}{\boldsymbol{h}_k^{\mathrm{H}}\underbrace{\left(\mathbf{I}_N + \boldsymbol{H}\boldsymbol{Q}\boldsymbol{H}^{\mathrm{H}}\right)}_{=:\boldsymbol{X}(\boldsymbol{q})}{}^{-1}\boldsymbol{h}_k} \quad \forall k \tag{A.61}$$

is indeed standard for $\mathcal{M} := \{\boldsymbol{\mu} \in \mathbb{R}^K \big| \mathbf{0} \leq \boldsymbol{\mu} < \mathbf{1}\}$ and $\boldsymbol{h}_k \neq \mathbf{0}\ \forall k$.

1. Since $\mu_k \leq 1$ holds for all k and $\boldsymbol{X}(\boldsymbol{q}) \succ \mathbf{0}$ for any $\boldsymbol{q} \geq \mathbf{0}$, it is easy to see that the nonnegativity condition $I_k^{\mathrm{FF}}(\boldsymbol{q}, \mu_k) \geq 0\ \forall k$ in Definition 9 holds.
2. Let $\boldsymbol{q} = \boldsymbol{q}' + \Delta\boldsymbol{q}'$ with $\boldsymbol{q}' \geq \mathbf{0}$ and $\Delta\boldsymbol{q}' \geq \mathbf{0}$ such that $\boldsymbol{q} \geq \boldsymbol{q}'$ is valid. Then, $\boldsymbol{X}(\boldsymbol{q})$ can be written as

$$\boldsymbol{X}(\boldsymbol{q}) = \boldsymbol{X}(\boldsymbol{q}') + \boldsymbol{H} \cdot \mathrm{diag}\{\Delta\boldsymbol{q}'\} \cdot \boldsymbol{H}^{\mathrm{H}} = \boldsymbol{X}(\boldsymbol{q}') + \boldsymbol{B}(\Delta\boldsymbol{q}')$$

with $\boldsymbol{B}(\Delta\boldsymbol{q}') \succeq \mathbf{0}$ for any $\Delta\boldsymbol{q}' \geq \mathbf{0}$. By means of the matrix inversion lemma from Appendix A.2, the denominator in (A.61) can be upper bounded via[6]

$$\begin{aligned}\boldsymbol{h}_k^{\mathrm{H}}\boldsymbol{X}^{-1}(\boldsymbol{q})\boldsymbol{h}_k &= \boldsymbol{h}_k^{\mathrm{H}}\boldsymbol{X}^{-1}(\boldsymbol{q}')\boldsymbol{h}_k - \boldsymbol{h}_k^{\mathrm{H}}\underbrace{\boldsymbol{X}^{-1}(\boldsymbol{q}')\boldsymbol{B}^{\frac{1}{2}}\big[\mathbf{I} + \boldsymbol{B}^{\frac{1}{2}}\boldsymbol{X}^{-1}(\boldsymbol{q}')\boldsymbol{B}^{\frac{1}{2}}\big]^{-1}\boldsymbol{B}^{\frac{1}{2}}\boldsymbol{X}^{-1}(\boldsymbol{q}')}_{\succeq\mathbf{0}}\boldsymbol{h}_k \\ &\leq \boldsymbol{h}_k^{\mathrm{H}}\boldsymbol{X}^{-1}(\boldsymbol{q}')\boldsymbol{h}_k\end{aligned}$$

which leads to $I_k^{\mathrm{FF}}(\boldsymbol{q}, \mu_k) \geq I_k^{\mathrm{FF}}(\boldsymbol{q}', \mu_k)$ for all k and any $\boldsymbol{q} \geq \boldsymbol{q}' \geq \mathbf{0}$.

3. For $\alpha > 1$, the denominator of $I_k^{\mathrm{FF}}(\alpha\boldsymbol{q})$ can be written as

$$\begin{aligned}\boldsymbol{h}_k^{\mathrm{H}}\boldsymbol{X}^{-1}(\alpha\boldsymbol{q})\boldsymbol{h}_k &= \boldsymbol{h}_k^{\mathrm{H}}\big(\mathbf{I} + \alpha\boldsymbol{H}\boldsymbol{Q}\boldsymbol{H}^{\mathrm{H}}\big)^{-1}\boldsymbol{h}_k \\ &= \alpha^{-1}\boldsymbol{h}_k^{\mathrm{H}}\big[\mathbf{I} + \boldsymbol{H}\boldsymbol{Q}\boldsymbol{H}^{\mathrm{H}} - (1-\alpha^{-1})\cdot\mathbf{I}\big]^{-1}\boldsymbol{h}_k \\ &= \alpha^{-1}\boldsymbol{h}_k^{\mathrm{H}}\big[\boldsymbol{X}(\boldsymbol{q}) - \beta\mathbf{I}\big]^{-1}\boldsymbol{h}_k\end{aligned}$$

with $0 < \beta := 1 - \alpha^{-1} < 1$. Using the matrix inversion lemma from Appendix A.2, we obtain

$$\begin{aligned}\boldsymbol{h}_k^{\mathrm{H}}\boldsymbol{X}^{-1}(\alpha\boldsymbol{q})\boldsymbol{h}_k &= \alpha^{-1}\boldsymbol{h}_k^{\mathrm{H}}\Big[\boldsymbol{X}^{-1}(\boldsymbol{q}) + \underbrace{\boldsymbol{X}^{-1}(\boldsymbol{q})\big[\beta^{-1}\mathbf{I} - \boldsymbol{X}^{-1}(\boldsymbol{q})\big]^{-1}\boldsymbol{X}^{-1}(\boldsymbol{q})}_{\succ\mathbf{0}}\Big]\boldsymbol{h}_k \\ &> \alpha^{-1}\boldsymbol{h}_k^{\mathrm{H}}\boldsymbol{X}^{-1}(\boldsymbol{q})\boldsymbol{h}_k\end{aligned}$$

where the strict inequality is due to the fact that $\boldsymbol{X}^{-1}(\boldsymbol{q})\big[\beta^{-1}\mathbf{I} - \boldsymbol{X}^{-1}(\boldsymbol{q})\big]^{-1}$ $\boldsymbol{X}^{-1}(\boldsymbol{q}) \succ \mathbf{0}$ and $\boldsymbol{h}_k \neq \mathbf{0}$ are valid for all k. The positive definiteness holds since $\beta^{-1}\mathbf{I} \succ \mathbf{I} \succeq \boldsymbol{X}^{-1}(\boldsymbol{q})$ for $\beta < 1$. Hence, we find for $\mu_k < 1$

$$I_k^{\mathrm{FF}}(\alpha\boldsymbol{q}, \mu_k) = \frac{1-\mu_k}{\boldsymbol{h}_k^{\mathrm{H}}\boldsymbol{X}^{-1}(\alpha\boldsymbol{q})\boldsymbol{h}_k} < \frac{1-\mu_k}{\alpha^{-1}\boldsymbol{h}_k^{\mathrm{H}}\boldsymbol{X}^{-1}(\boldsymbol{q})\boldsymbol{h}_k} = \alpha I_k^{\mathrm{FF}}(\boldsymbol{q}, \mu_k) \quad \forall k \in \mathcal{K}$$

which finally proves that the interference function $\boldsymbol{I}^{\mathrm{FF}}(\boldsymbol{q}, \boldsymbol{\mu})$ is *standard* for $\boldsymbol{\mu} \in \mathcal{M}$.

A.22 Proof of Lemma 9.4

Lemma 9.4 states that if the users in the two sets $\mathcal{I}_1$ and $\mathcal{I}_2$ attain their lower sum MMSE bounds

[6] The positive semi-definite square root of $\boldsymbol{B}(\Delta\boldsymbol{q}')$ is used here. Moreover, the dependency of $\boldsymbol{B}(\Delta\boldsymbol{q}')$ on $\Delta\boldsymbol{q}'$ is omitted for a clearer notation.

$$\sum_{k\in\mathcal{I}_i} \varepsilon_k = |\mathcal{I}_i| - \mathrm{rank}(\boldsymbol{\Phi}_{\mathcal{I}_i}) \quad \forall i \in \{1,2\} \tag{A.62}$$

then so do the users in the union $\mathcal{I}_1 \cup \mathcal{I}_2$ and the intersection $\mathcal{I}_1 \cap \mathcal{I}_2$. Summing up the two equations in (A.62), we obtain

$$\begin{aligned}\sum_{k\in\mathcal{I}_1} \varepsilon_k + \sum_{k\in\mathcal{I}_2} \varepsilon_k &= |\mathcal{I}_1| - \mathrm{rank}(\boldsymbol{\Phi}_{\mathcal{I}_1}) + |\mathcal{I}_2| - \mathrm{rank}(\boldsymbol{\Phi}_{\mathcal{I}_2}) \\ &= \underbrace{\sum_{k\in\mathcal{I}_1\cup\mathcal{I}_2} \varepsilon_k}_{\geq|\mathcal{I}_1\cup\mathcal{I}_2|-\mathrm{rank}(\boldsymbol{\Phi}_{\mathcal{I}_1\cup\mathcal{I}_2})} + \underbrace{\sum_{k\in\mathcal{I}_1\cap\mathcal{I}_2} \varepsilon_k}_{\geq|\mathcal{I}_1\cap\mathcal{I}_2|-\mathrm{rank}(\boldsymbol{\Phi}_{\mathcal{I}_1\cap\mathcal{I}_2})}\end{aligned} \tag{A.63}$$

where the two lower bounds result from the necessary conditions in (9.33). From (A.63), we conclude that

$$\mathrm{rank}(\boldsymbol{\Phi}_{\mathcal{I}_1\cup\mathcal{I}_2}) + \mathrm{rank}(\boldsymbol{\Phi}_{\mathcal{I}_1\cap\mathcal{I}_2}) \geq \mathrm{rank}(\boldsymbol{\Phi}_{\mathcal{I}_1}) + \mathrm{rank}(\boldsymbol{\Phi}_{\mathcal{I}_2}) \tag{A.64}$$

results from (A.62) due to $|\mathcal{I}_1| + |\mathcal{I}_2| = |\mathcal{I}_1 \cup \mathcal{I}_2| + |\mathcal{I}_1 \cap \mathcal{I}_2|$. However, we can show that the inequality

$$\mathrm{rank}(\boldsymbol{\Phi}_{\mathcal{I}_1\cup\mathcal{I}_2}) + \mathrm{rank}(\boldsymbol{\Phi}_{\mathcal{I}_1\cap\mathcal{I}_2}) \leq \mathrm{rank}(\boldsymbol{\Phi}_{\mathcal{I}_1}) + \mathrm{rank}(\boldsymbol{\Phi}_{\mathcal{I}_2}) \tag{A.65}$$

is true for any $\boldsymbol{\Phi}$, such that if (A.62) holds, then

$$\mathrm{rank}(\boldsymbol{\Phi}_{\mathcal{I}_1\cup\mathcal{I}_2}) + \mathrm{rank}(\boldsymbol{\Phi}_{\mathcal{I}_1\cap\mathcal{I}_2}) = \mathrm{rank}(\boldsymbol{\Phi}_{\mathcal{I}_1}) + \mathrm{rank}(\boldsymbol{\Phi}_{\mathcal{I}_2})$$

i.e., the two lower bounds in (A.63) must hold individually. In other words, the users of the two sets $\mathcal{I}_1 \cup \mathcal{I}_2$ and $\mathcal{I}_1 \cap \mathcal{I}_2$ achieve their lower bounds on the sum MMSEs as well, which proves Lemma 9.4.

It remains to show that (A.65) always holds. Let $\boldsymbol{U}$ be a basis for $\boldsymbol{\Phi}_{\mathcal{I}_1}$ with the smallest number of columns such that $\mathrm{rank}(\boldsymbol{\Phi}_{\mathcal{I}_1}) = \mathrm{rank}(\boldsymbol{U})$ and let $\boldsymbol{U}$ be partitioned into

$$\boldsymbol{U} = [\boldsymbol{U}_1, \boldsymbol{U}_2]$$

where $\boldsymbol{U}_1^{\mathrm{H}}\boldsymbol{U}_2 = \boldsymbol{0}$ and $\boldsymbol{U}_1^{\mathrm{H}}\boldsymbol{\Phi}_{\mathcal{I}_2} = \boldsymbol{0}$, i.e., $\boldsymbol{U}_1$ is not in the range of $\boldsymbol{\Phi}_{\mathcal{I}_2}$. The matrix $\boldsymbol{U}_2$, however, is in the range of $\boldsymbol{\Phi}_2$. Then, the rank of $\boldsymbol{\Phi}_{\mathcal{I}_1}$ is given by

$$\mathrm{rank}(\boldsymbol{\Phi}_{\mathcal{I}_1}) = \mathrm{rank}(\boldsymbol{U}_1) + \mathrm{rank}(\boldsymbol{U}_2). \tag{A.66}$$

The rank of the union matrix $\boldsymbol{\Phi}_{\mathcal{I}_1\cup\mathcal{I}_2}$ is equal to

$$\begin{aligned}\boldsymbol{\Phi}_{\mathcal{I}_1\cup\mathcal{I}_2} &= \operatorname{rank}([\boldsymbol{\Phi}_{\mathcal{I}_1}, \boldsymbol{\Phi}_{\mathcal{I}_2}]) \\ &= \operatorname{rank}([\boldsymbol{U}, \boldsymbol{\Phi}_{\mathcal{I}_2}]) \\ &= \operatorname{rank}([\boldsymbol{U}_1, \boldsymbol{\Phi}_{\mathcal{I}_2}]) \\ &= \operatorname{rank}(\boldsymbol{U}_1) + \operatorname{rank}(\boldsymbol{\Phi}_{\mathcal{I}_2})\end{aligned} \tag{A.67}$$

since $\boldsymbol{U}_2$ is in the range of $\boldsymbol{\Phi}_{\mathcal{I}_2}$ by construction and therefore can be dropped. Finally, the rank of the union matrix $\boldsymbol{\Phi}_{\mathcal{I}_1\cap\mathcal{I}_2}$ is upper bounded by

$$\operatorname{rank}(\boldsymbol{\Phi}_{\mathcal{I}_1\cap\mathcal{I}_2}) \le \operatorname{rank}(\boldsymbol{U}_2) \tag{A.68}$$

since $\boldsymbol{U}_1$ is not in the range of $\boldsymbol{\Phi}_{\mathcal{I}_2}$. Combining (A.66)–(A.68) leads to (A.65).

Note that the inequality in (A.68) may be strict which happens for example if

$$\boldsymbol{\Phi} = \begin{bmatrix} 1 & 1 & 2 \\ 0 & 1 & 0 \end{bmatrix}$$

with $\mathcal{I}_1 = \{1, 2\}$ and $\mathcal{I}_2 = \{2, 3\}$. In this case, $\boldsymbol{U}_2$ can be set to

$$\boldsymbol{U}_2 = \begin{bmatrix} 1 & 0 \\ 0 & 1 \end{bmatrix}$$

whereas $\boldsymbol{U}_1$ would have zero columns. Then, the intersection matrix (vector) $\boldsymbol{\Phi}_{\mathcal{I}_1\cap\mathcal{I}_2} = [1, 1]^{\mathrm{T}}$ has only rank one, whereas $\boldsymbol{U}_2$ has rank two. Due to the strict inequality in (A.68), inequality (A.64) can not hold which means that (A.62) must also be violated. For the given $\boldsymbol{\Phi}$, (A.62) would mean that $\varepsilon_1 = \varepsilon_2 = \varepsilon_3 = 0$, which obviously contradicts the lower bound on the total sum MMSE $\varepsilon_1 + \varepsilon_2 + \varepsilon_3 \ge 1$.

References

1. T.M. Cover, Broadcast channels. IEEE Trans. Inf. Theory **18**(1), 2–14 (1972)
2. D.P. Palomar, M.A. Lagunas, J.M. Cioffi, Optimum joint transmit receive linear processing for vectored DSL transmission with QoS requirements, in *Proceedings of the 36th Asilomar Conference on Signals, Systems, and Computers*, November 2002, pp. 388–392
3. V.L. Nir, M. Moonen, J. Verlinden, M. Guenach, Optimal power allocation for downstream xDSL with per-modem total power constraints: broadcast channel optimal spectrum balancing (BC-OSB). IEEE Trans. Signal Process. **57**(2), 690–697 (2009)
4. M. Coldrey, P. Bohlin, Training-based MIMO systems-part I: performance comparison. IEEE Trans. Signal Process. **55**(11), 5464–5476 (2007)
5. M. Coldrey, P. Bohlin, Training-based MIMO systems: part II-improvements using detected symbol information. IEEE Trans. Signal Process. **56**(1), 296–303 (2008)
6. G. Auer, I. Cosovic, Pilot design for multi-user MIMO, in *Proceedings of the IEEE International Conference on Acoustics, Speech, and Signal Processing*, April 2009
7. M. Guillaud, D.T.M. Slock, R. Knopp, A practical method for wireless channel reciprocity exploitation through relative calibration, in *Proceedings of the Eighth International Symposium on Signal Processing and Its Applications (ISSPA)*, August 2005, pp. 403–406

8. F. Rey, M. Lamarca, G. Vázquez, Optimal power allocation with partial channel knowledge for MIMO multicarrier systems, in *Proceedings of the IEEE 56th Vehicular Technology Conference* (*VTC Fall*), September 2002, pp. 2121–2125
9. F.A. Dietrich, R. Hunger, M. Joham, W. Utschick, Robust transmit wiener filter for time division duplex systems, in *Proceedings of the IEEE Symposium on Signal Processing and Information Technology* (*ISSPIT*), December 2003, pp. 415–418
10. R. Hunger, F. Dietrich, M. Joham, W. Utschick, Robust transmit zero-forcing filters, in *Proceedings of the ITG Workshop on Smart Antennas*, March 2004, pp. 130–137
11. F.A. Dietrich, P. Breun, W. Utschick, Robust Tomlinson-Harashima precoding for the wireless broadcast channel. IEEE Trans. Signal Process. **55**(2), 631–644 (2007)
12. K.R. Kumar, G. Caire, Channel state feedback over the MIMO-MAC, in *Proceedings of the International Symposium on Information Theory* (*ISIT*), June 2009, pp. 2361–2365
13. G. Caire, N. Jindal, M. Kobayashi, N. Ravindran, Multiuser MIMO achievable rates with downlink training and channel state feedback. IEEE Trans. Inf. Theory (2007), (Submitted). http://arxiv.org/abs/0711.2642
14. D. Li, Z.-Y. Wu, H.-W.J. Lee, X.-M. Yang, L.-S. Zhang, Hidden convex minimization. J. Global Optim. **31**(2), 211–233 (2005)
15. S. Shi, M. Schubert, MMSE transmit optimization for multiuser multiantenna systems, in *Proceedings of the IEEE International Conference on Acoustics, Speech, and Signal Processing* (*ICASSP*), March 2005
16. M. Schubert, S. Shi, E.A. Jorswieck, H. Boche, Downlink Sum-MSE transceiver optimization for linear multi-user MIMO systems, in *Proceedings of the Asilomar Conference on Signals systems and computers*, September 2005
17. S. Shi, M. Schubert, H. Boche, Downlink MMSE transceiver optimization for multiuser MIMO systems: duality and sum-MSE minimization. IEEE Trans. Signal Process. **55**(11), 5436–5446 (2007)
18. R. Hunger, M. Joham, W. Utschick, On the MSE-duality of the broadcast channel and the multiple access channel. IEEE Trans. Signal Process. **57**(2), 698–713 (2009)
19. S. Vishwanath, N. Jindal, A.J. Goldsmith, Duality, achievable rates, and sum-rate capacity of Gaussian MIMO broadcast channels. IEEE Trans. Inf. Theory **49**(10), 2658–2668 (2003)
20. H. Viswanathan, S. Venkatesan, H. Huang, Downlink capacity evaluation of cellular networks with known-interference cancellation. IEEE J. Sel. Areas Commun. **21**, 802–811 (2003)
21. R. Hunger, M. Joham, A general rate duality of the MIMO multiple access channel and the MIMO broadcast channel, in *Proceedings of the IEEE Global Conference on Communications* (*Globecom*) New Orleans, November 2008
22. R. Hunger, M. Joham, On the asymptotic optimality of block-diagonalization for the MIMO BC under linear filtering, in *Proceedings of the Workshop on Smart Antennas* (*WSA*), Berlin, February 2009
23. R. Hunger, M. Joham, An asymptotic analysis of the MIMO broadcast channel under linear filtering, in *Proceedings of the Conference on Information Sciences and Systems* (*CISS*), March 2009
24. R. Hunger, P. de Kerret, M. Joham, The geometry of the MIMO broadcast channel rate region under linear filtering at high SNR, in *Proceedings of the 43rd Asilomar Conference on Signals, Systems, and Computers*, November 2009
25. J. Zander, M. Frodigh, Comment on 'performance of optimum transmitter power control in cellular radio'. IEEE Trans. Veh. Technol. **43**(3), 636 (1994)
26. D. Gerlach, A. Paulraj, Base station transmitting antenna arrays for multipath environments. IEEE Trans. Signal Process. **54**(1), 59–73 (1996)
27. H. Boche, M. Schubert, Optimum SIR balancing using extended 1-Norm beamforming optimization, in *Proceedings of the International Conference on Acoustics, Speech, and Signal Processing* (*ICASSP*), May 2002, pp. 2945–2948
28. F. Rashid-Farrokhi, K.J.R. Liu, L. Tassiulas, Transmit beamforming and power control for cellular wireless systems. IEEE J. Sel. Areas Commun. **16**(8), 1437–1450 (1998)

29. E. Visotsky, U. Madhow, Optimum beamforming using transmit antenna arrays, in *Proceedings of the IEEE 49th Vehicular Technology Conference*, July 1999, pp. 851–856
30. H. Boche, M. Schubert, Solution of the SINR downlink beamforming problem, in *Proceedings of the Conference on Information Sciences and Systems* (*CISS*), March 2002
31. H. Boche, M. Schubert, A general duality theory for uplink and downlink beamforming, in *Proceedings of the IEEE 56th Vehicular Technology Conference* (*VTC Fall*), September 2002, vol. 1, pp. 87–91
32. P. Viswanath, D.N.C. Tse, Sum capacity of the vector Gaussian broadcast channel and uplink-downlink duality. IEEE Trans. Inf. Theory **49**(8), 1912–1921 (2003)
33. D. Tse, P. Viswanath, On the capacity of the multiple antenna broadcast channel, in *Multiantenna Channels: Capacity, Coding and Signal Processing, DIMACS Workshop American Mathematical Society 2003*, October 2002, vol. 62, pp. 87–105
34. D.A. Schmidt, M. Joham, R. Hunger, W. Utschick, Near maximum sum-rate non-zero-forcing linear precoding with successive user selection, in *Proceedings of the 40th Asilomar Conference on Signals, Systems, and Computers*, November 2006
35. M. Schubert, H. Boche, Solution of the multi-user downlink beamforming problem with individual SINR constraints. IEEE Trans. Veh. Technol. **53**(1), 18–28 (2004)
36. A. Wiesel, Y. Eldar, S. Shamai, Linear precoding via conic optimization for fixed MIMO receivers. IEEE Trans. Signal Process. **54**(1), 161–176 (2006)
37. A.M. Khachan, A.J. Tenenbaum, R.S. Adve, Linear processing for the downlink in multiuser MIMO systems with multiple data streams, in *Proceedings of the IEEE International Conference on. Communications*, June 2006, vol. 9, pp. 4113–4118
38. A. Mezghani, M. Joham, R. Hunger, W. Utschick, Transceiver design for multi-user MIMO systems, in *Proceedings of the ITG/IEEE Workshop on Smart Antennas*, March 2006
39. M. Schubert, S. Shi, MMSE transmit optimization with interference pre-compensation, in *Proceedings of the IEEE 62nd Vehicular Technology Conference* (*VTC Spring*), May 2005
40. E. Jorswieck, H. Boche, Transmission strategies for the MIMO MAC with MMSE receiver: average MSE optimization and achievable individual MSE region. IEEE Trans. Signal Process. **51**(11), 2872–2881 (2003)
41. R. Hunger, M. Joham, W. Utschick, Extension of linear and nonlinear transmit filters for decentralized receivers, in *Proceedings of the European Wireless*, Nicosia, April 2005, vol. 1, pp. 40–46
42. W. Utschick, M. Joham, On the duality of MIMO transmission techniques for multiuser communications, in *Proceedings of the European Signal Processing Conference* (*EUSIPCO*), September 2006
43. Y. Saad, *Iterative Methods for Sparse Linear Systems* (Society for Industrial and Applied Mathematics, Philadelphia, 2003)
44. A. Berman, R.J. Plemmons, *Nonnegative Matrices in the Mathematical Sciences* (Academic Press [Harcourt Brace Jovanovich Publishers], New York, 1979)
45. G.H. Golub, C.F.V. Loan, *Matrix Computations*, 2nd edn. (Johns Hopkins University Press, Philadelphia, 1991)
46. M. Codreanu, A. Tölli, M. Latva-aho, Weighted sum MSE minimization for MIMO broadcast channel, in *Proceedings of the Personal, Indoor and Mobile Radio Communications* (*PIMRC*), September 2006
47. S. Shi, M. Schubert, H. Boche, Downlink MMSE transceiver optimization with layer-MSE requirements, in *Proceedings of the Communications and Networking in China* (*ChinaCom*), October 2006
48. M. Codreanu, A. Tölli, M. Juntti, M. Latva-aho, Joint design of Tx-Rx beamformers in MIMO downlink channel. Trans. Signal Process. **55**(9), 4639–4655 (2007)
49. S. Shi, M. Schubert, H. Boche, Computational efficient transceiver optimization for multiuser MIMO systems: power minimization with user-MMSE requirements, in *Proceedings of the 40th Asilomar Conference on Signals, Systems, and Computers*, October 2006
50. C.E. Shannon, A mathematical theory of communication. Bell Syst. Tech. J. **27**, 1–54 (1948)
51. C.E. Shannon, Communication in the presence of noise. Proc. IRE **37**(1), 10–21 (1949)

52. E. Telatar, Capacity of multi-antenna Gaussian channels. Eur. Trans. Telecommun. **10**(6), 585–595 (1999)
53. W. Yu, T. Lan, Transmitter optimization for the multi-antenna downlink with per-antenna power constraints. IEEE Trans. Signal Process. **55**(6), 2646–2660 (2007)
54. S. Boyd, L. Vandenberghe, *Convex Optimization* (Cambridge University Press, New York, 2009), seventh printing with corrections 2009
55. B. Song, R.L. Cruz, B.D. Rao, Network duality for multiuser MIMO beamforming networks and applications. IEEE Trans. Commun. **55**(3), 618–630 (2007)
56. A. Liu, Y. Liu, H. Xiang, W. Luo, On the duality of the MIMO interference channel and its application to resource allocation, in *Proceedings of the IEEE Global Communications Conference* (*Globecom*), November 2009
57. S. Shi, M. Schubert, H. Boche, Weighted sum-rate optimization for multiuser MIMO systems, in *Proceedings of the 41st Conference on Information Sciences and Systems* (*CISS*), March 2007
58. S. Shi, M. Schubert, H. Boche, Rate optimization for multiuser MIMO systems with linear processing. IEEE Trans. Signal Process. **56**(8), 4020–4030 (2008)
59. M. Codreanu, A. Tölli, M. Juntti, M. Latva-aho, MIMO downlink weighted sum rate maximization with power constraints per antenna groups, in *Proceedings of the 65th Vehicular Technology Conference* (*VTC Spring*), April 2007, pp. 2048–2052
60. A.J. Tenenbaum, R.S. Adve, Linear processing and sum throughput in the multiuser MIMO downlink. IEEE Trans. Wireless Commun. **8**(5), 2652–2661 (2009)
61. R. Agarwal, S.S. Christensen, J.M. Cioffi, Beamforming design for the MIMO downlink for maximizing weighted sum-rate, in *Proceedings of the International Symposium on Information Theory and its Applications* (*ISITA*), December 2008
62. R. Böhnke, K.-D. Kammeyer, Weighted sum rate maximization for MIMO-OFDM systems with linear and dirty paper precoding, in *Proceedings of the 7th International ITG Conference on Source and Channel Coding* (*SCC*), January 2008
63. F. Boccardi, H. Huang, A near-optimum technique using linear precoding for the MIMO broadcast channel, in *Proceedings of the IEEE International Conference on Acoustics, Speech and, Signal Processing*, April 2007, pp. III-17–III-20
64. C. Guthy, W. Utschick, R. Hunger, M. Joham, Efficient weighted sum rate maximization with linear precoding. IEEE Trans. Signal Process. **58**(4), 2284–2297 (2010)
65. R. Hunger, D.A. Schmidt, M. Joham, A combinatorial approach to maximizing the sum rate in the MIMO BC with linear filtering, in *Proceedings of the 42nd Asilomar Conference on Signals, Systems, and Computers*, October 2008
66. T.M. Cover, J.A. Thomas, *Elements of Information Theory*, 2nd edn. (Wiley, New York, 2006)
67. S. Vishwanath, S.A. Jafar, A. Goldsmith, Optimum power and rate allocation strategies for multiple access fading channels, in *Proceedings of the IEEE 53rd Vehicular Technology Conference* (*VTC Spring*), May 2001, vol. 4, pp. 2888–2892
68. E. Biglieri, R. Calderbank, A. Constantinides, A. Goldsmith, A. Paulraj, H.V. Poor, *MIMO Wireless Communications* (Cambridge University Press, Cambridge, 2007)
69. B. Rimoldi, R. Urbanke, A rate-splitting approach to the Gaussian multiple-access channel. IEEE Trans. Inf. Theory **42**(2), 364–375 (1996)
70. W. Yu, J. M. Cioffi, Multiuser detection for vector multiple access channels using generalized decision feedback equalization, in *Proceedings of the 5th International Conference on Signal Processing* (*ICSP*), August 2000, vol. 3, pp. 1771–1777
71. W. Yu, W. Rhee, S. Boyd, J.M. Cioffi, Iterative water-filling for Gaussian vector multiple access channels, in *Proceedings of the International Symposium on Information Theory* (*ISIT*), June 2001, p. 322
72. A. Goldsmith, S.A. Jafar, N. Jindal, S. Vishwanath, Capacity limits of MIMO channels. IEEE J. Sel. Areas Commun. **21**(5), 684–702 (2003)
73. D.N.C. Tse, S.V. Hanly, Multiaccess fading channels-part I: polymatroid structure, optimal resource allocation and throughput capacities. IEEE Trans. Inf. Theory **7**, 2796–2815 (1998)

74. A.E. Gamal, T.M. Cover, Multiple user information theory. Proc. IEEE **68**(12), 1466–1483 (1980)
75. P.P. Bergmans, A simple converse for broadcast channels with additive white Gaussian noise. IEEE Trans. Inf. Theory **20**(2), 279–280 (1974)
76. M.H.M. Costa, Writing on dirty paper. IEEE Trans. Inf. Theory **29**(3), 439–441 (1983)
77. W. Yu, J.M. Cioffi, Sum capacity of Gaussian vector broadcast channels. IEEE Trans. Inf. Theory **50**(9), 1875–1892 (2004)
78. W. Yu, J.M. Cioffi, Trellis precoding for the broadcast channel, in *Proceedings of the IEEE Global Communications Conference* (*Globecom*), November 2001, pp. 1344–1348
79. H. Sato, An outer bound to the capacity region of broadcast channels. IEEE Trans. Inf. Theory **24**(3), 374–377 (1978)
80. G. Caire, S. Shamai, On achivable rates in a multi-antenna broadcast downlink, in *Proceedings of the 38th Annual Allerton Conference on Communications, Control, and, Computing*, October 2000
81. G. Caire, S. Shamai, On achivable rates in a multi-antenna Gaussian broadcast channel, in *Proceedings of the International Symposium on Information Theory* (*ISIT*), June 2001
82. S. Vishwanath, N. Jindal, A. Goldsmith, On the capacity of multiple input multiple output broadcast channels, in *Proceedings of the IEEE International Conference on Communications* (*ICC*), April 2002, vol. 3, pp. 1444–1450
83. W. Yu, J.M. Cioffi, Sum capacity of a Gaussian vector broadcast channel, in *Proceedings of the IEEE International Symposium on Information Theory* (*ISIT*), June 2002, p. 498
84. W. Yu, Uplink-downlink duality via minimax duality. IEEE Trans. Inf. Theory **52**(2), 361–374 (2006)
85. H. Weingarten, Y. Steinberg, S. Shamai, The capacity region of the Gaussian multiple-input multiple-output broadcast channel. IEEE Trans. Inf. Theory **52**(9), 3936–3964 (2006)
86. H. Weingarten, Y. Steinberg, S. Shamai, The capacity region of the Gaussian MIMO broadcast channel, in *Proceedings of the IEEE International Symposium on Information Theory* (*ISIT*), June 2004, p. 174
87. L.N. Trefethen, D. Bau, *Numerical Linear Algebra* (Society for Industrial Mathematics, Philadelphia, 1997)
88. S.A. Jafar, G.J. Foschini, A.J. Goldsmith, PhantomNet: exploring optimal multicellular multiple antenna systems. EURASIP J. Appl. Signal Process. **5**, 591–604 (2004)
89. J. Kim, J.M. Cioffi, Spatial multiuser access with antenna diversity using singular value decomposition, in *Proceedings of the IEEE International Conference on Communications* (*ICC*), June 2000, vol. 3, pp. 1253–1257
90. U. Erez, S. Shamai, R. Zamir, Capacity and lattice strategies for canceling known interference. IEEE Trans. Inf. Theory **51**(11), 3820–3833 (2005)
91. A.A. Goldstein, Convex programming in Hilbert space. Bull. Am. Math. Soc. **70**(5), 709–710 (1964)
92. E.S. Levitin, B.T. Polyak, Constrained minimization methods. USSR Comput. Math. Math. Phys. **6**, 1–50 (1966)
93. D.P. Bertsekas, On the Goldstein–Levitin–Polyak gradient projection method. IEEE Trans. Autom. Control **21**(2), 174–184 (1976)
94. M.S. Bazaraa, H.D. Sherali, C.M. Shetty, *Nonlinear Programming: Theory and Algorithms*, 3rd edn. (Wiley, New York, 2006)
95. J.W. Daniel, *The Approximate Minimization of Functionals* (Prentice-Hall, New York, 1971)
96. G.P. McCormick, R.A. Tapia, The gradient projection method under mild differentiability conditions. SIAM J. Control **10**(1), 93–98 (1972)
97. P.H. Calamai, J.J. Moré, Projected gradient methods for linearly constrained problems. Math. Programm. **39**, 93–116 (1987)
98. J.C. Dunn, Global and asymptotic convergence rate estimates for a class of projected gradient processes. SIAM J. Control Optim. **19**(3), 368–400 (1981)
99. D.A. Schmidt, March 2010, private communication

100. V.F. Demyanov, A.M. Rubinov, *Approximate Methods in Optimization Problems*, Modern Analytic and Computational Methods in Science and Mathematics, (Elsevier, New York, 1970)
101. L.M.G. Drummond, A. Iusem, A projected gradient method for vector optimization problems, Comput Optim Appl. **28**(1), 5–29 (2004), Kluwer Academic Publishers, Norwell
102. E.G. Birgin, J.M. Martínez, M. Raydan, Nonmonotone spectral projected gradient methods on convex sets. SIAM J. Optim. **10**(4), 1196–1211 (2000)
103. E.G. Birgin, J.M. Martínez, M. Raydan, *Spectral Projected Gradient Methods, in Encyclopedia of Optimization* (Springer, Berlin, 2009), pp. 3652–3659
104. C.E. Chidume, S.A. Mutangadura, An example on the Mann iteration method for Lipschitz pseudocontractions. Proc. Am. Math. Soc. **129**(8), 2359–2363 (2001)
105. R. Böhnke, K.-D. Kammeyer, Fast sum rate maximization for the downlink of MIMO-OFDM systems, in *Proceedings of the Canadian Workshop on Information Theory* (*CWIT*), June 2005
106. N.J. Higham, *Matrix Nearness Problems and Applications, in Applications of Matrix Theory* (University Press, Oxford, 1989), pp. 1–27
107. D. Bertsekas, A. Nedić, A.E. Ozdaglar, *Convex Analysis and Optimization* (Athena Scientific, Cambridge, 2003)
108. R. Hunger, D.A. Schmidt, M. Joham, W. Utschick, A general covariance-based optimization framework using orthogonal projections, in *Proceedings of the IEEE 9th Workshop on Signal Processing Advances in Wireless Communications* (*SPAWC*), July 2008, pp. 76–80
109. J. Liu, Y.T. Hou, S. Kompella, H.D. Sherali, Conjugate gradient projection approach for MIMO Gaussian broadcast channels, in *Proceedings of the IEEE International Symposium on Information Theory* (*ISIT*), June 2007, pp. 781–785
110. N. Jindal, S. Jafar, S. Vishwanath, A. Goldsmith, Sum power iterative water-filling for multi-antenna Gaussian broadcast channels, in *Proceedings of the 36th Asilomar Conference on Signals, Systems, and Computers*, November 2002, pp. 1518–1522
111. N. Jindal, W. Rhee, S. Vishwanath, S.A. Jafar, A. Goldsmith, Sum power iterative water-filling for multi-antenna Gaussian broadcast channels. IEEE Trans. Inf. Theory **51**(4), 1570–1580 (2005)
112. W. Yu, A dual decomposition approach to the sum power Gaussian vector multiple access channel sum capacity problem, in *Proceedings of the Conference on Information Sciences and Systems* (*CISS*), March 2003
113. M. Codreanu, M. Juntti, M. Latva-Aho, Low-complexity iterative algorithm for finding the MIMO-OFDM broadcast channel sum capacity. IEEE Trans. Commun. **55**(1), 48–53 (2007)
114. R. Hunger, D.A. Schmidt, W. Utschick, Sum-capacity and MMSE for the MIMO broadcast channel without eigenvalue decompositions, in *Proceedings of the IEEE International Symposium on Information Theory* (*ISIT*) Nice, June 2007
115. D.P. Palomar, M. Chiang, A tutorial on decomposition methods for network utility maximization. IEEE J. Sel. Areas Commun. **24**(8), 1439–1451 (2006)
116. D.P. Palomar, M. Chiang, Alternative distributed algorithms for network utility maximization: framework and applications. IEEE Trans. Autom. Control **52**(12), 2254–2269 (2007)
117. J. Brehmer, W. Utschick, Nonconcave utility maximization in the MIMO broadcast channel, EURASIP J. Adv. Signal Process. **2009**(1), 645041 (2009)
118. H. Viswanathan, S. Venkatesan, H. Huang, Downlink capacity evaluation of cellular networks with known-interference cancellation. IEEE J. Sel. Areas Commun. **21**(5), 802–811 (2003)
119. H. Boche, M. Wiczanowski, Optimization-theoretic analysis of stability-optimal transmission policy for multiple-antenna multiple-access channel. IEEE Trans. Signal Process. **55**(6), 2688–2702 (2007)
120. G. Wunder, T. Michel, Minimum rates scheduling for MIMO OFDM broadcast channels, in *Proceedings of the IEEE Ninth International Symposium on Spread Spectrum Techniques and Applications* (*ISSSTA*), August 2006, pp. 510–514
121. G. Wunder, T. Michel, Nonlinear downlink beamforming under QoS constraints: optimum precoding order and the need for time-sharing, in *Proceedings of the IEEE International Symposium on on Personal, September, Indoor and Mobile Radio Communications* (*PIMRC*), September 2007

122. J. Liu, Y.T. Hou, H.D. Sherali, On the maximum weighted sum-rate of MIMO Gaussian broadcast channels, in *Proceedings of the IEEE International Conference on Communications* (*ICC*), May 2008, pp. 3664–3668
123. M. Kobayashi, G. Caire, An iterative water-filling algorithm for maximum weighted sum-rate of Gaussian MIMO-BC. IEEE J. Sel. Areas Commun. **24**(8), 1640–1646 (2006)
124. R. Böhnke, K.-D. Kammeyer, Weighted sum rate maximization for the MIMO-Downlink using a projected conjugate gradient algorithm, in *Proceedings of the International Workshop on Cross Layer Design* (*IWCLD*), September 2007, pp. 82–85
125. S.I. Gel'fand, M.S. Pinsker, Coding for channel with random parameters. Prob. Control Theory **9**(1), 19–31 (1980)
126. U. Erez, S. ten Brink, A close-to-capacity dirty paper coding scheme. IEEE Trans. Inf. Theory **51**(10), 3417–3432 (2005)
127. J.H. Conway, N.J.A. Sloane, Voronoi regions of lattices, second moments of polytopes, and quantization. IEEE Trans. Inf. Theory **28**(2), 211–226 (1982)
128. U. Erez, R. Zamir, Achieving $\frac{1}{2}\log(1 + \text{SNR})$ on the AWGN channel with lattice encoding and decoding. IEEE Trans. Inf. Theory **50**(10), 2293–2314 (2004)
129. C. Guthy, W. Utschick, G. Dietl, P. Tejera, Efficient linear successive allocation for the MIMO broadcast channel, in *Proceedings of the 42nd Asilomar Conference on Signals, Systems, and Computers*, October 2008, pp. 336–340
130. J. Löfberg, YALMIP: a toolbox for modeling and optimization in MATLAB, in *Proceedings of the CACSD Conference*, 2004. http://control.ee.ethz.ch/joloef/yalmip.php
131. R. Hunger, W. Utschick, D.A. Schmidt, M. Joham, Alternating optimization for MMSE broadcasting, in *Proceedings of the International Conference on Acoustics, Speech, and Signal Processing* (*ICASSP*), May 2006, vol. IV, pp. IV-757–IV-760
132. R. Hunger, M. Joham, On the convexity of the MSE region of single-antenna users, in *Proceedings of the IEEE Global Conference on Communications* (*Globecom*), New Orleans, November 2008
133. T.H. Cormen, C.E. Leiserson, R.L. Rivest, C. Stein, *Introduction to Algorithms*, 2nd edn. (MIT Press, Cambridge, 2001)
134. S.S. Shitz, S. Verdú, The impact of frequency-flat fading on the spectral efficiency of CDMA. IEEE Trans. Inf. Theory **47**(4), 1302–1327 (2001)
135. A. Lozano, A.M. Tulino, S. Verdú, High-SNR power offset in multi-antenna communication, in *Proceedings of the IEEE International Symposium on Information Theory* (*ISIT*), June 2004, p. 287
136. A. Lozano, A.M. Tulino, S. Verdú, High-SNR power offset in multiantenna communication. IEEE Trans. Inf. Theory **51**(12), 4134–4151 (2005)
137. E. Gauthier, A. Yongaçoglu, J.-Y. Chouinard, Capacity of multiple antenna systems in Rayleigh fading channels, in *Proceedings of the Canadian Conference on Electrical and Computer Engineering*, March 2000, vol. 1, pp. 275–279
138. A. Grant, Rayleigh fading multi-antenna channels. EURASIP J. Appl. Signal Process. **3**, 316–329 (2002)
139. N. Jindal, High SNR analysis of MIMO broadcast channels, in *Proceedings of the International Symposium on Information Theory* (*ISIT*), September 2005, pp. 2310–2314
140. G. Caire, S. S. Shitz, On the achievable throughput of a multiantenna Gaussian broadcast channel. IEEE Trans. Inf. Theory **49**(7), 1691–1706 (2003)
141. J. Lee, N. Jindal, Dirty paper coding vs. linear precoding for MIMO broadcast channels, in *Proceedings of the 40th Asilomar Conference on Signals, Systems, and Computers*, October 2006, pp. 779–783
142. J. Lee, N. Jindal, High SNR analysis for MIMO broadcast channels: dirty paper coding versus linear precoding. IEEE Trans. Inf. Theory **53**(12), 4787–4792 (2007)
143. Z. Shen, R. Chen, J.G. Andrews, R.W. Heath, B.L. Evans, Sum capacity of multiuser MIMO broadcast channels with block diagonalization, in *Proceedings of the IEEE International Symposium on Information Theory* (*ISIT*), July 2006, pp. 886–890

144. M.R. McKay, I.B. Collings, A.M. Tulino, Achievable sum rate of MIMO MMSE receivers: a general analytic framework. IEEE Trans. Inf. Theory **56**(1), 396–410 (2010)
145. N. Jindal, MIMO broadcast channels with finite rate feedback, in *Proceedings of the IEEE Global Telecommunications Conference* (*Globecom*), November 2005, pp. 1520–1524
146. R.A. Horn, C.R. Johnson, *Matrix Analysis* (Cambridge University Press, Cambridge, 1990)
147. R.N. McDonough, A.D. Whalen, *Detection of Signals in Noise*, 2nd edn. (Academic Press, London, 1995)
148. R.S. Costas-Santos, On the elementary symmetric functions of a sum of matrices, arXiv.org, September 2009. http://arxiv.org/abs/math/0612464
149. R. Hunger, P. de Kerret, M. Joham, An algorithm for maximizing a quotient of two Hermitian form determinants with different exponents, in *Proceedings of the IEEE International Conference on Acoustics, Speech, and Signal Processing* (*ICASSP*), March 2010
150. J.A. Tague, C.I. Caldwell, Expectations of useful complex wishart forms. Multidimension. Syst. Signal Process. **5**, 263–279 (1994)
151. A.M. Tulino, S. Verdú, *Random Matrix Theory and Wireless Communications* (*Foundation and Trends in Communications and Information Theory*), vol. 1, no. 1. (Now Publishers Inc., Hanover, 2004)
152. A.K. Gupta, D.K. Nagar, *Matrix Variate Distributions (Monographs and Surveys in Pure and Applied Mathematics)* (Chapman & Hall/CRC, London, 1999)
153. P. Shaman, The inverted complex Wishart distribution and its application to spectral estimation. J. Multivar. Anal. **10**(1), 51–59 (1980)
154. M. Schubert, H. Boche, Iterative multiuser uplink and downlink beamforming under SINR constraints. IEEE Trans. Signal Process. **7**(53), 2324–2334 (2005)
155. R.D. Yates, A framework for uplink power control in cellular radio systems. IEEE J. Sel. Areas Commun. **13**(7), 1341–1347 (1995)
156. M. Schubert, H. Boche, A generic approach to QoS-based transceiver optimization. IEEE Trans. Commun. **55**(8), 1557–1566 (2007)
157. H. Boche, M. Schubert, Concave and convex interference functions-general characterizations and applications. IEEE Trans. Signal Process. **56**(10), 4951–4965 (2008)
158. H. Boche, M. Schubert, The structure of general interference functions and applications. IEEE Trans. Inf. Theory **54**(11), 4980–4990 (2008)
159. H. Boche, M. Schubert, A calculus for log-convex interference functions. IEEE Trans. Inf. Theory **54**(12), 5469–5490 (2008)
160. H. Boche, S. Naik, T. Alpcan, Characterization of convex and concave resource allocation problems in interference coupled wireless systems. IEEE Trans. Signal Process. **59**(5), 2382–2394 (2011)
161. M. Bengtsson, B. Ottersten, Optimal downlink beamforming using semidefinite optimization, in *Proceedings of the 37th Annual Allerton Conference on Communication, Control, and Computing*, September 1999, pp. 987–996
162. H. Boche, M. Schubert, SIR balancing for multiuser downlink beamforming–a convergence analysis, in *Proceedings of the IEEE International Conference on Communications* (*ICC*), April 2002, vol. 2, pp. 841–845
163. H. Boche, M. Schubert, A general theory for SIR balancing. EURASIP J. Wirel Commun. Netw. **2006**(2), 48–55 (2006)
164. H. Boche, M. Schubert, Multiuser interference balancing for general interference functions–a convergence analysis, in *Proceedings of the IEEE International Conference on Communications* (*ICC*), June 2007, pp. 4664–4669
165. H. Boche, M. Schubert, On the structure of the multiuser QoS region. IEEE Trans. Signal Process. **55**(7), 3484–3495 (2007)
166. G. Montalbano, D.T.M. Slock, Matched filter bound optimization for multiuser downlink transmit beamforming, in *Proceedings of the IEEE International Conference on Universal Personal Communications* (*ICUPC*), October 1998, vol. 1, pp. 677–681
167. P. Viswanath, V. Anantharam, D.N.C. Tse, Optimal sequences, power control, and user capacity of synchronous CDMA systems with linear MMSE multiuser receivers. IEEE Trans. Inf. Theory **45**(6), 1968–1983 (1999)

168. D.P. Palomar, M.A. Lagunas, J.M. Cioffi, Optimum linear joint transmit-receive processing for MIMO channels with QoS constraints. IEEE Trans. Signal Process. **52**(5), 1179–1197 (2004)
169. E. Jorswieck, A. Sezgin, B. Ottersten, A. Paulraj, Guaranteed performance region in fading orthogonal space-time coded broadcast channels. EURADIP J. Wirel. Commun. Netw. (2008)
170. R. Hunger, M. Joham, A complete description of the QoS feasibility region in the vector broadcast channel. IEEE Trans. Signal Process. **58**(7), 3870–3878 (2010)
171. M. Joham, R. Hunger, Feasible rate region of the MIMO broadcast channel with linear transceivers, in *Proceedings of the International ITG Workshop on Smart Antennas* (*WSA*), February 2010
172. R.S. Varga, *Matrix Iterative Analysis*, 2nd edn. (Springer, Berlin, 2000)
173. S.M. Kay, *Fundamentals of Statistical Signal Processing*, vol. 1 (Prentice Hall, Upper Saddle River, 1993)
174. R. Hunger, An introduction to complex differentials and complex differentiability, Technische Universität München, Technical Report TUM-LNS-TR-07-06, October 2007
175. P. Henrici, *Applied and Computational Complex Analysis*, vol. 1 (Wiley-Interscience, New York, 1974)
176. H. Trapp, Funktionentheorie einer Veränderlichen. Universitätsverlag Osnabrück bei (V&R unipress, Göttingen, 1996)
177. W. Wirtinger, Zur formalen theorie der funktionen von mehr komplexen veränderlichen. Mathematische Annalen **97**, 357–375 (1927)
178. P. Lancaster, On eigenvalues of matrices dependent on a parameter. Numerische Mathematik **6**, 377–387 (1964)
179. N.P. van der Aa, H.G.T. Morsche, R.R.M. Mattheij, Computation of eigenvalue and eigenvector derivatives for a general complex-valued eigensystem. Electron. J. Linear Algebra **16**, 300–314 (2007)

Index

DOI: 10.1007/978-3-642-31692-0,

MIX
Papier aus verantwortungsvollen Quellen
Paper from responsible sources
FSC® C105338

If you have any concerns about our products,
you can contact us on
ProductSafety@springernature.com

In case Publisher is established outside the EU,
the EU authorized representative is:
Springer Nature Customer Service Center GmbH
Europaplatz 3, 69115 Heidelberg, Germany

Printed by Libri Plureos GmbH
in Hamburg, Germany